Proceedings in Life Sciences

Metabolic Interconversion of Enzymes 1980

International Titisee Conference
October 1st – 5th, 1980

Organized by
E.J.M. Helmreich H. Holzer
H. Schroeder O.H. Wieland

Edited by Helmut Holzer

With 212 Figures

Springer-Verlag
Berlin Heidelberg New York 1981

Professor Dr. H. HOLZER
Albert-Ludwigs-Universität
Biochemisches Institut
Hermann-Herder-Str. 7
7800 Freiburg/FRG

ISBN 3-540-10979-X Springer-Verlag Berlin Heidelberg New York
ISBN 0-387-10979-X Springer-Verlag New York Heidelberg Berlin

Library of Congress Cataloging in Publication Data Main entry under title: Metabolic interconversion of enzymes, 1980. (Proceedings in life sciences) Proceedings of the 5th International Symposium on Metabolic Interconversion of Enzymes. Bibliography: p. Includes index. 1. Enzymes--Congresses. 2. Metabolic regulation--Congresses. I. Helmreich, Ernst. II. Holzer, Helmut, 1921-. III. International Symposium on Metabolic Interconversion of Enzymes (5th : 1980 : Titisee, Germany) IV. Title: International Titisee conference, October 1st-5th, 1980. V. Series. QP601.M467 599.01'927 81-14426 AACR2

Typesetting and printing: Beltz, Offsetdruck, Hemsbach/Bergstr.
Bookbinding: Brühlsche Universitätsdruckerei, Giessen
2131-3130-543210

Preface

Previous symposia on *Metabolic Interconversion of Enzymes* were held in 1970 (Santa Margherita-Ligure, Italy), 1971 (Rottach-Egern, Germany), 1973 (Seattle, USA) and 1975 (Arad, Israel). The present publication reports the proceedings of the 5th International Symposium on Metabolic Interconversion of Enzymes, held in Titisee/Black Forest (Germany) from October 1st-5th, 1980.

In the last few years, the number of enzymes for which control of activity by enzyme-catalyzed covalent modification, i.e., by interconversion, is of recognized metabolic importance has increased so much that is was not possible to have every such enzyme considered during a 3-day conference. The organizers therefore decided to devote only one of the three sections to "metabolic interconversion" per se, and to cover in the other two sections "enzyme regulation by proteolytic modification" and "novel aspects of regulation". According to the IUPAC-IUB Commission on Biochemical Nomenclature (cf. J. Biol. Chem. *252*, 5939-5941 (1977)), modification by proteolysis is not included in "metabolic interconversion". Considering, however, the close interrelationship of these two types of enzyme control, it has become a tradition, beginning with the conference in Rottach-Egern 1971, to include proteolytic modification in our conferences.

The program committe is greatly indebted to the Dr. Karl Thomae GmbH, subsidiary of C.H. Boehringer Sohn, Ingelheim/Rhein, which made this symposium possible as one of the traditional *Titisee Conferences*, organized and held in the Schwarzwald Hotel under the experienced direction of Dr. Hasso Schroeder and his staff, especially of Mrs. Ingeborg Wernicke and Mr. Norbert Frommann. We also thank our colleagues who were rapporteurs (E.-G. Afting, H. Betz, P. Bünning, W. Burgermeister, C. Gancedo, P.C. Heinrich, E. Siess, D.H. Wolf) and have summarized the discussion to the 36 papers presented at the meeting. Last but not least, we wish to thank all the participants who travelled from all continents to Titisee, for their presentations, and for making available the manuscripts which comprise this book. They deserve the credit for making this a successful meeting worthy of the preceding conferences.

Würzburg ERNST J.M. HELMREICH
Freiburg im Breisgau HELMUT HOLZER
München OTTO H. WIELAND

Contents

Metabolic Interconversion

Automodification of Poly (ADP-ribose) Synthetase and DNA Repair
O. Hayaishi, M. Kawaichi, N. Ogata, and K. Ueda (With 12 Figures) 3

cAMP-dependent Protein Kinase. Active Site Structure, Restricted Degradation, and Use in Assessing the Regulation of the Hormonal Response
S. Shaltiel, J.S. Jiménez, A. Kupfer, E. Alhanaty, M. Tauber-Finkelstein, Y. Zick, and R. Cesla (With 11 Figures). 10

Classification of Protein Phosphatases Involved in Cellular Regulation
P. Cohen, J.G. Foulkes, J. Goris, B.A. Hemmings, T.S. Ingebritsen, A.A. Stewart, and S.J. Strada (With 2 Figures) 28

The Mechanism of Action of Brain Cyclic AMP-dependent Protein Kinase: Structure of Active Site, Mechanism of Dissociation of the Holoenzyme, and Possible Biological Role of Its Subunits
E.S. Severin, S.N. Kochetkov, A.G. Gabibov, M.V. Nesterova, I.N. Trakht, L.P. Sashchenko, S.V. Shlyapnikov, and I.D. Grozdova (With 14 Figures). 44

Regulation of Skeletal Muscle Phosphatases by Mn^{2+} and Sulfhydryl-Disulfide Interchange
E.H. Fischer, L.M. Ballou, and D.L. Brautigan (With 4 Figures) 59

Regulation of Liver Phosphorylase Kinase
N.B. Livanova, I.E. Andreeva, T.B. Eronina, and G.V. Silonova (With 4 Figures). 68

Insulin Mediators and Their Control of Covalent Phosphorylation
J. Larner, Y. Oron, K. Cheng, G. Galasko, R. Cabelli, and L. Huang (With 11 Figures). 74

Regulation of Phosphofructokinase by Phosphorylation – Dephosphorylation – State of the Art
H.-D. Söling and I. Brand (With 5 Figures) 91

The Control of Liver Phosphofructokinase by Fructose 2,6-Bisphosphate
H.G. Hers, E. Van Schaftingen, and L. Hue (With 8 Figures) . . . 100

The Relationship Between Latent Phosphorylase Phosphatases and the Precursors of the Heat-stable Inhibitor in a Liver High-speed Supernatant
M.F. Jett and H.G. Hers (With 2 Figures) 108

Multimodulation of Phosphofructokinases in Metabolic Regulation
A. Sols, J.G. Castaño, J.J. Aragón, C. Domenech, P.A. Lazo, and A. Nieto (With 2 Figures) 111

Structure, Function, and Regulation of Mammalian Pyruvate Dehydrogenase Complex
L.J. Reed, F.H. Pettit, D.M. Bleile, and T-L. Wu (With 4 Figures) 124

Regulation of Adipose Tissue Pyruvate Dehydrogenase by Insulin: Possible Messenger Role of Peroxide(s)
O.H. Wieland and I. Paetzke-Brunner (With 6 Figures) 134

Proteolytic Modification

Mast Cell Proteases
R.G. Woodbury and H. Neurath (With 3 Figures). 145

Chemical Steps in the Selective Inactivation and Degradation of Glutamine Phosphoribosylpyrophosphate Amidotransferase in *Bacillus subtilis*
R.L. Switzer, D.A. Bernlohr, M.E. Ruppen, and J.Y. Wong 159

Inactivation and Turnover of Fructose-1,6-Bisphosphatase from *Saccharomyces cerevisiae*
M.J. Mazón, J.M. Gancedo, and C. Gancedo (With 4 Figures). . . 168

Catabolite Inactivation of Phosphoenolpyruvate Carboxykinase and Aminopeptidase I in Yeast
M. Müller and J. Frey (With 4 Figures). 174

Metabolic Interconversion of Yeast Fructose-1,6-Bisphosphatase
H. Holzer, P. Tortora, M. Birtel, and A.-G. Lenz (With 4 Figures) 179

Fructose-1,6-Bisphosphatase Converting Enzymes in Rabbit Liver Lysosomes: Purification and Effect of Fasting
S. Pontremoli, E. Melloni, and B.L. Horecker (With 10 Figures) . 186

Release and Modulation of Membrane-bound γ-Glutamyltranspeptidase by Limited Proteolysis
N. Katunuma, Y. Matsuda, A. Tsuji, and T. Miura
(With 6 Figures) 199

Inactivation of Cytosol Enzymes by a Liver Membrane Fraction
G.L. Francis and F.J. Ballard (With 2 Figures). 208

Intracellular Protein Topogenesis
G. Blobel (With 3 Figures). 217

Precursor Processing in the Biosynthesis of Rat Liver Cytochrome c Oxidase
P.C. Heinrich, E. Schmelzer, and T. Nagasawa (With 8 Figures) . 228

Characterization and Possible Pathophysiological Significance of Human Erythrocyte Proteinases
A. De Flora, E. Melloni, F. Salamino, B. Sparatore, M. Michetti, U. Benatti, A. Morelli, and S. Pontremoli (With 4 Figures). 239

Physiological Function and Catalytic Mechanism of Angiotensin Converting Enzyme
P. Bünning and J.F. Riordan (With 6 Figures) 248

Novel Aspects of Regulation

Regulation of Glutamine Synthetase Degradation
C.N. Oliver, R.L. Levine, and E.R. Stadtman (With 1 Figure). . . 259

Mono ADP-ribosylation and Poly ADP-ribosylation of Proteins
H. Hilz, P. Adamietz, R. Bredehorst, and K. Wielckens
(With 5 Figures) 269

ADP-ribosylation of Nonhistone Chromatin Proteins in Vivo and of Actin in Vitro and Effects of Normal and Abnormal Growth Conditions and Organ-specific Hormonal Influences
E. Kun, A.D. Romaschin, R.J. Blaisdell, and G. Jackowski
(With 9 Figures) 280

ADP-ribosylation of Ribonucleases
E. Leone, B. Farina, M.R. Faraone Mennella, and A. Mauro
(With 6 Figures) 294

Supramolecular Organization of Regulatory Proteins into *Calcisomes*: A Model of the Concerted Regulation by Calcium Ions and Cyclic Adenosine 3':5'-Monophosphate in Eukaryotic Cells
J. Haiech and J.G. Demaille (With 5 Figures) 303

Inhibition of Catalytic Subunit by Fragments of cAMP-dependent Protein Kinase Regulatory Subunit I and II
F. Hofmann, H. Wolf, R. Mewes, and M. Rymond (With 3 Figures) 314

Control of Mg^{2+} and Ca^{2+} Binding to Troponin C and Calmodulin by Phosphorylation of the Respective Holocomplexes Troponin and Phosphorylase Kinase
L.M.G. Heilmeyer, jr., B. Koppitz, K. Feldmann, J.E. Sperling, U. Jahnke, K.P. Kohse, and M.W. Kilimann (With 11 Figures) . . 321

Bacterial Prohistidine Decarboxylase: Kinetics of Conversion to the Active Enzyme
P.A. Recsei and E.E. Snell (With 9 Figures) 335

Sulfation of a Cell Surface Component Correlates with the Developmental Program During Embryogenesis of *Volvox carteri*
M. Sumper and S. Wenzl (With 10 Figures) 345

Structure, Possible Function, and Biosynthesis of VPg, the Genome-linked Protein of Poliovirus
C.J. Adler, B.L. Semler, P.G. Rothberg, N. Kitamura, and E. Wimmer (With 3 Figures) 356

The Acetylcholine Receptor: Control of Its Synthesis, Stability, and Cell Surface Distribution
H. Betz (With 2 Figures) 372

β-Catecholamine-Stimulated Adenylate Cyclase; an Associating-Dissociating System
A. Bakardjieva, R. Peters, M. Hekman, H. Hornig, W. Burgermeister, and E.J.M. Helmreich (With 9 Figures) 378

Subject Index 393

Contributors

You will find the adresses at the beginning of the respective contribution

Adamietz, P. 269
Adler, C.J. 356
Alhanaty, E. 10
Andreeva, I.E. 68
Aragón, J.J. 111
Bakardjieva, A. 377
Ballard, F.J. 208
Ballou, L.M. 59
Benatti, U. 239
Bernlohr, D.A. 159
Betz, H. 371
Birtel, M. 179
Blaisdell, R.J. 280
Bleile, D.M. 124
Blobel, G. 217
Brand, I. 91
Brautigan, D.L. 59
Bredehorst, R. 269
Bünning, P. 248
Burgermeister, W. 377
Cabelli, R. 74
Castaño, J.G. 111
Cesla, R. 10
Cheng, K. 74
Cohen, P. 28
De Flora, A. 239
Demaille, J.G. 303
Domenech, C. 111
Eronina, T.B. 68
Faraone Mennella, M.R. 294
Farina, B. 294
Feldmann, K. 321
Fischer, E.H. 59
Foulkes, J.G. 28
Francis, G.L. 208
Frey, J. 174
Gabikov, A.G. 44
Galasko, G. 74
Gancedo, C. 168
Gancedo, J.M. 168
Goris, J. 28
Grozdova, I.D. 44
Haiech, M. 303
Hayaishi, O. 3
Heilmeyer, L.M.G., Jr. 321
Heinrich, P.C. 228
Hekman, M. 377
Helmreich, E.J.M. 377
Hemmings, B. 28
Hers, B.L. 100, 108
Hilz, H. 269
Hofmann, F. 314
Holzer, H. 179
Horecker, B.L. 186
Hornig, H. 377
Huang, L. 74
Hue, L. 100
Ingebritsen, T.S. 28
Jackowski, G. 280
Jahnke, U. 321
Jett, M.F. 108
Jiménez, J.S. 10
Katunuma, N. 199
Kawaichi, M. 3
Kilimann, M.W. 321
Kitamura, N. 356
Kochetkov, S.N. 44
Kohse, K.P. 321
Koppitz, B. 321
Kun, E. 280
Kupfer, A. 10
Larner, J. 74
Lazo, P.A. 111
Lenz, A.-G. 179
Leone, E. 294
Levine, R.L. 259

Livanova, N.B. 68
Matsuda, Y. 199
Mauro, A. 294
Mazón, M.J. 168
Melloni, E. 186, 239
Mewes, R. 314
Michetti, M. 239
Miura, T. 199
Morelli, A. 239
Müller, M. 174
Nagasawa, T. 228
Nesterova, M.V. 44
Neurath, H. 145
Nieto, A. 111
Ogata, N. 3
Oliver, C.N. 259
Oron, Y. 74
Paetzke-Brunner, I. 134
Peters, R. 377
Pettit, F.H. 124
Pontremoli, S. 186, 239
Recsei, P.A. 335
Reed, L.J. 124
Riordan, J.F. 248
Romaschin, A.D. 280
Rothberg, P.G. 356
Ruppen, M.E. 159
Tymond, M. 314
Salamino, F. 239
Sashchenko, L.P. 44
Schmelzer, E. 228
Semler, B.L. 356
Severin, E.S. 44
Shaltiel, S. 10
Shlyapnikov, S.V. 44
Silonova, G.V. 68
Snell, E.E. 335
Söling, H.D. 91
Sols, A. 111
Saparatore, B. 239
Sperling, J.E. 321
Stadtman, E.R. 259
Stewart, A.A. 28
Strada, S.J. 28
Sumper, M. 345
Switzer, R.L. 159
Tauber-Finkelstein, M. 10
Tortora, P. 179
Trakht, I.N. 44
Tsuji, A. 199
Ueda, K. 3
Van Schaftingen, E. 100
Wenzl, S. 345
Wieland, O.H. 134
Wielckens, K. 269
Wimmer, E. 356
Wolf, H. 314
Wong, J.Y. 159
Woodbury, R.G. 145
Wu, T-L. 124
Zick, Y. 10

Metabolic Interconversion

Automodification of Poly(ADP-ribose) Synthetase and DNA Repair[1]

O. Hayaishi, M. Kawaichi, N. Ogata, and K. Ueda[2]

Poly (ADP-Ribose) is a unique homopolymer composed of a linear sequence of repeating ADP-ribose units that are derived from NAD, nicotinamide-adenine dinucleotide. The ADP-ribose moiety is linked together by ribose 1"→2' ribose glycosidic bonds and the terminal ribose is covalently attached to an acceptor protein (Hayaishi and Ueda, 1977).

Enzymatic synthesis of poly(ADP-ribosyl) protein involves at least two seemingly different reactions, namely (1) the initial attachment of the ADP-ribose unit of NAD to an acceptor protein and (2) subsequent elongation of the chain to produce a polymer and branching (Fig. 1). These two types of reaction seem to be catalyzed by an apparently homogeneous single enzyme, poly(ADP-ribose) synthetase (Kawaichi et al. 1980).

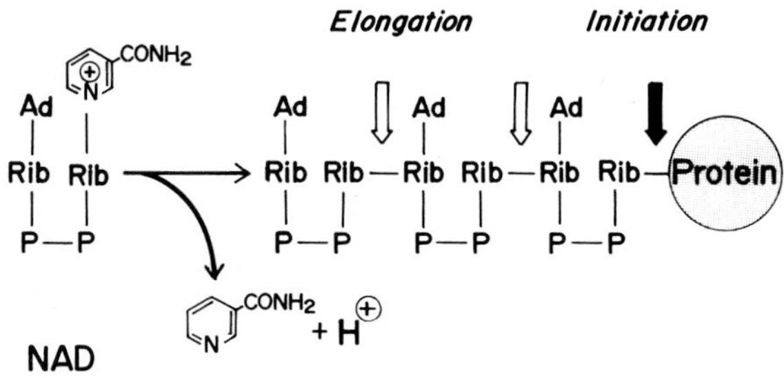

Fig. 1. Enzymatic synthesis of poly(ADP-ribosyl) protein

Purification of rat liver enzyme previously reported from our laboratory is shown in Table 1 (Okayama et al. 1977). The enzyme was purified from rat liver nuclei about 7000-fold, with an overall yield of 15%. The gel electrophoretic pattern of the final preparation is shown in Fig. 2. As shown here, the final preparation is apparently homogeneous. The molecular weight was estimated to be about 110,000 by gel filtration and ultracentrifugation.

One of the most interesting features of this enzyme is its requirement of DNA for activity. As shown in Table 2, the enzyme by itself was almost completely in-

---

1 This paper is dedicated to Dr. Helmut Holzer on the occasion of his 60th birthday

2 Department of Medical Chemistry, Kyoto University Faculty of Medicine, Yoshida, Sakyoku, Kyoto 606, Japan

Table 1. Purification of poly(ADP-ribose) synthetase

Step	Total activity	Specific activity
	nmol/h	nmol/h/mg protein
Rat liver nuclei	1150	1
Chromatin	3100	4
0.6M KCI extract	1780	7
Hydroxylapatite	730	37
Amm. sulfate	300	35
Sephadex G-150	320	660
P-cellulose	170	6900

Table 2. Requirements for poly(ADP-ribosyl)ation by purified enzyme

Addition	ADP-ribose incorporated
	pmol
None	1.1
Histone (H1)	0.3
DNA	27
DNA + Histone (H1)	105

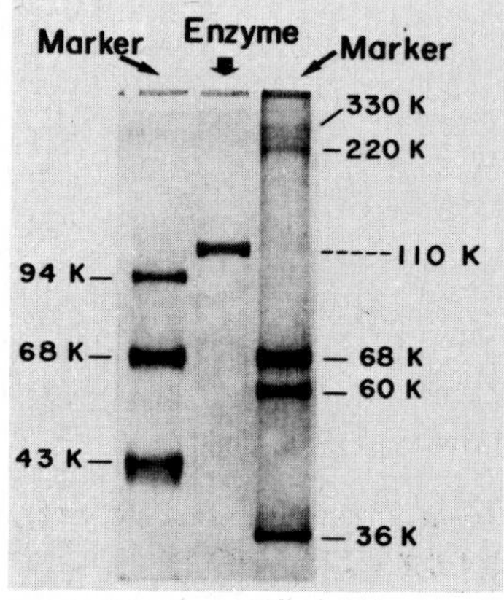

Fig. 2. SDS-gel electrophoresis of poly(ADP-ribose) synthetase

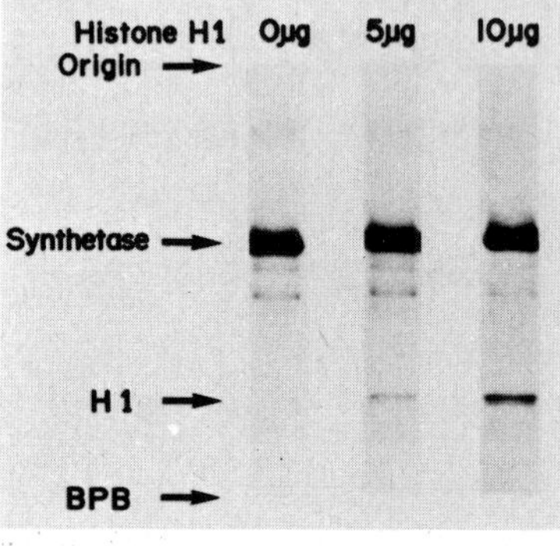

Fig. 3. Poly(ADP-ribose) acceptors in the reaction catalyzed by purified enzyme

active; the addition of histone alone was rather inhibitory. When DNA was added to the reaction mixture, the reaction was stimulated about 25-fold and further addition of histone increased the rate another 4-fold (Okayama et al. 1977).

When the products were analyzed by SDS gel electrophoresis, histone H1 as well as the synthetase was ADP-ribosylated. However, in the presence of DNA but without histone, the synthetase itself serves as the sole acceptor protein (Fig. 3).

When histone Hl was ADP-ribosylated, the results of fragmentation experiments and amino acid analysis revealed that glutamate residues No. 2, 14, 116 and the carboxyl group of the C-terminal lysine are ADP-ribosylated (Fig. 4) (Ogata *et al.* 1980b). Recently, Koide and co-workers at Rockefeller University also identified glutamate residues No. 2 and 116 as the ADP-ribosylation sites (Riquelme *et al.* 1979). It has been suggested that the N-terminal region and the carboxyl half of the molecule play an important role in the DNA binding while the apolar central region is required for other functions (Rattle et al. 1977). The well-defined apolar central region folds up with a precise globular conformation, while the N-terminal and C-terminal regions remain unstructured. All ADP-ribosylated amino acid residues are present in the so-called N-terminal and the C-terminal regions, and appear to play a role in the regulation of the binding of DNA with histones.

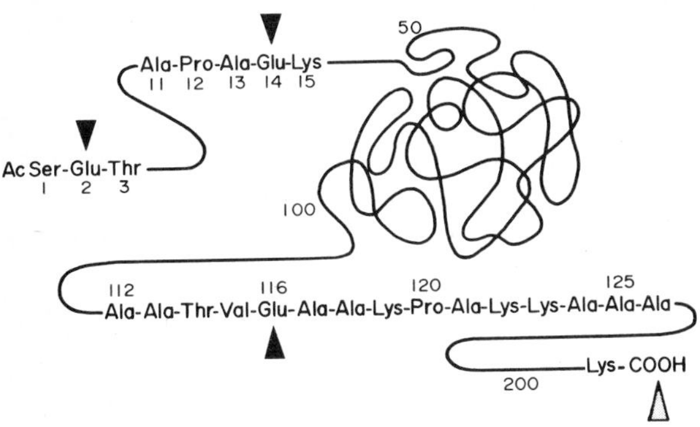

Fig. 4. Poly(ADP-ribosyl)ation sites of histone H1

As mentioned before, when the purified synthetase was incubated in the presence of DNA alone but in the absence of histone, the synthetase itself was ADP-ribosylated. The time course of chain elongation and increase in site number is illustrated in Fig. 5. After about 20 min, about 15 sites per enzyme molecule seem to be modified. Chain elongation, however, still continued. Since the molecular weight of ADP-ribosyl unit is approximately 540, after about 1 h, poly(ADP-ribose) moiety grows into a large molecule of nearly 650,000 daltons, that is about six times as big as the size of the native enzyme.

In Fig. 6 is shown a Lineweaver-Burk plot of the native enzyme, enzyme modified at 1 site with the average chain length of 3, enzyme modified at 3 sites with the average chain length of 24 and enzyme modified at 4 sites with the average chain length of 28. These results indicate that the modified enzyme is still catalytically active, although Vmax values gradually decrease and Km for NAD increases as the modification proceeds.

These observations were made with highly purified enzyme preparations in vitro. However, more recent experiments in our laboratory indicate that such automodification of the synthetase also takesplace in vivo and that it may play an important role in DNA repair (Ogata et al. 1980 a).

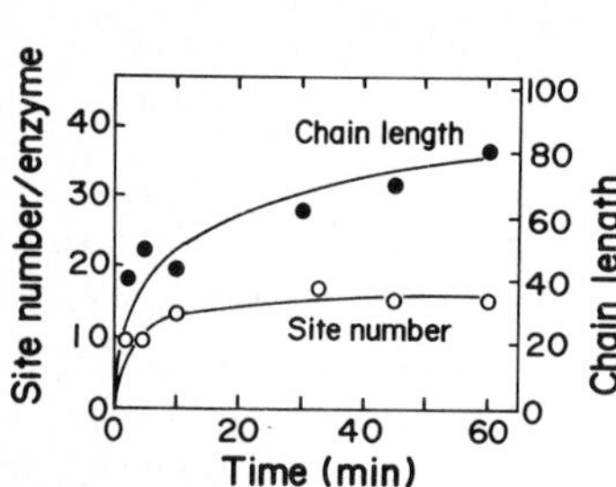

Fig. 5. Time course of chain elongation and increase in site number

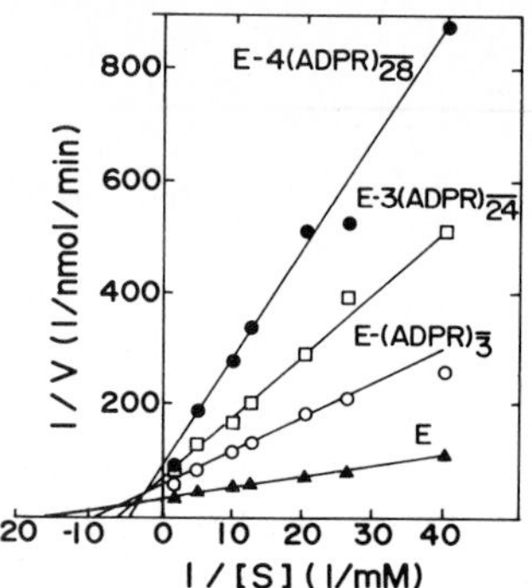

Fig. 6. Changes in Km and Vmax of automodified synthetase

Human lymphocytes were treated with N-methyl-N'-nitro-N-nitrosoguanidine (MNNG), a potent carcinogenic agent that damages DNA. The lymphocytes were then permeabilized and examined for the activities to synthesize poly(ADP-ribose) and DNA. Poly(ADP-ribose) synthetase activity started to increase almost immediately (Fig. 7). ^{3}H-thymidylate incorporation started to occur about 1 h after MNNG treatment. After a peak at 2 h DNA synthesis rapidly decreased while poly(ADP-ribose) synthetase activity continued to increase for several hours and then decreased.

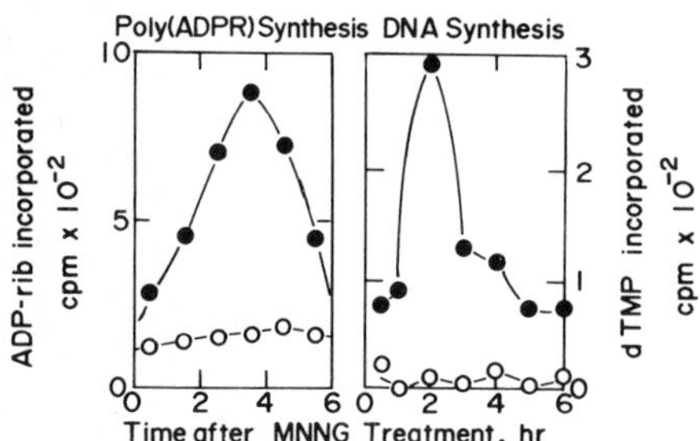

Fig. 7. Unscheduled DNA synthesis and poly-(ADP-ribose) synthesis in human lymphocytes after N-methyl-N'-nitro-N-nitrosoguanidine treatment

Similar results have been reported with other types of cells when DNA was damaged either by various alkylating agents or by physical agents such as X-ray and UV irradiation, indicating that poly(ADP-ribose) may be involved in DNA repair itself (Benjamin and Gill 1979; Berger et al. 1979; Juarez-Salinas et al. 1979; Durkacz et al. 1980). Further evidence to support this hypothesis was provided by the use of poly-(ADP-ribose) synthetase inhibitors.

3-Aminobenzamide (Fig. 8) is an analog of nicotinamide and is a specific and potent inhibitor of poly(ADP-ribose) synthetase (Purnell and Whish 1980). 3-Aminobenzamide at 3 mM concentration almost completely inhibits poly(ADP-ribose) synthesis. However, the incorporation of ^{3}H-thymidylate shown in Fig. 7 was not inhibited under identical conditions, indicating that poly(ADP-ribose) synthesis was not a prerequisite to unscheduled DNA synthesis.

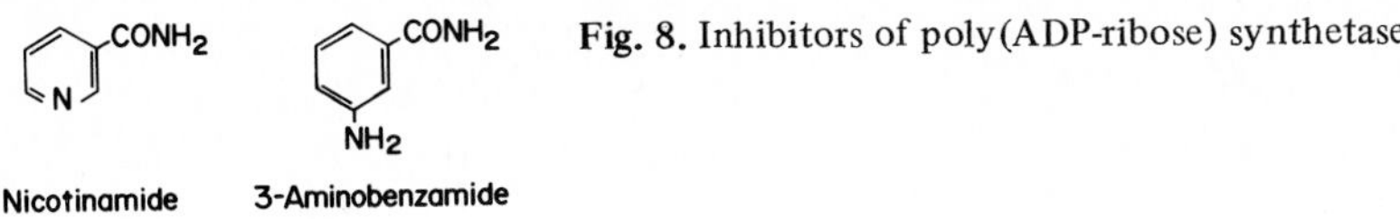

Fig. 8. Inhibitors of poly(ADP-ribose) synthetase

When L1210 mouse leukemic lymphoblasts were prelabeled with ^{3}H-thymidine and then exposed to MNNG for 20 min at 37°, DNA was fragmented and sedimented as a broad peak in the middle of the alkali sucrose gradient (Fig. 9). However, 1 and 4 h after removal of MNNG, DNA increased in size and a sizable portion was sedimented near the bottom of the gradient, the position at which undamaged DNA sedimented. However, when 3-aminobenzamide, a potent and specific inhibitor of the synthetase, was included in the culture medium, conversion of the low-molecular weight DNA to a fast-moving material was almost completely inhibited.

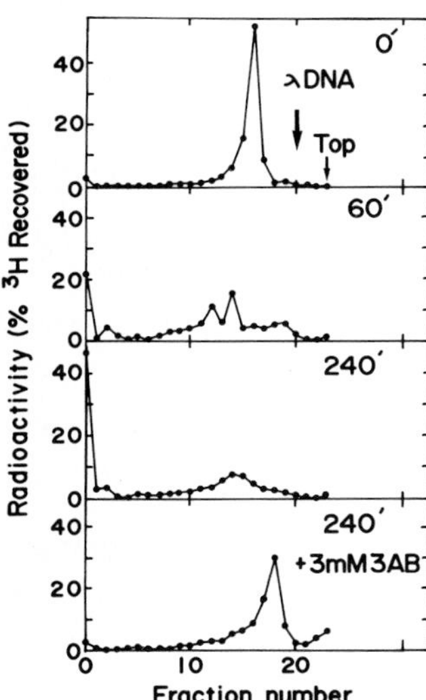

Fig. 9. Rejoining of damaged DNA in L1210 cells and inhibition by 3-aminobenzamide

These results, together with other experiments by Shall and co-workers (Durkacz et al. 1980) and Berger and others (Benjamin and Gill 1979; Berger et al. 1979; Juarez-Salinas et al. 1979), are consistent with the idea that poly(ADP-ribose) is closely linked with DNA excision repair, particularly in the rejoining of fragmented DNA.

When ADP-ribosylated lymphocytes were dissolved in SDS and analyzed by SDS-polyacrylamide gel electrophoresis, the major acceptor protein seems to form a broad band having the smallest molecular weight of 110,000 (Fig. 10) (Ogata et al. 1980 a). Since this molecular weight is exactly identical to that of poly(ADP-ribose) synthetase, we have examined its behavior under different NAD concentrations and incubation times. The results are shown in Figs. 11 and 12.

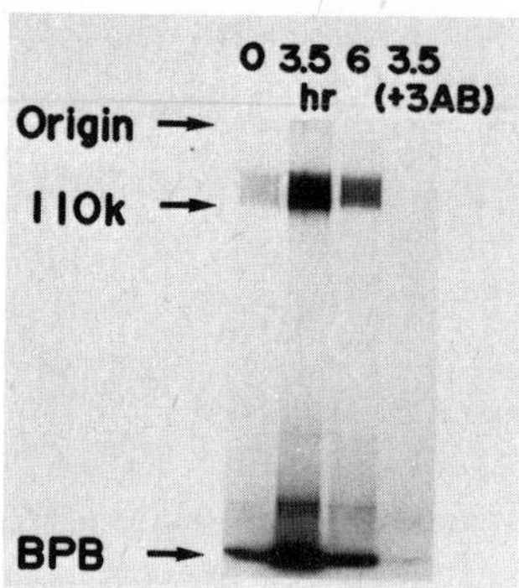

Fig. 10. Acceptor of poly(ADP-ribose) during DNA repair

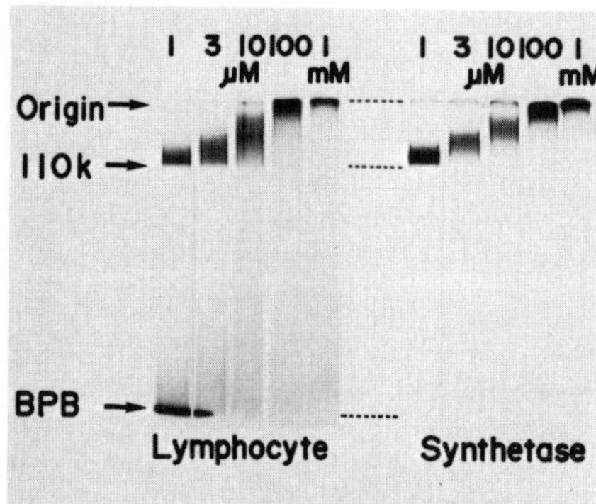

Fig. 11. Increase in molecular weight of 110K acceptor at increasing NAD concentrations

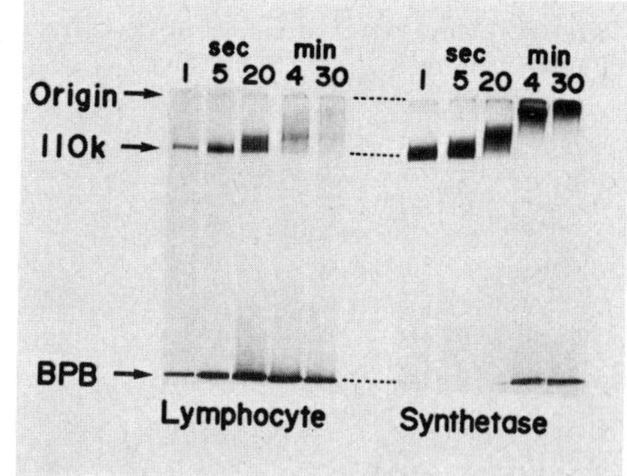

Fig. 12. Increase in molecular weight of 110K acceptor upon longer incubations

When lymphocytes and synthetase were incubated with increasing concentrations of NAD for 20 sec and the products were analyzed by SDS-gel electrophoresis, a tremendous increase in molecular size and heterogeneity was observed, as shown in Fig. 11.

Essentially identical results were obtained when incubation times were prolonged with a fixed NAD concentration of 3 μM (Fig. 12).

These results indicate that poly(ADP-ribose) synthetase is a unique enzyme and the enzyme itself serves as an excellent acceptor not only in the purified enzyme system but also in lymphocytes and cultured cells. Such automodification seems to play a role in DNA repair, in particular at the stage of DNA ligation.

References

Benjamin RC, Gill DM (1979) A connection between poly ADP-ribose synthesis and DNA damage. Fed Proc 38: 619

Berger NA, Sikorski GW, Petzold SJ, Kurohara KK (1979) Association of poly(adenosine diphosphoribose) synthesis with DNA damage and repair in normal human lymphocytes. J Clin Invest 63: 1164

Durkacz BW, Omidiji O, Gray DA, Shall S (1980) (ADP-ribose)$_n$ participates in DNA excision repair. Nature (London) 283: 593
Hayaishi O, Ueda K (1977) Poly(ADP-ribose) and ADP-ribosylation of proteins. Annu Rev Biochem 46: 95
Juarez-Salinas H, Sims JL, Jacobson MK (1979) Poly(ADP-ribose) levels in carcinogen-treated cells. Nature (London) 282: 740
Kawaichi M, Ueda K, Hayaishi O(1980) Initiation of poly(ADP-ribosyl) histone synthesis by poly(ADP-ribose) synthetase. J Biol Chem 255: 816
Ogata N, Kawaichi M, Ueda K, Hayaishi O (1980) Poly(ADP-ribosyl)ation of 110,000 dalton protein in human lymphocytes treated with *N*-methyl-*N*'-nitro-*N*-nitrosoguanidine. Biochem Int 1: 229
Ogata N, Ueda K, and Hayaishi O (1980 b) ADP-ribosylation of histone H1. J Biol Chem 255:7616
Okayama H, Edson CM, Fukushima M, Ueda K, Hayaishi O (1977) Purification and properties of poly(adenosine diphosphate ribose) synthetase. J Biol Chem 252: 7000
Purnell MR, Whish WJD (1980) Novel inhibitors of poly(ADP-ribose) synthetase. Biochem J 185: 775
Rattle HWE, Langan TA, Danby SE, Bradbury EM (1977) Studies on the role and mode of operation of the very-lysine-rich histones in eukaryote chromatin. Eur J Biochem 81: 499
Riquelme PT, Burzio LO, Koide SS (1979) ADP ribosylation of rat liver lysine-rich histone in vitro. J Biol Chem 254: 3018

Discussion[3]

Dr. Kun's question concerning the effect of 3-aminobenzamide on DNA polymerase β was answered by Dr. Hayaishi that it has practically no effect. This inhibitor seemed to be rather specific on poly(ADP-ribose) synthetase.

Dr. Hilz remarked that Berger, as well as Pitot, has just published papers showing that nicotinamide will stimulate DNA repair.

Dr. Hayaishi replied nicotinamide as a precursor of NAD is a vitamin and also one of the reaction products of the poly(ADP-ribose) synthesis. It may have many other biological effects. In his opinion the results with 3-amino-benzamide seemed to be more reliable.

3 Rapporteur: Ernst-Günter Afting

cAMP-dependent Protein Kinase. Active Site Structure, Restricted Degradation, and Use in Assessing the Regulation of the Hormonal Response

S. Shaltiel[2], J.S. Jiménez[3], A. Kupfer, E. Alhanaty, M. Tauber-Finkelstein, Y. Zick, and R. Cesla[1]

Within the last decade it became evident that cAMP-dependent protein kinase (c-AMPdpK) constitutes a major intracellular sensing device through which mammalian cells detect and respond to hormones that function via cAMP. Originally described by Walsh et al. (1968), this enzyme catalyzes the transfer of the γ-phosphoryl group of ATP onto the hydroxyl group of specific Ser or Thr residues of its target substrate proteins, often modulating their biological function (Krebs, 1972; Walsh and Krebs, 1973; Krebs and Beavo, 1979). cAMPdPK is known to be composed of two types of subunits, one catalytically active (C) and the other having a regulatory function (R) (Walsh and Krebs, 1973). These two subunits are assembled together to yield the inactive form of the enzyme (R_2C_2) which is now believed to be activated by cAMP according to the following equation:

$$R_2C_2 + 4\ cAMP \rightleftharpoons R_2(cAMP)_4C_2 \rightleftharpoons R_2(cAMP)_4 + 2C$$

(Corbin et al. 1978; Weber and Hilz 1979; Builder et al. 1980).

1 Structure, Function and Biorecognition of the C Subunit of the Enzyme

When the SH groups of C are allowed to react with an excess of Nbs_2 at neutral pH and an ionic strength of 0.22, the enzyme is inactivated and ~2 SH groups become modified, both the inactivation and the chemical modification taking place in a monophasic process (Fig. 1A). However, if the reaction is carried out at an ionic strength of 0.03, two clear-cut phases can be distinguished (Fig. 1B); a slower phase ($k_1 \doteq (2 \pm 0.3) \times 10^2 M^{-1} min^{-1}$) corresponding to the titration of 0.9 ± 0.1 mol of SH groups per mol of C (referred to as SH_I) and a faster phase ($k_2 = (3.4 \pm 0.5) \times 10^3 M^{-1} min^{-1}$, corresponding to the titration of 1.0 ± 0.1 mol SH groups per mol of C (referred to as SH_{II}). Kinetic analysis of a series of experiments in which the SH

Abbreviations: AMP-PNP, adenosine 5'-[β, γ-imido]triphosphate; cAMPdPK, cAMP-dependent protein kinase; C, catalytic subunit of cAMPdPK; Nbs_2, 5'5-dithiobis[2-nitrobenzoic acid]; R, regulatory subunit of cAMPdPK; TLCK, Nα-tosyl-L-lysine chloromethylketone

1 Department of Chemical Immunology, The Weizmann Institute of Science Rehovot, Israel

2 Incumbent of the Hella and Kleeman Chair in Biochemistry

3 On leave of absence from the Department of Physical Chemistry, University of Granada, Granada, Spain. Recipient of an EMBO post doctoral fellowship

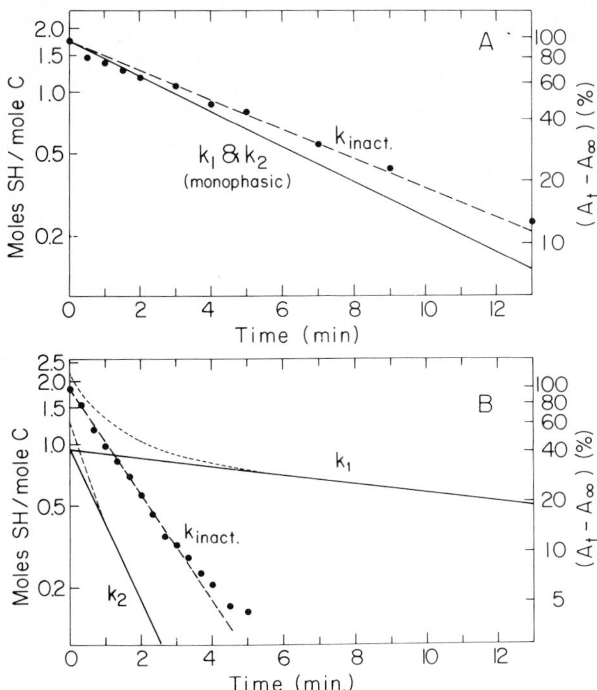

Fig. 1A, B. First order plots for the inactivation of C by Nbs_2 (●) and for the modification of the SH groups of the enzyme by the same reagent (——). Panel **A** $\mu = 0.22$; Panel **B** $\mu = 0.03$. Note that while the modification of the SH groups is monophasic at the higher ionic strength, it becomes biphasic upon lowering the ionic strength. Under the latter reaction conditions it is evident that the rate of inactivation of the enzyme (k_{inact}) is very similar to the rate of modification of SH_{II} (k_2) and not to the rate of modification of SH_I (k_1). For experimental details cf. Jiménez et al. (1980)

groups of C were titrated with Nbs_2 under identical conditions, except for the concentration of added NaCl (changing the ionic strength from 0.03 to 0.22) clearly showed (Jiménez et al., 1980) that the rate constants k_1 and k_2 change gradually (Fig. 2). However, this change occurs in opposing directions; while k_2 decreases from $3.4 \times 10^3 M^{-1} min^{-1}$ (at $\mu = 0.03$) to $9 \times 10^2 M^{-1} min^{-1}$ (at $\mu = 0.22$), i.e., ~3.8-fold, k_1 increases from $2 \times 10^2 M^{-1} min^{-1}$ to $9 \times 10^2 M^{-1} min^{-1}$, i.e., ~4.5-fold. Since the modification of the two SH groups is affected in opposite directions, it seems plausible to assume that the change in the relative reactivity of these sulfhydryls is a reflection of a simultaneous change in their microenvironment (steric availability, surrounding dielectric constant, dislocation of adjacent interacting groups, etc.). In other words, upon increasing the ionic strength of the medium from 0.03 to 0.22, the enzyme seems to undergo a pronounced conformational change.

The sulfhydryl group SH_{II} is structurally associated with the active site of the enzyme, specifically with the γ-P subsite of the ATP site in C (Jiménez et al. 1980; Shaltiel et al. 1980). This is indicated by the following:

1. The inactivation of C by Nbs_2 occurs concomitantly with the modification of SH_{II} and not with the modification of SH_I (Fig. 2).
2. There is a linear stoichiometric relationship between the inactivation of the enzyme and the modification of SH_{II}. In other words, there is a quantitative relationship between the fraction of SH_{II} modified at each stage of the reaction, and the percentage of enzyme inactivation (Fig. 3).
3. MgATP affords to the enzyme considerable protection from inactivation. This protection is effective at concentrations of MgATP ($\geqslant 10\ \mu M$) which correspond

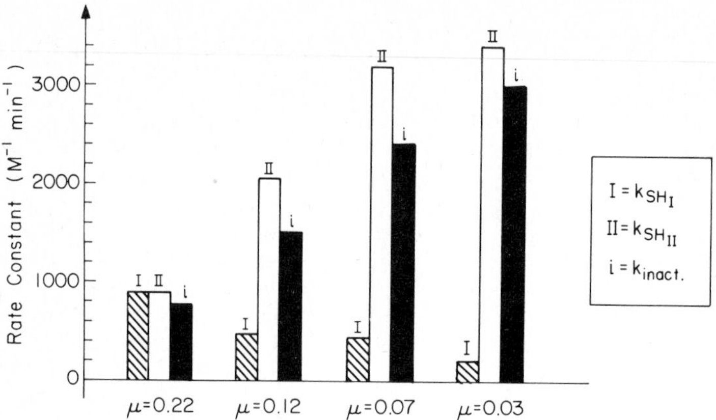

Fig. 2. Effect of ionic strength on the rate constants for the modification of SH_I (k_{SH_I}) and SH_{II} ($k_{SH_{II}}$) as well as the rate constant for the inactivation of the enzyme (k_{inact}). It is evident that: a upon lowering the ionic strength k_{SH_I} decreases while $k_{SH_{II}}$ increases; b the rate of inactivation of the enzyme (k_{inact}) increases concomitantly with the modification of SH_{II} (cf. Jiménez et al. 1980)

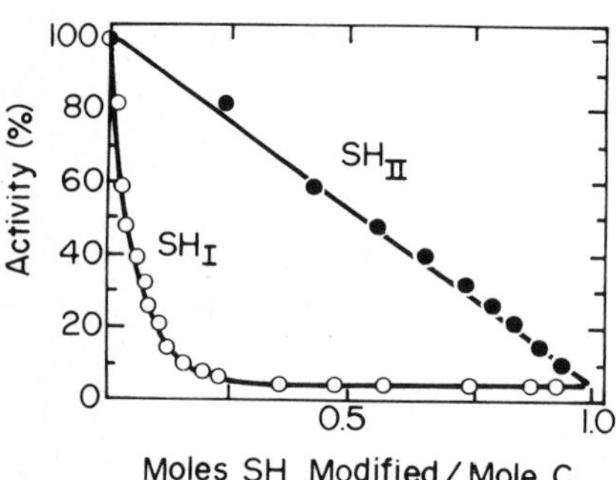

Fig. 3. Correlation between the extent of inactivation of C and the modification of SH_I (o) or SH_{II} (•). (data from Jiménez et al. 1980). Note the linear stoichiometric relationship between the fraction of SH_{II} modified at each stage of the reaction and the percentage of enzyme inactivation. No such relationship exists with regard to SH_I

to the K_m value of the enzyme for this nucleotide. In the series ATP, AMP-PNP, ADP and AMP (all in the presence of Mg^{2+}), maximal protection was observed with the two nucleotides containing a γ-P group (ATP and AMP-PNP). Upon removal of the γ-P group (as in ADP) the protection is considerably reduced, and in the absence of the β-P as well (AMP), no protection from inactivation occurs[4] (Fig. 4). It should be noted that the ability of the various MgATP analogs to afford protection against inactivation parallels the gradual decrease in the reactivity

4 The possibility that these differences in protection efficiency merely reflect the different affinities of the enzyme towards the various ATP analogs (Hoppe et al. 1978) could be ruled out here, since in all cases an appropriate excess of the analog (100 times its K_i value, which for such analogs is apparently close to the appropriate K_d values, Feramisco 1978; Armstrong et al. 1979) was used to ensure binding. In addition, the concentration of $Mg(CH_3COO)_2$ was in all cases such as to keep more than 50% of the ATP analog in the form of its magnesium complex (cf. Jiménez et al. 1980)

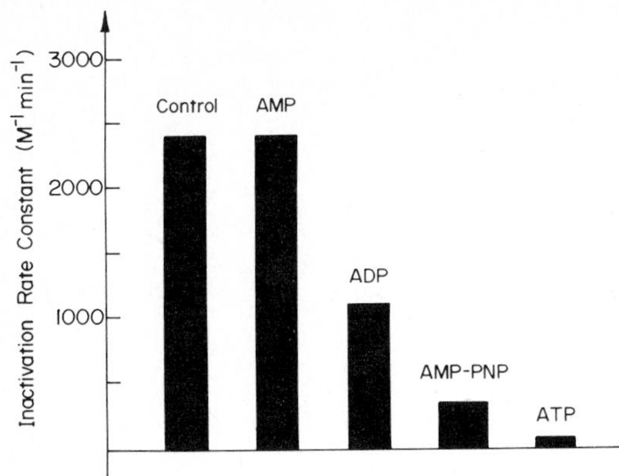

Fig. 4. Assessing the importance of the α, β and γ-P groups of ATP in affording a protection from inactivation of C by Nbs_2. The inactivation rate constants were measured for the various ATP analogs using in each case a concentration of 100 times its K_i value and keeping the Mg^{2+} concentration always such as to keep over 50% of the tested analog in the form of its magnesium complex. For experimental details cf. Jiménez et al. (1980)

of SH_{II} towards Nbs_2 whereas the reactivity of SH_I is not significantly altered. It can thus be concluded that the γ-P group of ATP is essential for affording optimal protection from inactivation by Nbs_2 and therefore that there is a structural connection between SH_{II} and the γ-P subsite within the ATP binding site in the enzyme. This sulfhydryl could either reside at this subsite or at a site intimately associated with it.

Since the catalytic function of this enzyme is to transfer the γ-P from ATP onto a target serine residue in the protein substrate, it would be reasonable to assume that the subsites accomodating these two substrates are vicinal in the three-dimensional structure of the enzyme and therefore that the presence of protein substrates could affect the reactivity of SH_{II} and thus also the inactivation of the enzyme by Nbs_2. This was found indeed to be the case. However, in contrast to ATP which dramatically attenuates the rate of inactivation of the enzyme by Nbs_2, histone H2b brings about an opposite effect, considerably accelerating (~5-fold) this inactivation (Kupfer et al. 1980b). As seen in Fig. 5, histone H2b also accelerates the rate of chemical modification of the enzyme by Nbs_2. But unlike ATP (which exerts its protective effect on SH_{II} only), histone H2b accelerates the rate of modification of both SH_I and SH_{II}, making the kinetics of the reaction monophasic ($k_1 = k_2$, cf. Kupfer et al. 1980b; Shaltiel et al. 1980). This acceleration occurs also with protamine, another well-known protein substrate of the kinase.

Although these two proteins were found to accelerate the rate of chemical modification of both SH_I and SH_{II}, it was assumed that the accelerated inactivation of the enzyme results mainly from the increase in the reactivity of SH_{II} rather than SH_I, as we had previously shown that modification of SH_{II} alone suffices to inactivate the enzyme. Furthermore, it could be argued that if the modification of either one of the two sulfhydryls by itself could cause a full inactivation of the enzyme, then the rate constant of inactivation (k_{inact}) should be equal to the sum of the rate constants for the modification of the two sulfhydryls ($k_1 + k_2$), i.e. k_{inact} should be $\sim 3 \times 10^4\ M^{-1}min^{-1}$. However, the value of k_{inact} as determined experimentally in the presence of histone H2b was $(1.6 \pm 0.2) \times 10^4 M^{-1} min^{-1}$, very close to the rate of modification of one of the sulfhydryls only.

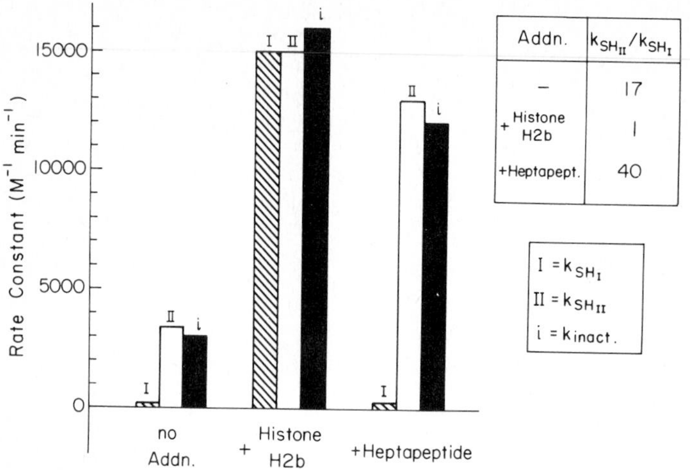

Fig. 5. Effect of histone H2b and of the heptapeptide Leu-Arg-Arg-Ala-Ser-Leu-Gly on the rate of inactivation of the enzyme (k_{inact}) as well as the rate of modification of SH_I (k_{SH_I}) and SH_{II} ($k_{SH_{II}}$). Note that the modification of the SH groups is monophasic in the presence of histone H2b and biphasic in the presence of the heptapeptide (cf. Kupfer et al. 1980b)

Since the protein substrate used in these experiments (histone H2b) has a considerable net positive charge at neutral pH, we attempted to find out whether the enhanced reactivity of the sulfhydryls arises from the fact that the enzyme-substrate complex, once formed, simply attracts the negatively charged Nbs_2. We therefore used several disulfide thiol reagents (analogs of Nbs_2; Brocklehurst 1979) which were either neutral or negatively charged in order to determine whether there is a correlation between the negative charge of the reagent and the enhanced reactivity of the sulfhydryls in the enzyme-substrate complex. When histone H2b was added to the reaction mixture and the inactivation of the enzyme was carried out in the presence of this protein substrate, it was found (Kupfer et al. 1980b) that the rate of inactivation is greatly accelerated for the two reagents – 6,6'-dithiodinicotinic acid, and 4,4'-dithiobis(3-nitrobenzoic acid) – (not illustrated) which like Nbs_2 have a negative net charge under the conditions of the experiment. However, thiol reagents which have no net charge under the same conditions (2,2'-dithiodipyridine; 2,2'-dithiobis [5-nitropyridine] and 4,4' dithiodipyridine, Brocklehurst and Little 1973; Grassetti and Murray 1969) inactivate the enzyme at a similar, if not a slightly slower rate in the presence of histone H2b.

As the accelerating effect of histone H2b does not occur with all the Nbs_2 analogs but only with those having a negative net charge, the major function of this protein substrate is not merely to promote an ionization of the thiols ($-SH \rightarrow -S^-$). Instead, it appears that the enzyme-histone complex, once formed, somehow brings into play positively charged residues (unavailable in the free enzyme) that would channel the negatively charged reagent to the vicinity of the SH groups and increase its local concentration around its site of action. In principle, these positively charged residues could either be provided by the histone, or become available in the enzyme itself around the sulfhydryls, as a result of a histone-induced conformational change.

In view of the fact that histone H2b is a relatively large substrate with many positively charged amino acid residues, it is not possible to conclude from the fact that this protein accelerates the rate of modification of both sulfhydryls that SH_I is necessarily vicinal to SH_{II} in the three-dimensional structure of the enzyme, and thus that SH_I (like SH_{II}) is also located near the ATP-binding site in C. Therefore, we attempted to restrict the accelerating effect of the protein substrate to the immediate vicinity of the ATP-binding site by using a considerably smaller substrate (with only a few basic amino acids) in order to find out whether the accelerated modification would then occur with both sulfhydryls or with one of them only. For this purpose we used a heptapeptide, Leu-Arg-Arg-Ala-Ser-Leu-Gly, which had been shown by Kemp et al. (1977) to constitute a good substrate for the enzyme and to have K_m and V_{max} values similar to those of histone H2b (Glass and Krebs 1979; Jiménez et al. 1980). In addition, this heptapeptide contains only two basic amino acid residues at positions n-2 and n-3 from its target serine (n), which were shown to provide major recognition elements at the active site of the enzyme (Hjelmquist et al. 1974; Huang et al.1974; Kemp et al. 1975; Zetterquist et al. 1976; Yeaman et al. 1976; Kemp et al. 1977).

As seen in Fig. 5, the binding of this heptapeptide (concentration tested 0.3 – 2.0 mM) accelerated the rate of inactivation of C by Nbs_2 to almost the same extent as the binding of the much larger and more basic protein substrate. However, while in the case of histone H2b the kinetics of modification of the sulfhydryls is monophasic (i.e., both SH_I and SH_{II} are modified at a similar rate, Fig. 5) in the case of the heptapeptide the modification trace is biphasic and the acceleration in the rate of modification occurs essentially with SH_{II} only (Fig. 5). As in the case of histone H2b, the net charge plays a crucial role: the sulfhydryl reagents with a negative net charge (Nbs_2 or 6,6'-dithiodinicotinic acid) inactivate the enzyme faster in the presence of the heptapeptide, while the rate of inactivation of the enzyme by the neutral reagent (2,2'-dithiodipyridine) is not affected by the presence of the heptapeptide (Kupfer et al. 1980b; Shaltiel et al. 1980).

The above results suggest that SH_{II} is located in the vicinity of the active site, in agreement with our findings (Jiménez et al. 1980), which showed that SH_{II} is intimately associated with the γ-P subsite in the ATP-binding site of the kinase. Furthermore, it can be inferred from the above that the channeling of the negatively charged reagents to the vicinity of SH_{II} may be attributed mainly to the positively charged amino acid residues which are considered essential for turning a given serine residue into a target for phosphorylation by this kinase. The fact that in the complex formed between C and its protein substrate the essential arginyl or lysyl residues of the substrate are close to SH_{II}, is in line with our previous observation that N^{α}-tosyl-L-lysine chloromethylketone (a chemically reactive derivative of lysine) acts as an affinity labeling reagent for C and inactivates this enzyme while reaching out with its chloromethyl-ketone moiety to become covalently linked to SH_{II} (Kupfer et al. 1979).

On the basis of the mapping of the ATP-binding site in R_2C_2 (Hoppe et al. 1977) and in C (Hoppe et al. 1978) we came to the conclusion that while the shielding of the ATP site in C by R may be partial only, it most likely does involve the ribose-triphosphate recognizing subsites in C (Hoppe et al. 1978). Therefore, if SH_{II} is located near the γ-P subsite of the ATP-binding site of C (Jiménez et al. 1980), one would expect that the reactivity of this sulfhydryl might be affected by the binding of R to C. This is indeed the case. Upon reaction with Nbs_2, the undissociated form of the enzyme (R_2C_2) loses its potential catalytic activity at a much slower rate than its dissociated form ($R_2(cAMP)_4$ + 2C). Fitting the time course of the inactivation to pseudo first-order reaction kinetics yielded in both cases monophasic reaction traces (Kupfer et al., 1980b). The rate constants for inactivation of the undissociated form of the enzyme (k_{inact} = 15 $M^{-1} min^{-1}$) was over 130-fold slower than that of the dissociated form k_{inact} = 2000 $M^{-1} min^{-1}$.

The higher rate of inactivation of the dissociated form of the kinase cannot be simply attributed to the presence of an excess of cAMP, since it was shown by Armstrong and Kaiser (1978) that the rate of inactivation (by Nbs_2) of the C subunit alone is not affected by addition of cAMP, and since it was found that the rate constant for inactivation of the dissociated enzyme (k_{inact} = 2 x $10^3 M^{-1} min^{-1}$) is close to that of the free catalytic subunit (k_{inact} = 3 x $10^3 M^{-1} min^{-1}$; Jiménez et al. 1980).

It is intersting to compare the rate of inactivation by Nbs_2 of the type I kinase with that of the type II enzyme. The rate of inactivation of the dissociated type I kinase (from rabbit muscle) is similar to, if not slower than, that of the pure C subunit. However, in the case of the type II kinase (from bovine heart) it was reported that the rate of inactivation in the presence of cAMP is considerably faster than that of the free subunit (Armstrong and Kaiser 1978). The distinct difference between the two types of kinase is consistent with our hypothesis regarding the structural aspects of the interaction between the C subunit and its protein substrates, since one of the major differences between the two isozymes is that while the R subunit of type II is phosphorylated by C, the R subunit of type I is not (Rosen and Erlichman 1975; Hofmann et al. 1975). Being a substrate for C, the type II regulatory subunit should contain structural features of a substrate, and indeed it has been proposed by Corbin et al. (1978) that arginyl residues are present in the vicinity of the serine residue which is phosphorylated in the type II R subunit. More recently, Potter and Taylor (1979 a, b) sequenced the phosphorylated site in the type II R subunit (from porcine skeletal muscle) and found it to have the sequence Asx-Arg-Arg-Val-Ser(P)-Val, i.e., to contain the essential positively charged amino acid residues, usually found in the substrates of this enzyme. These arginyl residues which are present in the type II (but not in the type I) R subunit, may very well channel the negatively charged Nbs_2 to SH_{II} and thus accelerate the rate of its modification and the subsequent inactivation of the enzyme. In other words, the accelerated inactivation of C in the presence of the type II regulatory subunit is due to the channelling effect created by the arginyl residues in the complex $R^{II}(cAMP)_4C_2$, which is analogous to the channeling effect created by the arginines of the heptapeptide mentioned above, when bound to C.

In conclusion, the results presented above suggest (cf. Fig. 6): (1) that SH_{II} resides in, or is structurally associated with, the γ-P subsite of the ATP-binding site in C (Jiménez et al. 1980); (2) that in the complex formed between the enzyme and its protein substrates this sulfhydryl group is located in the vicinity of one or more of the essential arginyl or lysyl residues in the substrate, which were previously shown to constitute major recognition elements at the active site of the enzyme. If we assume that the capability of the arginyl residues in the heptapeptide to channel

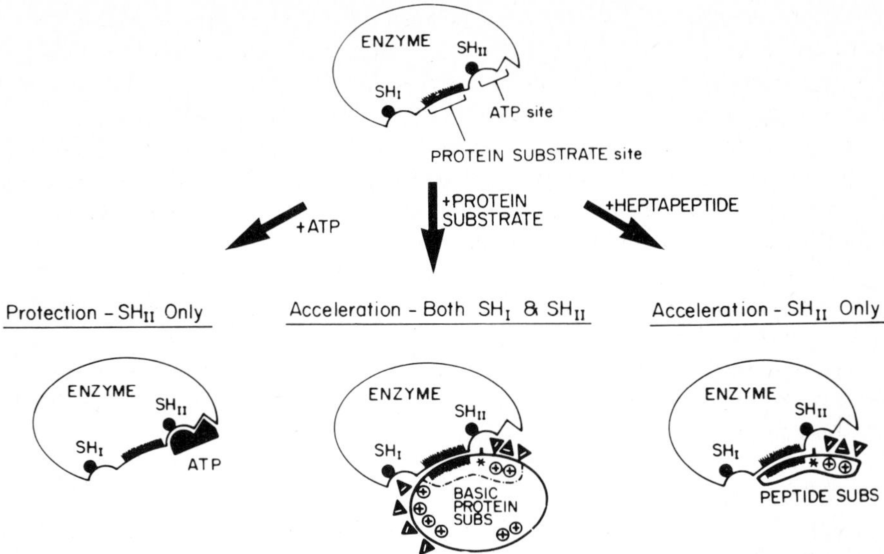

Fig. 6. A schematic representation of the active site of the catalytic subunit of cAMP-dependent protein kinase. This scheme illustrates a plausible spatial relationship between the ATP site, the protein (or peptide) accomodating site (target serine denoted) and the two sulfhydryls SH_I and SH_{II}. This proposed spatial relationship can account for the protective effect afforded by the γ-P of ATP as well as for the channeling effect of the basic amino acid residues (⊕) in the protein or peptide substrates. The negatively charged SH reagent is designated ▲

the negatively charged sulfhydryl reagents to SH_{II} is not fortuitous but reflects at least part of the mechanistic role of these basic amino acid residues, then it seems reasonable to propose (by analogy) that in the ternary complex formed between C, ATP, and the protein substrate, there is an important ionic interaction between one or more of these essential, positively charged amino acid residues and the negative charges of ATP. In fact, this would be in agreement with several pieces of evidence in the literature, which show that this ternary complex is formed through a sequential mechanism (Pomerantz et al. 1977; Jiménez et al. 1980). It should be emphasized, however, that a priori this suggestion would not favor an ordered (Feramisco 1978; Bolen et al. 1980) but rather a random (Matsuo et al. 1978) binding, since it is possible to visualize a channeling effect in both directions: the protein binding first and its arginyl or lysyl residues channeling and orienting the ATP molecule to its site, or vice versa.

Since a large protein substrate (histone H2b) considerably accelerates the rate of modification of both SH_I and SH_{II}, while a small heptapeptide substrate restricts the acceleration to SH_{II} only (Fig. 6), it becomes possible to use a series of peptide substrates containing the sequence of the heptapeptide mentioned above and additional basic amino acid residues at an increasing distance from the target serine and to determine at which stage the channeling effect extends to include SH_I. Such studies would shed light on the spatial relationship between the two sulfhydryls.

Finally, the channeling effect of the heptapeptide substrate provides an example of how the specificity of a chemical modification can be sharpened by a preferential (substrate-mediated) targeting of the reagent to a desired functional group in the enzyme. In a previous publication (Jiménez et al. 1980) we have shown that it is possible to change the relative rate of modification of SH_{II} and SH_I from a ratio of 1:1 to 16:1 by slightly modulating the ionic strength of the medium. By the additional resolution achieved here through the use of the heptapeptide, the relative rate of modification of the two sulfhydryls can be made to reach a ratio of 40:1. In other words, it become possible to essentially complete the modification of SH_{II} before SH_I starts to be modified. Furthermore, since Nbs_2 is a fully reversible label, it can also be used to temporarily mask SH_{II}, while SH_I is irreversibly tagged with an appropriate reagent, thus obtaining an enzyme exclusively labeled at SH_I. Once confirmed by sequence studies, this selective labeling approach may prove valuable for establishing structure-function relationships in this enzyme.

2 Specific Degradative Inactivation of the C Subunit of cAMPdPK and Its Possible Regulatory Implications

Brush border membranes from the rat small intestine prepared according to methods described in the literature (Schmitz et al. 1973, Hopfer et al. 1973) possess a cAMP-dependent, as well as a cAMP-independent, protein kinase activity. Upon preincubation of the membrane preparation with cAMP, the cAMP-dependent kinase activity vanished rapidly in a time-dependent process which did not affect the nondependent kinase activity. This was indicated by the fact that the time dependent inactivation triggered by cAMP leveled off at a basal value, very similar to the kinase activity remaining after inhibition by the specific cAMP-dependent kinase inhibitor characterized by Walsh et al. (1971). Attempting to find out whether the role played by cAMP in this process is that of triggering directly an inactivating factor (e.g., an enzyme) or merely dissociating the R_2C_2 form of the kinase to release free C, which only as such is susceptible to inactivation, we tried to reproduce a similar time-dependent inactivation with pure (exogenous) C obtained from rabbit muscle. Indeed, the preparation of brush border membranes was found to inactivate exogenous pure C without addition of any cAMP whatsoever (cf. Fig. 1 in Alhanaty and Shaltiel, 1979), strongly suggesting that the function of cAMP here is to release free C. At the same time, these experiments demonstrated that the inactivation of the kinase which was originally observed in the membrane preparation itself is not a unique characteristic to the membrane-bound kinase only, as it occurs also with pure (cytosolic) C obtained from rabbit muscle.

The time-dependent inactivation of pure C seems to be associated with a specific, degradative (apparently proteolytic) activity found so far in membranes from the brush border of the small intestine and of the kidney (both in the rat). As seen in Fig. 7, when aliquots of the reaction mixture are removed at different times and subjected to polyacrylamide gel electrophoresis in the presence of sodium dodecylsulfate, it was found that the brush border membrane preparations degrade C (M.W. ~40,000) with concomitant quantitative appearance of a clipped protein (M.W.

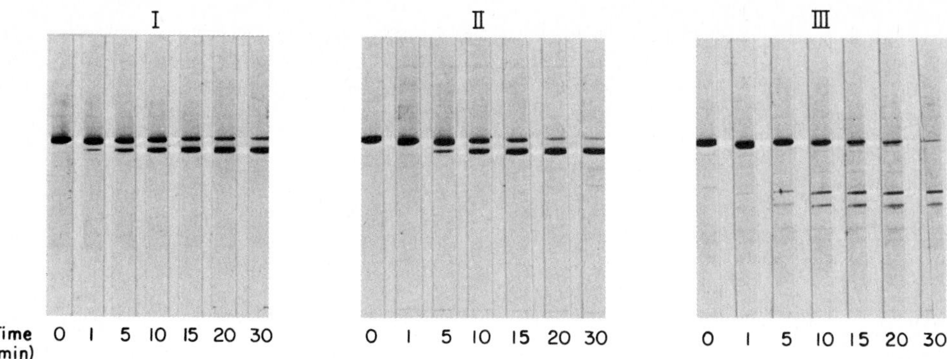

Fig. 7. Time course of the degradation of C by the membranal brush border enzyme from the rat small intestine (*I*) and from the rat kidney (*II*) as illustrated by polyacrylamide gel electrophoresis in the presence of sodium dodecylsulfate. The different pattern of degradation obtained under the same conditions with trypsin (*III*) is given for comparison. (cf. also Alhanaty and Shaltiel 1979)

~34,000) which was shown to be devoid of protein kinase activity as measured with the substrate histone H2b (Alhanaty and Shaltiel 1979). Under the conditions of the experiment depicted in Fig. 7, no intermediary degradation products were detected and the degraded protein (C') was not degraded any further. It should be noted that while the free C added to the membrane preparation is clipped, the other endogenous proteins in the membrane are not degraded in the course of this reaction (cf. Fig. 2 in Alhanaty and Shaltiel 1979).

In principle, it could be argued that the lack of degradation (proteolysis) of the other endogenous proteins of the membrane might be due to a special resistance of these proteins, essential for their coexistence with the degrading enzyme in the membrane, or that these proteins represent resistant "cores" of proteins which had been already degraded in the course of the isolation of the membranes. On the other hand, the membranal preparation failed to degrade (under the same conditions) any of six other (arbitrarily chosen) proteins from various sources. Several additional observations illustrate the specifivity of action of the membranal enzymatic activity. For example, the restricted and limited action of the membranal degrading enzyme could not be simulated (under the same reaction conditions) by other proteolytic enzymes such as trypsin, clostripain, chymotrypsin or papain. In addition, the inactivation of the C subunit by the brush border membranal enzyme is blocked in the presence of MgATP, the nucleotide substrate of the enzyme (cf. Fig. 3 in Alhanaty and Shaltiel 1979). This complete protection from inactivation and from degradation could not be achieved by either the nucleotide or the metal ion alone, but was very effective when both were present.

Solubilization (with additonal purification) of the membranal degradative enzyme could be achieved with octyl-β-D-glucopyranoside (~0.8%). Using this purified fraction, the selectivity of action of the enzyme was further illustrated in the experiment depicted in Fig. 8. Pure undissociated cAMP-dependent protein kinase (in its R_2C_2 Form) was incubated with the solubilized enzyme preparation. The kinase retained its molecular integrity as well as its potential catalytic activity (ex-

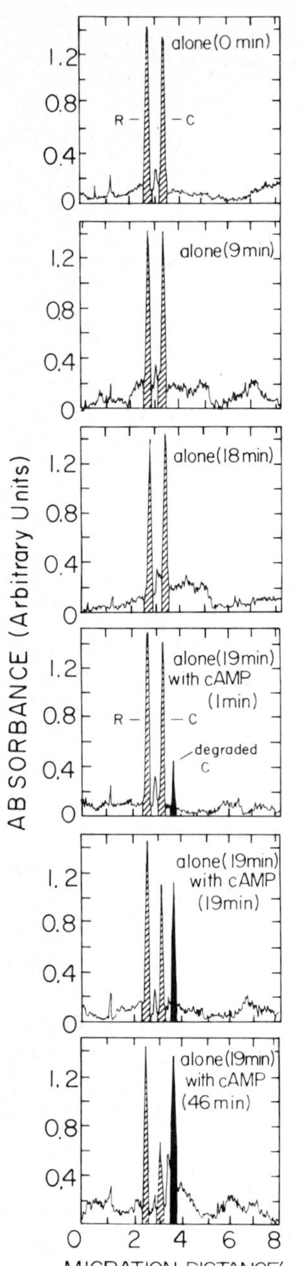

Fig. 8. Illustration of the protection from degradation afforded by the regulatory subunits (*R*) to the catalytic subunits (*C*) of cAMPdPK. The undissociated form of the enzyme (R_2C_2) was incubated with the membranal degrading enzyme (pH 7.5, 23°). No degradation occurred for 18 min as seen from the densitometric scans of the electrophoresis gels. Upon addition of cAMP to dissociate the R_2C_2 form of the enzyme, the *C* subunit is selectively degraded with concomitant formation of degraded *C*. For experimental details cf. Alhanaty and Shaltiel 1979

pressed upon assay in the presence of cAMP). However, upon addition of cAMP to the incubation mixture the kinase becomes readily inactivated with concomitant degradation of the C subunit, but not of the regulatory protein R (Fig. 8 and Fig. 4A in Alhanaty and Shaltiel 1979). This preferential degradation of C over R and R_2C_2

emphasized the selectivity of action of the membranal degrading enzyme, since in general the R subunits of cAMP-dependent protein kinase are known to be very susceptible to degradation by other proteases (Potter and Taylor, 1979 a, b).

The physiological significance of the specific inactivation of C through degradation is not clear yet. In principle, it might be the first step initiating its digestion (Schimke and Doyle 1970; Holzer et al. 1975) or constitute a safety device to prevent phosphorylation of proteins where and when such phosphorylation may have undesired consequences. Alternatively, the specific degradation of C might be a means of transferring a unique physiological stimulus if, for example, one or more of the degradation products formed will be found to possess an important function (inhibition, activation, amplification, etc.). In any case, the fact that the membranal enzyme exhibits such a specific, restricted and limited action on C, the fact that the biodegradative inactivation takes place when C is in its free (active) form but not in its R_2C_2 (inactive or "stored") form, the fact that the substrate MgATP protects C from inactivation, and finally the fact that the enzyme is found in a membrane whose function (fluid and electrolyte secretion) is known to be definitely affected by cAMP (Schafer et al. 1970; Kimberg 1974; DeJonge 1976) make it quite probable that the biorecognition between the enzyme described here and the free catalytic subunit of cAMPdPK is not fortuitous but indicates a distinct physiological (possibly regulatory) assignment. A most interesting possibility will arise, of course, if the degraded form of C will be found to posses a new (yet unknown) catalytic activity, in response to interaction with either a modulator or an intracellular metabolite. The specific degradation described above may then constitute the means by which the cell diverts the action of the enzyme from one assignment to another.

3 cAMPdPK as an Intracellular Sensor for the Detection of Extracellular Stimuli

Mouse thymocytes constitute a very good model cellular system for the study of the mechanism of the hormonal response and the role played by cAMPdPK as a major (if not the sole) intracellular sensor for the detection of extracellular stimuli. This model system has the following useful features:

1. The cells can be isolated without the use of digestive enzymes which may alter or damage the cell surface.
2. They can be kept in a viable state for several hours, a period during which their hormonal response (including the onset of cellular refractoriness) is completed.
3. They can be stimulated by β-adrenergic agonists and PGE_1 (Nordeen and Young 1978; Zick et al. 1978) as well as by specific humoral factors (Trainin 1974; Scheid et al. 1978).
4. They can be ruptured rapidly and under mild conditions (Zick et al. 1979) which allow the follow-up of the biochemical consequences of the hormonal stimulation.

The response of viable mouse thymocytes to stimulation by β-agonists, PGE_1 and the specific humoral factor denoted THF (Trainin 1974) is characterized by a rapid rise (within 30 sec) in the intracellular level of cAMP followed by a slower decline in the

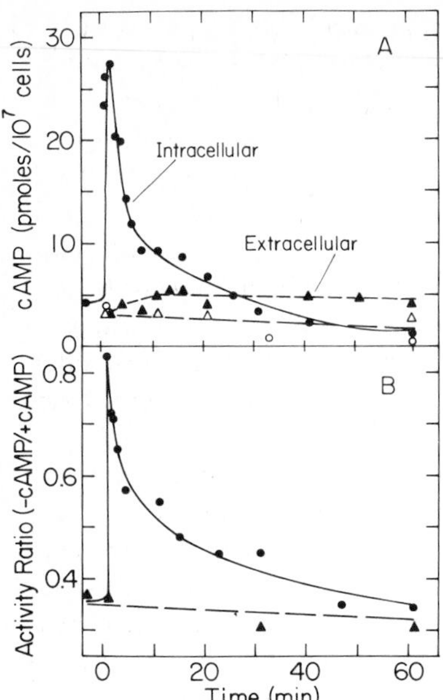

Fig. 9 A, B. The hormonal response of freshly prepared viable thymocytes to isoproterenol as monitored by the rise and fall in: **A** the level of intracellular and extracellular cAMP; **B** the activity ratio (state of activation) of intracellular cAMP-dependent protein kinase. Panel **A** ●, ▲ cAMP in isoproterenol-treated cells; ○, △ cAMP in control (buffer treated) cells: Panel B activity ratio of cAMPdPK in isoproterenol-treated (●) or control (buffer treated) cells (▲). For experimental details see Zick et al. (1980)

level of this nucleotide. As seen in Fig. 9, the spike-like response to isoproterenol which is reflected in the intracellular level of cAMP is also monitored with fidelity by the activity ratio of cAMPdPK, i.e. by the activity of the enzyme measured in the absence of cAMP divided by its activity in the presence of an excess of added cyclic nucleotide.

In order to ascertain that under the experimental conditions used by us to rupture the thymocytes (Zick et al. 1979) the activity ratio does represent a true measure of the state of activation of cAMP-dependent protein kinase in vivo, while the cells were still intact, it was necessary to establish that the dissociation of the enzyme measured in vitro was not due to a release of previously unavailable cAMP (cf. Palmer et al. 1980 for recommended critical controls). Indeed, when the rupture and subsequent handling of the cells were carried out in the presence of charcoal (which was shown with [^{3}H] cAMP to have the capacity of adsorbing under these conditions over ten fold the amount of intracellular cAMP) no significant change in the activity ratio was observed (the values differed by no more than 10%). Furthermore, upon addition of exogenous undissociated cAMP-dependent protein kinase (type I) to the ruptured cells, no evidence for a measurable dissociation of the added enzyme was found, as will be shown in detail elsewhere. The activity ratio of cAMP-dependent protein kinase can thus be taken here as a built-in sensitive indicator of the level of intracellular cAMP (at least to a first approximation). This intracellular detecting device will function. of course, only within a certain range of levels of cAMP, a range determined by the intracellular concentration of cAMP-dependent protein kinase. However, since most, if not all, the intracellular effects of cAMP appear to be mediated through the cAMP-dependent kinase, then this intracellular sensor for cAMP is not only sensitive and convenient but also a very meaningful measure from the physiological point of view, as it represents a major, if not the sole, device through which the cell detects intracellularly an extracellular hormonal stimulus. In a series of experiments to be described elsewhere it was found that the decline in intracellular cAMP and in the activity ratio of the kinase is not due to a time-dependent excretion of cAMP, nor to a disappearance of available hormone (.e.g., through destruction or internalization by the cell), since after spinning down the stimulated cells the supernatant is capable of fully triggering "fresh" cells. In addition, if the isoproterenol-triggered cells are challenged by a hormone with a different specificity (e.g., prostaglandin E_1) say 90 min after the peak response in cAMP and in the activity ratio, they do respond to it (not illustrated) raising the activity ratio of cAMPdPK > 0.75. Therefore, the above mentioned decline in the activity ratio represents the onset of a hormone-specific refractory state.

Using cAMPdPK as an intracellular sensor of cAMP we have been studying the mechanism responsible for the onset of the hormone-induced refractory state (desensitization). Our approach was to look for a chemical reagent that would irreversibly inhibit the desensitization under physiological conditions, preferably in intact cells (Zick et al. 1980). If this inhibition occurs without loss of cell viability it may be assumed to result from the selective labeling of a key component of the system responsible for the desensitization. This component may then be identified and localized within the cell. Indeed, under appropriate, mild conditions N^{α}-tosyl-L-lysine chloromethyl ketone (TLCK), originally designed to label the active site of trypsin (Shaw 1970), inhibits selectively and irreversibly the onset of the hormone-induced refractory state in intact thymocytes, while preserving their viability.

The onset of the refractory state in our thymocyte system is characterized by a high activity ratio of cAMP-dependent protein kinase (usually > 0.8) reached within 0.5 min after stimulation, followed by a drop in this ratio (usually down to < 0.45) which occurs within 30 min after stimulation (cf. Fig. 10). Therefore, in screening for mild and selective reagents to block and thus identify the molecular species responsible for desensitization, we looked for reagents that would not affect the immediate cellular response to the hormonal stimulus (elevation of activity ratio within 0.5 min) but would prevent the desensitization, i.e. the subsequent drop in the activity ratio (Zick et al. 1980). TLCK fulfills these requirements (Fig. 10). Preincubation of thymocytes with this reagent for up to 90 min does not affect at all their immediate peak response to isoproterenol (activity ratio of cAMPdPK measured after reaction with TLCK remains constant at ~0.85). However, the same reagent prevents the subsequent drop in activity ratio to levels below 0.45 observed at 30 min in cells which were not treated with TLCK. The longer the preincubation time with TLCK (up to ~30 min) the less is the drop in activity ratio. This time-dependent prevention of desensitization presumably reflects either the rate at which the reagent penetrates into (or through) the cell membrane during the contact period between the reagent and the cells, or the rate of chemical reaction between TLCK and its cellular target(s).

The specificity and selectivity of action of TLCK is quite impressive. It does not seem to be due to a reversible non covalent binding, since a close analog of TLCK in which the chemically reactive $-COCH_2Cl$ group is replaced by the nonreactive $-COCH_2SCH_2-CH_2OH$ derivative (Scheme I) fails to inhibit under the same conditions the desensitization of the cells (Table 1). On the other hand, it is unlikely that

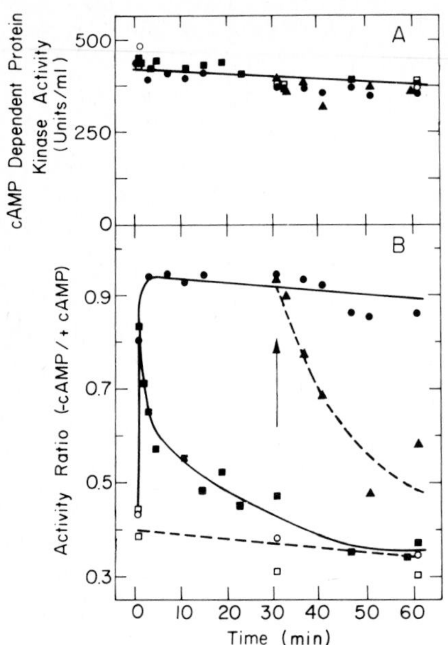

Fig. 10 A, B. Effect of TLCK on the course of the hormonal response of thymocytes as monitored by: **A** the activity of cAMP-dependent protein kinase measured in the presence of cAMP (5 μM); **B** the activity ratio of cAMP-dependent protein kinase. The incubation with TLCK (●, ○) or with buffer alone (■, □) lasted 60 min before washing and stimulation with isoproterenol (time zero) (●, ■) or, as a control, incubation with buffer alone (○, □). At the time indicated by the *arrow* (Panel **B**) some of the TLCK-treated samples (which had already been stimulated by isoproterenol) were challenged with propranolol (final concentration 10^{-5} M) (▲). For experimental details see Zick et al. (1980)

Table 1. Structural features of TLCK responsible for the selective inhibition of desensitization

Reagent	Final concentration (M)	Activity ratio of cAMPdPK after incubation for t min with stimulant[a]			
		Isoproterenol-treated cells		Buffer-treated cells	
		t = 0.5 min	t = 60 min	t = 0.5 min	t = 60 min
Control (no reagent added)	–	0.74	0.32	0.33	0.27
TLCK	1×10^{-4}	0.80	0.85	0.43	0.40
TLCK-ME[b]	1×10^{-4}	0.72	0.45	0.43	0.30
$Cl\text{-}CH_2CONH_2$	1×10^{-4}	0.70	0.40	0.40	0.36
	1×10^{-3}	0.75	0.38	0.38	0.34

[a] For experimental details cf. Zick et al. (1980)

[b] A chemically unreactive analog of TLCK obtained by allowing it to react with a 100-fold excess of 2-mercaptoethanol

TLCK blocks desensitization through a non specific alkylation, since it does not take place with chloroacetamide even at a ten fold higher molar excess (Table 1). The latter can be regarded as an adequate analog of TLCK as far as chemical reactivity goes, because it has a similar reactive group ($-COCH_2Cl$) and it is also known to react with nucleophiles (e.g., SH groups) at a rate comparable with, if not higher than, the rate of reaction of TLCK with the same nucleophiles (Whitaker and Perez-Villasenor 1968). It seems, therefore, that TLCK draws its enhanced reactivity towards its cellular target(s) from a specific affinity associated somehow with the molecular features of the N^{α}-tosyl-L-lysyl residue.

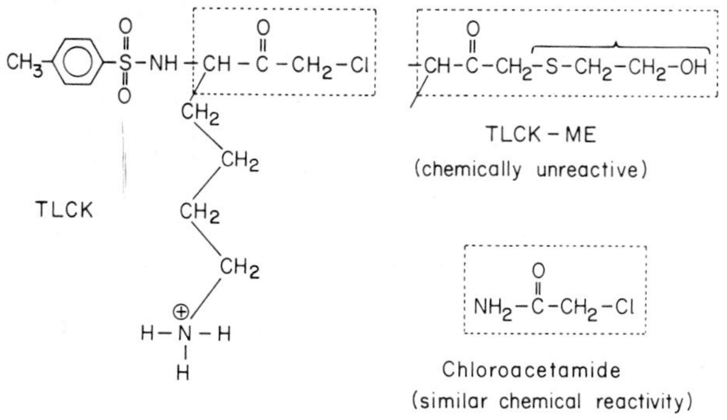

CH_3-C$_6$H$_4$-SO$_2$-NH-CH-CO-CH_2-Cl ; side chain $(CH_2)_4$-NH_3^+

TLCK

-CH-CO-CH_2-S-CH_2-CH_2-OH

TLCK – ME
(chemically unreactive)

NH_2-CO-CH_2-Cl

Chloroacetamide
(similar chemical reactivity)

Fig. 11. Structures of TLCK, TLCK-ME and chloroacetamide

The inhibition of the onset of the refractory state by TLCK is quite selective. This is indicated by the following:

1. After reaction with TLCK the majority of the cells (> 90%) remain viable;
2. Upon incubation (60 min, pH 7, 37°) the TLCK-treated cells and the control (buffer treated) cells maintain a similar basal activity ratio of cAMP-dependent protein kinase, the same total kinase activity measured in the presence of cAMP, and similar basal valus of intracellular cAMP (1-2 pmol cAMP/10^7 cells, Zick et al. 1980);
3. The TLCK-treated cells preserve the ability to recognize and respond to isoproterenol with the same immediacy as untreated cells, and reach a similar state of activation in their cAMP-dependent protein kinase (> 0.8).
4. The cells preserve the ability to recognize and respond to the β-antagonist propranolol (Fig. 10), strongly indicating that TLCK does not impair the biorecognition site of the hormone receptor. This is further supported by the finding that the effect of TLCK, though hormone-dependent, is not hormone-specific and is expressed upon challenge of the cells with different hormones (e.g., isoproterenol as well as prostaglandin E_1, data not shown). It seems therefore that after treatment with TLCK the cells are in many respects indistinguishable from non treated cells and that the consequences of the TLCK treatment become apparent only upon addition of hormone.

Together with some other studies on the inhibition of cellular refractoriness at 4° (Zick, Cesla and Shaltiel, submitted for publication) the above results imply that it is possible to dissect in a selective irreversible manner between hormonal stimulation (coupling) and hormonal desensitization (uncoupling) – two processes which are consecutive and inseparable events under normal physiological conditions. This leads us to hypothesize that the signal transfer which enables activation of adenylate cyclase is, at least in part, distinct from the signal which brings about the onset of the refractory state and to conclude that these two processes are partially autonomous and regulated by either two different proteins or two different sites on the same protein. Furthermore, these results set the stage for the identification and characterization of the molecular species which are involved in the onset of cellular refractoriness.

Acknowledgement. We thank Mrs. Elana Friedman for excellent secretarial assistance.

References

Alhanaty E, Shaltiel S (1979) Biochem Biophys Res Commun 89: 323

Armstrong RN, Kaiser ET (1978) Biochemistry 17: 2840

Armstrong RN, Kondo H, Kaiser ET (1979) Proc Natl Acad Sci USA 76: 722

Bolen DW, Stingelin J, Bramson HN, Kaiser ET (1980) Biochemistry 19: 1170

Brocklehurst K (1979) Int J Biochem 10: 259

Brocklehurst K, Little G (1973) Biochem J 133: 67

Builder SE, Beavo JA, Krebs EG (1980) J Biol Chem 255: 2350

Corbin JD, Sugden PH, West L, Flockhart DA, Lincoln TM, McCarthy D (1978) J Biol Chem 253: 3997

DeJonge HR (1976) Nature (London) 262: 590

Feramisco JR (1978) Fed Proc 37: 1328

Grassetti DR, Murray JF (1969) J Chromatogr 41: 121

Glass DB, Krebs EG (1979) J Biol Chem 254: 9728

Hjelmquist G, Andersson J, Edlund B, Engström L (1974) Biochem Biophys Res Commun 61: 559

Hofmann F, Beavo JA, Bechtel PJ, Krebs EG (1975) J Biol Chem 250: 7795

Holzer H, Betz H, Ebner E (1975) In: Horecker BL, Stadtman ER (eds) Current topics in cellular regulation, vol 9. Academic Press, London New York, p 103

Hofper U, Nelson K, Perrotto J, Isselbacher KJ (1973) J Biol Chem 248: 25

Hoppe J, Marutzky R, Freist W, Wagner KH (1977) Eur J Biochem 80: 369

Hoppe J, Friest W, Marutzky R, Shaltiel S (1978) Eur J Biochem 90: 427

Huang TS, Bylund DB, Stull JT, Krebs EG (1974) FEBS Lett 42: 249

Jiménez JS, Kupfer A, Gani V, Shaltiel S (1980) Biochemistry (submitted)

Kemp BE, Bylund DB, Huang TS, Krebs EG (1975) Proc Natl Acad Sci USA 72: 3448

Kemp BE, Graves DJ, Benjamini E, Krebs EG (1977) J Biol Chem 252: 4888

Kimberg DB (1974) Gastroenterology 67: 1023

Krebs EG (1972) Curr Top Cell Regul 5: 99

Krebs EG, Beavo JA (1979) Annu Rev Biochem 48: 923

Kupfer A, Gani V, Jiménez JS, Shaltiel S (1979) Proc Natl Acad Sci USA 76: 3073

Kupfer A, Jiménez JS, Shaltiel S (1980a) Biochem Biophys Res Commun 96: 77

Kupfer A, Jiménez JS, Gottlieb P, Shaltiel S (1980b) Biochemistry (submitted)

Matsuo M, Huang C, Huang LC (1978) Fed Proc 37: 1328

Nordeen SK, Young DA (1978) J Biol Chem 253: 1234

Palmer WK, McPherson JM, Walsh DA (1980) J Biol Chem 255: 2663

Pomerantz AH, Allfrey VG, Merrifield RB, Johnson EM (1977) Proc Natl Acad Sci USA 74: 4261

Potter RL, Taylor SS (1979a) J Biol Chem 254: 2413

Potter RL, Taylor SS (1979b) J Biol Chem 254: 9000

Rosen OM, Erlichman J (1975) J Biol Chem 250: 7788

Schafer DE, Lust WD, Sircar B, Goldberg ND (1970) Proc Natl Acad Sci USA 67: 851

Scheid MP, Goldstein G, Boyse EA (1978) J Exp Med 147: 1727

Schimke RT, Doyle D (1970) Annu Rev Biochem 39: 929

Schmitz J, Preiser H, Maestracci D, Ghosh BK, Cerda JJ, Grane RK (1973) Biochim Biophys Acta 323: 98

Shaltiel S, Alhanaty E, Cesla R, Jiménez JS, Kupfer A, Zick Y (1980) In: Rosen O, Krebs EG (eds) Protein phosphorylation. Cold Spring Harbor Conferences on Cell Proliferation pp. 83-101

Shaw E (1970) Physiol Rev 50: 244

Trainin N (1974) Physiol Rev 54: 272

Walsh DA, Krebs EG (1973) Enzymes 8: 555
Walsh DA, Perkins JP, Krebs EG (1968) J Biol Chem 243: 3763
Walsh DA, Ashby CD, Gonzalez C, Calkins D, Fischer EH, Krebs EG (1971) J Biol Chem 246: 1977
Weber W, Hilz H (1979) Biochem Biophys Res Commun 90: 1073
Whitaker JR, Perez-Villasenor J (1968) Arch Biochem Biophys 124: 70
Yeaman SJ, Cohen P, Watson DC, Dixon GH (1976) Biochem Soc Trans 4: 1027
Zetterquist O, Ragnarsson U, Humble E, Berglund L, Engström L (1976) Biochem Biophys Res Commun 70: 696
Zick Y, Cesla R, Shaltiel S (1978) FEBS Lett 90: 239
Zick Y, Cesla R, Shaltiel S (1979) J Biol Chem 254: 879
Zick Y, Cesla R, Shaltiel S (1980) Proc Natl Acad Sci USA 77: 5967

Discussion[5]

Dr. Helmreich asked Dr. Shaltiel to comment on the experiments showing a relationship between receptor desensitization and cAMP-dependent protein kinase activity.

Dr. Shaltiel answered that N^{α}-tosyl L-lysine chloromethyl ketone (TLCK)-treated mouse thymocytes retain their capacity to recognize and respond to hormonal stimuli, yet they selectively lose their ability to become desensitized upon persistent triggering by a hormone. TLCK inhibits the regulatory process which normally uncouples the adenylate cyclase system without interfering with previous or subsequent molecular events connected with the transfer of the hormonal signal across the cell membrane.

Dr. Walter put the question whether the degraded C-subunit retained catalytic activity.

Dr. Shaltiel replied that the degraded C was inactive toward some classical substrates of the kinase e.g. histone H2b.

Dr. Demaille remarked that the SH_{II} group may be either close to or in the active site and that it may perhaps be involved in the catalytic mechanism. This could not be reconciled with percyanylation experiments done in his laboratory in which the –SCN derivative of the catalytic subunit still exhibited ~60% of the activity of the native enzyme.

Dr. Shaltiel replied that in the ternary complex formed between the enzyme, ATP and the protein substrate, SH_{II} is structurally associated with the γ-P subsite of the ATP binding site in the catalytic subunit of this kinase. It may reside at this subsite without participating directly in the catalytic event. Moreover it seemed to him that Dr. Demaille's findings with the –SCN derivate would exclude the possibility that SH_{II} is involved in the catalytic mechanism as an $-S^-$ or as a proton donor but it would not exclude other (catalytic or recognition) roles in which the –SCN derivative might replace the sulfhydryl with a ~60% efficiency.

5 Rapporteur: Ernst-Günter Afting

Classification of Protein Phosphatases Involved in Cellular Regulation

P. Cohen[1], J.G. Foulkes[1], J. Goris[2], B.A. Hemmings[1], T.S. Ingebritsen[1], A.A. Stewart[1], and S.T. Strada[3]

1 Protein Phosphorylation and the Coordinated Control of Cellular Metabolism

The discovery that glycogen phosphorylase was controlled by a phosphorylation-dephosphorylation mechanism [1] was the first example of enzyme regulation by reversible covalent modification, and phosphorylase kinase [2] and glycogen synthase [3] were the second and third enzymes in which this phenomenon was identified. However, enzyme regulation by phosphorylation-dephosphorylation is not confined to glycogen metabolism and over 20 enzymes are now known to be controlled in this manner [4]. An important generality that seems to be emerging is that enzymes in biodegradative pathways are activated by phosphorylation whereas enzymes in biosynthetic pathways are inactivated by phosphorylation [5, 6] (Table 1). Two implications of this finding are that different enzymes may be regulated by the same protein kinases and protein phosphatases, or that different protein kinases and protein phosphatases may respond to the same effector molecules. These ideas are already established in the case of cyclic AMP-dependent protein kinase (cAMP-PrK) and Ca^{2+}-calmodulin dependent protein kinases.

cAMP-PrK activates phosphorylase kinase and inactivates glycogen synthase allowing coordinated control of the pathways of glycogen breakdown and synthesis in response to adrenaline (muscle) or glucagon (liver). This enzyme also activates triglyceride lipase [7] and inactivates glycerol phosphate acyl transferase [8], providing coordinated control of trigylceride breakdown and synthesis in adipose tissue by adrenaline. The inactivation of glycogen synthase and acetyl-CoA carboxylase [9, 10] by cAMP-PrK enables the rates of both glycogen and fatty acid synthesis in the liver to respond to glucagon.

When the concentration of Ca^{2+} in the cytoplasm of mammalian cells is elevated in response to neural or hormonal stimuli, this divalent cation binds to calmodulin, or to structurally related proteins, which are the major intracellular Ca^{2+} receptors. The Ca^{2+}-calmodulin complex then binds to and activates a number of different enzymes thereby changing cellular activities. Three different calmodulin-dependent protein kinases have been identified, namely myosin kinase [11], phosphorylase kina-

1 Department of Biochemistry, University of Dundee, Dundee DD1 4HN, Scotland

2 Present address: Katholieke Universitet, Leuven, Faculteit der Geneeskunde, Afdeling Biochimie, Campus Gasthuisberg, 3000 Leuven, Belgium

3 Present address: Department of Pharmacology, University of Texas, Health Science Center, Houston, Texas 77025, USA

Table 1. Enzymes found in the cytoplasm of mammalian cells that are regulated by phosphorylation [1]

	Types of protein kinase involved			
	cAMP	Ca^{2+}-calmodulin	Other	
Activation by phosphorylation				*Biodegradative pathway*
Glycogen phosphorylase	–	+	–	Glycogenolysis
Phosphorylase kinase	+	+	–	Glycogenolysis
Myosin	–	+	–	ATP hydrolysis
Triglyceride lipase	+	–	–	Triglyceride breakdown
Cholesterol esterase	+	–	–	Cholesterol ester hydrolysis
Inactivation by phosphorylation				*Biosynthetic pathway*
Glycogen synthase	+	+	+	Glycogen synthesis
Acetyl-CoA carboxylase	+	–	+	Fatty acid synthesis
Glycerol phosphate acyl transferase	+	–	–	Triglyceride synthesis
HMG-CoA reductase	–	–	+	Cholesterol synthesis
Initiation factor eIF-2	–	–	+	Protein synthesis

se [12, 13], and tryptophan hydroxylase kinase [14], but many more are likely to exist, such as those identified in the cell membranes of various tissues [15]. The Ca^{2+}-calmodulin system therefore illustrates the concept of different protein kinases being controlled by the same regulator molecule.

Many enzymes involved in biosynthetic pathways appear to be regulated by protein kinases which are not stimulated by cyclic nucleotides or by Ca^{2+}-calmodulin (Table 1). It is not yet known whether these protein kinases regulate a number of different enzymes, or whether they are specific for one substrate. Since their mechanisms of regulation are in the main unknown, the question of whether they respond to a common regulator has not yet been answered.

In contrast to work on protein kinases, the nature and regulation of the protein phosphatases involved in cellular regulation has been much more controversial, but the questions that need to be answered are essentially the same. This article will review recent work from this laboratory which is aimed at solving these problems. These studies have led to the conclusion that many of the protein phosphatases described in the literature can be divided into two major classes, termed type-1 and type-2. There is now strong evidence that type-1 protein phosphatases are under hormonal control.

2 Phosphatases Involved in Glycogen Metabolism in Mammalian Skeletal Muscle

2.1 Preparation of Phosphoprotein Substrates

A major problem in studying protein phosphatases is the preparation of the substrates used to assay these enzymes. This arises from the phenomenon of "multisite phosphorylation", which is shown by many enzymes including phosphorylase kinase and glycogen synthase.

Phosphorylase kinase is phosphorylated preferentially on two serine residues by cAMP-PrK when incubations are carried out at low Mg^{2+} (1.0 mM). One serine is located on the β-subunit and the other on the α-subunit (Table 2) and both sites are phosphorylated in vivo in response to adrenaline [24]. Phosphorylation of these two sites increases the V_{max} 15-fold at saturating Ca^{2+}, and decreases the $A_{0.5}$ for Ca^{2+} 15-fold from 23 μM to 1.5 μM [25]. The activation of phosphorylase kinase is caused by the phosphorylation of the β-subunit, while the phosphorylation of the α-subunit does not appear to influence the activity [26, 27].

On the other hand, incubation of phosphorylase kinase with cAMP-PrK for prolonged periods at high Mg^{2+} (10 mM) leads to the incorporation of 7-9 mol of phosphate rather than 2, and activates the enzyme to a level two to three fold higher than is obtained after phosphorylation to 2 mol at low Mg^{2+}. Nearly all this additional phosphate is incorporated into the α-subunit [28]. Phosphorylase kinase will also phosphorylate and activate itself in vitro at a slow rate (autophosphorylation) and 7-9 molecules of phosphate can be incorporated during this reaction, again mostly into the α-subunit [29]. There is no evidence that the large amounts of phosphate that can be introduced in vitro either by cAMP-PrK at high Mg^{2+} or by autophos-

Table 2. Amino acid sequences at the phosphorylation sites of the enzymes of glycogen metabolism [16-23]

cAMP-PrK, cyclic AMP-dependent protein kinase
PhK, phosphorylase kinase
Ph, phosphorylase
GS, glycogen synthase
GSK-3, glycogen synthase kinase-3
I_I, inhibitor-1

Protein substrate	Sequence	Protein kinase
PhK (α)	phe-arg-arg-leu-ser(P)-ile-ser-thr-glu-ser-gln-pro-asx-gly-gly-his	cAMP-PrK
PhK (β)	arg-thr-lys-arg-ser-gly-ser(P)-val-tyr-glu-pro-leu-lys	cAMP-PrK
	1a 1b	
GS	arg-arg-ala-ser(P)-cys-thr-ser-ser-ser-gly-gly-ser-lys-arg-ser-asn-ser(P)-val-asp	cAMP-PrK
	2 ?	
GS	pro-leu-ser-arg-thr-leu-ser(P)-val-ser-ser-leu-pro-gly-leu-glu-	cAMP-PrK, Phk
	3a 3b 3c	
GS	arg-tyr-pro-arg-pro-ala-ser(P)-val-pro-pro-ser(P)-pro-ser-leu-ser(P)-arg	GSK-3
I_I	ile-arg-arg-arg-arg-pro-thr(P)-pro-ala-thr	cAMP-PrK
Ph	glu-lys-arg-lys-gln-ile-ser(P)-val-arg-gly-leu-ala-gly-val-glu	PhK

phorylation occur in vivo [30, 31]. Only phosphorylase kinase labelled specifically in the two sites phosphorylated by cAMP-PrK at 1.0 mM Mg^{2+} has been used in this laboratory as a substrate for protein phosphatases.

Glycogen synthase is phosphorylated in vitro by several protein kinases, each of which decrease the activity of the enzyme. The peptides phosphorylated by three glycogen synthase kinases have been isolated and sequenced (Table 2), and this analysis demonstrated that each kinase phosphorylates distinct sites. cAMP-PrK phosphorylates sites 1a and 1b [17], and also site 2 [32]. Sites 1a and 1b are separated by only 13 residues in the primary sequence, near the C-terminus of the molecule [19], whereas site 2 is located 7 residues from the N-terminus of the polypeptide chain [20]. Phosphorylase kinase preferentially phosphorylates glycogen synthase at site 2, whereas glycogen synthase kinase 3 phosphorylates three serines, sites 3a, 3b and 3c, all situated in the same nine amino acid segment of the polypeptide chain [21]. The only overlapping specificity between these enzymes therefore occurs at site 2.

Three glycogen synthase substrates have been used in the work described below. Glycogen synthases b_1, b_2 and b_3 refer to phosphorylated forms of glycogen synthase prepared by phosphorylation with cAMP-PrK, phosphorylase kinase and glycogen synthase kinase-3 respectively.

2.2 Separation of Two Phosphorylase Kinase Phosphatases Specific for the Dephosphorylation of the α- and β-subunits

In our initial work, extracts of rabbit skeletal muscle were fractionated by precipitation with 30% ethanol at −10°C, chromatography on DEAE-cellulose, precipitation with 45% ammonium sulphate, and gel filtration on Sephadex G200. This procedure co-purified two protein phosphatases which were separated by the final step. One phosphatase (apparent M_r = 170K) dephosphorylated the α-subunit of phosphorylase kinase much faster than the β-subunit (α phosphorylase kinase phosphatase), while a second enzyme (apparent M_r = 75-80K) dephosphorylated the β-subunit much faster than the α-subunit (β phosphorylase kinase phosphatase) [33].

Subsequently, the same purification procedure was used to examine a number of different protein phosphatase activities [34]. The substrates used were phosphorylase *a*, glycogen synthase b_1, glycogen synthase $b_{2,3}$ (phosphorylated by a fraction now known to have contained both phosphorylase kinase and glycogen synthase kinase-3), and histones H1 and H2B. These experiments showed that phosphorylase phosphatase, β-phosphorylase kinase phosphatase, and glycogen synthase phosphatase activities were co-purified. The α-phosphorylase kinase phosphatase also had some glycogen synthase phosphatase activity and slight phosphorylase phosphatase activity. Both enzymes were very active histone phosphatases [34].

2.3 Protein Phosphatase-1 and Protein-Phosphatase-2

The finding that phosphorylase phosphatase, β phosphorylase kinase phosphatase, and most of the glycogen synthase phosphatase in rabbit skeletal muscle had co-purified through a several 100-fold purification suggested that a single major enzyme might catalyse all the dephosphorylation reactions which inhibit glycogenolysis or activate glycogen synthesis. This idea has been substantiated by many subsequent findings.

1. The elution of β-phosphorylase kinase phosphatase from the final gel filtration step was found to vary from preparation to preparation. Sometimes it was eluted as a species of M_r = 75-80K, sometimes as a species of M_r = 45K and sometimes as a mixture of the two components. However despite this unusual behaviour, the phosphorylase phosphatase and glycogen synthase phosphatase activities always followed the β-phosphorylase kinase phosphatase, and this provided strong evidence that all three activities were catalysed by the same enzyme [34]. This enzyme was therefore termed protein phosphatase-1 (PrP-1) and the two forms of M_r = 75-80K and M_r = 45K were termed PrP-1a and PrP-lb respectively. The α-phosphorylase kinase phosphatase was termed protein phosphatase-2 (PrP-2) [34].

2. PrP-1 was found to be potently inhibited by two heat-stable proteins from skeletal muscle termed inhibitor-1 (I_1) and inhibitor-2 (I_2) [35-37]. Both proteins have been purified to homogeneity from rabbit skeletal muscle and characterised extensively in this laboratory [38]. I_1 is only an inhibitor of PrP-1 after it has been phosphorylated on a unique threonine residue by cAMP-PrK (Table 2). Both inhibitor proteins were found to inactivate the phosphorylase phosphatase, β-phosphorylase kinase phosphatase, and glycogen synthase phosphatase activites of PrP-1 in an identical manner, and at concentrations in the nanomolar range. This again demonstrated that all three activities were catalysed by the same enzyme, since I_1 and I_2 appear to be specific inhibitors of protein phosphatase-1. They do not inhibit protein phosphatase-2 [36, 37], a specific histone phosphatase [36], or pyruvate dehydrogenase phosphatase [39].

3. 80%-90% of the phosphorylase phosphatase and glycogen synthase phosphatase activity in skeletal muscle extracts can be inhibited by I_2 [36], and greater than 90% of the β-phosphorylase kinase phosphatase activity can also be blocked by this protein (Foulkes, Ingebritsen, and Cohen, unpublished work). These experiments again suggested that PrP-1 was the major protein phosphatase involved in glycogen metabolism in skeletal muscle, a view strengthened by the finding that a major proportion of the PrP-1 is bound to the protein-glycogen complex that can be isolated from skeletal muscle [31, 37]. In contrast, PrP-2 is not closely associated with this complex [31, 37].

4. PrP-1 has recently been purified by a different procedure, to a specific activity 20-fold higher than that obtained previously [40]. This preparation was also resolved into two components by the final gel filtration on Sephadex G100, which are presumably identical to PrP-1a and PrP-1b, although their apparent molecular weights were found to be 62K and 35K rather than the values of 75-80K and 45K obtained previously. The enzymatic properties of these two components were indistinguishable from each other and from the less highly purified PrP-1a and PrP-1b described earlier. Both enzymes dephosphorylated the β-subunit of phosphorylase kinase at a 40-fold greater rate than the α-subunit, were inhibited by very similar concentrations of I_1 and I_2, and had the same relative activities against phosphorylase, phosphorylase kinase, and glycogen synthase. Glycogen synthases b_1, b_2, and b_3 were found to be dephosphorylated at similar rates by PrP-1 (Table 3). In the case of glycogen synthase b_1, which is phosphorylated on three serine residues, the relative rates of de-

phosphorylation were found to be:- site 2 (100) > site 1a [15] ≫ site 1b (< 1). The activity of glycogen synthase is determined by the state of phosphorylation of site 2 and site 1a. No effect of the phosphorylation of site 1b on the activity has so far been detected [17, 32].

3 Role of Inhibitor-1 in the Regulation of Glycogen Metabolism

I_1 is phosphorylated by cAMP-PrK in vitro at similar rates to phosphorylase kinase and glycogen synthase [22]. It is only active in its phosphorylated state, and it is not phosphorylated by phosphorylase kinase or glycogen synthase kinase-3 [41]. It inhibits each of the activities of PrP-1 in a very similar manner, at concentrations in the nanomolar range [37]. The concentration of I_1 in rabbit skeletal muscle (1.8 μM) is higher than the concentration of PrP-1, which must be less than 0.7 μM [40]. All of these results are consistent with the idea that I_1 could be important in the regulation of PrP-1 in vivo. Recently it has been demonstrated that I_1 is phosphorylated in vivo and that its degree of phosphorylation is markedly enhanced by adrenaline [42, 43]. This suggests that the activation of phosphorylase kinase by cAMP-PrK is only one of the two mechanisms by which the level of phosphorylase *a* is elevated in response to adrenaline. Equally as important may be the activation of I_1 by cAMP-PrK, and the inhibition of PrP-1 (Fig. 1).

I_1 provides a mechanism for amplifying the effects of cAMP and for exerting tight control for a number of protein kinase-protein phosphatase cycles. Furthermore, since PrP-1 dephosphorylates sites that are phosphorylated by protein kinases other than cAMP-PrK, I_1 may represent a mechanism for introducing control by cAMP into proteins which are phosphorylated by cyclic AMP-*independent* protein kinases.

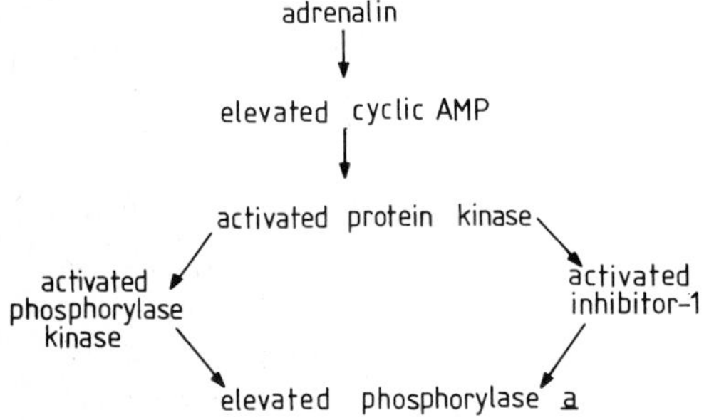

Fig. 1. Elevation of phosphorylase *a* through the simultaneous activation of inhibitor-1 and phosphorylase kinase

4 The Dephosphorylation of Inhibitor-1

A feature of the inhibition of PrP-1 by I_1 is that I_1 does not inhibit its own dephosphorylation by PrP-1 even at concentrations, 1,000-fold higher than those which inhibit the dephosphorylation of other substrates [37]. This phenomenon is observed with the most highly purified preparations of PrP-1, double reciprocal plots being linear from 0.4 to 5.0 μM I_1 [40]. I_1 is dephosphorylated by PrP-1 with kinetic constants similar to those observed for the dephosphorylation of other substrates (Table 3). The dephosphorylation of I_1 can be inhibited by I_2 [37] although higher concentrations are needed than those required to inhibit the dephosphorylation of other substrates (Foulkes, Strada, and Cohen, unpublished work). These properties of the system mean there is no need to invoke the existence of a separate enzyme for the dephosphorylation of I_1, and the state of phosphorylation of I_1 in vivo may merely be a reflection of the relative activities of cAMP-PrK and PrP-1. The existence of other I_1 phosphatases is however not excluded, and PrP-2 has also been shown to dephosphorylate I_1 effectively in vitro [37].

5 Dephosphorylation of Inhibitor-1 in Vivo in Response to Insulin

I_1 is one of the simplest examples of an interconvertible protein. It is phosphorylated on a single residue (Table 2) by one protein kinase, and it may be dephosphorylated by a single protein phosphatase (PrP-1). Furthermore, since I_1 appears to possess little ordered structure [31, 44], it is unlikely to bind allosteric effectors which might alter the rates at which it is phosphorylated or dephosphorylated. The ability of insulin to influence the state of phosphorylation of I_1 was therefore examined using a perfused hind limb system [45]. This hormone was found to decrease the activity of inhibitor-1 almost twofold [45]. This inactivation of I_1 by insulin occurs with a half-time of 10-15 min (Foulkes, Jefferson, and Cohen, unpublished work), which is similar to the time required for the activation of glycogen synthase by insulin.

The dephosphorylation of I_1 by insulin is likely to be important in the stimulation of glycogen synthesis by this hormone. Furthermore, the results demonstrate that insulin must act by either inhibiting cAMP-PrK or by activating an inhibitor-1 phosphatase (PrP-1?).

6 Evidence for the Involvement of Protein Phosphatase-1 in the Regulation of Other Metabolic Pathways

PrP-1 clearly has a broad substrate specificity. It can dephosphorylate all ten phosphorylation sites involved in glycogen metabolism (Table 2), although the relative rates of dephosphorylation of these sites vary considerably. It dephosphorylates threonine as well as serine residues, and it dephosphorylates sites phosphorylated by several different protein kinases (Table 2). These results raise the possibility that its specificity is even broader and that it may dephosphorylate enzymes involved in metabolic pathways other than glycogen metabolism.

One of the first pieces of evidence which implicated cAMP-PrK in the regulation of metabolic processes other than glycogen metabolism was the finding that its concentration was similar in different cell types, and that it was present at high levels even in tissues where the enzymes of glycogen metabolism were almost absent [46]. The tissue distribution of PrP-1 was therefore investigated by measuring the phosphorylase phosphatase activity that was sensitive to I_1 and I_2 in different tissue ex-

Table 3. Comparison of the substrate specificities of protein phosphatase-1 (PrP-1) from skeletal muscle and ATP-Mg-dependent protein phosphatase (F_C) from liver [40]

Protein substrate	Concentration in assay mg/ml	μM	Relative activities[a] PrP-1	F_C
Phosphorylase *a*	1.0	10.2	100	100
Phosphorylase kinase *a*	0.8	2.4	54	59
Glycogen synthase b_1 [b]	0.2	2.3	22	15
Glycogen synthase b_2 [c]	0.2	2.3	30	30
Glycogen synthase b_3 [d]	0.2	2.3	15	20
Inhibitor-1	0.04	2.0	10	5
Acetyl-CoA carboxylase	0.22	0.9	13	10
ATP-citrate lyase	0.50	4.2	6	6
Initiation factor eIF-2	0.002	0.01	70	63
Troponin-I[e] (thr-11)	0.04	2.0	290	270

[a] Assays were carried out in the absence of divalent cations and activities are expressed as % phosphate released in the assays
[b] Prepared by phosphorylation with cyclic AMP-dependent protein kinase (0.84 mol/subunit)
[c] Prepared by phosphorylation with phosphorylase kinase (0.52 mol/subunit)
[d] Prepared by phosphorylation with glycogen synthase kinase-3 (0.90 mol/subunit)
[e] Prepared by phosphorylation with phosphorylase kinase

tracts. These results indicated that the concentration of PrP-1 only varied about two-fold in all tissues tested (skeletal muscle, cardiac muscle, brain, mammary gland, adipose tissue, liver, kidney, and lung). In contrast, the activity of phosphorylase and phosphorylase kinase varied 300 to 500-fold. The concentration of PrP-1 in adipose tissue was 50% of that found in skeletal muscle although the enzymes of glycogen metabolism were virtually absent [47]. Thus the tissue distribution of PrP-1 resembles that of cAMP-PrK and supports the idea of a wider role for this enzyme in cellular regulation.

PrP-1 has now been shown to dephosphorylate all of the enzymes listed in Table 1 [8, 9, 48-51], except for hormone-sensitive triglyceride lipase which has not yet been tested. The enzymes involved in biodegradative pathways are inactivated and those involved in biosynthetic pathways are activated by this enzyme. These experiments do not however prove that PrP-1 is the *major* protein phosphatase acting on these enzymes in vivo. There is currently only evidence for this in the case of acetyl-CoA carboxylase in lactating rabbit mammary gland [52].

7 The ATP-Mg-dependent Protein Phosphatase and Protein Phosphatase-1 Have Identical Substrate Specificities

Merlevede and Riley [53] reported in 1966 that the activity of phosphorylase phosphatase in bovine adrenal cortex was increased by incubation with ATP-Mg, and similar observations were made subsequently in dog liver [54], pigeon skeletal muscle [55], and Neurospora crassa [56]. Kalala et al. [57] showed that a distinct phosphorylase phosphatase was activated by ATP-Mg. This enzyme could be separated from other phosphatases, which were fully active in the absence of ATP-Mg, by gel filtration or sucrose density gradient centrifugation. Goris et al. [58] subsequently resolved the ATP-Mg-dependent phosphorylase phosphatase into two components by chromatography on DEAE-cellulose, termed F_C and F_A, both of which were required for activity. F_C, which has been partially purified from dog liver, is an inactive phosphatase, whereas F_A is the activating factor [59]. The ATP-Mg-dependent phosphorylase phosphatase has been identified in liver, heart, and skeletal muscle from rats and rabbits, and similarly resolved into the F_C and F_A components [60]. The F_C and F_A components from different tissues can be used interchangeably [60].

Recently we have demonstrated that the ATP-Mg-dependent phosphatase is not specific for the dephosphorylation of phosphorylase *a*, and that it has a broad substrate specificity [40]. Moreover its specificity is identical to PrP-1. It has the same relative activity against all phosphoprotein substrates tested (Table 3), it dephosphorylates the β-subunit of phosphorylase kinase 40-fold faster than the α-subunit, and it is inhibited by very similar concentrations of I_1 and I_2 [40]. The ATP-Mg-dependent protein phosphatases from rabbit muscle and dog liver have identical properties.

In view of these striking similarities it is tempting to speculate that PrP-1 and the ATP-Mg dependent protein phosphatase possess the same catalytic subunit. However, the two enzymes must clearly differ in structure since the ATP-Mg-dependent protein phosphatase is inactive until incubated with F_A and ATP-Mg, whereas PrP-1 is fully active in the absence of these agents. The requirement for ATP-Mg and a protein factor (F_A) to activate F_C might suggest that F_C is activated by a phosphorylation mechanism, but there is currently no evidence to support such a conclusion [59]. The apparent molecular weight of partially purified F_C from rabbit skeletal muscle, determined on a calibrated Sephadex G100 column, was found to be 64K [40]. The molecular weights of highly purified PrP-1a and PrP-1b, determined on the same columns, were 62 ± 2K and 34 ± 1K respectively [40]. It is inknown whether PrP-1b is a fragment of PrP-1a, or whether PrP-1a is a dimer of PrP-1b, or a complex between PrP-1b and another protein. The structural relationship between the different forms of PrP-1, and between PrP-1 and F_C, can only be answered when homogeneous preparations of these enzymes are available.

Previous estimations of protein phosphatase activities in gel filtered tissue extracts would not have included F_C since it is only active in the presence of ATP-Mg [36, 47]. However, preliminary measurements suggest that the concentration of F_C is 30%-50% of the concentration of PrP-1 in liver and muscle of normal fed rats and rabbits. Of course, if PrP-1 and F_C are interconvertible forms of the same enzyme, their relative amounts may vary considerably under different metabolic or hormonal conditions.

8 Separation of Protein Phosphatase C from Mammalian Liver into Two Distinct Enzymes

Over the past 5 years several laboratories have isolated a protein phosphatase of M_r = 35K from liver, skeletal muscle, and heart muscle, which has a broad specificity and which is capable of reversing phosphorylation reactions catalysed by both cAMP-PrK and other types of protein kinase [61-66]. This enzyme, which has been termed protein phosphatase C [67], has been obtained in a highly purified form from each of these tissues using an 80% ethanol precipitation at room temperaturee at an early stage of the preparation. This treatment not only denatures many contaminating proteins, but is also believed to dissociate the 35K catalytic subunit from higher molecular weight complexes which may contain regulatory subunits [67]. For several years we believed that protein phosphatase C must be identical to protein phosphatase-1b, but recent work has indicated that this is not the case.

Protein phosphatase C was isolated from rat and rabbit livers using phosphorylase *a* as a substrate, up to and including the second chromatography on DEAE-Sephadex [68]. These preparations, which were purified 1000 to 3000-fold behaved as a single component when subjected to gel filtration on Sephadex G100 Superfine. However, when I_1 and I_2 were included in the assays, and phosphorylase kinase was also used as a substrate, it became clear that each protein phosphatase C preparation was a mixture of two enzymes. One activity dephosphorylated the β-subunit of phosphorylase kinase much faster than the α-subunit, was potently inhibited by I_1 and I_2, and appeared to be identical to PrP-1b from skeletal muscle. The second activity dephosphorylated the α-subunit of phosphorylase kinase five-fold faster than the β-subunit and was insensitive to I_1 and I_2. The properties of this enzyme suggested that it was similar to PrP-2 from skeletal muscle.

All preparations of protein phosphatase C from rabbit liver and some preparations from rat liver were largely type-2 protein phosphatase with small amounts of type-1 protein phosphatase. Other preparations from rat liver contained similar amounts of type-1 and type-2 [68].

The results demonstrate that two distinct protein phosphatase catalytic subunits are present in mammalian liver. Both enzymes have very similar physical properties and broad substrate specificities, but they can clearly be distinguished by the use of I_1, I_2 and phosphorylase kinase. The type-2 catalytic subunit is much more sensitive than PrP-1 to inhibition by ATP, and is eluted from DEAE-cellulose at slightly higher salt concentrations (0.17 vs 0.15 M) and slightly earlier on Sephadex G100 (M_r = 34K vs M_r 33K).

9 Separation of Protein Phosphatase-2 into Several Subclasses

A difference between the type-2 catalytic subunit present in preparations of protein phosphatase C from liver (M_r = 34K) and the PrP-2 described previously from rabbit skeletal muscle (M_r = 170K) was that the former had a much higher activity ratio phosphorylase phosphatase/phosphorylase kinase phosphatase. In view of this difference a more detailed comparison of the phosphorylase kinase phosphatase and phosphorylase phosphatase activities in rabbit liver and rabbit skeletal muscle has been carried out.

Only 50% of the phosphorylase phosphatase activity in rabbit liver extracts can be inhibited by I_2 when assays are carried out in the absence of Mn^{2+}, whereas 90% of the activity in rabbit muscle can be inhibited by I_2. Although this difference implied that the concentration of PrP-2 was five-fold higher in liver than in muscle, experiments using phosphorylase kinase as a substrate revealed that the α-phosphorylase kinase phosphatase activities in liver and muscle extracts were very similar. The explanation for this discrepancy became apparent when the extracts were chromatographed on DEAE-cellulose, and assayed at high concentrations of I_2 to inhibit PrP-1. Under these conditions three peaks of α-phosphorylase kinase phosphatase activity and two peaks of phosphorylase phosphatase activity were detected (Fig. 2). The first peak of α-phosphorylase kinase phosphatase activity (termed PrP-2b) had very little phosphorylase phosphatase activity, whereas the second and third peaks (termed PrP-2a$_1$ and PrP-2a$_2$) were active against both substrates. PrP-2b was the dominant α-phosphorylase kinase phosphatase in skeletal muscle and this enzyme is the PrP-2 described originally [33]. This enzyme is, however, a very minor activity in rabbit liver. The existence of large amounts of PrP-2a$_1$ and PrP-2a$_2$ in liver explains why only 50% of the phosphorylase phosphatase in this tissue extract can be inhibited by I_2. In rabbit liver 50% of the phosphorylase phosphatase measured under optimal conditions in the absence of Mn^{2+} is catalysed by PrP-1 and 50% by PrP-2a$_1$ and PrP-2a$_2$, whereas these percentages are 90% and 10% respectively in rabbit skeletal muscle.

The phosphorylase phosphatase activity of PrP-2a$_1$ and PrP-2a$_2$ is activated about two-fold by the inclusion of 1.0 mM Mn^{2+} in the assays, whereas PrP-1 is slightly inhibited. Accordingly, only 25% of the phosphorylase phosphatase in

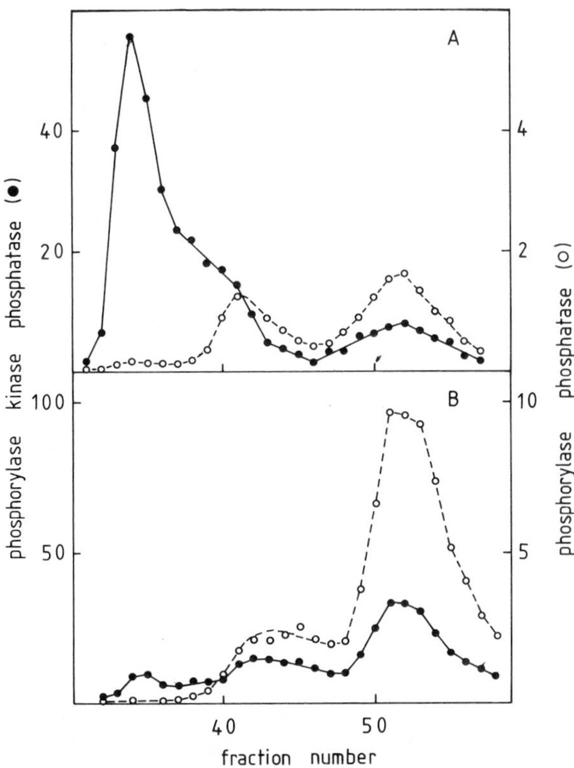

Fig. 2 A, B. Chromatography of rabbit skeletal muscle (**A**) and liver (**B**) cytosol on DEAE-cellulose (12 x 2 cm). The columns were equilibrated with 10 mM Tris-HCl pH 7.0 – 0.1 mM EGTA – 0.1% v/v 2-mercaptoethanol (solution A). Cytosol (35 ml) was applied to the columns which were then washed with solution A containing 50 mM NaCl (100 ml). The columns were then developed with a 300 ml linear gradient from 0.05-0.5M NaCl in solution A. The fractions (5 ml) were assayed for phosphorylase phosphatase (o---o) and phosphorylase kinase phosphatase (●——●) activities in the presence of I_2 (100 units) and 2.0 mM $MnCl_2$ as in [68]. Phosphorylase phosphatase activity is expressed as U/ml, and phosphorylase kinase phosphatase as % counts released per min by 0.02 ml of each fraction

rabbit liver extracts and 80% of the phosphatase in skeletal muscle extracts can be inhibited by I_2, if assays contain Mn^{2+}.

PrP-2b, PrP-2a$_1$, and PrP-2a$_2$ are eluted from DEAE-cellulose at 0.12, 0.20, and 0.28M NaCl respectively (Fig. 2). PrP-1 and the ATP-Mg protein phosphatase both elute at 0.16M NaCl. The elution positions of all five phosphatase activities are identical whether muscle or liver extracts are chromatographed on DEAE-cellulose.

10 The Subunit Structures on PrP-2a$_1$ and PrP-2a$_2$

The resolution of two phosphorylase phosphatases on DEAE-cellulose, which correspond to PrP-2a$_1$ and PrP-2a$_2$, has been described previously in rat liver by Tamura et al. [69]. These workers have now purified both of these enzymes to homogeneity [70, 71]. PrP-2a$_1$ is composed of three subunits, α, β, and γ, whose molecular weights are 35K, 69K, and 59K respectively, whereas PrP-2a$_2$ contains only two subunits, α and β, with molecular weights of 35K and 69K respectively. Both enzymes can be dissociated either by freezing and thawing in 0.2M mercaptoethanol or by treatment with ethanol at room temperature, and these procedures activate the enzyme four to five fold. Gel filtration of the dissociated enzymes demonstrates that the α-subunit (M_r = 35K) is the catalytic subunit. PrP-2a$_1$ therefore appears to differ from PrP-2a$_2$ by the presence of one additional subunit.

The identification of the two protein phosphatases isolated by Tamura et al. as PrP-2a$_1$ and PrP-2a$_2$ is based on their elution behaviour on DEAE-cellulose and Sephadex G200, and on their dissociation to type-2 catalytic subunits following freezing and thawing in 0.2M mercaptoethanol.

Lee and co-workers [72] have also isolated a protein phosphatase from rat liver which contains three subunits with molecular weights 65K, 55K, and 35K respectively. However, the 55K component was present in variable amounts in contrast to PrP-2a$_1$ isolated by Tamura and Tsuiki [71]. The enzyme purified by Lee et al. is undoubtedly PrP-2a, and is likely to be a mixture of PrP-2a$_1$ and PrP-2a$_2$. The tendency of the γ-subunit (M_r = 55K) to dissociate from PrP-2a$_1$ has been noted by Tamura and Tsuik [71].

PrP-2b from skeletal muscle is not activated by freezing and thawing in 0.2M mercaptoethanol, or by treatment with 80% ethanol at room temperature, and this treatment does not dissociate the enzyme to the M_r = 35K catalytic subunit. These properties and its much lower phosphorylase phosphatase/α-phosphorylase kinase phosphatase activity ratio distinguish PrP-2b from PrP-2a$_1$ and PrP-2a$_2$. The α-subunit of PrP-2a$_1$ and PrP-2a$_2$ is clearly the type-2 catalytic subunit present in protein phosphatase C preparations from liver (Sec. 8). PrP-2b does not contain the same catalytic subunit as PrP-2a$_1$ and PrP-2a$_2$ (unpublished work).

11 Classification of Other Protein Phosphatases

Crouch and Safer [73] have purified to homogeneity a protein phosphatase, active on initiation factor eIF-2, from rabbit reticulocytes. This enzyme is composed of one α-subunit (M_r = 38K) and one β-subunit (M_r = 69K). We have demonstrated that this enzyme is a type-2 protein phosphatase based on its pecificity for the α-subunit of phosphorylase kinase and its insensitivity to I_1 and I_2 [51]. Its phosphorylase phosphatase/α-phosphorylase kinase phosphatase activity ratio is similar to the α-subunit (M_r = 35K) derived from PrP-$2a_1$ or PrP-$2a_2$. The subunit structure of this enzyme appears to be $\alpha\beta$, compared to $\alpha\beta_2$ for PrP-$2a_2$ [72, 73].

Pato and Adelstein [74] have purified to homogeneity a protein phosphatase active on both myosin light chains and myosin light chain kinase from turkey gizzard smooth muscle. This enzyme is composed of three subunits, a, β, and γ, with molecular weights of 38K, 60K, and 55K respectively. We have demonstrated that this enzyme is also a type-2 protein phosphatase based on its specificity for the α-subunit of phosphorylase kinase and insensitivity to I_1 and I_2. Its phosphorylase phosphatase/α-phosphorylase kinase phosphatase activity ratio is also similar to the α-subunit (M_r = 35K) derived from PrP-$2a_1$ or PrP-$2a_2$. Its subunit compositon, $\alpha\beta\gamma$, demonstrates that it is very similar to PrP-$2a_1$ which is reported to have the subunit composition $\alpha_2\beta\gamma$ [71, 74].

12 Summary

Many of the protein phosphatases in the literature appear to fall into two groups termed type-1 and type-2. Type-1 protein phosphatases dephosphorylate the β-subunit of phosphorylase kinase and are potently inhibited by I_1 and I_2, whereas type-2 protein phosphatases dephosphorylate the α-subunit of phosphorylase kinase and are insensitive to I_1 and I_2.

Two type-1 protein phosphatases have been identified, namely PrP-1 and the ATP-Mg-dependent protein phosphatase. The properties of these two enzymes are so similar that they may be interconvertible forms of the same enzyme.

Three type-2 protein phosphatases have been identified, namely PrP-2b, PrP-$2a_1$, and PrP-$2a_2$. The latter two enzymes share the same catalytic subunit but differ in that PrP-$2a_2$ lacks one of the two additional subunits. PrP-2b has a much lower activity ratio phosphorylase phosphatase/α-phosphorylase kinase phosphatase. Its subunit composition is unknown.

Several potential methods of regulating type-1 protein phosphatases have already been identified. The regulation of type-2 protein phosphatases is not understood, but the existence of two subunits, β and γ, which suppress the activity of the catalytic α-subunit, suggests that they may be involved in the regulation of the enzyme.

The relative amounts of PrP-1, PrP-2b, and PrP-$2a_1$ + $2a_2$ differ considerably in liver and skeletal muscle from normal fed rabbits, and these enzymes can be quantitated accurately in crude extracts by the use of phosphorylase kinase and phosphorylase as substrates and by the inclusion of I_1 and I_2 in the assays. Studies of the effects of dietary and hormonal manipulations on the relative activities of these phosphatases should clarify their metabolic roles.

Acknowledgements. This work was supported by a Programme grant from the Medical Research Council London and the British Diabetic Association (to PC). Jozef Goris acknowledges an EMBO Long-Term Fellowship and a NATO Research Fellowship, and Thomas Ingebritsen a NATO Post-doctoral fellowship from the National Science Foundation, U.S.A. Alexander Stewart is a post-graduate student of the medical Research Council, London. Samuel Strada was the recipient of a Fogarty International Scholarship.

Note Added in Proof. Since this paper was submitted, a third type-2 protein phosphatase (termed PrP-2c) has been identified. This enzyme (Mr app = 46,000) is Mg^{2+} dependent, and explains several reports in the literature of a glycogen synthase phosphatase which does not dephosphorylate glycogen phosphorylase.

References

1. Krebs E, Fischer EH (1956) Biochim Biophys Acta 20: 150-157
2. Krebs EG, Graves DJ, Fischer EH (1959) J Biol Chem 234: 2867
3. Friedman DL, Larner J (1963) Biochemistry 2: 669-675
4. Krebs EG, Beavo JA (1979) Annu Rev Biochem 48: 23-59
5. Cohen P, Embi N, Foulkes JG, Hardie DG, Nimmo GA, Rylatt DB, Shenolikar S (1979) Miami Winter Symp 16: 463-481
6. Cohen P (1980) In: Cohen P (ed) Molecular aspects of cellular regulation, vol I. Elsevier/North Holland, Amsterdam, pp 255-269
7. Steinberg D (1976) Adv Cyclic Nucleotide Res 7: 157-198
8. Nimmo HG (1980) In: Cohen (ed) Molecular aspects of cellular regulation, vol I. Elsevier/North Holland, Amsterdam, pp 135-152
9. Hardie DG, Guy PS (1980) Eur J Biochem 110: 167-177
10. Brownsey RW, Hardie DG (1980) FEBS Lett 120: 67-70
11. Yagi K, Yazawa M, Kakiuchi S, Oshimo M, Uenishi K (1978) J Biol Chem 253: 1338-1340
12. Cohen P, Burchell A, Foulkes JG, Cohen PTW, Vanaman TC, Nairn AC (1978) FEBS Lett 92: 287-293
13. Cohen P, Klee CB, Picton C, Shenolikar S (1980) Ann NY Acad Sci 356: 151-161
14. Yamauchi T, Fujisawa H (1980) FEBS Lett 116: 141-144
15. Schulman H, Greengard P (1978) Proc Nat Acad Sci USA 75: 5432-5436
16. Yeaman SJ, Cohen P, Watson DC, Dixon GH (1977) Biochem J 162: 411-421
17. Proud CG, Rylatt DB, Yeaman SJ, Cohen P (1977) FEBS Lett 80: 435-442
18. Huang TS, Feramisco JR, Glass DB, Krebs EG (1979) Miami Winter Symp 16: 449-461
19. Parker PJJ, Embi N, Cohen P (1981) FEBS Lett 123: 332-336
20. Embi N, Rylatt DB, Cohen P (1979) Eur J Biochem 100: 339-347
21. Rylatt DB, Aitken A, Bilham T, Condon GD, Embi N, Cohen P (1980) Eur J Biochem 107: 529-537
22. Cohen P, Rylatt DB, Nimmo GA (1977) FEBS Lett 76: 182-186
23. Nolan C, Novoa WB, Krebs EG, Fischer EH (1964) Biochemistry 3: 542
24. Yeaman SJ, Cohen P (1975) Eur J Biochem 51: 93-104
25. Cohen P (1980) Eur J Biochem 111: 563-574
26. Cohen P (1973) Eur J Biochem 34: 1-14
27. Cohen P (1980) FEBS Lett 119: 301-306
28. Singh TJ, Wang JH (1977) J Biol Chem 252: 625-632
29. Wang JH, Stull JT, Huang TS, Krebs EG (1976) J Biol Chem 251: 4521-4527
30. Nimmo HG, Cohen P (1977) Adv Cyclic Nucleotide Res 8: 145-226
31. Cohen P (1978) Curr Top Cell Regul 14: 117-196
32. Embi N, Parker PJJ, Cohen P (1981) Eur J Biochem 115: 405-413
33. Antoniw JF, Cohen P (1976) Eur J Biochem 68: 45-54
34. Antoniw JF, Nimmo HG, Yeaman SJ, Cohen P (1977) Biochem J 162: 423-433
35. Huang FL, Glinsmann WH (1976) Eur J Biochem 70: 419-426
36. Cohen P, Nimmo GA, Antoniw JF (1977) Biochem J 162: 435-444
37. Nimmo GA, Cohen P (1978) Eur J Biochem 87: 341-351, 353-365
38. Foulkes JG, Cohen P (1980) Eur J Biochem 105: 195-203
39. Reed LJ, Pettit FH, Yeaman SJ, Teague WM and Bleile DM (1980) FEBS Symp 60: 47-56
40. Stewart AA, Hemmings B, Cohen P, Goris J, Merlevede W (1981) Eur J Biochem 115: 197-205
41. Embi N, Rylatt DB, Cohen P (1980) Eur J Biochem 107: 519-527
42. Foulkes JG, Cohen P (1979) Eur J Biochem 97: 251-256

43. Khatra BS, Chiasson JL, Shikama H, Exton JH, Soderling TR (1980) FEBS Lett 114: 253-256
44. Cohen P, Nimmo GA, Shenolikar S, Foulkes JG (1978) Krause EG, Pinna L, Wollenberger A (eds) FEBS Symp vol 54, Pergamon Press, Oxford, pp 161-169
45. Foulkes JG, Jefferson LS, Cohen P (1980) FEBS Lett 112: 21-24
46. Kuo JF, Greengard P (1969) Proc Nat Acad Sci USA 64: 1349-1355
47. Burchell A, Foulkes JG, Cohen PTW, Condon GD, Cohen P (1978) FEBS Lett 92: 68-72
48. Ingebritsen TS, Gibson DM (1980) In: Cohen P (ed) Molecular aspects of cellular regulation, vol. I, Elsevier/North Holland, Amsterdam, pp 63-94
49. Boyd GS, Gorbon AM (1980) In: Cohen P (ed) (1980) Molecular aspects of cellular regulation, vol. I. Elsevier/North Holland, Amsterdam, pp 95-134
50. Hardie DG, Cohen P (1979) FEBS Lett 103: 333-338
51. Stewart AA, Crouch D, Cohen P, Safer B (1980) FEBS Lett 118: 16-19
52. Hardie DG (1980) In: Cohen P (ed) Molecular aspects of cellular regulation, vol. I. Elsevier/North Holland, Amsterdam, pp 33-62
53. Merlevede W, Riley GA (1966) J Biol Chem 241: 3521-3524
54. Merlevede W, Goris J, DeBrandt C (1969) Eur J Biochem 11: 499-502
55. Chelala CA, Torres HN (1970) Biochim Biophys Acta 198: 504-513
56. Tellez-Inon MT, Torres HN (1973) Biochim Biophys Acta 297: 399-412
57. Kalala LR, Goris J, Merlevede W (1973) FEBS Lett 32: 284-288
58. Goris J, Defreyn G, Merlevede W (1979) FEBS Lett 99: 279-282
59. Goris J, Dopere F, Vandenheede JR, Merlevede W (1980) FEBS Lett 117: 117-121
60. Yang SD, Vandenheede JR, Goris J, Merlevede W (1980) FEBS Lett 111: 201-204
61. Brandt H, Capulong ZL, Lee EYC (1975) J Biol Chem 250: 8038-8044
62. Chou CK, Alfano J, Rosen OM (1977) J Biol Chem 252: 2855-2859
63. Li HC, Hsiao KJ, Chan WWS (1978) Eur J Biochem 84: 215-225
64. Killilea SD, Aylward JK, Mellgren RL, Lee EYC (1978) Arch Biochem Biophys 191: 638-646
65. Khandelwahl RL (1979) Can J Biochem 57: 1337-1343
66. Titanji VPK, Zetterqvist O, Engstrom L (1980) FEBS Lett 111: 209-213
67. Lee EYC, Mellgren RL, Killilea SD, Aylward JT (1978) Esman V (ed) FEBS Symp, vol 43. Pergamon Press, Oxford, pp 327-346
68. Ingebritsen TS, Foulkes JG, Cohen P (1980) FEBS Lett 119: 9-15
69. Tamura S, Kikuchi K, Hiroya A, Kikuchi H, Hasokawa M, Tsuiki S (1978) Biochem Biophys Acta 524: 349-356
70. Tamura S, Kikuchi H, Kikuchi K, Kiroya A, Tsuiki S (1980) Eur J Biochem 104: 347-355
71. Tamura S, Tsuiki S (1980) Eur J Biochem 111: 217-224
72. Lee EYC, Aylward JH, Mellgren RL, Killilea SD (1979) Miami Winter Symp 16: 483-499
73. Crouch D, Safer B (1980) J Biol Chem 255: 7918-7924
74. Pato MD, Adelstein RS (1980) J Biol Chem 255: 6535-6538

Discussion[4]

Dr. Sparrow asked whether there is any Ca^{2+}-regulation of the phosphatases.

Dr. Cohen answered that very recent work had demonstrated that PrP-2b is a Ca^{2+} dependent enzyme that is stimulated 10-fold by calmodulin. PrP-1 and PrP-2a are unaffected by Ca^{2+} and calmodulin in the presence or absence of I_1 and I_2.

4 Rapporteur: Ernst-Günter Afting

The Mechanism of Action of Brain Cyclic AMP-dependent Protein Kinase: Structure of Active Site, Mechanism of Dissociation of the Holoenzyme, and Possible Biological Role of Its Subunits

E.S. Severin, S.N. Kochetkov, A.G. Gabibov, M.V. Nesterova, I.N. Trakht, L.P. Sashchenko, S.V. Shlyapnikov, and I.D. Grozdova[1]

In recent years great attention has been devoted to the investigation of the biological role of cyclic AMP-dependent protein kinases. Intensive studies have been carried out on the effect of cyclic nucleotides and protein phosphorylation on the processes of proliferation and differentiation of cells. In this connection the role of protein kinases in transcription, biosynthesis of proteins, and regulation of various enzymes is being currently investigated. In this paper some new data on the mechanism of action of cAMP-dependent pig brain protein kinase are presented and the biological role of phosphorylation in the cell cycle is discussed.

1 Mechanism of Phosphotransferase Reaction

The catalytic subunit of cAMP-dependent pig brain protein kinase is a protein of molecular weight 38,000. It transfers the terminal phosphate of ATP ($K_{M_{ATP}} = 1.2 \cdot 10^{-5}$M) to serine residues of lysine-rich histones, namely: Ser 38 in histone H1, Ser 18 in H2a, Ser 14 and Ser 36 in histone H2b (Severin et al. 1978).

The kinetic mechanism of the phosphotransferase reaction was studied by the method of Cleland. Evidence for the Ping-Pong Bi Bi mechanism of the phosphotransferase reaction can be clearly seen from Fig. 1. The Lineweaver-Burk plot of the reaction rate vs concentration of substrates (ATP and histone H1) is a series of parallel lines. The reaction products ADP and phosphohistone H1 are competitive with histone H1 and ATP, respectively. Accordingly, ADP does not compete with ATP, and phosphohistone H1 is not competitive with histone H1 [Gabibov et al. to be published (a)]. The kinetics of the phosphotransferase reaction suggests the formation of a relatively labile intermediate phosphoryl protein.

As seen in Fig. 2, incubation of the catalytic subunit with excess [γ-^{32}P] ATP in the absence of the second substrate, i.e., histone H1, led to incorporation of one equivalent of the radioactive phosphate. The individual phosphoform of the catalytic subunit was isolated by Sephadex G-25 gel filtration. Thereupon, pH dependence of stability of the phospho-protein bond was studied and the constant of hydrolysis of phosphoryl protein at pH 4.1 was calculated.

Abbreviations used: cAMP, adenosine 3′, 5′-monophosphate; R, regulatory subunit; C, catalytic subunit; oATP, sodium periodate oxidized dialdehyde derivative of ATP

1 Institute of Molecular Biology, USSR Academy of Sciences, Moscow, USSR

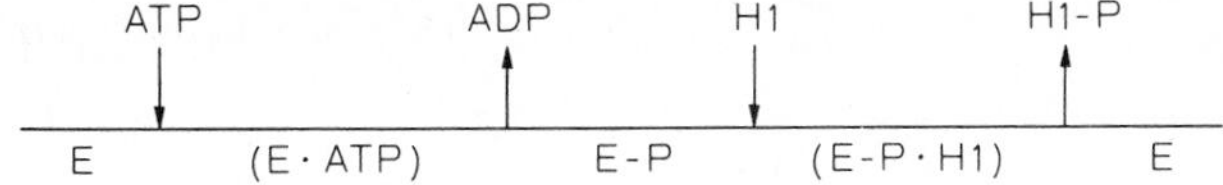

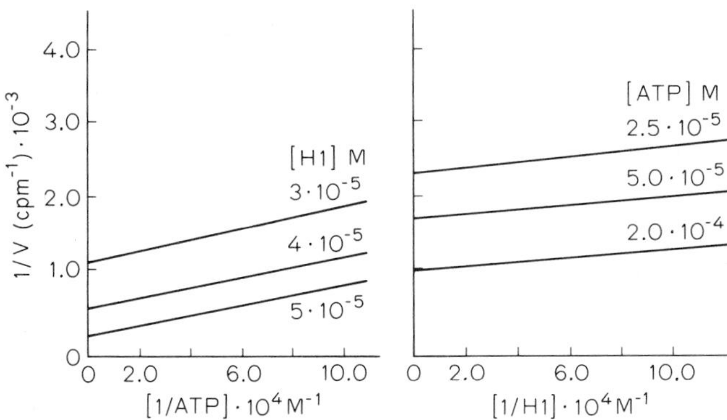

Fig. 1. Lineweaver-Burk plot of the phosphotransferase reaction vs concentration of substrate. *Top,* scheme of mechanism of Ping-Pong Bi Bi type of the reaction. *Right,* reaction rate as a function of ATP concentration; *left,* reaction rate as a function of histone H1 concentration. Enzyme concentration $1.0 \cdot 10^{-7}$ M

As a result of total alkaline hydrolysis of the phosphoform, 3-N-phosphohistidine residue was identified. The [^{32}P]-phosphohistidine-containing peptide was isolated after tryptic digestion (Kochetkov et al. 1976; Kochetkov et al. 1977). The data obtained permitted us to assume that the histidine residue in the active site acts as the catalytic group of protein kinase which is directly involved in ATP cleavage.

In accordance with the established kinetic mechanism of the Ping-Pong type, the phosphotransferase reaction can be subdivided into two half-reactions:

1. formation of the enzyme phosphoform

$$E + ATP \rightleftarrows E \cdot ATP \rightleftarrows E - P, \text{ and}$$

2. transfer of phosphate to the second substrate, namely histone H1

$$E - P + H1 \rightleftarrows E - P \cdot H1 \rightleftarrows E + H1 - P.$$

When this reaction proceeds in the absence of the second substrate, the phosphoform obtained is subjected to hydrolysis, and thus the ATPase activity of protein kinase is observed.

Using the quenched flow method, we have determined the relevant constants (K_S, k_2, k_3, and k_4) of the phosphotransferase reaction which are shown in the scheme below [Gabibov et al. to be published (b)].

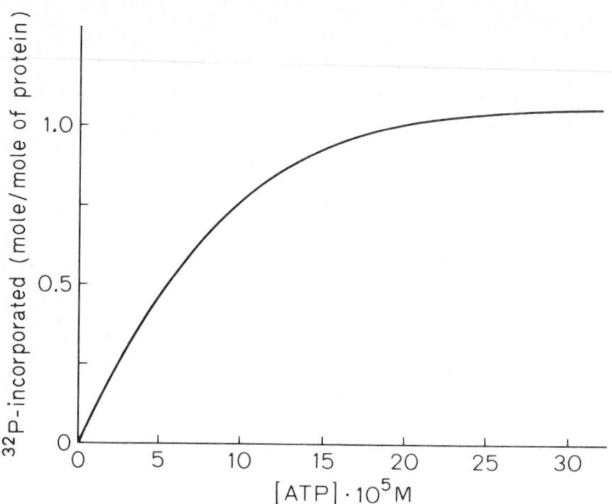

Fig. 2. Incorporation of $[^{32}P]$-phosphate into catalytic subunit upon incubation with $[\gamma\text{-}^{32}P]$ ATP. Enzyme concentration $2.5 \cdot 10^{-5}$ M (10 mM Tris · HC1 buffer pH 7.5 containing 10 mM $MgC1_2$; incubation time 15 min, at 30°C)

$$E + ATP \underset{}{\overset{K_S = 5 \cdot 10^{-5}M}{\rightleftarrows}} E \cdot ATP \xrightarrow[ADP]{k_2 = 27\ sec^{-1}} E - P \xrightarrow{k_4 = 42\ sec^{-1},\ H1} E + H1 - P.$$

$$E - P \xrightarrow{H_2O,\ k_3 = 7.6\ sec^{-1}} E + P_i$$

As follows from the scheme, the constant of hydrolysis of phosphoryl enzyme (k_3 = 7.6 sec^{-1}) is lower than the constant of formation of the enzyme phosphoform (k_2 = 27 sec^{-1}). Such a ratio of constants in the absence of the second substrate should lead to accumulation of the phosphoryl enzyme, and this enabled us to obtain it in the individual state. It is also interesting to mention that the constant of rate of phosphate transfer to the histone (k_4 = 42 sec^{-1}) is approximately six times higher than the constant of hydrolysis of the phosphoform (k_3 = 7.6 sec^{-1}), which is indicative of a highly specific character of the second half-reaction probably due to protein-protein interactions. Besides, the ratio k_4/k_3 indicates that formation of the enzyme phosphoform can be detected only in the absence of the second substrate.

2 Functional Groups of the Active Site of Protein Kinase

In order to study the structure of the ATP-site and to elucidate the nature of functional groups of the active site of protein kinase, we investigated the interaction of the enzyme with various reversible and irreversible inhibitors, modified ATP analogs.

The comparison of the affinity for the ATP-site of various nucleotide and nucleoside derivatives (Table 1) permitted us to conclude that the main role in recogni-

Table 1. Interaction of catalytic subunit of protein kinase with nucleotide and nucleoside derivatives

Compound	$k_i \cdot 10^{-4}$ M
ATP	0.12 (K_mApp)
ADP	0.80
AMP	1.12
Ado	0.74
Ara-Ado	0.96
2'-d-Ado	1.65
3', 5'-cAMP	6.80
2', 3'-cAMP	56.0
GTP	15.0
UTP	5.5
CTP	2.4
1,N^6-etheno-ADP	32.0
IMP	145.0
8-Br AMP	26.0

tion and binding of the substrate by the enzyme was played by the heterocyclic base since consecutive "removal" of the phosphate groups did not lead to significant decrease in the affinity (Kochetkov et al. 1978).

The heterocyclic base is likely to be bound in the hydrophobic cleft due to the stacking interactions. The presence of a free amino group in position 6 of the base plays an important role in this binding. There are no strict limitations as to the conformation of the ribose residue (Kochetkov et al. 1978). We also synthesized various modified ATP analogs containing reactive groups and capable of covalent blocking of functional groups of the active site (Table 2).

One of these analogs, namely adenosine chloromethane pyrophosphonate, appeared to be an irreversible inhibitor of the enzyme; ATP protected the enzyme from irreversible inhibition. Once the [^{14}C]-radioactive analog was used, incorporation of 1.2 equivalent of the radioactive label per one mol of the enzyme was observed. pK of the modified group of the protein was about 6.5. On the basis of these findings one can suggest that the histidine residue of the active site undergoes modification (see Fig. 3) (Gulyaev et al. 1976).

Another ATP derivative, namely fluorosulfophenyl analog of adenylic acid, also irreversibly inhibited the enzyme. In this case, pK of the modified group was about 4.7 which might indicate the formation of an anhydride bond with the γ-carboxyl group of glutamic or aspartic acids of the active site (see Fig. 4) (Gulyaev et al. 1976). Treatment of the enzyme-inhibitor complex with the [^{14}C]-labeled O-methylhydroxyamine resulted in specific incorporation of the [^{14}C]-label into the protein that was characteristic of derivatives of similar type.

Finally, treatment of the catalytic subunit with o-ATP followed by sodium borohydride reduction brought about irreversible inhibition of the enzyme activity (Fig. 5). This reaction proceeded strictly in the ATP-site of the active center (protection from inhibition with excess ATP, incorporation of one equivalent of the [^{14}C]-labeled inhibitor, isolation of the individual [^{14}C]-labeled peptide after tryptic digestion).

Table 2. Interaction of catalytic subunit of protein kinase with synthesized ATP analogs

No.	ATP analogs	Protein kinase Muscle	Brain
1	$(BrCH_2CH_2)pA$	$2 \cdot 10^{-4}$	$2 \cdot 10^{-4}$
2	$(BrCH_2CH_2)ppA$	$0.7 \cdot 10^{-4}$	$1.7 \cdot 10^{-4}$
3	$(BrCH_2CH_2)pppA$	$10 \cdot 10^{-4}$	–
4	$(ClCH_2)pA$	$1 \cdot 10^{-4}$	$4 \cdot 10^{-4}$
5	$(ClCH_2)ppA$	0.017	0.05
6	$(ClCH_2)pppA$	$7.9 \cdot 10^{-4}$	–
7	$(ClCH_2CH_2O)pA$	$0.7 \cdot 10^{-4}$	0.03
8	$(ClCH_2CONHCH_2)pA$	–	–
9	$(BrCH_2CONHCH_2CH_2O)ppA$	$5.3 \cdot 10^{-4}$	–
10	$pp(CH_2)pA$	$0.023 \cdot 10^{-4}$	–
11	p(CH)pA \\ $BrCH_2CONH$	0.013	–
12	p(CH)pA-2′, 3′-O-isopropylidene | $HNCOCH_2Br$	–	–
13	$p(CH_2)ppA$	$5 \cdot 10^{-4}$	–
14	p(CH)ppA \\ $BrCH_2CONH$	0.027	–
15	$p(CH_2CH_2)ppA$	$14 \cdot 10^{-4}$	–
16	$(FSO_2C_6H_4)pA$	0.043	0.067
17	$(FSO_2C_6H_4CO)A$	0.019	–
18	$p(CH_2)ppA\text{-}8\text{-}NH(CH_2)_6NHCOCH_2Cl$	–	–
19	$pppA\text{-}8\text{-}NH(CH_2)_6NHCOCH_2Cl$	$1.2 \cdot 10^{-4}$	–
20	$pppA\text{-}8\text{-}NH(CH_2)_2NHCOCH_2Cl$	–	–
21	pppA-8-Br	$4 \cdot 10^{-4}$	–

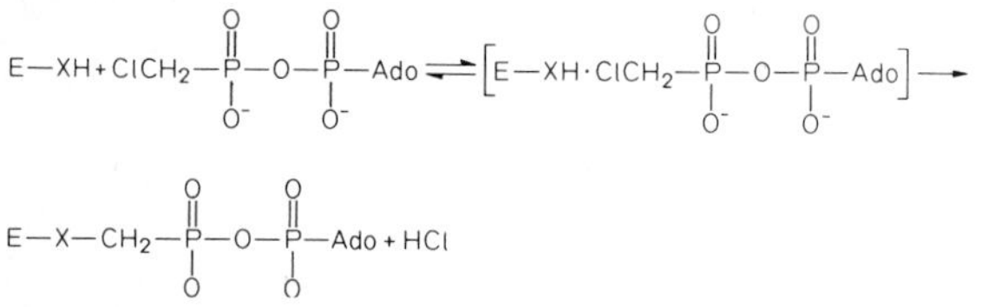

$$E{-}XH + ClCH_2{-}\overset{O}{\overset{\|}{\underset{O^-}{\underset{|}{P}}}}{-}O{-}\overset{O}{\overset{\|}{\underset{O^-}{\underset{|}{P}}}}{-}Ado \rightleftharpoons \left[E{-}XH \cdot ClCH_2{-}\overset{O}{\overset{\|}{\underset{O^-}{\underset{|}{P}}}}{-}O{-}\overset{O}{\overset{\|}{\underset{O^-}{\underset{|}{P}}}}{-}Ado\right] \longrightarrow E{-}X{-}CH_2{-}\overset{O}{\overset{\|}{\underset{O}{\underset{|}{P}}}}{-}O{-}\overset{O}{\overset{\|}{\underset{O}{\underset{|}{P}}}}{-}Ado + HCl$$

Fig. 3. Interaction of catalytic subunit with adenosine 5′-chloromethylpyrophosphonate

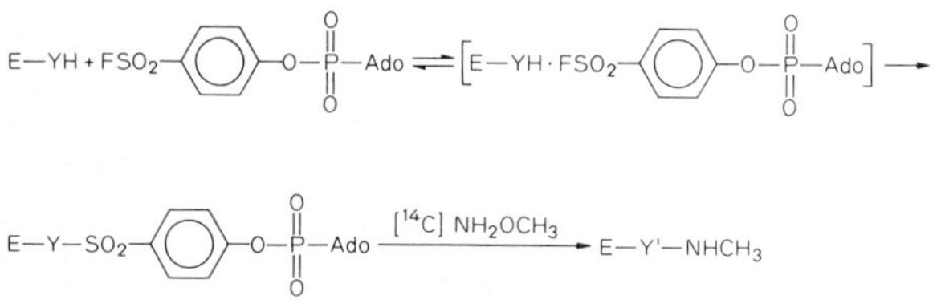

$$E{-}YH + FSO_2{-}C_6H_4{-}O{-}\overset{O}{\overset{\|}{\underset{O}{\underset{\|}{P}}}}{-}Ado \rightleftharpoons \left[E{-}YH \cdot FSO_2{-}C_6H_4{-}O{-}\overset{O}{\overset{\|}{\underset{O}{\underset{\|}{P}}}}{-}Ado\right] \longrightarrow$$

$$E{-}Y{-}SO_2{-}C_6H_4{-}O{-}\overset{O}{\overset{\|}{\underset{O}{\underset{\|}{P}}}}{-}Ado \xrightarrow{[^{14}C]\ NH_2OCH_3} E{-}Y'{-}NHCH_3$$

Fig. 4. Interaction of the catalytic subunit with adenosine 5′-(p-fluorosulfophenylphosphate)

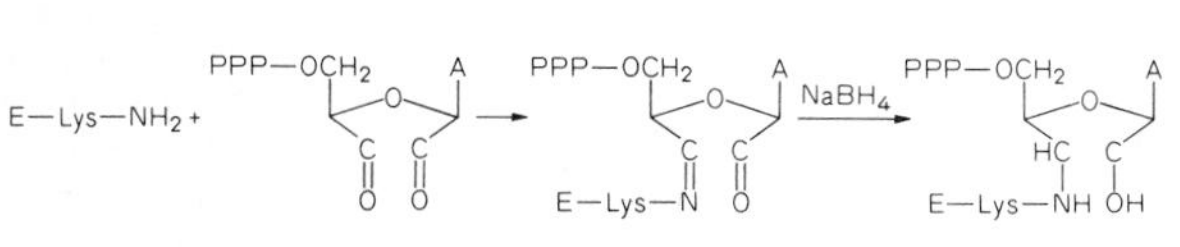

$$E{-}Lys{-}NH_2 + PPP{-}OCH_2\text{(ribose dialdehyde, A; C=O, C=O)} \longrightarrow PPP{-}OCH_2\text{(A; C=N{-}Lys{-}E, C=O)} \xrightarrow{NaBH_4} PPP{-}OCH_2\text{(A; HC{-}NH{-}Lys{-}E, C{-}OH)}$$

Fig. 5. Interaction of the catalytic subunit with oATP

These studies have shown that the lysine residue of the active site is subjected to selective modifications. Besides, the following amino acid sequence for the isolated [^{14}C]-labeled peptide was established: Val – Gly – Pro – Lys – Arg.

An assumption can be made that this peptide is a part of the cationic locus of the active site which is involved in binding of the negatively charged polyphosphate chain of ATP (Kochetkov et al. 1977).

Thus, the ATP-binding site of the active center of protein kinase consists of at least three loci (Fig. 6):

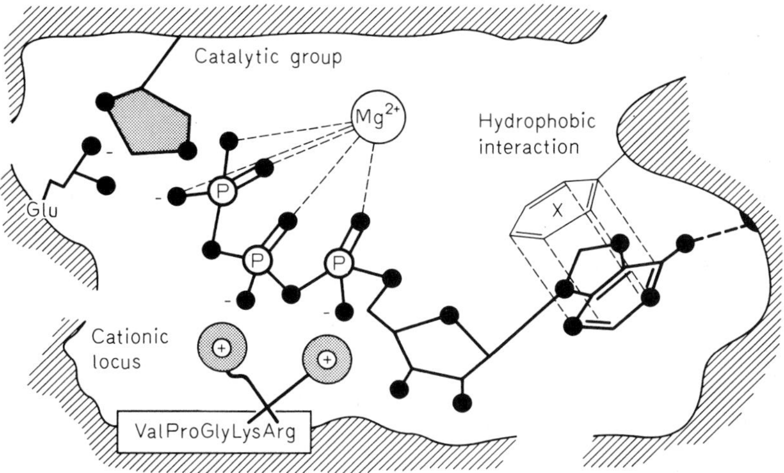

Fig. 6. Structure of the ATP-binding site of the active center of catalytic subunit of protein kinase

1. the hydrophobic cleft of the adenine moiety binding;
2. the cationic locus that binds the polyphosphate chain of ATP (Lys, Arg);
3. the catalytic site (His).

3 A Possible Model for Activation of cAMP-dependent Protein Kinase. Reaction of Autophosphorylation of the Regulatory Subunit and Mechanism of cAMP Binding

Two main types of protein kinases – type I and type II – can be distinguished by certain details in the dissociation mechanism, in particular with regard to modulation of this process by the reaction of autophosphorylation (Corbin et al. 1975).

This process was discovered and described by Rosen et al. several years ago for bovine muscle protein kinase (Rosen and Erlichman 1975).

In this paper we summarize some data obtained on the role of the autophosphorylation reaction in the mechanism of dissociation of pig brain enzyme-protein kinase type II (Ulmasov et al. 1980a, Ulmasov et al. 1980b).

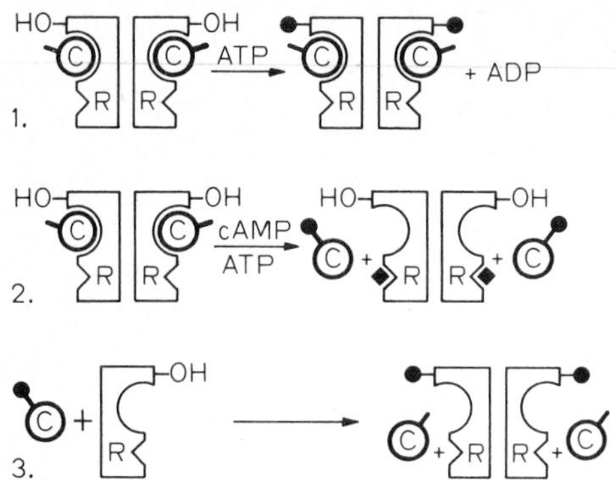

Fig. 7. Autophosphorylation of protein kinase. *1* Autophosphorylation of holoenzyme; *2* autophosphorylation of holoenzyme in the presence of cAMP; *3* transfer of phosphate residue from catalytic subunit to regulatory one

In the course of our investigations we have detected the following regularities characteristic of the autophosphorylation reaction of the holoenzyme of cAMP-dependent protein kinase (Fig. 7):

1. in the presence of [γ-^{32}P] ATP, 1.6 mol of phosphate is incorporated per one mol of the holoenzyme;
2. the reaction of autophosphorylation occurs only within a tetrameric complex of the holoenzyme (R_2C_2) and is inhibited by addition of cAMP;
3. incubation of the isolated [^{32}P]-labeled phosphoform of the catalytic subunit with the regulatory subunit under the conditions of the autophosphorylation reaction leads to the transfer of the radioactive phosphate to the regulatory subunit;
4. there is no cooperativity in the process of cAMP binding, and phospho- and dephosphoforms of the holoenzyme bind 2 mol of cAMP per one mol of the holoenzyme (R_2C_2);
5. and finally, the phosphoholoenzyme is more sensitive to cAMP (Kd_{cAMP} = $1.0 \cdot 10^{-9}$M) than dephosphoholoenzyme ($Kd_{cAMP} = 7.5 \cdot 10^{-9}$M).

On the whole, the autophosphorylation reaction can be subdivided into a number of discrete stages (Fig. 8): at stage I, binding of ATP in the active site of the catalytic subunit within the holoenzyme takes place; then hydrolysis of the phosphoether bond of the ATP molecule proceeds with the formation of an intermediate phosphoform of the enzyme which contains a phosphohistidine residue in the catalytic subunit (stage II); and, finally, the proper autophosphorylation reaction of the regulatory subunit occurs which consists in the transfer of the phosphate residue from the catalytic subunit to the regulatory one (Ser or Thr residues being phosphorylated) (stage III).

It is important to note that phosphorylation of both the protein kinase substrate (for example, histone H1) and the regulatory subunit involves one and the same active site of the catalytic subunit, and phosphorylation of the regulatory subunit takes place only within the holoenzyme through formation of the intermediate phosphoform of the catalytic subunit.

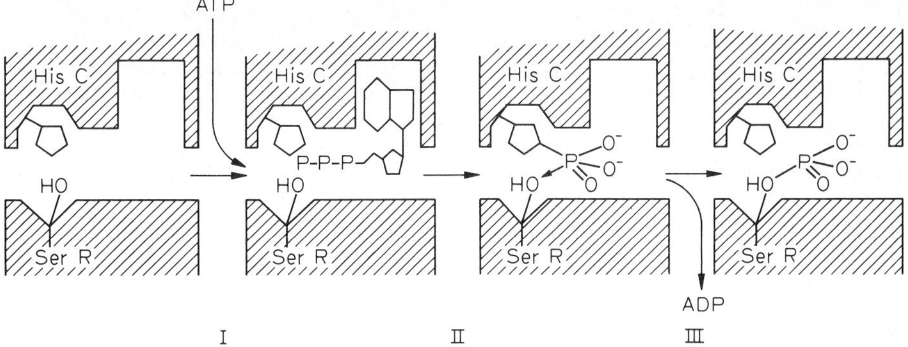

Fig. 8. Mechanism of autophosphorylation reaction of protein kinase. *I*, Formation of noncovalent complex of holoenzyme with ATP; *II*, transfer of γ-phosphate from ATP to histidine residue of catalytic subunit; *III*, transfer of phosphate from histidine residue of catalytic subunit to serine residue of regulatory subunit

An important step in the mechanism of protein kinase activation is binding of cAMP in the allosteric site of the enzyme. What is the mechanism of allosteric regulation of cAMP, which elements of the cAMP molecule are responsible for binding in the allosteric site of cAMP? These questions are being thoroughly and intensively investigated in various laboratories though they are still far from being finally solved.

The cAMP molecule in solution is known to have preferentially *anti*-conformation (data of NMR-spectroscopy, conformational analysis and circular dichroism). At the same time, the 8-substituted cAMP analogs which have capacity to activate the holoenzyme are believed to exist in solution preferentially in *syn*-conformation (Tunitskaya et al. 1977).

Difference in conformation can be observed in circular dichroism spectra (see Fig. 9A) of cAMP (*anti*-conformation in solution, negative maximum in the region of 260 nm) and 8-Br-cAMP (*syn*-conformation, slight positive maximum in the region of 260 nm and negative maximum at 280 nm).

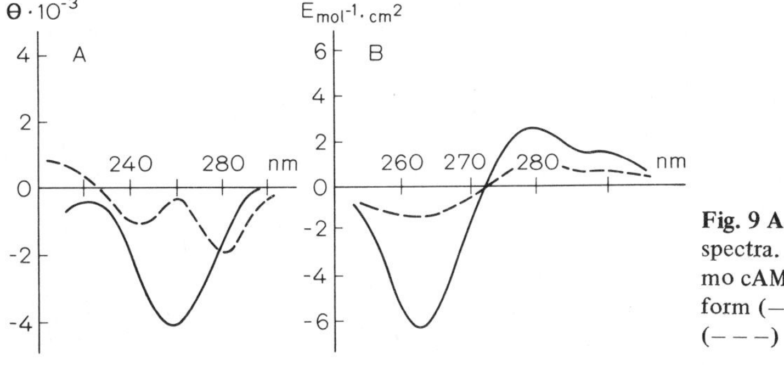

Fig. 9 A, B. Circular dichroism spectra. **A**, cAMP (——), 8-bromo cAMP (– – –); **B**, Phosphoform (——), dephosphoform (– – –)

In the studies of circular dichroism spectra of the holoenzyme, addition of cAMP was observed to cause significant changes in the region of 265 nm (Fig. 9B).

On the whole, binding of cAMP to protein kinase can be represented as consisting as a minimum of two stages (Fig. 10). First, the initial recognition by the protein of the cyclophosphate ring and the ribose moiety of the cAMP molecule; conformation of the purine base of the nucleotide is not determining at this stage. At the first stage, the electrostatic interaction of the negatively charged cyclophosphate system of cAMP with the cationic locus of the enzyme is most important for binding.

At the second stage, fixation of the adenine moiety of cAMP in the hydrophobic cleft takes place. Probably, at this stage certain conformation of cAMP in the allosteric site of the regulatory subunit becomes fixed.

The above considerations allowed one to assume that the interaction of cAMP with the phosphorylated form of the cAMP-dependent protein kinase is accompanied by induced changes both in the cyclic nucleotide and the holoenzyme and leads to the complete dissociation of the protein kinase phosphoform into subunits.

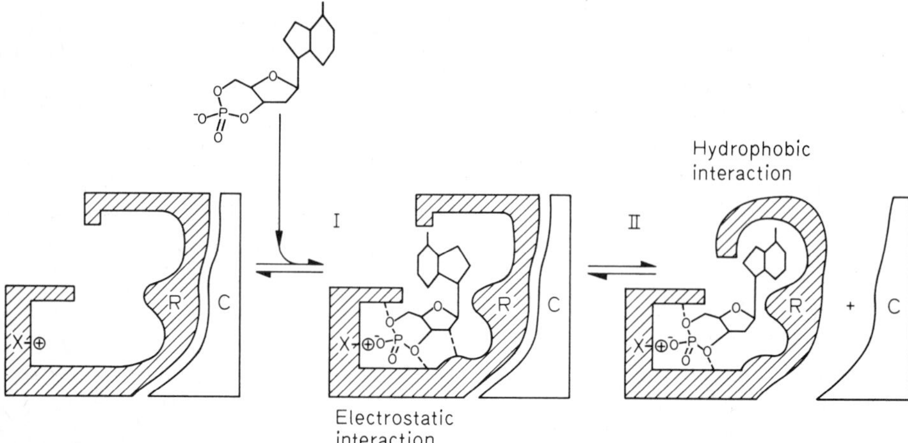

Fig. 10. Scheme of protein kinase dissociation under cAMP action. *I*, Binding of cAMP to regulatory subunit within holoenzyme; *II*, holoenzyme dissociation

4 Study of the Dynamics of Phosphorylation Processes Throughout the Cell Cycle

Using the plasmodial stage of a true slime mould *Physarum polycephalum* and a number of model systems, we studied phosphorylation of histones and structure-functional significance of this modification. Investigation of the protein kinase activity throughout the cell cycle of *Physarum polycephalum* revealed two types of histone phosphorylation: cAMP-dependent and cAMP-independent, and discrete time periods for these processes in the cell were established (Trakht et al. 1980a) (Fig. 11).

Cyclic AMP-dependent protein kinase has the maximum activity in S phase; the protein kinases which do not depend on cyclic nucleotides are most effective prior to

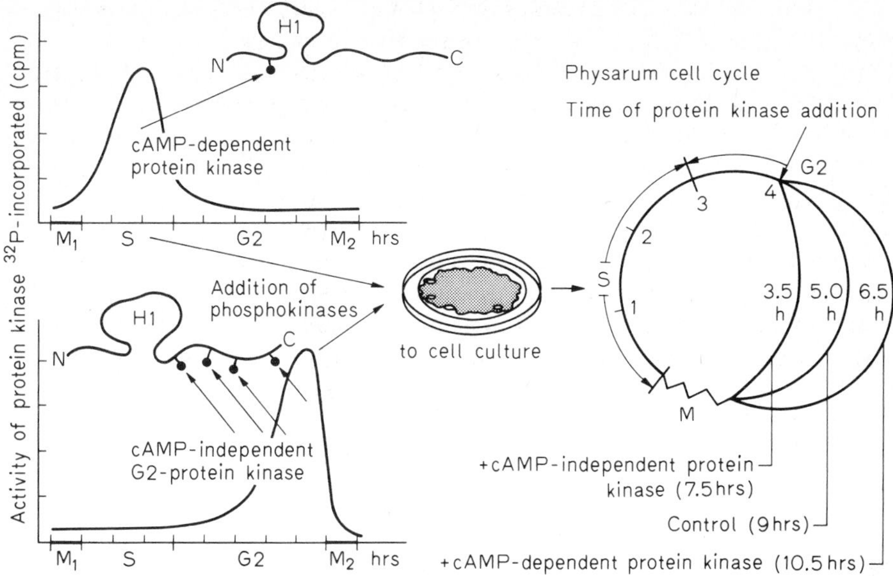

Fig. 11. Regulatory role of cAMP-dependent and cAMP-independent protein kinases throughout the cell cycle of *Physarum polycephalum*

mitosis, i.e., in the period of the most intensive histone phosphorylation. These enzymes are likely to play a certain role in the processes of chromatin compactization (Trakht et al. 1980a). For *Physarum polycephalum,* histone H1 was shown to be phosphorylated in vivo at all stages of the cell cycle; the degree of phosphorylation gradually increasing on approaching mitosis. Prior to mitosis, the number of phosphate groups attached to the molecule of histone H1 increased to as high as 7. The addition of cAMP-dependent pig brain protein kinase which possesses the ability of N-terminal phosphorylation (like protein kinase from S-stage of the cell cycle of *Physarum polycephalum*), as well as of the catalytic subunit of the enzyme, caused a delay of mitosis of about 1.5 h (Trakht et al. 1980b). On the other hand, addition of cAMP-independent protein kinase which is capable of phosphorylating the C-terminal part of histone H1 led to an advancement of mitosis by 1.5 h (Trakht et al. 1980b).

Thus for this case, the two types of phosphorylation considered above differ in their biological role:

1. cAMP-dependent phosphorylation that activates the process of transcription and controls the enzyme activity (during G1 and S stages of the cell cycle);
2. cAMP-independent phosphorylation that affects structural organization of chromatin fibers (G2 and M stages of the cell cycle) (Trakht et al. 1980c).

Besides, in our experiments the cAMP content has also been shown to change throughout the cell cycle; the lowest cAMP concentration being characteristic of mitotic cells, and the highest, of interphase (quiescent) cells.

It was interesting to find out whether these regularities are true for the whole organism in the process of embryonic development, regeneration, etc. As an object for our investigations, planaria *Dugesia tigrina* was chosen; one of its peculiarities being the unusual ability of regenerating the whole organism from 1/100 part of the body.

Normally, the cAMP content in the planaria is about 35 pmol/mg of the protein. During the first 30 min after sectioning of the planaria, cAMP level sharply decreases (Fig. 12A). In 3 h, the content begins to recover, and in a day achieves the initial level.

On the other hand, treatment of the sectioned planaria with dibutyryl-cAMP of different concentrations significantly inhibits regeneration (Fig. 12B).

These findings suggest that the possible role of cAMP in living organisms consists in inhibition of proliferation processes.

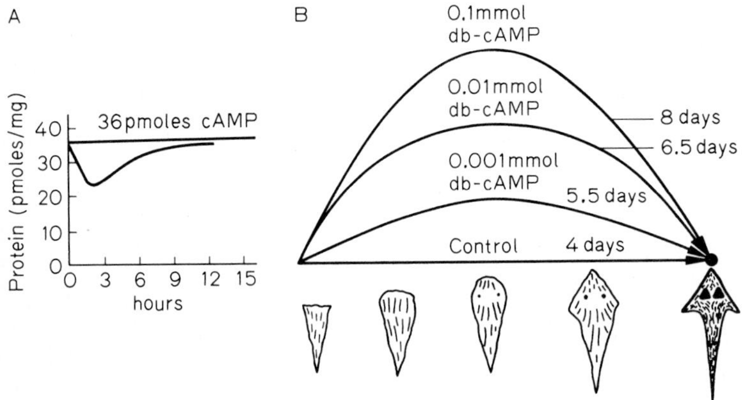

Fig. 12 A, B. Regeneration of planaria *Dugesia tigrina.* **A** Endogenous concentration of cAMP in the sectioned planaria; **B** Regeneration time of planaria after sectioning

5 Nuclear Translocation and Effect of the Regulatory Subunit of cAMP-dependent Protein Kinase on Chromatin Template Activity

While the main function of the catalytic component of protein kinase as a factor modifying properties of proteins is more or less clear, the biological role of the regulatory subunit of protein kinase in the nucleus still remains obscure.

In order to check the assumption that the regulatory subunit can affect the level of transcription by interacting directly with the template like the cAMP-binding protein of bacteria, we have studied changes in the template activity of chromatin after treatment with the regulatory subunit in vitro. It appeared that addition of the homogenous regulatory subunit of the pig brain protein kinase into the RNA-polymerase reaction mixture causes an increase in the template activity of chromatin (Nesterova et al. 1980) (Fig. 13). A high specificity of such effect of the regulatory subunit on chromatin is also evidenced by the data obtained by us on the phospho-

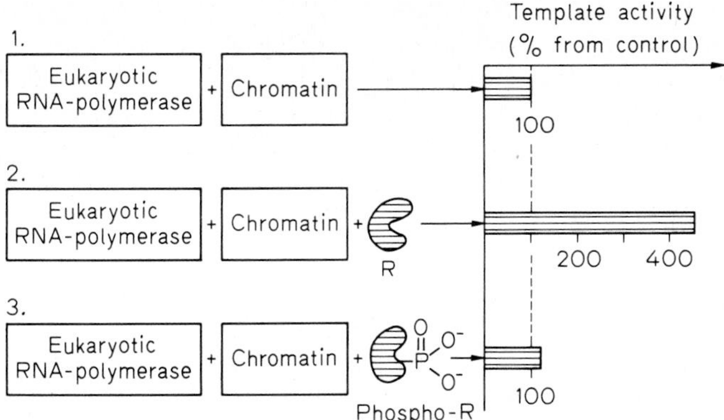

Fig. 13. Effect of regulatory subunit of protein kinase on chromatin template activity. *1*, Control template activity; *2*, template activity after addition of regulatory subunit; *3*, template activity after addition of phosphoform of regulatory subunit

form of the regulatory subunit. The latter has been found not to affect the level of transcription in this system at all. Another support in favor of specific interaction of the regulatory subunit with the template of RNA synthesis is the constant of binding of this protein to brain chromatin which equals 10^{-10} M.

Further experiments have shown that the template activity of chromatin after treatment of the latter with the regulatory subunit changes due to the increase in the number of initiation sites of RNA synthesis (Nesterova et al. 1980).

In the light of these findings, particular interest acquires the investigation of the mechanisms of nuclear translocation and acception of protein kinase by structural elements of genome in normal and tumor cells.

To study the mechanism of nuclear translocation of protein kinase and acception by chromosomes of the tritium-labeled protein kinase and its subunits, the autoradiography method was used for different cell cultures in the normal state and at neoplastic transformation (Nesterova et al. 1980). In normal cells, like Chinese hamster fibroblasts (CHO) or mouse fibroblasts (3T3), we observed a pronounced translocation of exogenous protein kinase and its subunits separately into interphase nuclei and acception of these proteins by metaphase chromosomes (Fig. 14a). Whereas nuclear translocation of the [^{3}H]-regulatory subunit and acception by chromosomes of the SV40 transformed 3T3 cells were practically not detected (Fig. 14b). Addition into the system of the regulatory subunit-cAMP complex led to essential recovery of nuclear translocation of the labeled protein and its acception by metaphase chromosomes (Fig. 14c).

The results obtained demonstrate that transformation of 3T3 cells with SV40 virus does not lead to any changes in specific sites for acception of the regulatory subunit by chromatin, but affects the system of cAMP synthesis that is directly connected with transport of cAMP-binding proteins through the membrane. These data also suggest that nuclear translocation of the regulatory subunit occurs irrespective of the presence of the catalytic subunit and depends on the specific interaction with cAMP molecules.

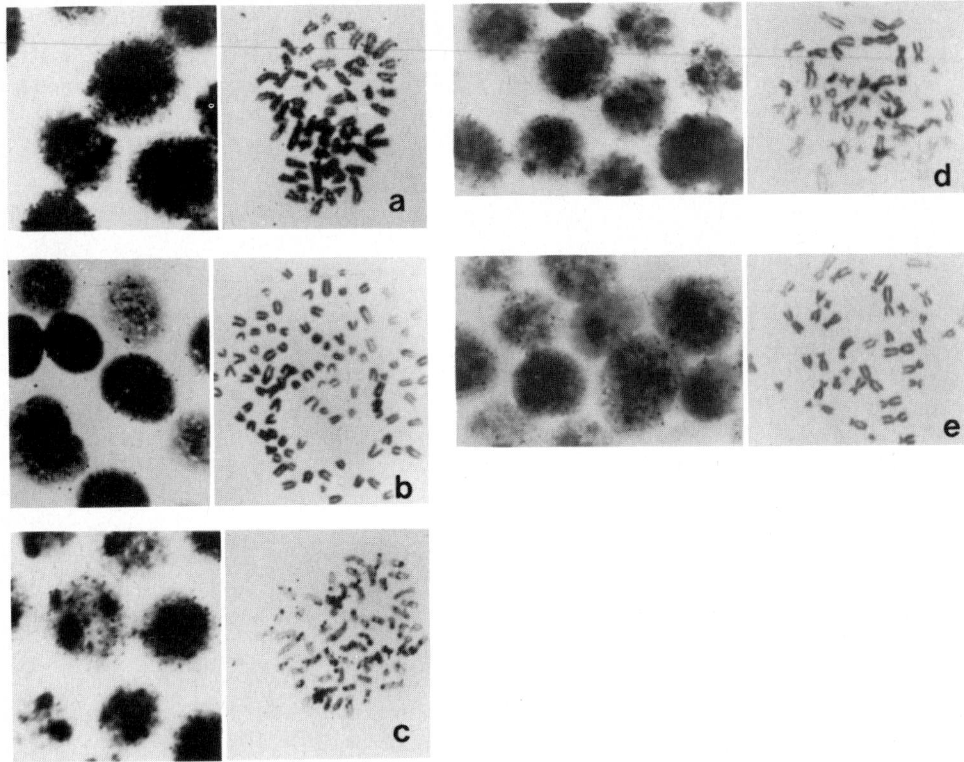

Fig. 14 a-e. Nuclear translocation and acception by chromosomes of ^{3}H-regulatory subunit (R*). **a** 3T3 cells + R*; **b** 3T3 (SV40) + R*; **c** 3T3 (SV40) + cAMP + R*; **d** KB + R*; **e** KB + R* + cAMP

At the next step of our studies we compared the previous results with nuclear translocation and acception by chromosomes of the [^{3}H]-regulatory subunit for the cells which underwent neoplastic transformation of another type. These experiments were carried out with the cell culture of human spontaneous cancer-KB cells. For this cell culture, we found that the regulatory subunit penetrated the nucleus but was not accepted by chromosomes (Fig. 14d). Addition of cAMP did not produce any effect. The results allowed us to suppose that the character of interaction of the regulatory subunit with the genetic material of cancer cells differs from that observed for normal and SV40 transformed cells (Fig. 14e).

Further investigations of phosphorylation processes in various cell cultures will undoubtedly contribute to elucidation of the molecular mechanisms of action of protein kinases and their role in the regulation of enzyme activity and cell division processes both in the normal state and at neoplastic transformation.

References

Corbin JD, Keely SL, Park CR (1975) The distribution and dissociation of cyclic adenosine 3′, 5′-monophosphate-dependent protein kinases in adipose, cardiac and other tissues. J Biol Chem 250: 218-225

Gabibov AG, Kochetkov SN, Sashchenko LP, Severin ES [to be published (a)] Mechanism of phosphotransferase reaction catalysed by cAMP-dependent protein kinase from pig brain. I. Steady state kinetics. Isolation of phosphorylated intermediate. Bioorg Khim (Rus)

Gabibov AG, Kochetkov SN, Smirnov IV, Sashchenko LP, Severin ES [to be published (b)] Mechanism of phosphotransferase reaction catalysed by cAMP-dependent protein kinase from pig brain. II. Transient phase kinetics. Application of quenched flow technique. Bioorg Khim (Rus)

Gulyaev NN, Tunitskaya VL, Baranova LA, Nesterova MV, Murtuzaev IM, Severin ES (1976) Investigation of the active site of catalytic subunit of histone kinase. Biokhimiya 41: 1241-1249

Kochetkov SN, Bulargina TV, Sashchenko LP, Severin ES (1976) Role of histidine residue in the active site of cyclic AMP-dependent histone kinase. FEBS Lett 71: 212-214

Kochetkov SN, Bulargina TV, Sashchenko LP, Severin ES (1977) Studies on the mechanism of action of histone kinase dependent on adenosine 3′, 5′-monophosphate. Evidence for involvement of histidine and lysine residues in the phosphotransferase reaction. Eur J Biochem 81: 111-118

Kochetkov SN, Khachatryan LL, Bagirov EM, Sashchenko LP, Severin ES (1978) Mechanism of action of cyclic AMP-dependent histone kinase. Substrate specificity of enzyme's catalytic subunit. Biokhimiya 43: 150-155

Nesterova MV, Barbashov SF, Aripdzhanov AA, Abdukarimov A, Severin ES (1980) Nuclear translocation and effect of cAMP-dependent protein kinase on the process of transcription. Biokhimiya 45: 979-991

Rosen OM, Erlichman J (1975) Reversible autophosphorylation of a cyclic 3′, 5′-AMP-dependent protein kinase from bovine cardiac muscle. J Biol Chem 250: 7788-7794

Severin ES, Sashchenko LP, Kochetkov SN, Kurochkin SN (1978) Structure and function of protein kinase from pig brain. In: George WJ, Ignarro LJ (eds) Advances in cyclic nucleotide research, vol 9. Raven Press, New York, pp 171-184

Trakht IN, Grozdova ID, Severin ES (1980a) Study of activity of protein kinases and phosphodiesterases in the cell cycle of *Physarum polycephalum*. Biokhimiya 45: 636-643

Trakht IN, Grozdova ID, Gulyaev NN, Severin ES, Gnuchev NV (1980b) Effects of some protein kinases, cyclic nucleotides and specific inhibitors of phosphorylation on the mitotic cycle of *Physarum polycephalum*. Biokhimiya 45: 788-793

Trakht IN, Grozdova ID, Vasiliev VYu, Severin ES (1980c) Evolutionary aspects of the biological role of cyclic nucleotides. Biosystems 12: 305-316

Tunitskaya VL, Gulyaev NN, Poletaev AI, Severin ES (1977) Study of conformation of cyclic adenosine 3′, 5′-monophosphate and some of its derivatives by the circular dichroism method. Biokhimiya 42: 746-753

Ulmasov KhA, Nesterova MV, Severin ES (1980a) Autophosphorylation of cAMP-dependent pig brain protein kinase. Biokhimiya 45: 661-668

Ulmasov KhA, Nesterova MV, Severin ES (1980b) Cyclic AMP-dependent protein kinase from pig brain: subunit structure, mechanism of autophosphorylation and dissociation into subunits under cAMP action. Biokhimiya 45: 835-844

Discussion[2]

Questions came from Drs. Wimmer, Fischer, Hilz, and Shaltiel.

The information given by Dr. Severin was as follows:

Addition of pure c-AMP dependent protein kinase to *Physarum* cells causes a delay of mitosis for about 1.5 h, addition of c-AMP-independent protein kinase leads to an advancement of mitosis by about 1.5 h, indicating that the enzymes have been picked up by the cells. The protein was introduced by using a spray technique. In the experiments on the influence of the regulatory subunit on the template activity a binding constant of 10^{-10} M for the pig brain regulatory subunit for brain chromatin was determined. Regulatory subunit was used in a concentration used for acidic nonhistone proteins [see Biokhymia 45: 979-991 (1980)].

The kinetic experiments were done with pigeon muscle catalytic subunit. For this enzyme the phosphotransferase reaction exhibits a random mechanism.

Using γ-^{32}P-ATP about one mol radioactive phosphate into the catalytic subunit of the brain enzyme was incorporated. After tryptic digestion one radioactive histidine containing peptide was isolated and sequenced.

2 Rapporteur: Dieter H. Wolf

Regulation of Skeletal Muscle Phosphatases by Mn^{2+} and Sulfhydryl-Disulfide Interchange

E.H. Fischer, L.M. Ballou, and D.L. Brautigan[1]

1 Introduction

In 1945 Cori and Cori described the first interconversion of an enzyme, namely that of active phosphorylase *a* from rabbit skeletal muscle into phosphorylase *b*, inactive in the absence of AMP. At that time it was thought that phosphorylase *a* contained AMP bound as a prosthetic group and, therefore, the enzyme catalyzing the *a* to *b* conversion was named "prosthetic group removing" (or PR) enzyme. The enzyme was partially purified by the procedures available at that time; these relied mainly on precipitation with ammonium sulfate or solvents such as alcohol or acetone. Ten years later, when it was found that phosphorylase was activated by the introduction of covalently bound phosphate derived from the γ position of ATP, it was recognized that PR enzyme was in fact a phosphorylase *a* phosphatase (EC 3.1.3.17) (Fischer and Krebs 1955; Wosilait and Sutherland 1956). In the next 25 years, protein kinases and phosphatases were extensively investigated. In fact, as of today, ca. 30 enzymes have been shown to undergo reversible phosphorylation (Krebs and Beavo 1979) and this process is now known to play a fundamental role in the regulation of various metabolic pathways.

Much detailed information has been accumulated on the structure of protein kinases, their regulation by cyclic nucleotides or calcium, and their intracellular localization within the cytoplasm, membranes, nuclei, mitochondria, or other subcellular organelles. By contrast, much less information has been gathered on the protein phosphatases that carry out the reverse reactions (Lee et al. 1980). Preparations capable of dephosphorylating a number of phosphoproteins have led to the conclusion that the phosphatase might be a "multifunctional" enzyme (Nakai and Thomas 1973, 1974; Killilea et al. 1976; Khandelwal et al. 1976; Antoniw et al. 1977; Chou et al. 1977). By contrast, other preparations appear to be quite specific for either phosphorylase *a*, the α- or β-subunit of phosphorylase kinase, glycogen synthase, or troponin I (Ray and England 1976; Antoniw and Cohen 1976; Gratecos et al. 1977; Kikuchi et al. 1977; Tan and Nuttall 1978; Laloux et al. 1978). The basis for these discrepancies is not yet understood. The enzymes appear in tissue extracts as high molecular weight complexes but none have been adequately purified to allow a comparison of their structures and substrate specificities. A M_r ca. 35,000 form of the phosphatase, presumably a catalytic component of the multisubunit complex, has

1 Department of Biochemistry SJ-70, University of Washington Medical School, Seattle, WA 98195, USA

been purified in different laboratories (Brandt et al. 1974, 1975; Khandelwal et al. 1976; Gratecos et al. 1977). When this enzyme is obtained by alcohol precipitation (Brandt et al. 1975), it displays broad substrate specificity. In the presence of divalent metal ions it also hydrolyzes p-nitrophenol phosphate, an activity attributed to contamination by an "alkaline-type" phosphatase (Li et al. 1979). Although these preparations appear to be homogeneous, the multifunctional nature of the M_r = 35,000 enzyme may indicate that it consists of an admixture of different phosphatases of similar molecular weight.

Even more complex has been the matter of regulation of the protein phosphatases. In a kinetic analysis of phosphorylase interconversion following electrical stimulation in intact frog muscle, Danforth et al. (1962) proposed that the major regulation occurred through control of phosphorylase kinase with no evidence for regulation of the phosphatase. Thus, the simplest mechanism would be for the phosphatase to be always active without any control. Such a mechanism seemed theoretically feasible and even plausible at that time since potential kinase activity was thought to exceed by far phosphatase activity as calculated on the basis of the approximate concentrations and activities of the two enzymes in muscle. However, more recent data indicate that this is not the case and that, in fact, the activities of phosphorylase kinase and phosphatase are fairly well balanced. Therefore, it would be more rational to expect a mechanism by which both enzymes are simultaneously regulated, perhaps in an inverse manner, and there have been several recent reports suggesting such a control. For example, regulation by substrate competition (e.g., phosphorylase kinase and glycogen synthase competing with phosphorylase *a*), by modifiers of substrate (e.g., inhibition of phosphorylase *a* dephosphorylation by AMP or IMP, or activation by glucose, G1c6-P or caffeine) or by modifiers of the phosphatase itself (such as Mn^{2+} or thiols) have all been reported. Direct regulation of the phosphatase could occur by a number of mechanisms: activation of phosphorylase phosphatase by a Mg-ATP-dependent protein cofactor (Goris et al. 1979; Yang et al. 1980) or inhibition by two heat-stable proteins, one (inhibitor I) being active only in its phosphorylated form while the other (inhibitor II) is not phosphorylated (Cohen 1978). Finally, the phosphatase reaction might be affected by "chemical compartmentation" of the enzyme or protein substrates through their differential attachments to glycogen particles or sarcoplasmic membranes, for instance (Haschke et al. 1970; Varsanyi and Heilmeyer 1979), or the thin filaments containing troponin I, another substrate of the enzyme. However, while evidence suggesting interconvertible forms of phosphorylase phosphatase has been presented in the past in other organs and species such as the adrenal cortex (Riley and Haynes 1963; Merlevede and Riley 1966) pigeon breast muscle (Torres and Chelala 1970 a, b) and *Neurospora crassa* (Tellez de Inon and Torres 1973), no direct covalent modification of the rabbit muscle enzyme has been demonstrated as yet. Considering the complexitiy and multiplicity of possible interactions outlined above, it is clear that no simple scheme will fully account for the physiological mechanism by which protein phosphatases are regulated.

2 Purification of a Phosphorylase Phosphatase Complex

Rabbit skeletal muscle coarsely ground and then extracted with buffer containing EDTA and protease inhibitors yields a single form of the enzyme that elutes from Sephacryl 6B with a molecular size similar to that of bovine liver catalase (250,000). No protein phosphatase activity appears at molecular weights of less than about 100,000. This form of the enzyme has been purified by a combination of ion exchange (DEAE-Sephadex 50) and gel filtration (Sepharose 6B-CL) chromatography. The final product (M_r 250,000), contains three major polypeptide components which, when examined by SDS gel electrophoresis, displays three polypeptides of M_r 83,000, 72,000, and 32,000 in a stoichiometry of approximately 1:6:1. It is not yet known how these are arranged to give the high molecular weight form of the enzyme but the 83,000 and the 32,000 molecular weight proteins that are present in stoichiometric amounts are both catalytically active phosphatases. The function of the 72,000 molecular weight protein is yet unknown (Fig. 1).

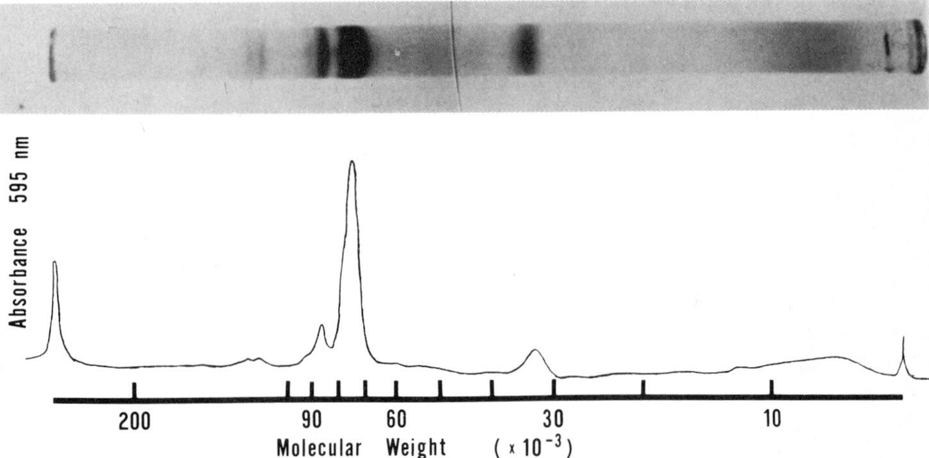

Fig. 1. SDS gel electrophoresis of sarcoplasmic phosphorylase phosphatase complex of M_r = 250,000 obtained after Sepharose 6B-CL chromatography. Fractions with the highest activities were pooled, concentrated and subjected to electrophoresis on a 8.75% polyacrylamide gel in the presence of reducing agent and 0.1% SDS. Staining with Coomassie blue revealed three major bands of M_r 83,000, 72,000, and 32,000 in relative molar proportions of 1:6.4:1

3 Manganese Stimulation

The complex form of the enzyme is stimulated more than tenfold by Mn^{2+}, with a $K_{0.5}$ of 10^{-5} M. The reaction is not mimicked by Mg^{2+} or Ca^{2+}; it is abolished by trypsin. This activation reaction is unusual in that it takes several minutes and the activation is not reversed by gel filtration or prolonged incubation with 1 mM EDTA. The activation by Mn^{2+} does not involve binding of the metal ion to the protein since addition of radioactive $[^{54}Mn]^{2+}$ followed by gel filtration gave an enzyme which was more than fivefold activated even though it contained less than 1 part per 1000 of manganese. The activity generated by Mn^{2+} treatment represented between two-thirds and three-fourths of the total phosphatase activity (Fig. 2).

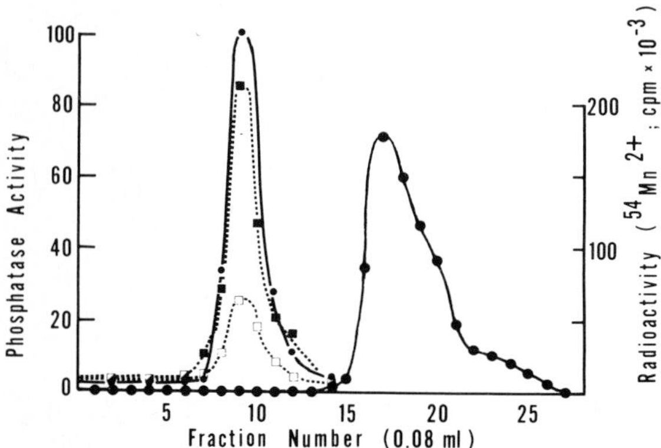

Fig. 2. $[^{54}Mn]^{2+}$ binding to the activated phosphatase complex. Sarcoplasmic phosphatase complex (100 μL) was desalted on a 0.5 x 7 cm column of Sephadex G25-Fine in 20 mM imidazole-acetate pH 7.5, 0.1 mM DTT, 5% glycerol containing 0.1 M NaCl. Each fraction was diluted with 200 μL of salt-free buffer and the phosphorylase phosphatase activity measured after 15 min incubation with (●—●) and without (□ · · · □) 1 mM Mn^{2+}

4 Solvent Precipitation

Ethanol (80% at room temperature) has been used for the past 5 years to purify a phosphatase of molecular weight 35,000 (Brandt et al. 1975). This procedure only provided us with poor yields of activity when we started with the complex form of the enzyme. On the other hand, 50% acetone at room temperature led to quantitative recovery of activity. Extraction of the acetone precipitate with buffer and passage through columns of Sephacryl S200 yielded two peaks of phosphatase activity (Fig. 3). A large peak of activity (*closed circles*) contained enzyme that was dependent on manganese treatment and was completely trypsin sensitive. This phosphatase exhibited a molecular weight of 83,000 by SDS gel electrophoresis and accounted for the bulk of the total activity. The second enzyme (*open circles*) contained activity independent of manganese and insensitive to trypsin. It is similar to the M_r 33,000 enzyme previously characterized (Brandt et al. 1974, 1975; Khandelwal et al. 1976; Gratecos et al. 1977). Thus the sarcoplasmic phosphatase complex contains at least two active subunits that can be dissociated and recovered in lower-molecular weight forms. These two phosphatases can be distinguished on the basis of three criteria (Table 1): (1) the larger one exhibits a high dependence on manganese activation while the activity of the M_r 32,000 enzyme is independent of manganese; (b) the 83,000 phosphatase is not inhibited by sulfhydryl reagents while that of M_r 32,000 is strongly inhibited, and (3) the larger species is sensitive to trypsin while the smaller one is not.

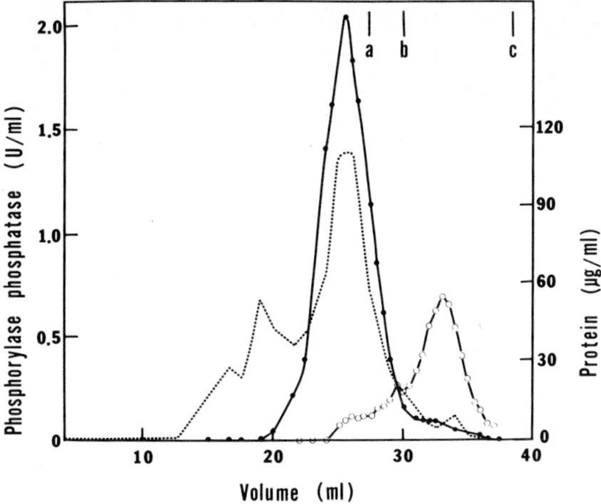

Fig. 3. Recovery of two catalytic subunits following acetone precipitation of sarcoplasmic phosphatase complex. The complex was precipitated without bovine serum albumin by 50% acetone at +20°C in the presence of 1 mM Ca^{2+}. After centrifugation for 2 min at 8500 x g, the pellet was extracted and 500 µL applied to a 1.5 x 30 cm column of Sephacryl S-200 equilibrated in 15 mM morpholinopropanesulfonic acid buffer, pH 7.5, 15 mM 2-mercaptoethanol, and 50 mM NaCl. Fractions of 0.5 mL were collected at 8 mL/h and analyzed for protein concentration (· · ·) and phosphorylase phosphatase activity after 15 min incubation with 1 mM Mn^{2+} (●—●). The major peak of activity (V_e = 25 mL) corresponded to a molecular weight of ca. 90,000 and was poorly active in the absence of Mn^{2+}. Virtually no M_r 35,000 enzyme was recovered due to the absence of albumin in the column buffer. In a separate experiment an identical sample of phosphatase complex was precipitated and chromatographed in the presence of 0.1 mg/mL bovine serum albumin. The phosphorylase phosphatase activity measured without Mn^{2+} (○—○) appeared at V_e = 33 mL corresponding to a M_r of 35,000. This activity was the same when measured with or without Mn^{2+}. Standards were: *a*, bovine serum albumin (M_r 68,000); *b*, ovalbumin (45,000) and *c*, ribonuclease (14,700)

Table 1. Comparative properties of skeletal muscle protein phosphatases

Phosphatase preparations	Mn^{2+}[a]		DTNB[b]		Sensitivity to	
	–	+	–	+	acetone[c]	trypsin[d]
	Per cent relative activities					
M_r 83,000	10	100	100	160	no	yes
M_r 32,000	100	100	100	15	no	no
*p*NP-Pase[e]	0	100	?	?	yes	?

[a] Incubated for 15 min with 1 mM Mn acetate
[b] Incubated for 5 min with 1 mM DTNB
[c] Following precipitation with 50% solvent at 20°C
[d] After digestion with 5% w/w TPCK-treated trypsin for 30 min at 30°C
[e] *p*NP-Pase: *p*-nitrophenyl-phosphate phosphatase

5 Interaction of the M_r 32,000 Subunit with Disulfides

As previously reported from this laboratory, the 32,000 molecular weight form of the enzyme is not inhibited by iodoacetamide but rapidly inhibited by a number of disulfide compounds. To further investigate this reaction, gel filtered muscle extracts were incubated with radioactive oxidized glutathione ([^{14}C]-GSSG). After desalting by gel filtration it was found that a significant amount of glutathione had been incorporated into trichloroacetic acid-precipitable protein. Reduction with 2-mercaptoethanol released most of the bound radioactivity and, at the same time, nearly doubled the phosphatase activity of these extracts (Table 2). When the distribution of the radioactive glutathione among the soluble muscle proteins was studied by SDS gel electrophoresis, only two major peaks were found corresponding to proteins of molecular weight of ca. 43,000 and 35,000. Further examination of the reaction of GSSG, as well as of other structurally analogous disulfides with muscle proteins of subunit molecular weights in the range of 35,000, indicated that glutathione most probably labeled the phosphatase. Various disulfides were chosen to compare both charge effects and the structural dependence on the reactivity at the sulfhydryl groups present in skeletal and cardiac muscle lactate dehydrogenase, muscle glyceraldehyde 3-phosphate dehydrogenase, and the 32,000 form of the phosphatase. These include hydroxyethyl disulfide, aminoethyl disulfide, carboxyethyl disulfide, glutathione, and dithio-5,5′-bis(2-nitrobenzoate) (DTNB). The effects are shown in Table 3. Lactate dehydrogenase, one of the major muscle proteins, was not inhibited, nor was it labeled by any of the disulfides even though this enzyme contains a well recognized and essential cysteinyl residue at position 165 in a highly conserved peptide sequence. Reaction of skeletal or cardiac muscle lactate dehydrogenase with radioactive GSSG showed no incorporation. Glyceraldehyde 3-phosphate dehydrogenase is another muscle soluble protein which contains in its active site an essential sulfhydryl group directly involved in catalysis. Though this enzyme is the target of iodoacetate poisoning of glycolysis and reacts strongly with some disulfides such as DTNB and hydroxyethyl disulfide, it reacted quite poorly with oxidized glutathione. The 32,000 form of the posphatase was also strongly inhibited by DTNB. By contrast, the order of its reactivity toward the other disulfides was virtually the reverse of that exhibited by glyceraldehyde-3-phosphate dehydrogenase. In particular, oxidized glutathione showed the strongest reaction with the phosphatase while hydro-

Table 2. Inhibition of sarcoplasmic phosphorylase phosphatase by oxidized [^{14}C]-glutathione

	Trichloroacetic acid		Phosphorylase phosphatase
	Soluble	Precipitated	
	^{14}C cpm/mL		mU/mL
GSSG-treated sarcoplasm	6,040	29,800	4,348
+50 mM 2-mercaptoethanol	33,590	4,650	7,795

Table 3. Effects of sulfhydryl reagents

	Inhibition of enzyme activity %		
Reagent	Phosphatase $M_r = 32{,}000$	*LDH*	*G3PDH*
Iodoacetamide	1	2	>99
Mercuribenzoate	95	>99	>99
Disulfides			
Hydroxyethyl-	18	0	>99
Aminoethyl-	20	0	80
Carboxyethyl-	5	0	73
Glutathione-	40	2	20
5-(2-nitrobenzoate)-(DTNB)	80	5	>99

xyethyl disulfide was one of the weakest. These results indicate considerable specificity in the reaction of oxidized glutathione for the sulfhydryl groups present in these three enzymes and suggest that reaction with the disulfide may be an important mechanism for the regulation of the phosphatases involved in glycogen metabolism and protein synthesis.

6 Conclusion

In summary, the scheme illustrated in Fig. 4 shows how the high-molecular weight, M_r 250,000 form of the phosphatase that is present in muscle extracts can be resolved by room temperature acetone treatment into two active species. The first is a M_r 83,000 phosphatase that is activated by a reaction with Mn^{2+} that does not involve binding of the metal ion to the protein, that is insensitive to sulfhydryl reagents and sensitive to trypsin. It is possible that manganese is indeed the physiological activator of this enzyme since estimates of the intramuscular concentration of the ion are between 10 and 20 μM, precisely the half-effective concentration found in the in vitro assay. The second phosphatase of M_r 32,000 is active in the absence of manganese, inhibited by sulfhydryl reagents and very stable to trypsin; in particular, this enzyme showed an unusual selectivity for reaction with GSSG. Despite the low intracellular concentrations of oxidized glutathione, this reaction may be of physiological importance. If so, one could predict that a phosphatase reductase might be involved in the reversal of the disulfide formation reaction, with concomitant activation of the enzyme.

The mechanism of manganese activation of the M_r 83,000 enzyme is still obscure. The metal ion does not seem to be incorporated into the protein, and there is no evidence that the phosphatase is a metalloenzyme. Therefore, the simplest explantation is that the enzyme has undergone a change in conformation, but a number of other possibilities exist. Since we have not been able to reverse the reaction as yet, the phenomenon has been difficult to study. If the conversion reaction were

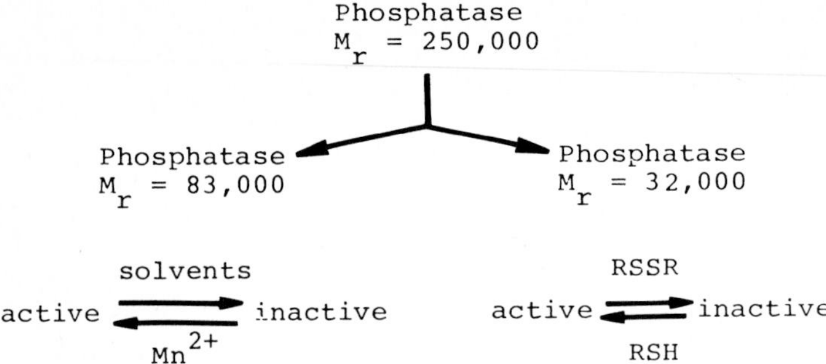

Fig. 4. Interconversion of sarcoplasmic phosphatases

really irreversible, it could result from the elimination of a bound inhibitor. The enzyme also could have sustained a covalent modification, e.g., the cleavage of an inhibitory peptide by a Mn^{2+}-dependent peptidase or of a phosphoryl group by a separate Mn^{2+}-dependent phosphatase. However, no rephosphorylation of the enzyme has been demonstrated as yet.

Likewise, the reason for the native phosphatase to exist in a high molecular weight complex is unknown. No major differences in regulatory and/or catalytic behavior between the "heavy" enzyme and its free subunits have been ascertained. The complex could represent a functional unit in which several phosphatases of varying specificities are held together for a concerted action on different substrates, or on a single substrate possessing multiple sites phosphorylated by kinases of different specificities. Alternatively, it could serve to partition the enzyme on specific structural elements within the cell.

The changes in activity of the individual catalytic subunits described herein are but a few aspects of the regulation of the protein phosphatases. The overall physiological regulation is more likely to result from several interactive mechanisms affecting the substrate as well as the enzyme.

Acknowledgements. The authors thank R.B. Olsgaard and C. Diltz for their technical assistance in these studies and Ms. Alice Palacio for typing the manuscript.

This research was supported by grants from the National Institutes of Health (AM7902) and the Muscular Dystrophy Association. D.L.B. was a recipient of a postdoctoral fellowship from the MDA and a National Research Service Award (NIAMDD-NIH AM06023-01). L.M.B. was supported by a Molecular and Cellular Biology training grant to the University of Washington Medical School (NIH GM07270).

References

Antoniw JF, Cohen P (1976) Eur J Biochem 68: 45-54
Antoniw JF, Nimmo HG, Yeaman SJ, Cohen P (1977) Biochem J 162: 423-433
Brandt H, Killilea SD, Lee EYC (1974) Biochem Biophys Res Commun 61: 598-604
Brandt H, Capulong ZL, Lee EYC (1975) J Biol Chem 250: 8038-8044
Chou CK, Alfano J, Rosen OM (1977) J Biol Chem 252: 2855-2859
Cohen P (1978) Curr Top Cell Regul 14: 117-196

Danforth WH, Helmreich E, Cori CF (1962) Proc Natl Acad Sci USA 48: 1191-1196
Cori CF, Cori GT (1945) J Biol Chem 158: 341
Fischer EH, Krebs EG (1955) J Biol Chem 216: 121
Goris J, Defreyn G, Merlevede W (1979) FEBS Lett 99: 279-282
Gratecos D, Detwiler TC, Hurd S, Fischer EH (1977) Biochem 16: 4812-4817
Haschke RH, Heilmeyer LMG Jr, Meyer F, Fischer EH (1970) J Biol Chem 245: 6657-6663
Khandelwal RL, Vandenheede JR, Krebs EG (1976) J Biol Chem 251: 4850-4858
Kikuchi K, Tamura S, Hiraga A, Tsuiki S (1977) Biochem Biophys Res Commun 75: 29-37
Killilea SD, Brandt H, Lee EYC, Whelan WJ (1976) J Biol Chem 251: 2363-2368
Krebs EG, Beavo JA (1979) Ann Rev Biochem 48: 923-959
Laloux M, Stalmans W, Hers H-G (1978) Eur J Biochem 92: 15-24
Lee EYC, Silberman SR, Ganapathi MK, Petrvic S, Paris H (1980) Adv Cyclic Nucleotide Res 13:95-131
Li H-C, Hsiao K-J, Sampathkumar S (1979) J Biol Chem 254: 3368-3374
Merlevede W, Riley GA (1966) J Biol Chem 241: 3517-3524
Nakai C, Thomas JA (1973) Biochem Biophys Res Commun 52: 530-536
Nakai C, Thomas JA (1974) J Biol Chem 249: 6459-6467
Ray KP, England PJ (1976) Biochem J 157: 369-380
Riley GA, Haynes RC Jr (1963) J Biol Chem 238: 1563-1570
Tan AWH, Nuttall FQ (1978) Biochim Biophys Acta 522: 139-150
Tellez de Inon MT, Torres HM (1973) Biochim Biophys Acta 297: 399-412
Torres HN, Chelala CA (1970a) Biochim Biophys Acta 198: 495-503
Torres HN, Chelala CA (1970b) Biochim Biophys Acta 198: 504-513
Varsanyi M, Heilmeyer LMG Jr (1979) Biochemistry 18: 4869-4875
Wosilait WD, Sutherland EW (1956) J Biol Chem 218: 469-481
Yang S-D, Vandenheede JR, Goris J, Merlevede W (1980) FEBS Lett 111: 201-205

Discussion[2]

Questions were raised by Drs. Shaltiel, Söling, Cohen, Stadtman, and Larner

Information presented by Dr. Fischer was as follows:

Neither manganese activation or G-S-S-G inhibition was reproducible by looking at the effects of a series of oxido-reduction systems with varying redox potential. Mn^{2+} activation can be demonstrated on the free 83,000 subunit obtained by acetone treatment of the complex before metal activation and on the high molecular weight complex itself.

G-S-S-G reacts with the high molecular weight enzyme. Activity is inhibited by G-S-S-G and can be restored by excess-β-mercaptoethanol. K_a is in a reasonable concentration range comparable with the physiological concentration of G-S-S-G in the cell. Manganese could be the physiological activator of the catalytic subunit. Half maximum activation is 10 μM while intramuscular concentration is estimated at 10-20 μM.

In crude muscle no large effects of Mn^{2+} are seen. As soon as the complex is obtained through gel filtration it becomes highly sensitive to manganese displaying an approx ten-fold activation. This activation is not likely to be due to procedures used during the purification. There is no evidence that the enzyme itself is a metal enzyme.

No separate 83,000 and 33,000 molecular weight subunits can be obtained under conditions where homogenization is avoided. The 83,000 molecular weight component cannot be resolved further on SDS gels.

The presence of two independent phosphatase forms can be concluded from the inactivation of only one form by DTNB, the sensitivity of only one form against Mn^{2+}, and the fact that only one form is destroyed by trypsin while the other is not.

2 Rapporteur: Dieter H. Wolf

Regulation of Liver Phosphorylase Kinase

N.B. Livanova, I.E. Andreeva, T.B. Eronina, and G.V. Silonova[1]

1 Introduction

The structure, physical and kinetic properties as well as the regulatory mechanisms of skeletal muscle phosphorylase kinase have been studied extensively. In contrast to our current understanding of muscle phosphorylase kinase, the structure and regulatory properties of liver enzyme have not yet been fully clarified. Results obtained with crude and partially purified preparations of liver phosphorylase kinase suggest two main mechanisms of regulation: firstly, covalent modification through a phosphorylation reaction catalyzed by cyclic AMP-dependent protein kinase in response to glucagon and epinephrine (Vandenheede et al. 1979), and secondly, Ca^{2+} has been proposed as a possible candidate for an intracellular mediator in the hormonal control by cyclic AMP-independent hormones (e.g., vasopressin, angiotensin and α adrenergic agents) (Van de Werve et al. 1977); besides that, the control of kinase reaction can occur due to binding of effectors to the substrate, phosphorylase *b*, to the enzyme, phosphorylase kinase, or both.

Since it has already been shown that muscle phosphorylase kinase is very sensitive to slight conformational changes of its phosphorylase *b* substrate (Bot et al. 1970; Tu and Graves 1973; Morange and Buc 1979), we have tested in the present paper the effect of AMP and Glc-6-P on the conversion rate from phosphorylase *b* to *a* using the preparations of muscle and liver phosphorylase kinase and a substrate phosphorylase *b* from the same as well as from the other source hoping to find out whether the effect of these ligands is caused by binding to the substrate, or to the enzyme, or both. We have been interested, too, in possible effect of Ca^{2+} on liver phosphorylase kinase. It is known that muscle phosphorylase kinase is almost totally dependent on Ca^{2+}; the results of Shenolicar et al. (1979) strongly suggest that the component which confers Ca^{2+} sensitivity to the phosphorylase kinase reaction is the integral component (δ subunit) of muscle phosphorylase kinase, identical to the calcium binding protein, termed calmodulin. Despite the fact that all preparations of muscle phosphorylase kinase contained stoichiometric quantities of calmodulin, these preparations could be additionally stimulated by pure calmodulin. In contrast to the muscle enzyme, the liver phosphorylase kinase is only partially dependent on calcium; its inhibition by EGTA is no more than 60%-70% (Van de Werve et al. 1977;

Abbreviations: Gle-6-P, glucose-6-phosphate; EGTA, ethylenglycol bis (β-aminoethyl)-N,N,-N',N'-tetraacetic acid; Phosphorylase *b*, E.C. 2.4.1.1.; Phosphorylase kinase, E.C. 2.7.1.38

1 The A.N. Bakh Institute of Biochemistry, USSR Academy of Sciences, Moscow 117071, USSR

Sakai et al. 1979). There are no data on liver calmodulin and its role in the regulation of liver phosphorylase kinase.

2 Materials and Methods

Crystalline rabbit skeletal muscle phosphorylase *b* was prepared according to Fischer and Krebs (1958). Purified rabbit muscle phosphorylase kinase was isolated by the procedure of Krebs et al. (Hayakawa et al. 1973; De Lange et al. 1968). Phosphorylase *b* from rabbit liver was purified as described in our paper (Livanova et al. 1976). Partially purified rabbit liver phosphorylase kinase was prepared from the glycogen pellet using the chromatography on the column of ω-aminohexyl-Sepharose. The specific activity of the preparations of liver phosphorylase kinase was about 30 units; the molecular weight estimated using a Sepharose 4B column was equal to 310000 ± 10000. One unit of phosphorylase *b* kinase was the amount of enzyme which produced one unit of phosphorylase *a*/min at 30°C under the assay conditions; one unit of phosphorylase *a* produced 1 μmol of phosphate/min at 30°C. Phosphorylase kinase was assayed in the following medium: 100 μl of tris-glycerophosphate buffer 0.125M pH 8.6, 100 μl of phosphorylase *b* (300-500 μg) dissolved in glycerophosphate buffer 0.025M pH 6.8, 50 μl of ATP 18 mM and magnesium acetate 60 mM; the reaction was started by addition of a variable quantity of phosphorylase kinase in a volume of 50 μl. The preparations of muscle and liver phosphorylase kinase were used without previous activation. AMP, Glc-6-P, EGTA or $CaCl_2$ (0.6 mM) were included into the tris-glycerophosphate buffer, calmodulin was previously mixed with phosphorylase kinase. The final pH was 8.2. After incubation for 5 min at 30°C the conversion was stopped by dilution in 0.05M ice-cold glycerophosphate buffer pH 6.8 and was followed by measuring the increment of activity of phosphorylase assayed in the absence of AMP, with respect to time. Phosphorylase *a* activity was assayed following Illingworth and Cori (1953). In all experiments we determined the initial rate of conversion from phosphorylase *b* to phosphorylase *a* by phosphorylase kinase, when the percent of conversion was less than 40. The preparation of brain calmodulin homogeneous by the criteria of polyacrylamide gel electrophoresis was kindly provided by Dr. V.A. Tkachuk from the Biochemistry Department of Moscow State University. Partially purified calmodulin from rabbit liver was isolated by a modification of the procedure described by Dedman et al. (1977).

3 Influence of AMP and Glc-6-P on the Phosphorylase *b* to Phosphorylase *a* Conversion Rate

It was already known that AMP at 0°C and Glc-6-P at different temperatures have a strong inhibitory effect upon the muscle phosphorylase *b* to *a* conversion rate by phosphorylase kinase from muscle. Bot et al. (1970) and Morange and Buc (1979) gave strong arguments in favor of an indirect effect of AMP on muscle phosphorylase kinase. Tu and Graves (1973) using the alternative substrate tetradecapeptide in kinase reaction received direct evidence in favor of substrate-directed effect of Glc-6-P.

But these experiments do not permit to establish whether the effect of AMP and Glc-6-P is due to the change of the substrate conformation or to local interference of the effector, bound on phosphorylase *b*, with kinase action, or to both effects.

We have studied the dependence of phosphorylase *b* to *a* conversion rate towards AMP and Glc-6-P concentration for muscle and liver phosphorylase kinase using a proper substrate for each enzyme as well as an alternative substrate – phosphorylase *b* from the other source, hoping to differentiate between these possibilities.

The results on the effect of AMP are plotted on Fig. 1. At the temperature 30°C one can see slight slowing effect of AMP only in the system muscle phosphorylase kinase + muscle phosphorylase *b*. In all other cases AMP has an activating effect. The studies of the effect of AMP upon the conversion rate from phosphorylase *b* to *a*, utilizing for each type of phosphorylase kinase the native and alternative substrate, confirmed the results previously obtained by Bot et al. (1970) and Morange and Buc (1979) that the action of this ligand is substrate-directed. But the effect of AMP is different also in the cases when phosphorylase kinases from muscle and from liver act on the same substrate. This provides evidence that the effect of AMP depends not only on the conformational state of the substrate but also on the modifying action of this ligand on the complex of phosphorylase *b* with phosphorylase kinase.

These data on the action of AMP upon the rate of phosphorylase *b* to *a* conversion have to be taken into consideration if one wants to compare the interaction of covalent and noncovalent regulation in muscle and in liver. In the skeletal muscle AMP by binding on phosphorylase *a* inhibits phosphorylase phosphatase (Martensen et al. 1973) and phosphorylase kinase (Morange and Buc 1979); i.e., in the presence of AMP the system of covalent modification of phosphorylase is less efficient. In the liver AMP on the one hand inhibits phosphorylase phosphatase (Brandt et al. 1975), and on the other activates phosphorylase kinase. Therefore, it appears that in the liver the increase of the AMP concentration causes the increase of the rate of active glycogen phosphorylase formation. It seems reasonable, because in the muscle the activity of phosphorylase *b* in the presence of AMP is almost equal to the activity of phosphorylase *a*, but the liver phosphorylase *b* is inactive in the presence of AMP, so the covalent modification is the only way of active phosphorylase formation.

Glc-6-P was already known to have a strong inhibitory effect upon the rate of phosphorylase kinase reaction in muscle by acting on the protein substrate (Tu and Graves, 1973). By changing the liver and muscle phosphorylase kinase and phosphorylase *b* in this system we have shown that phosphorylase kinase from liver has less specificity to the protein substrate than muscle enzyme, because the liver kinase makes no difference between the conformational change induced by Glc-6-P in liver and in muscle phosphorylase, while the muscle kinase does differentiate it (Fig. 2). One possible explanation for such different behavior of the kinases from two sources is that the enzyme from liver has lower molecular weight and, probably, less complex quaternary structure (perhaps through partial proteolysis) than the enzyme from muscle. The fact that the degree of the inhibitory action of Glc-6-P on the conversion rate of phosphorylase *b* to phosphorylase *a* catalyzed by muscle phosphorylase kinase strongly changes when phosphorylase *b* from muscle is replaced by the substrate from liver confirms the substrate-directed kind of Glc-6-P operation.

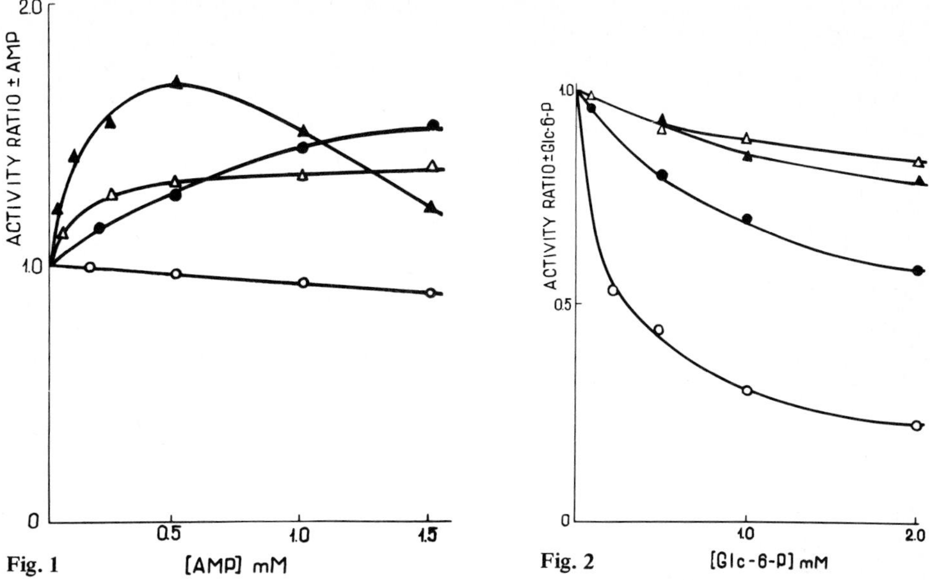

Fig. 1 Fig. 2

Fig. 1. Dependence of the phosphorylase *b* to *a* conversion rate towards AMP concentration. The activity ratio is defined as the activity in the presence of the ligand divided by the activity in its absence. The experiment is performed as described in Materials and Methods. The conversion probe contains muscle (○, ●) or liver (△, ▲) phosphorylase kinase with muscle (○, ▲) or liver (△, ●) phosphorylase *b*

Fig. 2. Dependence of the phosphorylase *b* to *a* conversion rate towards Glc-6-P concentration. The conversion probe contains muscle (○, ●) or liver (△, ▲) phosphorylase kinase with muscle (○, ▲) or liver (△, ●) phosphorylase *b*

4 Effect of Calmodulin on the Activity of Liver Phosphorylase Kinase

In addition to the study of the effect of AMP and Glc-6-P on the phosphorylase kinase reaction partial purification of liver calmodulin and the study of its effect on liver phosphorylase kinase were undertaken. Previous check of the influence of partially purified preparation of calmodulin from rabbit liver on the activity of Ca^{2+}-dependent phosphodiesterase (Table 1) showed that liver calmodulin is as effective as highly purified brain calmodulin and this activation is fully depdent on Ca^{2+}.

The effect of liver calmodulin on the activity of phosphorylase kinase from liver is illustrated in Fig. 3. The maximal extent of activation by liver calmodulin is about threefold. Apparently identical extent of activation by liver calmodulin was obtained with phosphorylase kinase from muscle (data not presented).

What is most surprising is that the activation of liver phosphorylase kinase by calmodulin does not depend on the presence of Ca^{2+}. A typical experiment is illustrated in Fig. 4. As shown in this figure, in the presence of 5 mM EGTA, liver phosphorylase kinase was inhibited by about 30% of the original activity. In the presence of EGTA the degree of activation by calmodulin from brain and from liver was the same as in the presence of Ca^{2+} without EGTA.

Table 1. Influence of brain and liver calmodulin on the activity of cyclic AMP-dependent phosphodiesterase from rabbit heart[a]

Additions to assay	pmol cyclic AMP/mg min	
	Ca^{2+} 10^{-5} M	EGTA 10^{-3} M
None	845 ± 41	801 ± 92
Calmodulin from brain	14,280 ± 2100	400 ± 8
Calmodulin from liver	12,770 ± 1050	439 ± 21

Standard assay contained: tris 20 mM, $MgCl_2$ 5 mM, cyclic AMP 0.01 mM, phosphodiesterase 18 μg, brain (7 μg) or liver (15 μg) calmodulin in the final volume 50 μl; pH 8.0, 37°C, 10 min incubation

[a] The data presented in Table 1 were received by Dr. V.A. Tkachuk of the Department of Biochemistry of Moscow State University; the authors are greatly indebted to him

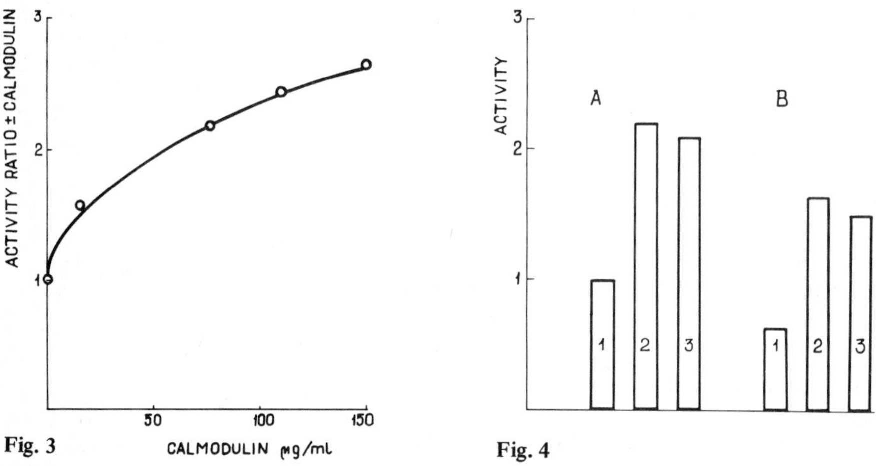

Fig. 3 **Fig. 4**

Fig. 3. Influence of liver calmodulin on the activity of liver phosphorylase kinase

Fig. 4 A, B. Influence of calmodulin from brain and liver on the activity of liver phosphorylase kinase in the absence (**A**) and in the presence (**B**) of EGTA. The *ordinate* shows the activity in arbitrary units. *Channel 1*, no additions; *channel 2*, acivity in the presence of saturating concentration of brain calmodulin; *channel 3*, activity in the presence of saturating concentration of liver calmodulin. Phosphorylase kinase was 30 μg/ml, in the assays, EGTA 5 mM, $CaCl_2$ 0.2 mM

Till now the role of calmodulin in enzymes regulation was considered to be that of functional Ca^{2+}-receptor (Means and Dedman 1980). Yamaki and Hidaka (1980) reported that the stimulation by calmodulin of cyclic GMP-dependent protein kinase was Ca^{2+}-independent. Futher studies are necessary to find out the mechanism of this Ca^{2+}-independent stimulation by calmodulin.

References

Bot G, Kovács EF, Pólyik EN (1970) The influence of allosteric effectors on the conversion of phosphorylase *b*. Acta Biochim Biophys Acad Sci Hung 5: 9-18

Brandt H, Capulong ZL, Lee EYC (1975) Purification and properties of rabbit liver phosphorylase phosphatase. J Biol Chem 250: 8038-8044

Dedman JR, Potter JD, Jackson RL, Johnson JD, Means AR (1977) Physicochemical properties of rat testis Ca^{2+}-dependent regulator protein of cyclic nucleotide phosphodiesterase. J Biol Chem 252: 8415-8422

Fischer EH, Krebs EG (1958) The isolation and crystallization of rabbit skeletal muscle phosphorylase *b*. J Biol Chem 231: 65-71

Hayakawa T, Perkins JP, Walsh DA, Krebs EG (1973) Physicochemical properties of rabbit skeletal muscle phosphorylase kinase. Biochemistry 12: 567-573

Illingworth B, Cori GT (1953) Crystalline muscle phosphorylase. In: Snell E (ed) Biochemical preparations, vol 3. John Wiley and Sons Inc New York, pp 1-9

De Lange RJ, Kemp RG, Riley WD, Cooper RA, Krebs EG (1968) Activation of skeletal muscle phosphorylase kinase by ATP and adenosine 3'-5'-monophosphate. J Biol Chem 243: 2200-2208

Livanova NB, Eronina TB, Silonova GV, Ramensky EV (1976) ω-Aminohexyl-sepharose in purification of liver glycogen phosphorylase *b*. FEBS Lett 69: 95-98

Martensen TM, Brotherton JE, Graves DJ (1973) Kinetic studies of the inhibition of muscle phosphorylase phosphatase. J Biol Chem 248: 8323-8328

Means AR, Dedman JR (1980) Calmodulin – an intracellular calcium receptor. Nature (London) 285: 73-77

Morange M, Buc H (1979) The interplay between covalent and non-covalent regulation of glycogen phosphorylase. Biochimie 61: 633-643

Sakai K, Matsumura S, Okimura Y, Yamamura H, Nishizuka Y (1979) Liver glycogen phosphorylase kinase. J Biol Chem 254: 6631-6637

Shenolicar S, Cohen PTW, Cohen P, Nairn AC, Perry SV (1979) The role of calmodulin in the structure and regulation of phosphorylase kinase from rabbit skeletal muscle. Eur J Biochem 100: 329-337

Tu J-I, Graves DJ (1973) Inhibition of the phosphorylase kinase catalyzed reaction by glucose-6-P. Biochem Biophys Res Commun 53: 59-65

Vandenheede JR, De Wulf H, Merlevede W (1979) Liver phosphorylase *b* kinase. Eur J Biochem 101: 51-58

Van de Werve G, Hue L, Hers H-G (1977) Hormonal and ionic control of the glycogenolytic cascade in rat liver. Biochem J 162: 135-142

Yamaki T, Hidaka H (1980) Ca^{2+}-independent stimulation of cyclic GMP-dependent protein kinase by calmodulin. Biochem Biophys Res Commun 94: 727-733

Discussion[2]

Questions were raised by Drs. Cohen and Fischer. Information emerging from this discussion was as follows: The liver calmodulin preparation used was not pure calmodulin. The calciuminsensitivity and the decreased specificity against glucose-6-phosphate of the liver phosphorylase kinase preparation used may be due to partial proteolysis. Dr. Fischer's explanation for the calcium insensitivity is the presence of small acidic proteins in Dr. Livanova's calmodulin preparation, which lead to an unspecific increase in phosphorylation.

2 Rapporteur: Dieter H. Wolf

Insulin Mediators and Their Control of Covalent Phosphorylation

J. Larner[1,2], Y. Oron[3], K. Cheng[3], G. Galasko[3], R. Cabelli[2], and L. Huang[2]

1 Introduction

We have detailed two separate mechanisms for insulin to activate glycogen synthase. In mechanism I, insulin acts without glucose present via a mediator to convert the cyclic AMP-dependent protein kinase to a desensitized holoenzyme, effectively lowering the response of the cell machinery to existing concentrations of cyclic AMP. Mechanism II, originally discovered in rat adipocytes, is seen in the presence of glucose, or of a hexose whose transport is accelerated by insulin. Enhanced phosphorylation of the 6 position is also required, since the hexose-6-phosphate acts informationally to activate the phosphoprotein phosphatase to convert glycogen synthase to its dephospho state. Today I should like to discuss recent findings of our laboratory which demonstrate the presence of mechanism II in muscle, namely, mouse diaphragm, pointing out the generality of this mechanism and discuss our recent studies on the identification of insulin mediators which control protein phosphorylation state.

2 Results and Discussion

Since this paper is in the nature of a review, the materials and methods are not separately described. Instead, the reader is referred to the material and methods section of the appropriate original papers.

2.1 Background on Glycogen Synthase and Its Control by Insulin

We have used this enzyme to investigate the mechanism of insulin action for more than 20 years. Perhaps I could be forgiven for citing a few "firsts" discovered with this enzyme (Table 1). In 1960, it was first established as the enzyme site at which insulin regulates glycogenesis. In 1963, it was shown to be the first enzyme activated by dephosphorylation – to be followed several years later by pyruvate dehydrogenase. In 1964, it was the first enzyme shown to be interconverted under the aegis of cyclic AMP by a protein kinase. In 1971, it was shown that the enzyme had multiple sites of phosphorylation,

1 Established Investigator of the American Diabetes Association
2 Department of Pharmacology, University of Virginia, Charlottesville, VA 22908, USA
3 Department of Pharmacology, University of Witwatersrand Medical School, Johannesburg, South Africa

Table 1. Glycogen synthase – A "Hallmark" enzyme

First enzyme site by which insulin stimulates glycogen synthesis. Villar-Palasi and Larner (1960).
First enzyme activated by dephosphorylation and inactivated by phosphorylation. Friedman and Larner (1963).
First demonstration of a cyclic AMP-dependent protein kinase. Rosell-Perez and Larner (1964).
First enzyme with multiple phosphorylation sites in a single chain. Smith et al. (1971); Takeda et al. (1975).
First enzyme site of dephosphorylation with insulin. Roach et al. (1977).
First enzyme system used as assay for insulin mediator. Larner et al. (1976).

6 per 85,000 dalton subunit subsequently shown to be present on a single polypeptide chain in 1975. In 1977, it was the first enzyme where dephosphorylation with insulin action was shown by analysis of covalent P. In 1975, it was the first enzyme system used to demonstrate successfully an insulin generated mediator, namely an inhibitor of the cyclic AMP-dependent protein kinase.

2.2 Control by Insulin of Metabolism by Dephosphorylation-Mechanism I

Insulin acts anabolically to enhance macromolecular synthesis – often called the pleiotypic effect. To demonstrate the generality of the control by insulin of metabolism including glycogen, fat, cholesterol synthesis, the initiation of protein synthesis, and the inhibition of gluconeogenesis, the action of insulin on key enzyme is shown in Table 2.

Table 2. Insulin action on metabolism related to dephosphorylation

Enzyme	Insulin effect	Phosphorylation dephosphorylation control
Glycogen synthase ↑	1960, Villar-Palasi and Larner	1963, Friedman and Larner
Pyruvate dehydrogenase ↑	1970, Jungas	1969, Linn, Pettit and Reed
Acetyl-CoA carboxylase ↑	1974, Halestrap and Denton	1973, Carlson and Kim
Glycogen phosphorylase ↓ (liver only)	1967, Bishop and Larner	1956, Wosilait and Sutherland
Phosphorylase *b* kinase ↓	1968, Torres et al., BBA	1965, Posner, Stern and Krebs
Hormone sensitive lipase ↓	1973, Khoo et al.	1970, Huttunen et al., Corbin et al.
HMG CoA reductase ↑	1979, Ingebritsen et al.	1973, Beg, Allman and Gibson
HMG CoA reductase kinase ↓	1979, Ingebritsen et al.	1978, Ingebritsen et al.
Cholesterol esterase ↓		1974, Trzeciak and Boyd
Pyruvate kinase ↑		1974, Ljungstrom et al.
Phosphofructo kinase ↑	1972, Taunton et al.	1975, Brand and Soling
Fructose bisphosphatase ↓	1972, Taunton et al.	
Phosphoprotein phosphatase ↓ Inhibitor Type I	1980, Chang and Huang, Foulkes, Jefferson and Cohen	1976, Huang and Glinsman
Protein synthesis initiation		1975, Levin et al.

There are now nine instances where insulin acts to control metabolism by dephosphorylation of key enzymes. Usually the biosynthetic enzymes such as glycogen synthase, pyruvate dehydrogenase, acetyl CoA carboxylase, HMG CoA reductase, are activated, whereas the degradative enzymes including phosphorylase, phosphorylase *b* kinase, hormone sensitive lipase, cholesterol esterase, are inactivated. In the case of glycogen synthase, the effect of insulin to activate the enzyme was discovered 3 years before the mechanism of its control by covalent phosphorylation-dephosphorylation. In most other instances, covalent phosphorylation control was appreciated before the effects of hormones were initiated. Thus, the use of insulin as a hormonal "probe" led to the discovery of interconvertable forms of glycogen synthase.

2.3 Mechanism II

In a previous symposium we detailed our studies on the two separate mechanisms by which insulin activates glycogen synthase (Lawrence and Larner 1978). We demonstrated in rat adipocytes that, in the presence of glucose, a separate mechanism was present, termed Mechanism II. For this mechanism to obtain, both a transportable hexose as well as its phosphorylation in the 6 position were necessary. The hexose 6-phosphate allosterically activated the phosphoprotein phosphatase to convert the phospho or *D* form of glycogen synthase to the dephospho or *I* form. We next studied muscle to determine whether this was a general mechanism or simply limited to fat.

The mouse diaphragm was developed as a system because it was very sensitive to insulin in terms of glycogen synthase response (Oron and Larner 1980). As shown in Fig. 1,

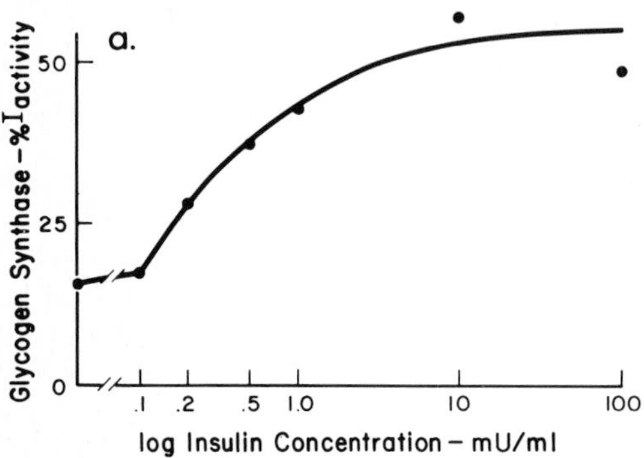

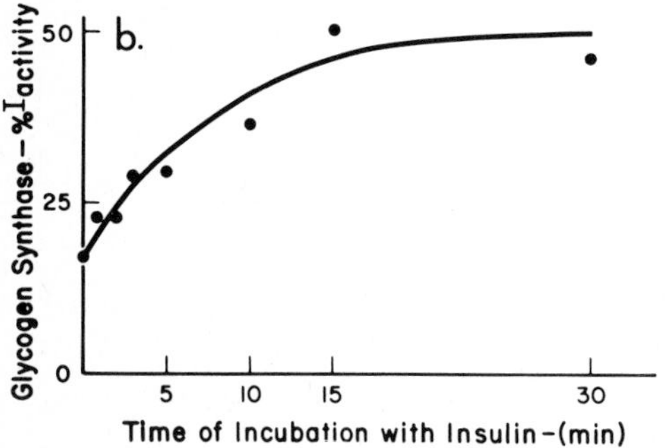

Fig. 1a, b. The response of mouse diaphragm GS to insulin. **a** Intact mouse diaphragms were incubated in Gey and Gey solution for 15 min, then transferred to fresh medium containing the indicated concentrations of insulin and incubated for 30 min. **b** Intact mouse diaphragms were incubated for 15 min, then transferred to fresh medium for the indicated time with and without insulin (10 mu/ml). Diaphragms were processed and GS assayed by standard methods

the diaphragm responds to insulin by activating glycogen synthase in a dose- and time-dependent manner with stimulations of two- to fivefold over control. The tissue is less sensitive to the hormone than the rat adipocyte responding at 0.2 mu/ml with a maximal response at about 10 mu/ml in contrast to the rat adipocyte which is maximal at 100 uu/ml. Activation of the enzyme by insulin is observed as early as after 1 min of incubation with the hormone. The enzyme is not activated appreciably with glucose alone, in contrast to the rat adipocyte (Lawrence and Larner 1978), but in keeping with our previous experience in rat diaphragm (Villar-Palasi and Larner 1961; Ng and Larner 1976), under conditions where it clearly responds to insulin. Therefore, we turned to the synthetic analog 2-deoxyglucose, whose phosphorylation product 2-deoxyglucose 6-P is stable and accumulates in tissue in contrast to glucose 6-P whose concentrations vary according to metabolic demands. In contrast to glucose, 2-deoxyglucose activates the enzyme significantly with increasing time of incubation with the sugar. When 2-deoxyglucose 6-P concentrations are determined in the tissue as a function of increasing 2-deoxyglucose concentration in the medium (Fig. 2), the increase matches the increase in activation of glycogen synthase. Under these conditions, ATP concentrations were shown to be essentially unchanged. As control hexose, we used 3-0-methyl glucose whose transport is also increased by insulin. However, it is not phosphorylated and does not activate glycogen synthase, again demonstrating that phosphorylation is required. The direct activation of the glycogen synthase phosphoprotein phosphatase by 2-deoxyglucose 6-P in the absence and presence of fluoride which inhibits the phosphatase is shown in Fig. 3. Analogous data with glucose 6-P, a more sensitive activator, have demonstrated similar results. The concentrations required to activate the phosphatase in the test tube match the calculated concentrations in the tissue measured (2 deoxyglucose 6-P) and estimated from the literature (glucose 6-P). Thus, from this work it is clear that mechanism II is operative in diaphragm skeletal muscle as well as fat (Lawrence and Larner 1978), and the question of its presence in other insulin-sensitive tissues is now open. Since glucose alone does activate glycogen synthase in adipocytes but not effectively in muscle, the physiological role of the concentration of blood glucose in regulating synthase in muscle is uncertain. Irrespective of the physiological role, the mechanistic role is of clear interest. The importance of glucose 6-P as a messenger molecule has been reported recently in two other systems unrelated to glycogen metabolism. Levin et al. (1975) demonstrated in elegant experiments that it acts as a messenger to control the initiation step in protein synthesis, another system regulated by covalent phosphorylation and dephosphorylation. Most recently Brautigan and co-workers in 1980 demonstrated that it increased Ca^{2+} uptake into "skinned" skeletal muscle fibers. Thus, the generality of the dual roles of hexose 6-P as a metabolite and as a messenger molecule seems to be increasing in importance.

2.4 Control by Insulin of Protein Phosphorylation

In rat adipocytes prelabeled with [^{32}P], Benjamin and Singer (1975), and Avruch et al. (1976) demonstrated that insulin acted to enhance dephosphorylation, but surprisingly to enhance phosphorylation of certain polypeptides as well. As shown in Table 3, an enzyme whose phosphorylation is enhanced in adipocytes and in liver is ATP-citrate lyase.

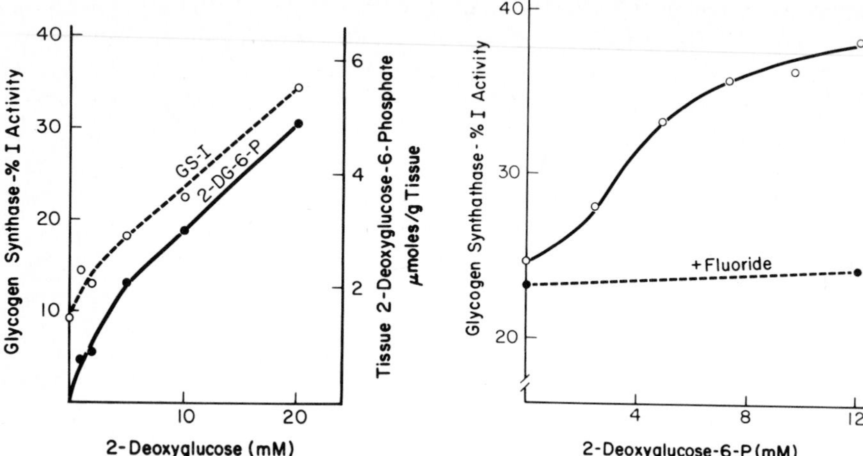

Fig. 2. Dose-dependent increase of diaphragm GS to 2-deoxyglucose. Intact mouse diaphragms were incubated for 15 min, then transferred to tubes containing [^{3}H]-2-deoxyglucose (1-20 mM, approximately 0.3-3.0 m Ci/mmol) in 3 ml Gey and Gey solution. For GS assay diaphragms were incubated in the same way without labeled 2-deoxyglucose. After 30 min the incubation was terminated by blotting and washing the diaphragms six consecutive times for 10 sec in 50 ml 2 mM 2-deoxyglucose in cold (4°C) Gey and Gey buffer. Diaphragms were blotted, frozen, chipped and weighed, then processed for GS assay and quantitation of 2-deoxyglucose 6-phosphate

Fig. 3. 2-deoxyglucose 6-phosphate stimulation of GS phosphoprotein phosphatase. Frozen diaphragm fragments were homogenized in 6 volumes of 10 mM MOPS, 5 mM EDTA pH 7.0 for 15 sec in a Polytron. The homogenate was rapidly filtered through a short column of hydrated Sephadex G-25 preequilibrated with the homogenization buffer. The effluent was incubated for 6 min at 30°C with increasing concentrations of 2-deoxyglucose 6-phosphate. The phosphatase reaction was terminated by adding 2 volumes of stopping solution (Tris-HCl 10 mM, EDTA 5 mM, KF100 mM, glycogen 1 mg/ml, pH 7.6). Controls contained stopping solution during the incubation. Aliquots were filtered through two consecutive columns of Sephadex G-25 preequilibrated with stop solution in order to remove 2-deoxyglucose 6-phosphate, and assayed for GS activity

Table 3. Increased phosphorylation with insulin sarcolemnal protein 16K daltons (seconds) Walaas et al. (1977)

ATP-Citrate lyase 123-130K daltons	Ramakrishna and Benjamin Alexander et al.
Ribosomal protein – S6 13K daltons	Lastick and McConkey Smith et al.

Two other polypeptides have also been shown to demonstrate increased phosphorylation with insulin, namely ribosomal S6-protein originally shown by Lastick and McConkey (1978) and confirmed by Smith et al. (1979), and a muscle sarcolemnal protein of 31K daltons first shown by Walaas et al. (1977). Here insulin acted very rapidly to enhance phosphorylation in seconds. Thus, there appears to be a complex picture of both decreased and increased phosphorylation with insulin.

2.5 Altered Phosphorylation in Extracts from Control and Insulin-pretreated Diaphragms

When extracts are prepared from mouse or rat diaphragms of control and insulin-treated tissue, incubated with [^{32}P]-ATP, and examined for protein phosphorylation, an effect is observed as early as after 1 min of insulin treatment. As shown in Fig. 4 (Oron et al. 1980), decreased phosphorylation is observed with insulin-pretreated tissue extracts under all conditions tested, namely: without (*a*) or with added cyclic AMP (*b*), with added histone substrate (*c*) or with added histone and cyclic AMP (*d*). As shown in Fig. 5, the decrease in phosphorylation occurs rapidly, remains decreased for about 5 min and then

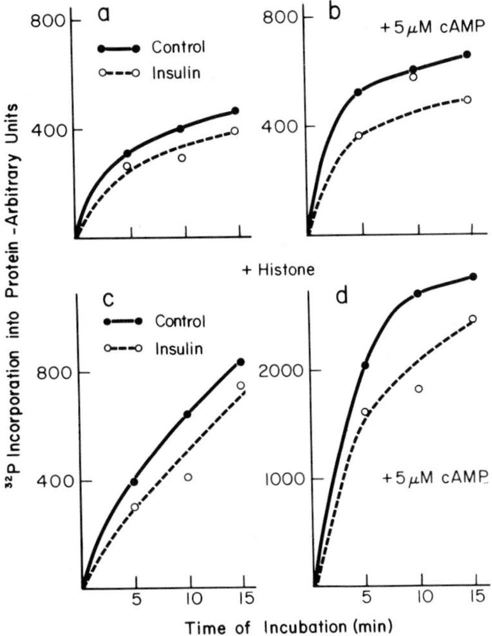

Fig. 4 a-d. Decreased phosphorylation of endogenous proteins in extracts of insulin-treated mouse diaphragms. Intact mouse diaphragms were incubated alone or with 10 mu/ml insulin for 20 min. Diaphragm extratcs (1/10 final dilution wt/vol) were incubated for the indicated times and analyzed for trichloracetic acid insoluble radioactivity. The concentration of cyclic AMP was 5 μM, and of histone 0.5 mg/ml. Results are expressed as [^{32}P] incorporation relative to the total activity of glycogen synthase in each extract. Mean results of three experiments

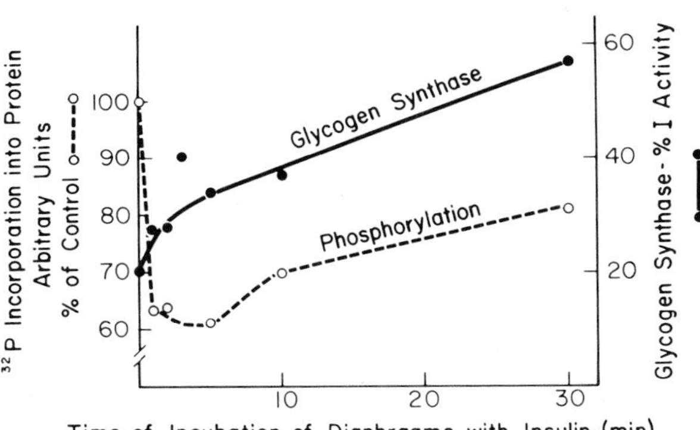

Fig. 5. Time course of insulin inhibition of protein phosphorylation. Intact diaphragms were incubated for the indicated times with insulin 10 mu/ml. Extracts were assayed for protein phosphorylation at 1:10 final dilution (wt/vol). Results are expressed as % of controls (without insulin) calculated relative to total glycogen synthase activity. *Closed symbols* represent the activity of GS-I. Each point is the mean of three separate experiments

"relaxes" over the next 25 min – but is still diminished even at 30 min. The maximal rate of glycogen synthase activation follows the maximal decrease in phosphorylation capacity.

When polyacrylamide gels are examined for radioactivity (Fig. 6), an interesting result is observed. Although decreased phosphorylation is observed in two bands of 94 and

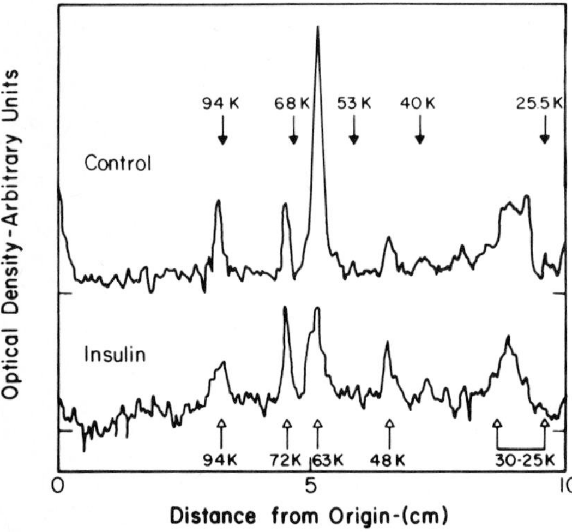

Fig. 6. [^{32}P] incorporation into individual polypeptides. Dried electropherograms of phosphorylated diaphragm extracts were autoradiographed for 48 h at −90°C. The developed autoradiograms were scanned with a densitometer. Densitometric profiles of extracts of control (*upper*) and insulin treated (6 min 10 mu/ml) diaphragms (*lower*) are shown. *Arrows* mark the calculated molecular weights based on standards

63K, increased phosphorylation is also observed in two other bands of 72 and 48K. Thus even where tissue was exposed to insulin for 6 min, the time of major decrease in phosphorylation, a mixed picture of decreased as well as increased phosphorylation is observed. Similar experiments have been obtained with rat diaphragms as well, and these experiments are similar to the results in adipocytes and hepatocytes by Benjamin and Singer (1975), and by Avruch et al. (1976). A main advantage of the present system is its stability and its availability to analysis and manipulation.

In a carefull examination of the conditions required to observe the decreased phosphorylation with insulin, another very interesting and surprising result is obtained. Namely, a key point experimentally is the concentration of frozen tissue powder to extracting fluid. When the tissue is extracted in a more concentrated way, the inhibition is observed. When the tissue is extracted in a more dilute way, a stimulation of phosphorylation of the insulin versus the control is observed. Although these results can be interpreted in several ways, it is clear that insulin has altered the phosphorylation capacity of extracts prepared from treated tissues. Insulin thus rapidly and dramatically influences the phosphorylation of cellular proteins.

2.6 Altered Cyclic AMP-dependent Protein Kinase with Insulin Action

As shown in Table 4, starting in the late 1960s and continuing through the present time, evidence has accumulated mainly from our laboratory, but from others as well, that insulin acts to maintain the cyclic AMP-dependent protein kinase in its holoenzyme form –

Table 4.

Date	Holoenzyme increased with insulin	Tissue	Assay
1967	Villar-Palasi and Wenger	Rat skeletal Muscle (in vivo)	Glycogen synthase *I* Kinase-holoenzyme increased
1970	Shen et al.	Rat diaphragm	Glycogen synthase *I* Kinase-holoenzyme increased
1972	Walaas et al.	Rat diaphragm	Histone Kinase-holoenzyme increased
1972	Miller and Larner	Rat diaphragm	Glycogen Synthase *I* Kinase-holoenzyme increased
1973	Miller and Larner	Perfused rat liver	Glycogen Synthase *I* Kinase-holoenzyme increased
1978	Guinovart et al.	Rat adipocyte	Type 2 holoenzyme increased-gel binding
1978	Walkenbach et al.	Rat diaphragm	Type 1 holoenzyme desensitized to cAMP-histone kinase and gel binding
1979	Northrup et al.	Rat kidney slice	Histone kinase desensitized to cAMP-insulin vs glucagon
1980	Walkenbach et al.		Histone kinase desensitized to cAMP-insulin vs epinephrine

but in a cyclic AMP-desensitized state. This has now been shown with muscle, liver, and fat. It has been shown with glycogen synthase and histone as substrates. It has been shown by enzyme assays and by visualizing the enzyme on polyacrylamide gels by [^{3}H] cyclic AMP binding. And finally, it has been shown by the altered phosphorylation or activity of two cyclic AMP-dependent protein kinase substrates, namely phosphorylase *b* kinase and phosphoprotein phosphatase inhibitor I. Our most recent experiments published this year (Walkenbach et al. 1980) are shown in Figs. 7 and 8. Here, in rat diaphragms glycogen synthase activity, cyclic AMP concentrations, and protein kinase activity states were measured with either suboptimal epinephrine and increasing insulin, or optimal insulin and increasing epinephrine. As can be seen, in both cases, insulin caused a decrease in protein kinase activity state (holoenzyme favored) under conditions where the concentration of cyclic AMP was unaltered by insulin. For this reason we decided to use the protein kinase as bioassay for the presumptive insulin mediator.

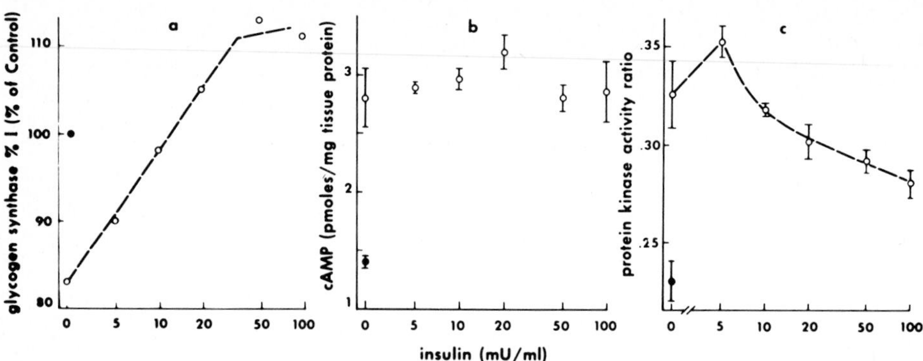

Fig. 7 a-c. Dose-response relationship of insulin on glycogen synthase, protein kinase and cyclic AMP in the presence of epinephrine. Rat hemidiaphragms were exposed to the concentration of insulin indicated for 15 min. Epinephrine (0.25 μM) was added during the last 6 min of insulin exposure. Tissues were then assayed for **(a)** GS, **(b)** cyclic AMP, and **(c)** protein kinase activity ratio. *Closed circles* indicate enzyme activities of hemi diaphragms receiving no hormonal treatment

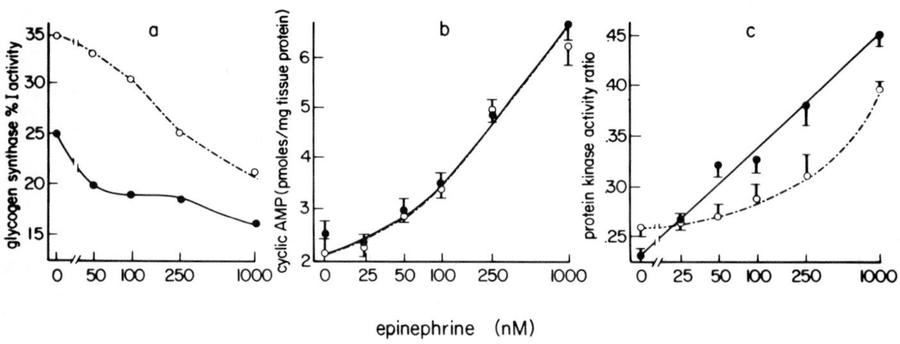

Fig. 8 a-c. The effect of insulin on the dose-response relationship of epinephrine on **(a)** GS, **(b)** cyclic AMP, and (c) protein kinase activity ratio. Rat hemi diaphragms were exposed to the concentrations of epinephrine indicated without (●—●) or with (○—○) concurrent insulin (50 mμ/ml)

2.7 Justification for an Insulin Mediator

Over a period of 10 years we have obtained two types of evidence for a chemical mediator in insulin action.

1. The dissociation which has been observed in our own and in other laboratories in the action of insulin on several metabolic parameters. For example, Huijing et al. (1969) observed that, in the perfused rat heart, insulin was able to enhance glucose transport under conditions where it no longer stimulated glycogen synthase conversion to the active or *I* form. Conversely, in France, Eboue'-Bonis et al. (1967) demonstrated in rat diaphragm that with the use of NEM, an SH blocking agent, the effect of insulin on glucose transport was abolished, under conditions where the effect of insulin to activate glycogen synthase was retained. This two-way dissociation argued strongly for a chemical mediator,

although the argument was not, of course, conclusive. We did in fact propose a mediator in 1972 (Larner 1972).

2. Two proteins have been shown in our laboratory to interact with the insulin mechanism. First, we demonstrated that a pituitary peptide of 30,000 daltons (Miller and Larner 1972) blocked all of the actions of insulin on glucose uptake, glycogen synthesis, and glycogen synthase activation. In addition, it blocked the effect of insulin to inactivate the cyclic AMP-dependent protein kinase. This block by the pituitary peptide was overcome by high concentrations of insulin. This suggested that the pituitary peptide as well as insulin acted on the outside of the cell membrane.

More recently we studied the anti-insulin receptor antibody isolated from a patient with insulin-resistant diabetes. This insulin receptor blocking antibody (150,000 daltons) curiously has insulin-like properties. In rat adipocytes, it fully reproduced the actions of insulin on glycogen synthase activation (Lawrence et al. 1979). This also demonstrated that a molecule much larger than insulin was able to set into motion from the cell membrane the intracellular events, and suggested a chemical messenger as the mechanism rather than the entry of insulin or of the antibody itself.

2.8 Identification and Isolation of an Insulin Mediator

Our plan was to prepare from control and insulin-treated muscle tissue a protein-free fraction by heart or by acid denaturation and to search for the insulin mediator by its effect to inactivate the protein kinase. As shown in Table 5, we were able to detect the

Table 5.

Experimental condition	*% Inhibition of protein kinase* Control	Insulin	% Net effect due to insulin
A) Cyclic AMP-dependent protein kinase			
– cyclic AMP	70.0	93.0	33
+ cyclic AMP	38.0	74.0	95
B) Cyclic AMP-independent protein kinase			
PC 0.4	70.0	64.0	–
PC 0.7	82.0	79.0	–
Phosphorylase *b* kinase	69.0	68.0	–

From Larner et al. (1979)

presence of increased protein kinase inhibitory material in the insulinized as compared to the control muscle (Larner et al. 1976). We then began to fractionate the material.

With two additional steps (Table 6) we identified and partially purified the insulin mediator, which was shown to carry out three of the intracellular actions of insulin to

Table 6. Purification of Insulin Mediator

Step	Procedure
1	Frozen powdered muscle.
2	Heat deproteinized.
3	Paper chromatography.
4	Sephadex G-25 chromatography of nucleotide "free" fraction.
5	High-voltage electrophoresis at pH 1.9 of fraction II.
6	High-voltage electrophoresis at pH 3.5 of fraction 1 and 3.

control protein phosphorylation and dephosphorylation, namely inhibition of the cyclic AMP-dependent protein kinase, activation of pyruvate dehydrogenase phosphoprotein phosphatase (Jarett and Seals 1979), and activation of glycogen synthase phosphoprotein phosphatase (Larner et al. 1979a; Cheng et al. 1980). After the removal of nucleotides by paper chromatography or more recently by charcoal adsorption, molecular sieving on Sephadex G-25 was carried out at acid pH as shown in Fig. 9. Peak 2 is the only peak with ninhydrin reactive material and is the peak with the major 230 nm absorption. Peak 2 contains the insulin mediator in terms of protein kinase inhibition (Table 3), pyruvate dehydrogenase activation (Jarett and Seals 1979) and glycogen synthase phosphoprotein phosphatase activation (Fig. 10). Peak 2 is also altered by insulin treatment, such that the material is redistributed to higher molecular weight without alteration in mass (Cheng et al. 1980). The characteristics of the mediator in peak 2 are that it is not a nucleotide, but rather a peptide of a molecular weight of approximately 1000-1500. It is partially destroyed by subtilisin – not by trypsin, and it is completely destroyed by hydrolysis in 6 N HCl at 100°C for 18 h.

It is of considerable interest that in order to understand further the bell-shaped curve of the concentration dependence of activation and inhibition of glycogen synthase phosphoprotein phosphatase by peak 2, we have further fractionated the material by high-voltage electrophoresis (Table 6, steps 4 and 5). We have obtained two separate fractions. Fraction *1→6* in Fig. 11 (Cheng et al. 1980), which is more basic, appears to be the insulin mediator by its characteristics effects on the cyclic AMP-dependent protein kinase and on synthase phosphoprotein phosphatase. In addition, we have obtained a second peptide or peptide like material (*1→4*) which activates the protein kinase in the absence of cyclic AMP, and which inhibits the phosphoprotein phosphatase. This peptide is more acidic than *1→6*. We have fractionated these two peptides by at least two independent methods, HVE and thin layer chromatography. The latter compound is also of interest

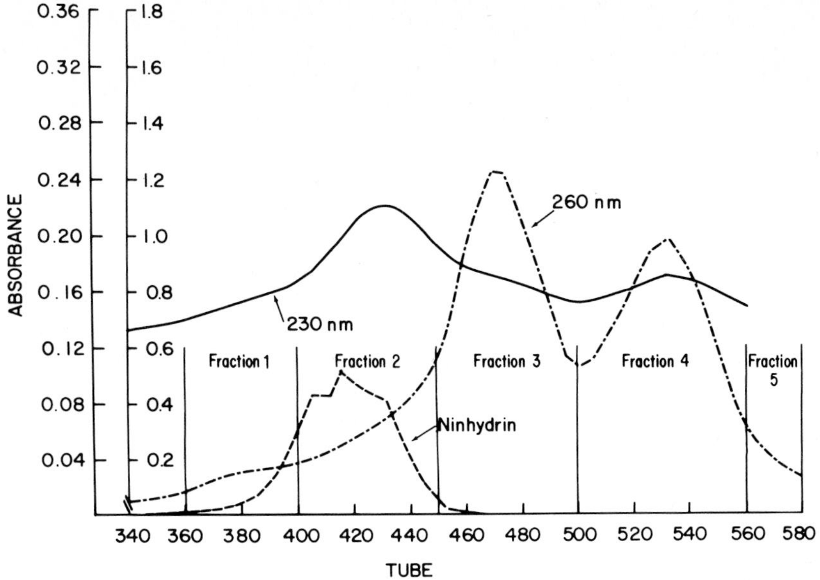

Fig. 9. Sephadex G-25 chromatogram of a skeletal muscle extract that had been deproteinized and from which the major nucleotides had been removed by paper chromatography. Frozen muscle powder (25 g) was heat treated, extracted, and chromatographed on paper. After elution, lyophilization, and reconstitution, 5 ml of the purified paper eluate was applied to the column; the column was washed twice with equal volumes of 0.05 N formic acid and then developed with the same solution; 2.5 ml fractions were collected at a flow rate of 15 drops/min in a fraction collector and then analyzed for absorbancy at 230 and 260 nm and for ninhydrin reactivity

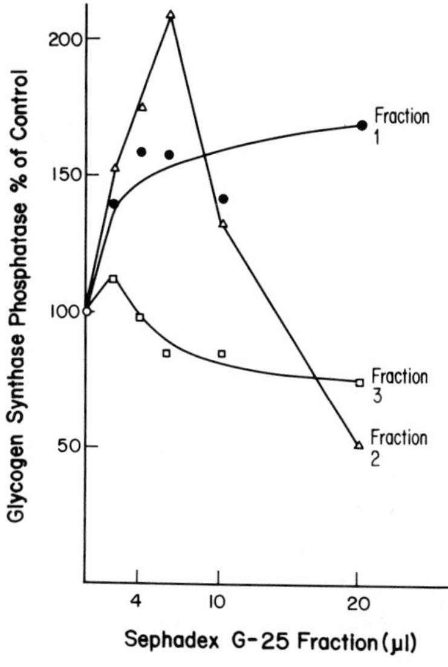

Fig. 10. Activation of glycogen synthase phosphoprotein phosphatase by Sephadex G-25 column fractions. Varying amounts of reconstituted fractions 1,2, and 3 from insulin-treated rat skeletal muscle extract were lyophilized and redissolved in 0.2 ml of 140 mM MES (pH 7.0) containing glycogen synthase D (0.2 mg/ml) and a crude glycogen synthase phosphoprotein phosphatase (2.8 mg/ml). After incubation at 21°C for 20 min, 20 μl portions were removed for GS assay in the absence and presence of glucose 6-phosphate. The reaction mixtures for the GS assay contained (total volume 90 μl) 4.4 mM UDPG, 13.3 mM EDTA, 100 mM KF, 6.7 mg/ml glycogen, 7.2 mM glucose 6-phosphate, if present, and 33 mM tris pH 7.8. Mixtures were incubated 10 min at 30°C and analyzed

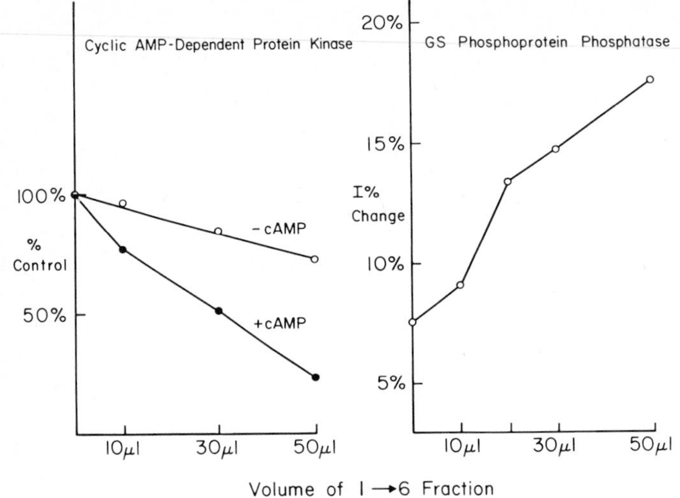

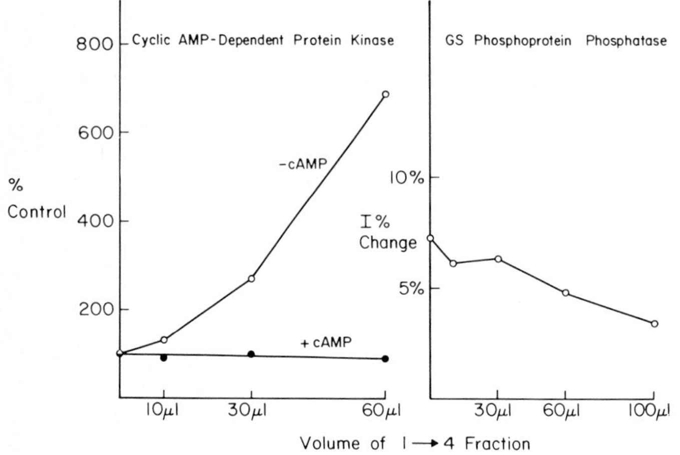

Fig. 11. Effect of fractions *1→4*, and *1→6* on cyclic AMP-dependent protein kinase and glycogen synthase phosphoprotein phosphatase. The control activities of protein kinase in the absence and presence of cyclic AMP were 0.06 and 1.5 nmol/min/mg protein respectively. The final concentration of cyclic AMP is 2.5 μM, when present

in terms of its regulatory potential in the control of increased protein phosphorylation observed with insulin. Both materials appear to be peptides in nature since they are destroyed by hydrolysis in 6 N HCl and contain amino acids after acid hydrolysis

2.9 Hypothesis and Summary

Because of these recent results it is tempting to hypothesize that insulin acts via a receptor at the cell surface to activate a cell membrane-bound protease to catalyze a limited proteolytic event, leading to the liberation of the peptide mediator intracellularly. The precursor of the mediator could conceivably be present in the membrane as an integral protein with a cytoplasmic tail, or it could be present loosely attached to the cell membrane. Certainly models exist in the erythrocyte cell membrane of integral proteins with cytoplasmic tails and of membrane bound proteases, which are attractive to consider. Also attractive to consider are the close links between the anabolic pleiotypic actions of insulin and the pleiotypic actions of proteases in tissue culture cell systems. I refer especially to the work of Dennis Cunningham and collaborators with thrombin acting on quiescent chick embryo cells (Carney and Cunningham 1978a, b). Furthermore, the structural relationship of insulin to pancreatic proteolytic enzymes has been shown by computer analysis, (DeHaen et al. 1976). The insulin-like actions of very limited tryptic action in various eukaryotic cells has been recognized for some time (Rieser and Rieser 1964). Therefore, it is important now to recognize a new form of hormonal communication via intracellular peptides (Seals and Czech 1980) with insulin action and which control phosphorylation and dephosphorylation. Thus, in addition to nucleotides and ions we must now add peptides to the list of intracellular messengers.

Further work will be required to establish the structure of these two mediators, their possible relationship to each other, as well as their sites of origin and mechanism of action on the protein kinases and phosphatases.

Acknowledgment. Supported in part by NIH grant #AM14334 and by the University of Virginia Diabetes Research and Training Center grant #AM22125.

References

Alexander MC, Kowaloff EM, Witters LA, Denniby DT, Avruch J (1979) Purification of a hepatic 123,000-dalton hormone-stimulated ^{32}P-peptide and its identification as ATP-citrate lyase. J Biol Chem 254:8052-8056

Avruch J, Leone GR, Martin DB (1976) Effect of epinephrine and insulin on phosphopeptide metabolism in adipocytes. J Biol Chem 251:1511-1515

Beg ZH, Allman DW, Gibson DM (1973) Modulation of 3-hydroxy-3-methyl glutaryl coenzyme A reductase activity with cAMP and with protein fractions from rat liver cytosol. Biochem Biophys Res Commun 54:1362-1369

Benjamin WB, Singer I (1975) Actions of insulin, epinephrine, and dibutyryl cyclic adenosine 3'-5'-monophosphate on fat cell protein phosphorylations. Biochemistry 14:3301-3309

Bishop JS, Larner J (1967) Rapid activation-inactivation of liver uridine diphosphate glucose-glycogen transferase and phosphorylase by insulin and glucagon in vivo. J Biol Chem 242:1355-1356

Brand IA, Soling HD (1976) Activation and inactivation of rat liver phosphofructokinase by phosphorylation-dephosphorylation. FEBS lett 57:163-168

Brautigan DL, Kerrick WGL, Fischer EH (1980) Insulin and glucose 6-phosphate stimulation of Ca^{2+} uptake by skinned muscle fibers. Proc Natl Acad Sci USA 77:936-939

Carlson CA, Kim KH (1973) Regulation of hepatic acetyl coenzyme A carboxylase by phosphorylation and dephosphorylation. J Biol Chem 248:378-380

Carney DH, Cunningham DD (1978a) Cell surface action of thrombin is sufficient to indicate division of chick cells. Cell 14:811-823

Carney DH, Cunningham DD (1978b) Role of specific cell surface receptors in thrombin stimulated cell division. Cell 15:1341-1349

Chang LY, Huang LC (1980) Effects of insulin treatment on the activities of phosphoprotein phosphatase and its inhibitors. Acta Endocrinol 95:427-432

Cheng K, Galasko G, Huang L, Kellogg J, Larner J (1980) Studies on the insulin mediator II Separation of two antagonistic biologically active materials from fraction II. Diabetes 29:659-661

Corbin JD, Reimann EM, Walsh DA, Krebs EG (1970) Activation of adipose tissue lipase by skeletal muscle cyclic adenosine 3'5'-monophosphate stimulated protein kinase. J Biol Chem 245:4849-4851

DeHaen C, Swanson E, Teller DC (1976) The evolutionary origin of proinsulin. J Mol Biol 106:639-661

Eboué-Bonis D, Chambout AM, Volfin P, Clauser H (1967) Selective action of N-ethylmaleimide on the stimulation by insulin of the metabolism of surviving diaphragms. Biol Soc Chim Biol 49: 415-431

Ernst V, Levin DH, London IM (1978) Evidence that glucose 6-phosphate regulates protein synthesis initiation in reticulocyte lysates. J Biol Chem 253:7163-7172

Foulkes JG, Jefferson LS, Cohen P (1980) The hormonal control of glycogen metabolism: dephosphorylation of protein phosphatase inhibitor I in vivo in response to insulin. FEBS Lett 112: 21-24

Friedman DL, Larner J (1963) Studies on UDPG-α-glucan transglucosylase III. Interconversion of two forms of muscle UDPG-α-glucan transglucosylase by a phosphorylation-dephosphorylation reaction sequence. Biochemistry 2:669-675

Guinovart J, Lawrence JC, Larner J (1978) Hormonal effects on fat cell adenosine 3'5'-monophosphate-dependent protein kinase. Biochim Biophys Acta 539:181-194

Halestrap AP, Denton RM (1974) Hormonal regulation of adipose tissue acetyl-coenzyme A carboxylase by changes in polymeric state of the enzyme. Biochem J 142:365-377

Huang FL, Glinsmann WH (1976) Phosphorylase phosphatase inhibitors from rabbit skeletal muscle. Fed Proc 35:1410

Huijing F, Nuttall FQ, Villar-Palasi C, Larner J (1969) A dissociation in the action of insulin on transport and transferase conversion. Biochim Biophys Acta 177:204-212

Huttunen JK, Steinberg D, Mayer SE (1970) ATP-dependent and cyclic AMP-dependent activation of rat adipose tissue lipase by protein kinase from rabbit skeletal muscle. Proc Natl Acad Sci USA 67:290-295

Ingebritsen TS, Lee HS, Parker RA, Gibson DM (1978) Reversible modulation of the activities of both liver microsomal HMG-CoA reductase and its activating enzyme. Evidence for regulation by phosphorylation-dephosphorylation. Biochem Biophys Res Commun 81:1268-1277

Ingebritsen TS, Geelen MJH, Parker RA, Evenson KJ, Gibson DM (1979) Modulation of hydroxymethyl glutaryl-CoA reductase activitiy, reductase kinase and cholesterol synthesis in rat hepatocytes in response to insulin and glucagon. J Biol Chem 254:9986-9989

Jarett L, Seals JR (1979) Pyruvate dehydrogenase activation in adipocyte mitochondria by an insulin-generated mediator from muscle. Science 206:1407-1408

Jungas RL (1970) Effect of insulin on fatty acid synthesis from pyruvate, lactate, or endogenous sources in adipose tissue. Evidence for the hormonal regulation of pyruvate dehydrogenase. Endocrinology 86:1368-1375

Khoo C, Steinberg D, Thompson B, Mayer SE (1973) Hormonal regulation of adipocyte membranes: The effects of epinephrine and insulin on the control of lipase, phosphorylase kinase, phosphorylase and glycogen synthase. J Biol Chem 248:3823-3830

Larner J (1972) Insulin and glycogen synthase. Diabetes 21 Suppl 2, 428-438

Larner J, Takeda Y, Brewer HB, Huang LC, Hazen R, Brooker G, Murad F, Roach P (1976) In: Shaltiel S (ed) Studies on glycogen synthase and its control by hormones in metabolic interconversion of enzymes 1975. Springer, Berlin Heidelberg New York, pp 71-85

Larner J, Galasko G, Cheng K, DePaoli-Roach AA, Huang L, Daggy P, Kellogg J (1979a) Science 206:1408-1410

Larner J, Lawrence JC, Roach PJ, DePaoli-Roach AA, Walkenbach RJ, Guinovart J, Hazen RJ (1979b) About insulin and glycogen, hormones and cell culture. Cold Spring Harbor Conf Cell Prolifer 6:95-112

Lastick SM, McConkey EH (1978) In: Prescott DM, Fox CF (eds) Ribosomal protein phosphorylation and control of cell growth in cell reproduction. Academic Press London New York, pp 61-69

Lawrence JC, Larner J (1978) Activation of glycogen synthase in rat adipocytes by insulin involves increased glucose transport and phosphorylation. J Biol Chem 253:2104-2113

Lawrence JC Jr, Larner J, Kahn CR, Roth J (1979) Autoantibodies to the insulin receptor activate glycogen synthase in rat adipocytes. Mol Cell Biochem 22:153-157

Levin DH, Ranu RS, Ernst V, Fefer MA, London IM (1975) Association of a cyclic AMP-dependent protein kinase with a purified translational inhibitor isolated from hemin-deficient rabbit reticulocyte lysates. Proc Natl Acad Sci USA 72:4849-4853

Linn TC, Pettit FH, Reed LJ (1969) a-keto acid dehydrogenase complexes X regulation of the activity of the pyruvate dehydrogenase complex from beef kidney mitochondria by phosphorylation and dephosphorylation. Proc Natl Acad Sci USA 62:234-241

Ljungstrom O, Hjelmquist G, Engstrom L (1974) Phosphorylation of purified rat liver pyruvate kinase by cyclic 3',5'-AMP stimulated protein kinase. Biochem Biophys Acta 358:289-298

Miller TB, Larner J (1972) Anti-insulin actions of a bovine pituitary diabetogenic peptide on glycogen synthesis. Proc Natl Acad Sci USA 69:2774-2777

Miller TB, Larner J (1973) Mechanisms of control of hepatic glycogenesis by insulin. J Biol Chem 248:3483-3488

Ng FM, Larner J (1976) Actions of insulin-potentiating peptides on glycogen synthesis. Diabetes 25:413-419

Northrup JE, Krezowski PA, Palumbo PJ, Kim JK, Hui YSF, Dousa TP (1979) Insulin inhibition of hormone-stimulated protein kinase systems in rat renal cortex. Am J Phys 236:E649-654

Oron Y, Larner J (1980) Insulin action in intact mouse diaphragm I Activation of glycogen synthase through stimulation of sugar transport and phosphorylation. Mol Cell Biochem 32:153-160

Oron Y, Galasko G, Larner J (1980) Insulin action in intact mouse diaphragm II Inhibition of endogenous protein phosphorylation. Mol Cell Biochem 254:161-167

Posner JB, Stern P, Krebs EG (1965) Effects of electrical stimulation and epinephrine in muscle phosphorylase, phosphorylase b kinase and adenosine 3',5'-phosphate. J Biol Chem 240:982-985

Ramakrishna S, Benjamin WB (1979) Fat cell protein phosphorylation identification of phosphoprotein – 2 as ATP citrate lyase. J Biol Chem 254:9232-9236

Rieser P, Rieser CH (1964) Anabolic responses of diaphragm muscle to insulin and to other pancreatic proteins. Proc Soc Exp Biol Med 116:669-671

Roach PJ, Rosell-Perez M, Larner J (1977) Muscle glycogen synthase in vivo state. Effects of insulin administration on the chemical and kinetic properties of the purified enzyme. FEBS Lett 80: 95-98

Rosell-Perez M, Larner J (1964) Studies on UDPG-α-glucan transglucosylase V. Two forms of the enzyme in dog skeletal muscle and their interconversion. Biochemistry 3:81-88

Seals JR, Czech MP (1980) Evidence that insulin activates an intrinsic plasma membrane protease in generating a secondary chemical mediator. J Biol Chem 255:6529-6531

Shen LC, Villar-Palasi C, Larner J (1970) Hormonal alteration of protein kinase sensitivity to 3',5'-cyclic AMP. Physiol Chem Phys 2:536-544

Smith CH, Brown NE, Larner J (1971) Molecular characteristics of the totally dependent and independent forms of glycogen synthase of rabbit skeletal muscle. Biochim Biophys Acta 242:81-88

Smith CJ, Wejksnora PJ, Warner JR, Rubin CS, Rosen OM (1979) Insulin-stimulated protein phosphorylation in 3T3-L1 preadipocytes. Proc Natl Acad Sci USA 76:2725-2729

Takeda Y, Brewer HB, Larner J (1975) Structural studies on rabbit muscle glycogen synthase. J Biol Chem 250:8943-8950

Taunton OD, Shifel FB, Greene HL, Herman RH (1972) Rapid reciprocal changes of rat hepatic glycolytic enzymes and fructose-1, 6-diphosphatase following glucagon and insulin injection in vivo. Biochem Biophys Res Commun 48:1663-1670

Torres HN, Marechal LR, Bernard E, Belocopitow E (1968) Control of muscle glycogen phosphorylase activity by insulin. Biochim Biophys Acta 156:206-209

Trezeciak WH, Boyd GS (1974) Activation of cholesteryl esterase in bovine adrenal cortex. Eur J Biochem 46:201-207

Villar-Palasi C, Larner J (1960) Insulin mediated effect on the activity of UDPG-glycogen transglucosylase activity of muscle. Biochim Biophys Acta 39:171-173

Villar-Palasi C, Larner J (1961) Insulin treatment and increased UDPG-glycogen transglucosylase activity in muscle. Arch Biochem Biophys 94: 436-442

Villar-Palasi C, Wenger JI.(1967) In vivo effect of insulin on muscle glycogen synthetase. Identification of the action pathway. Fed Proc 26:563

Walaas O, Walaas E, Gronnerod O (1972) Effect of insulin and epinephrine on cyclic AMP-dependent protein kinase in rat diaphragm. Isr J Med Sci 8:353-357

Walaas O, Walaas E, Lystad E, Alertsen AR, Horn RS, Fossum S (1977) A stimulatory effect of insulin on phosphorylation of a peptide in sarcolemma-enriched membrane preparation from rat skeletal muscle. FEBS Lett 80:417-422

Walkenbach RJ, Hazen R, Larner J (1978) Reversible inhibition of cyclic AMP-dependent protein kinase by insulin. Mol Cell Biochem 19:31-41

Walkenbach RJ, Hazen R, Larner J (1980) Hormonal regulation of glycogen synthase. Insulin decreases protein kinase sensitivity to cyclic AMP. Biochim Biophys Acta 629:421-430

Wosilait WD, Sutherland EW (1956) The relationship of epinephrine and glucagon to liver phosphorylase: Enzymatic inactivation of liver phosphorylase. J Biol Chem 218:469-481

Regulation of Phosphofructokinase by Phosphorylation-Dephosphorylation-State of the Art

H.-D. Söling and I. Brand[1]

The possibility of a "futile cycling" of fructose-6-phosphate (F-6-P) through the phosphofructokinase (PFK) and fructose-1,6-bisphosphatase (FDPase) catalyzed reactions in liver has been discussed first by Gevers and Krebs (1966). Later on, several groups (Clark et al. 1974; Rognstadt and Katz 1976; Van Schaftingen et al. 1980a) could indeed demonstrate the occurence of such a cycling, which is increased in the presence of a high concentration of glucose but will decrease in the presence of glucagon.

In view of a glucagon-dependent regulation of L-type pyruvate kinase by phosphorylation-dephosphorylation it seemed reasonable to speculate that the glucagon-mediated decreases in the rate of F-6-P phosphorylation might also result from a covalent modification of either PFK or FBPase (or both) by cAMP-dependent phosphorylation mechanisms.

Brand and Söling (1975) have described in 1975 a rapid inactivation of PFK in the pH 5.5 pellet from a rat liver homogenate in the presence of 20 mM $MgCl_2$. This could be reversed by addition of equilimolar concentrations of ATP and the final activity after ATP addition exceeded even the initial activity. When Mg-inactivated PFK from a rat liver homogenate was subjected to gel chromatography on Sephadex G-100 (Söling and Brand 1976), it appeared at an apparent molecular weight of about 120,000 which is higher than the molecular weight of the PFK protomer ($\approx$ 82,000) but lower than the dimer. It was, therefore, concluded that inactive PFK exists mainly as a complex of a protomer with a PFK-kinase.

Brand and Söling (1975) could indeed show that addition of ATP-γ-[^{32}P] to such a complex leads to a phosphorylation of PFK together with an increase in V_{max} activity of PFK. It was shown further that the phosphate incorporated was covalently bound to serine in the PFK protein. [^{32}P]-labeled PFK prepared in this way could be dephosphorylated in the presence of Mg^{2+} by a protein factor which was partially purified by DEAE-cellulose chromatography. Dephosphorylation was associated with loss of enzymatic activity (Brand and Söling 1975).

Further experiments showed that the reactivation of inactive PFK in the presence of ATP was completely inhibited by 100 mM $(NH_4)_2SO_4$ while inactivation of PFK was inhibited completely by 20 mM P_i. A buffer containing these concentrations of $(NH_4)_2SO_4$ and P_i was therefore thought to serve the function of a "stopping medium" by the aid of which the amount of active and inactive PFK could be fixed to the intracellular state during tissue homogenization. Using this stopping medium it was found that in fed animals most of the PFK in liver exists in the active state, whereas during star-

1 Abteilung für Klinische Biochemie, Zentrum Innere Medizin der Universität Göttingen, Humboldtallee 1, 3400 Göttingen, FRG

vation active PFK contributes only 20%-30% to total PFK activity. Refeeding in vivo or perfusion of isolated livers from starved rats in the presence of 30 mM glucose converted almost all PFK back into the active form (Brand et al. 1976). Glucagon or dibutyryl-3',5'-cAMP inhibited or abolished this in vitro effect of glucose in experiments with isolated perfused rat livers (Söling et al. 1978a). On the basis of these and other experiments it was assumed that PFK exists in two active states: a stable phosphorylated and a labile dephosphorylated state. The enzyme in the labile form dissociates rapidly into its inactive subunits. Phosphorylation of the subunits induces a reassociation to the active tetramer and higher aggregates (Brand and Söling 1975). However, more recent experiments performed in our laboratory (Brand I. unpublished results) have shown that the situation is far more complex. When livers from starved rats were homogenized in the stopping medium and immediately spun in an Airfuge^R centrifuge in such a way that the time elapsing between obtaining the tissue and the assay of enzyme activity was less than 8 min, the extracts showed the same high PFK activity as extracts from livers of fed animals. Liver extracts prepared from fed animals showed no difference in V_{max} activity of PFK irrespective of whether they were rapidly prepared by the Airfuge method or by the conventional ultracentrifugation method which lasted about 60 min. These results mean that the low PFK activity found in livers from starved rats did not reflect the true situation in the liver cell but rather indicates a greater lability of PFK in extracts from livers of starved animals. Whether this is the result of a different pattern of small-molecular weight ligands during starvation or resides in a different state of PFK itself is not completely clear yet. There exist experimental data in support of either explanation: The incorporation of [^{32}P] into PFK in isolated liver cells from starved rats is very low, but it increases significantly when glucose (30 mM) is added to the medium (Brand et al. unpublished data; manuscript in preparation) (Fig. 1). This effect of glucose cannot be

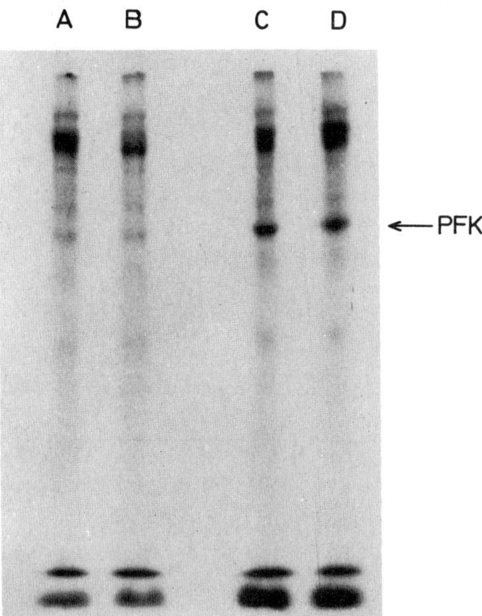

Fig. 1. Autoradiographs of [^{32}P]-labeled phosphofructokinase from isolated rat liver cells. Isolated liver cells from 24 h starved male Wistar rats were incubated in the absence (*lanes A* and *B*) or presence (*lanes C* and *D*) of 30 mM glucose. At the end of incubation, the liver cells were sedimented, washed and suspended in a buffer containing 4 μg digitonin/mg wet weight. The extract was centrifuged for 5 min at 2000 g and the supernatant first treated with normal rabbit γ-globulin. The resulting unspecific precipitate was removed by centrifugation and the supernatant incubated with specific anti-PFK γ-globulin. The precipitate was sedimented by centrifugation, washed and subjected to SDS-PAA electrophoresis

explained by changes in the specific radioactivity of the intracellular ATP-pool. Whether the increased phosphorylation of glucose is responsible for the greater stability of PFK is an as yet unanswered question.

We know, on the other hand, that the metabolite pattern in homogenates from livers of fed animals or livers from starved animals which had been perfused with high concentrations of glucose is completely different from that found in liver extracts from starved rats. Various glycolytic metabolites, notably F-1, 6-BP, are much higher in the fed state and their concentrations increase considerably during the conventional processing of liver extracts. We found F-1, 6-BP concentrations of up 250 μM in the supernatant after ultracentrifugation of extracts from livers of fed animals, while extracts from livers of starved rats showed concentrations below 10 μM. F-6-P and F-1,6-BP can stabilize PFK against inactivation by dissociation. The newly detected metabolite fructose-2,6-bisphosphate (F-2,6-BP), the concentration of which increases also with the supply of glucose to liver metabolism (Van Schaftingen et al. 1980b), may also contribute to the higher stability of PFK under these circumstances.

The effects of ligand binding on PFK stability complicates the interpretation of phosphorylation experiments. This became evident during an experiment in which we incubated partially purified inactive PFK from rat liver with ATP-γ-[^{32}P] in the presence and absence of 3',5'-cAMP. As can be seen from Fig. 2, phosphorylation of PFK took place only in the presence of 3',5'-cAMP. Induction of catalytic PFK activity by ATP on the other hand occurred to the same extent in the presence and absence of 3',5'-cAMP.

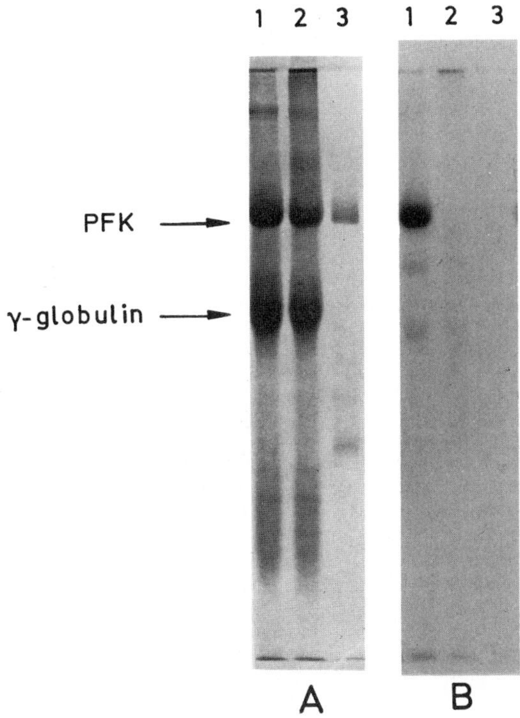

Fig. 2A, B. Effect of 3',5'-cAMP on [^{32}P]-incorporation into rat liver phosphofructokinase. Liver tissue from starved rats was homogenized in "stopping medium" and the resulting S_3 subjected to DEAE-cellulose chromatography as described by Brand and Söling (1975). The fraction containing the inactive PFK was incubated in the presence of 0.6 mM ATP-γ-[^{32}P] and the change in V_{max} activity and [^{32}P] incorporation into PFK measured as described in the legend to Fig. 1. **A** Coomassie Blue stain; **B** Autoradiograph; *Lane 1,* Incubation in the presence of 0.1 mM cAMP; *Lane 2,* Incubation in the absence of cAMP; *Lane 3,* Purified rat liver PFK. The V_{max} activity of PFK increased almost tenfold after the addition of ATP in the presence and absence of cAMP

This indicates that under these circumstances the reactivation of PFK resulted from ATP binding rather than from PFK phosphorylation.

The reason why we had not detected 3',5'-cAMP dependency in our earlier experiments (Brand and Söling 1975) may have been that we had sufficient amounts of free protein kinase catalytic subunits present to allow for phosphorylation of PFK in the absence of exogenous 3',5'-cAMP.

These results demonstrate that a change in the V_{max} activity or kinetic properties of an enzyme concomitantly with an increased incorporation of [^{32}P] from ATP-γ-[^{32}P] into the enzyme protein need not necessarily result from the phosphorylation of the enzyme but can as well be a consequence of a binding of ATP per se.

After the first description of phosphorylation of rat liver PFK, several reports on phosphorylation of PFK have appeared, but most studies are confined to skeletal muscle PFK (Hofer and Fürst 1976; Hussey et al. 1977; Riquelme et al. 1978; Uyeda et al. 1978; Hofer and Sørensen-Ziganke 1979) for the obvious reason that muscle tissue contains almost 100 times more PFK than liver tissue. As can be seen from Table 1, the amounts

Table 1. Amounts of covalently bound phosphate found in purified skeletal muscle PFK in different laboratories

State of phosphorylation of PFK (mol P_i/mol tetramer)		Reference
0.32 0.81	Skeletal muscle	Uyeda et al. (1978a)
0.60 0.16	Skeletal muscle	Hussey et al. (1977)
 1.12-1.36 0.80-0.96	Skeletal muscle (fresh muscle) (frozen muscle)	Riquelme et al. (1978)
 1.93 4.78 7.92	Skeletal muscle (Resting muscle, post mortem) (Resting muscle, in vivo) (Contracting muscle, in vivo)	 Hofer and Sørensen-Ziganke (1979) ∅
0.44-2.2	Skeletal muscle	Own results

of covalently bound phosphate shows a considerable variation even when PFK from the same tissue was used. Uyeda et al. (1978) and Hussey et al. (1977) were able to separate species of skeletal muscle PFK with different amounts of covalently bound phosphate, but were unable to detect differences in the kinetic properties or the specific activities between the separated species.

One reason for the different amounts of phosphate found in purified PFK may be due to the fact that the phosphorylatable serine in the PFK protomer is situated very close to the carbonyl terminus (position 6 from the carboxyl terminus; Kemp personal communication), and is cleaved off easily by proteases (Riquelme and Kemp 1980) with little detectable change of the apparent molecular weight and no measurable change in V_{max} activity or kinetic properties.

Recently several groups have described a change in the kinetic properties of crude rat liver PFK following in vivo (Kagimoto and Uyeda 1979; Nieto and Castano 1980) or in vitro (isolated liver cells, isolated perfused livers; Pilkis et al. 1979; Castano et al. 1979) treatment with glucagon. We could confirm these findings. However, as shown in Fig. 3, the removal of small-molecular weight compounds from the liver extract by gel filtration abolished the difference.

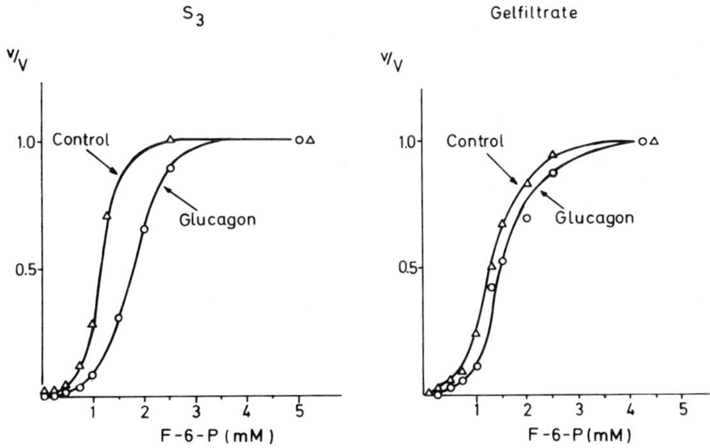

Fig. 3. Effect of glucagon on the kinetic behavior of phosphofructokinase from isolated hepatocytes obtained from fed rats. The hepatocytes which had been isolated by standard procedures, were incubated in the presence or absence of 10^{-6} M glucagon. Ten min after addition of glucagon the cells were separated from the incubation medium centrifugation and treated as described by Castaño et al. (1979). The dependence of the reaction velocity on the concentration of F-6-P was measured at 25°C and pH 7.5, as described by Castaño et al. (1979) (*Left side* of figure). Aliquots 0.75 ml of the centrifuged liver extract were chromatographed on a 0.9 × 18 cm column of Biogel P-6 which was equilibrated with (final concentrations): 50 mM Hepes, 100 mM KF, 15 mM EGTA, 0.1 mM F-6-P, 0.35 mM G-6-P pH 7.4. The PFK containing fractions were combined and used for the assay of PFK activity as described for the unfiltered S_3 (*right side* of figure)

This is in accordance with similar observations of Van Schaftingen et al. (1980b), but in contrast to a report of Castaño et al. (1979) who found the difference even after a more than 100-fold purification of PFK.

While Van Schaftingen et al. (1980b) and ourselves should like to relate the effect of glucagon mainly to a shift in the pattern of small-molecular weight ligands, Kagimoto and Uyeda (1979, 1980) feel that it results from an increased phosphorylation of PFK by a cAMP-dependent protein kinase. These authors found that glucagon increased phosphorylation of liver PFK in vivo and in isolated rat livers, and induced an increased sensitivity of PFK against inhibition by ATP and citrate (Kagimoto and Uyeda 1980). We observed also a stimulation by 10^{-7} M glucagon of PFK phosphorylation in isolated liver cells from fed rats in the presence of 16 mM L-lactate and 4 mM pyruvate (Fig. 4). But the same enhancement of PFK phosphorylation was produced by addition of glucose (Fig. 4). The effect of glucose was also observed in liver cells from starved rats (Fig. 1).

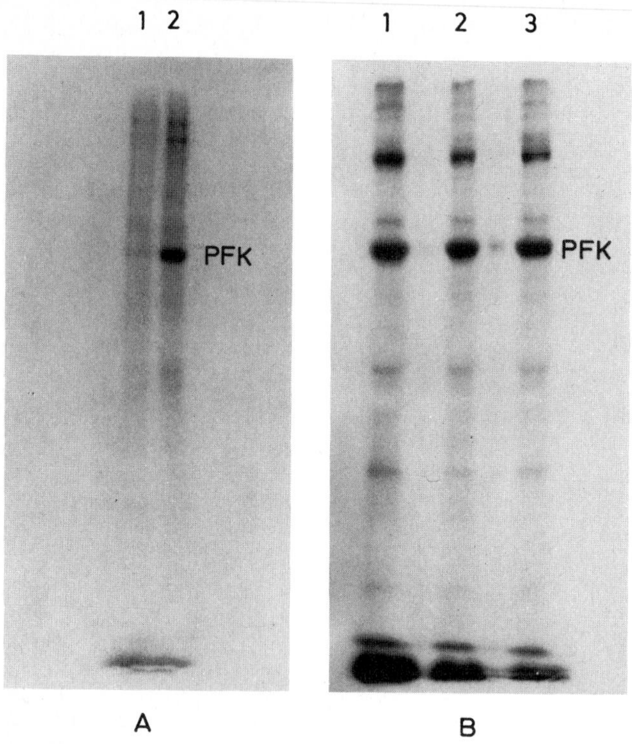

Fig. 4A, B. Effects of glucagon and glucose on phosphorylation of PFK in isolated hepatocytes from fed rats. **A** Effect of glucagon on PFK phosphorylation in isolated hepatocytes from a fed rat in the presence of 16 mM L-lactate and 4 mM pyruvate in the medium. *Lane 1,* No glucagon; *lane 2,* $2 \cdot 10^{-7}$ M glucagon. **B** Effects of glucose plus and minus glucagon on PFK phosphorylation in isolated hepatocytes from a fed rat. *Lane 1,* 5 mM glucose in the medium. *Lane 2,* 30 mM glucose + $2 \cdot 10^{-7}$ M glucagon in the medium. *Lane 3,* 30 mM glucose in the medium. The processing of the isolated liver cells for the electrophoretic separation of phosphorylated PFK was performed as given in the legend to Fig. 1

We cannot decide yet whether phosphorylation of PFK induced by glucose occurs at the same site as when phosphorylated in the presence of glucagon.

It is possible to achieve some phosphorylation of crude PFK in the presence of external cAMP or of purified PFK in the presence of cAMP-dependent protein kinase or its catalytic subunit. But in our hands the maximum incorporated under such conditions was 0.2 mol of phosphatate per mol of PFK tetramer.

It is not clear yet how dephosphorylation of phosphorylated PFK is catalyzed. In our hands, PFK which had been labeled by incubating isolated liver cells with [^{32}P] phosphate was not dephosphorylated to any measurable extent by purified rabbit skeletal muscle phosphorylase phosphatase, nor was PFK activity affected by this treatment. We have described previously (Söling et al. 1978b) an enzyme which catalyzed the rapid inactivation of rat liver as well as rat skeletal muscle PFK, and has been purified to homogeneity from rat liver. As shown in Fig. 5, this enzyme leads to an inactivation of rat liver PFK by promoting the dissociation of highly aggregated PFK into smaller-molecular

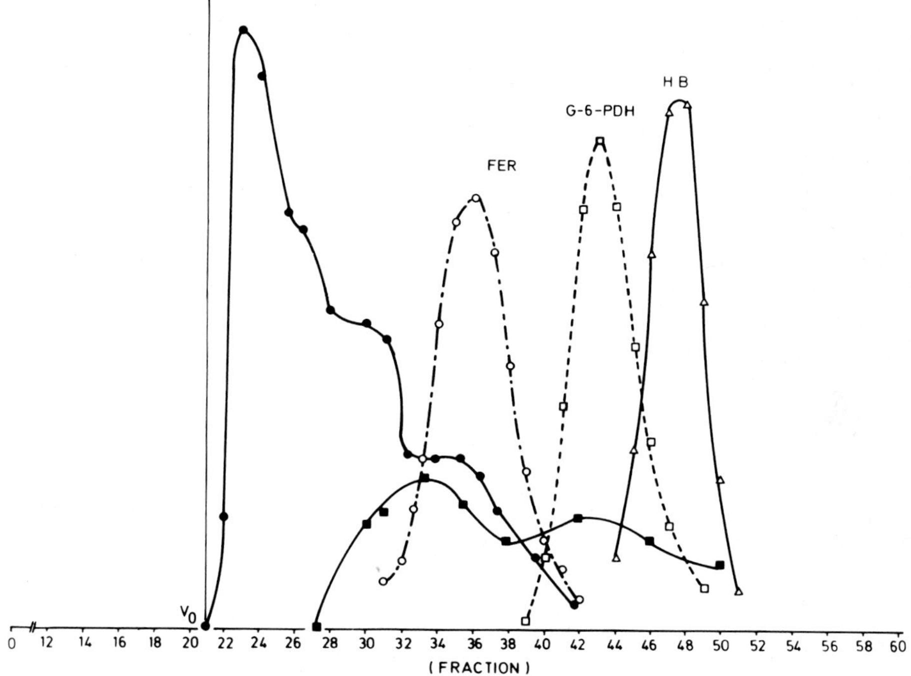

Fig. 5. Effect of incubation of purified rat liver phosphofructokinase with purified inactivating enzyme on the elution pattern of phosphofructokinase on a Sepharose Cl-6 B column. Purified PFK (9 units) were incubated with 3.5 units of inactivating enzyme in 1 ml buffer. The buffer consisted (final concentration) of: 16.5 mM Tris-Cl, 10 mM fructose-6-phosphate, 1.7 mM 2-mercaptoethanol and 35 vol% glycerol. The incubation was carried out at 25°C for 30 min. In the control experiment, the same amount of PFK was incubated under the same conditions except that the inactivating enzyme was omitted. ■—■ with inactivating enzyme; ●—● control incubation. Marker enzymes: *FER* = Ferritin; *G-6-PDH* = glucose-6-phosphate dehydrogenase, *HB* = Hemoglobine

weight species and inactive protomers. We do not know yet how this inactivation is brought about. We cannot exclude the removal of small-molecular weight ligands from the PFK molecule. One has to consider this possibility since it is known that treatment of PFK with FBPase leads by removal of PFK-bound F-1,6-BP to a rapid dissociation and inactivation of PFK (Söling et al. 1977). We could not, however, detect any FBPase-like activity in the purified inactivating enzyme. Whether this enzyme dephosphorylates PFK, which has been phosphorylated intracellularly or with protein kinase catalytic subunit, is currently under investigation. This role of small-molecular weight compounds in stabilizing PFK is underlined by the fact that several metabolites can inhibit or abolish inactivation of PFK by the inactivating enzyme at rather low concentrations (Table 2). It

Table 2. Low-molecular weight inhibitors of PFK inactivation by the PFK inactivating enzyme

Inhibitor	App. K_i
Fructose-1,6-bisphosphate	$9 \cdot 10^{-7}$ M
Fructose-6-phosphate	$6 \cdot 10^{-6}$ M
ATP	$2.5 \cdot 10^{-5}$ M
Inorganic phosphate	$1 \cdot 10^{-3}$ M

seems unlikely that PFK inactivation results from proteolytic effects since we could not detect a measurable decrease in the molecular weight of the PFK protomer after inactivation, and since protease inhibitors failed to inhibit PFK inactivation by the inactivating enzyme. The inactivating enzyme exhibits rather the properties of a protein phosphatase in that it is not only dependent on high concentrations of Mg^{2+}, but is inhibited by phosphate and fluoride as well as by polyarginine and polyamines.

In summary, we have to state that phosphorylation of liver and muscle PFK is now well established. But we still know very little about the regulation of phosphorylation and dephosphorylation of PFK and we know almost nothing about the functional consequence of PFK phosphorylation for its catalytic properties or its stability.

Acknowledgment. We should like to mention, that Hans-Joachim Lück and Dr. Christopher Proud have contributed to the work as far as results from our own laboratory are concerned.

Note Added in Proof. It has been shown in the meantime that the glucose induced phosphorylation of liver PFK is specific for PFK and does not affect pyruvate kinase. (Brand and Söling, manuscript submitted). PFK-inactivation by the inactivating factor is inhibited by fructose-2,6-bisphosphate with an apparent K_i below $5 \cdot 10^{-8}$ M (Söling, Kuduz and Brand, FEBS Letters, in the press).

References

Brand I, Söling HD (1975) FEBS Lett 57:163-168
Brand I, Müller MK, Unger C, Söling HD (1976) FEBS Lett 68:271-274
Castaño JG, Nieto A, Feliu DE (1974) J Biol Chem 254:5575-5579
Clark MG, Kneer NM, Bosch AL, Lardy HA (1974) J Biol Chem 249:5695-5703
Gevers W, Krebs HA (1966) Biochem J 98:720-735
Hofer HW, Fürst M (1976) FEBS Lett 62:118-122
Hofer HW, Sørensen-Ziganke B (1979) Biochem Biophys Res Commun 90:199-203
Hussey Ch R, Liddle PF, Ardron D, Kellett GL (1977) Eur J Biochem 80:497-506
Kagimoto T, Uyeda K (1979) J Biol Chem 254:5584-5587
Kagimoto T, Uyeda K (1980) Arch Biochem Biophys 203:792-799
Nieto A, Castano JG (1980) Biochem J 186:953-957
Pilkis S, Schlumpf J, Pilkis J, Claus TH (1979) Biochem Biophys Res Commun 88:960-967
Riquelme PT, Kemp RG (1980) J Biol Chem 255:4367-4371
Riquelme PT, Mosey MM, Marcus F, Kemp RG (1978) Biochem Biophys Res Commun 85:1480-1487
Rognstadt R, Katz J (1976) Arch Biochem Biophys 177:337-345
Söling HD, Brand I (1976) Interconversion of rat liver phosphofructokinase by phosphorylation and dephosphorylation, In: Shaltiel S (ed) Metabolic interconversion of enzymes 1975. Springer, Berlin Heidelberg New York, pp 203-212

Söling HD, Bernhard G, Kuhn A, Lück HJ (1977) Arch Biochem Biophys 182:563-572
Söling HD, Brand I, Whitehouse S, Imesch E, Unger C, Lück HJ, Kuhn A (1978a) Regulation of hepatic glycolysis and gluconeogenesis at the step of phosphofructokinase/FDPase, In: Esmann V (ed) Regulatory mechanisms of carbohydrate metabolism. FEBS 11th Meet Copenhagen 1977, vol 42. Pergamon Press, Oxford New York, pp 261-283
Söling HD, Brand I, Imesch E, Unger C, Lück HJ (1978b) Hormonal control of hepatic glycolysis and gluconeogenesis with special reference to the role of enzyme interconversion. In: Dumont J, Numez J (eds) Hormones and cell regulation, vol 2. North Holland Publishing Company, Amsterdam New York Oxford, pp 209-225
Uyeda K, Miyatake A, Luby LJ, Richards EG (1978a) J Biol Chem 253:8319-8327
Uyeda K, Miyatake A, Luby LJ, Richards EG (1978b) J Biol Chem 253:8319-8327
Van Schaftingen E, Hue L, Hers HG (1980a) Biochem J 192:263-271
Van Schaftingen E, Hue L, Hers HG (1980b) Biochem J 192:897-901

Discussion[2]

Regarding the nucleotide specificity Dr. Söling said that GTP but not App (NH)p can replace ATP as stimulating factor of PFK in the extracts after starvation and refeeding. Questioned about the nature of PFK-inactivase Dr. Söling pointed out that the inactivase apparently has no ATPase or FPPase activity. The addition of ATP alone to muscle or liver PFK treated with the inactivase does not reactivate the inactivated PFK unless a heat-labile compound from the 100,000 *g* supernatant is added. Furthermore, a heat-stable inhibitor of this inactivase has been highly purified in his laboratory with a molecular weight of 21,000.

2 Rapporteur: Elmar A. Siess

The Control of Liver Phosphofructokinase by Fructose 2,6-Bisphosphate

H.G. Hers, E. Van Schaftingen, and L. Hue[1]

1 Introduction

One major role of the liver is to form glucose from lactate and other precursors through gluconeogenesis, a process which operates both in the fed and the fasted animal. The reverse conversion is catalyzed by glycolysis and the simultaneous operation of these two metabolic pathways would be responsible for a futile recycling of metabolites.

There is general agreement that glycolysis is regulated at the level of phosphofructokinase which antagonizes the action of hexose diphosphatase. It is only recently that an appropriate methodology has been developed by Rognstadt and Katz (1976) to measure metabolite recycling between fructose 6-phosphate and triose phosphates in the liver and we have applied this methodology to both anesthetized rats (Van Schaftingen et al. 1980a) and isolated hepatocytes (Van Schaftingen et al. 1980b). This work led to the discovery of a new physiological stimulator of phosphofructokinase which was identified as fructose 2,6-bisphosphate. The concentration of this effector in the liver is regulated by glucose and by glucagon. In this presentation we tell the story of its discovery.

2 The Fructose 6-Phosphate/Fructose 1,6-Bisphosphate Cycle and Its Control by Glucose and Glucagon

We have measured in a semiquantitative way the recycling of metabolites between fructose 6-phosphate and fructose 1,6-bisphosphate by the formation of [6-^{14}C]glucose from [1-^{14}C]galactose as recommended by Rognstadt and Katz (1976). Figure 1 shows that

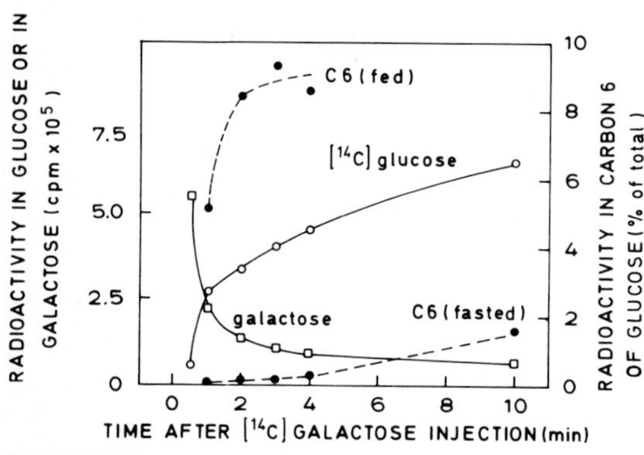

Fig. 1. Time course of the conversion of [1-^{14}C]galactose to [1,6-^{14}C]glucose by anaesthetized rats. 2.5 × 10^6 cpm of [1-^{14}C]-galactose was injected into the portal vein and the blood was collected from the aorta. The radioactivity present at 10 min in C-6 of glucose in the fasted rat can be attributed to the activity of the Cori cycle

1 Laboratoire de Chimie Physiologique, Université de Louvain, UCL 75 39, Faculté de Médecine 75 Avenue Hippocrate, 1200 Brussels, Belgium

there was no recycling in the liver of fasted rats in vivo. Since gluconeogenesis is maximally active during fasting, these results indicate that phosphofructokinase was inactive under this condition. In contrast, a rather large although variable recycling was observed in the livers of fed rats, and also of fasted rats in which glucose had been administered 10 min previously. In these animals, recycling was completely inhibited by glucagon. We have estimated that as a mean, as much as 15% of the glucose formed from galactose had been recycled through the triose phosphates in the fed animal (Van Schaftingen et al. 1980a). Recycling was also observed with hepatocytes isolated from the livers of fed rats although not with hepatocytes obtained from fasted animals, incubated in the presence of physiological concentrations of glucose. In the latter preparation, however, a large recycling was observed in the presence of higher concentrations of glucose (Fig. 2). As already observed by Rognstadt and Katz (1976), this recycling was completely cancelled by glucagon (Van Schaftingen et al. 1980b). All these data indicate that the activity of phosphofructokinase in the liver is greatly increased by glucose and decreased by glucagon.

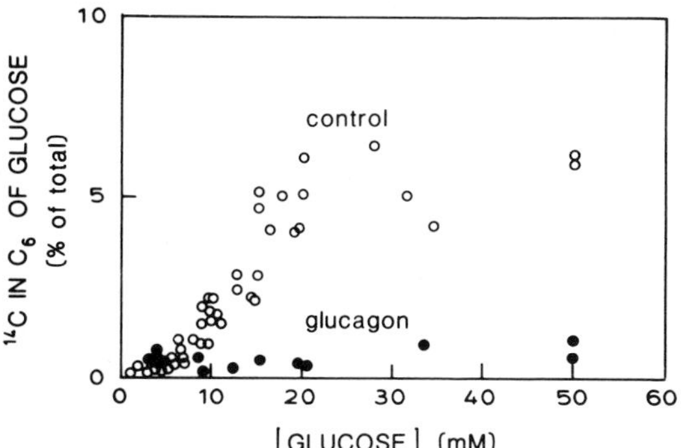

Fig. 2. The effect of glucose concentration on the randomization of carbon atoms by hepatocytes isolated from the livers of fasted rats. The cells were incubated at 37°C during 20 min at the concentration of glucose indicated, in the presence or in the absence of 10^{-6} M-glucagon. [From Van Schaftingen et al. (1980b)]

The activity of phosphofructokinase from liver or from other tissues is known to be regulated by several effectors. Indeed, the saturation curve for the substrate fructose 6-phosphate is sigmoidal, whereas the other substrate, ATP, when present in excess, inhibits the reaction. Various positive effectors like fructose 1,6-bisphosphate, the product of the reaction, and also AMP, increase the affinity of the enzyme for fructose 6-phosphate and counteract the inhibition by ATP (for a review, see Hofmann 1976). Another potential mechanism of regulation could be by phosphorylation and dephosphorylation of the enzyme, as has been suggested by the fact that a decrease in the affinity of phosphofructokinase for fructose 6-phosphate and an increased inhibition by ATP were observed upon incubation of hepatocytes in the presence of glucagon (Castano et al. 1979; Kagimoto and Uyeda 1979; Pilkis et al. 1979). These changes were indeed reported to be stable upon gel filtration as well as upon a 400-fold purification of the enzyme (Castano et al. 1979; Uyeda and Kagimoto 1979). They are conveniently expressed by the ratio of the activities measured at low (0.25-0.5 mM) and saturating (5 mM) concentration of fructose 6-phosphate.

In order to clarify the mechanism by which glucose and glucagon modify the activity of phosphofructokinase in the liver, we have measured the concentration of the known effectors as well as the activity ratio of phosphofructokinase in the liver in vivo under various experimental conditions (Table 1). The results clearly indicate that a change in the concentration of nucleotides could not be responsible for the observed modification of activity (results not shown). The concentration of fructose 6-phosphate was sometimes elevated when recycling was minimal. The concentration of fructose 1,6-bisphosphate was elevated in all conditions in which phosphofructokinase was very active. It was however difficult to decide if this change of concentration was the cause or the result of the change in activity. The activity ratio of phosphofructokinase also varied in parallel with the randomization of carbon and could therefore be an explanation for it. Similar results were also observed with isolated hepatocytes (Van Schaftingen et al. 1980b).

Table 1. Effect of the nutritional condition and of glucagon on the randomization of carbon by the livers of anesthetized rats, on the concentration of several metabolites in these livers, and on the activity ratio of phosphofructokinase. General procedures as in Fig. 1. The analysis was performed 3 min after the injection of [1-^{14}C]galactose. [From Van Schaftingen et al. (1980a)]

	^{14}C in C_6 of Glucose (%)	[F6P] nmol/g	[FDP] nmol/g	PFK $\frac{v_{0.25}}{v_5}$
Fed	4.9	46	32	0.64
Fasted	0.1	17	5	0.31
Fasted + Glucose	2.6	43	47	0.51
Fasted + Glucose + Glucagon	0.3	90	8	0.13

3 The Control of Phosphofructokinase by a Low-Molecular-Weight Effector, Different from Fructose 1,6-Bisphosphate

We show in Fig. 3 that, in confirmation of the work by others (Castano et al. 1979; Kagimoto and Uyeda 1979; Pilkis et al. 1979), phosphofructokinase present in an extract of hepatocytes incubated in the presence of glucagon had less affinity for its substrate fructose 6-phosphate than the enzyme present in a control extract. However, and in contrast with previous reports (Castano et al. 1979; Kagimoto and Uyeda 1979), this difference was cancelled after gel filtration of the two extracts. In other words, gel filtration could mimic the effect of glucagon, suggesting that the hormonal effect was to remove a positive effector. In confirmation of this hypothesis it was found that the activity of the filtered extract could be restored to the level of the control by the addition of an ultrafiltrate prepared from the homogenate of a control liver. Similar results were obtained by addition of a liver extract heated at 80°C and centrifuged. The addition of fructose 1,6-bisphosphate was without effect as was expected, since under our assay conditions this reaction product is removed by aldolase and converted finally to glycerol-

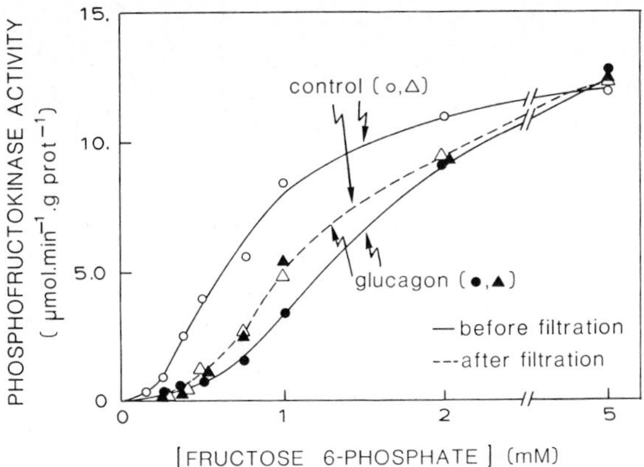

Fig. 3. Saturation curve of phosphofructokinase for fructose 6-phosphate. The enzymic activity was assayed in an extract of hepatocytes obtained from the liver of a fed rat and incubated in the presence or in the absence of 10^{-6} M-glucagon. The enzymic activity was also measured after filtration of the extracts on a column of Sephadex G-25 (*dotted line*). [From Van Schaftingen et al. (1980b)]

phosphate. From these experiments we concluded that the liver contains a positive effector of phosphofructokinase which is different from fructose 1,6-bisphosphate and the concentration of which is greatly decreased by glucagon (Van Schaftingen et al. 1980b).

The increase in activity ratio of phosphofructokinase present in a liver filtrate could be used as a bioassay for the measurement of this newly discovered stimulator. We show in Fig. 4 that the concentration of this stimulator in the isolated hepatocytes was greatly

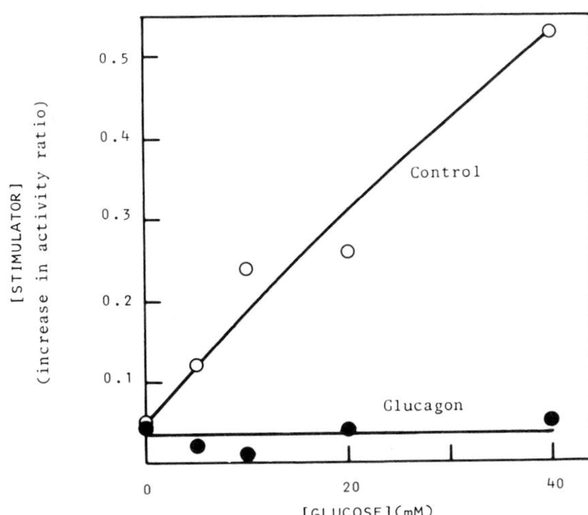

Fig. 4. Effect of the concentration of glucose in the incubation medium and of glucagon on the concentration of the stimulator of phosphofructokinase in hepatocytes. Hepatocytes obtained from the liver of a fasted rat were incubated for 10 min in the presence of glucose at the concentrations indicated with or without 10^{-6} M-glucagon. [From Van Schaftingen et al. 1980b)]

increased by increasing concentrations of glucose, and greatly decreased by glucagon. Figure 5 shows the time sequence of its formation under the effect of glucose and of its disappearance under the action of glucagon.

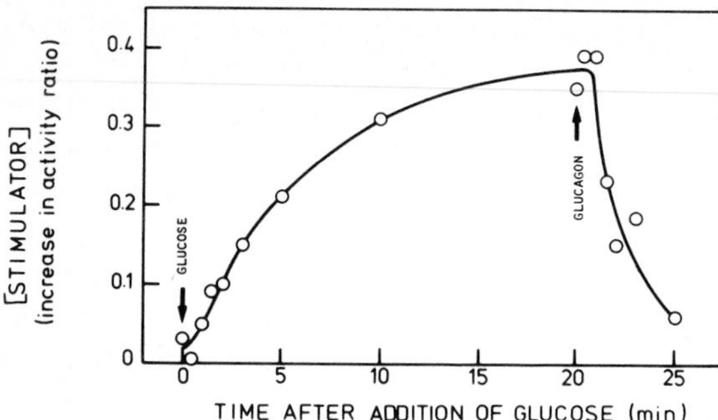

Fig. 5. Time course of the effect of glucose and of glucagon on the concentration of the stimulator of phosphofructokinase in isolated hepytocytes. Hepatocytes isolated from the liver of a fasted rat were preincubated for 30 min in the absence of glucose. Glucose was then added at a final concentration of 20 mM; glucagon was added 20 min later at a final concentration of 10^{-6} M. [From Van Schaftingen et al. (1980b)]

4 Identification of the Stimulator of Phosphofructokinase as Fructose 2,6-Bisphosphate (Van Schaftingen et al. 1980c)

The nature of the stimulator was progressively recognized thanks to the following analyses performed on a heated extract obtained from a rat liver perfused in the presence of 50 mM-glucose. The stimulator was destroyed by alkaline phosphatase, although not by hexosediphosphatase, and was not adsorbed by charcoal. Its molecular weight, as estimated by gel filtration, appeared to be similar to that of fructose 1,6-bisphosphate. It was not found in a trichloroacetic acid extract. It was precipitated by barium and, upon anion exchange chromatography, was eluted in close association with fructose 1,6-bisphosphate. All these properties indicate that the stimulator is a highly acid-labile, non-nucleotidic, diphosphoric ester.

The stimulator was purified by barium precipitation and Dowex chromatography. The acid lability of the purified compound was then investigated in detail. Figure 6 shows

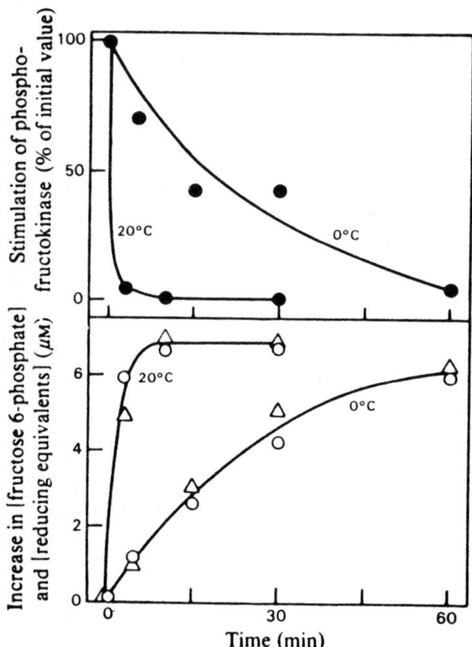

Fig. 6. Acid hydrolysis of the stimulator. The stimulator was incubated in the presence of 0.01 M-HCl either at 0°C or at 20°C. One reducing equivalent corresponds to the reducing power of 1 mol of fructose 6-phosphate. At time zero, fructose 6-phosphate was not detectable and the reducing equivalents were 3.5 μM. (○) fructose 6-phosphate; (△) increase in (reducing equivalents). [From Van Schaftingen et al. (1980c)]

that, when incubated in the presence of 0.01 M-HCl, the stimulator was completely destroyed within a few minutes at 20°C and at a slower rate at 0°C (half-life equal to 15 min). We then tried to identify the products formed in the course of this acid hydrolysis and we could show that fructose 6-phosphate was formed in parallel with the disappearance of the stimulatory property and in an amount equimolecular to the acid-labile phosphate present in the preparation. The formation of fructose 6-phosphate was measured not only enzymatically (reduction of NADP in the presence of phosphoglucoisomerase and glucose 6-phosphate dehydrogenase) but also by the appearance of its reducing power. We therefore concluded that fructose 6-phosphate is linked to phosphate by its anomeric carbon and that, most likely, the structure of the stimulator is fructose 2,6-bisphosphate (Fig. 7). We could not be sure, however, that this structure was complete and nothing was known about the anomeric form.

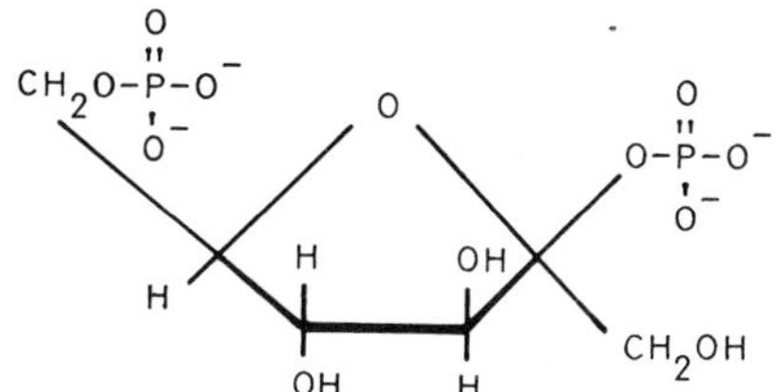

Fig. 7. Fructose 2,6-bisphosphate. The beta form of fructose 2,6-bisphosphate is shown although there is no definite proof that this anomer is the natural one

5 Synthesis of a Stimulator of Phosphofructokinase from Phosphoric Acid and Fructose 6-Phosphoric Acid (Van Schaftingen and Hers 1980)

Considering the remarkable acid lability of the stimulator, even in the cold, we expected that the reverse reaction (i.e., the synthesis of a small amount of fructose 2,6-bisphosphoric acid from phosphoric acid and fructose 6-phosphoric acid) could also occur in a rather specific way under similar conditions of pH and temperature. Table 2 shows that it was indeed the case. A small amount of stimulator, which was also measured as acid-revealed fructose 6-phosphate, could be separated from phosphate and fructose 6-phosphate by anion exchange chromatography. The amount formed was already maximal upon 1 min of incubation at 20°C. Fructose 1,6-bisphosphate was also formed, although at a much slower rate but in an amount which was finally at least 30 times that of fructose 2,6-bisphosphate. These data allow the conclusion that no constituent other than fructose 6-phosphate and phosphate enter into the structure of the stimulator.

6 Dose-response Curve of Phosphofructokinase to Fructose 2,6-Bisphosphate

The dose response curve of liver phosphofructokinase to the synthetic and to the natural stimulator is shown in Fig. 8. It appears that the two preparations were almost equally active, displaying a half-maximal effect at a concentration close to 0.1 μM. The fact that the synthetic stimulator was as effective as the natural one suggests that only the natural anomer was formed in the course of the chemical synthesis.

Table 2. Synthesis of the stimulator of phosphofructokinase and of fructose 1,6-bisphosphate. One vol of 85% phosphoric acid was mixed with 0.2 vol of 1.8 M-fructose 6-phosphate. The reaction was run for the time and at the temperature indicated and stopped by neutralization. The mixture was chromatographed on Dowex AG1 and the fraction containing the diphosphoric esters were pooled and analyzed. Acid-revealed fructose 6-phosphate corresponds to the fructose 6-phosphate measured after 10 min of incubation at pH 2.0 at 20°C. [From Van Schaftingen and Hers (1980)]

Temperature	Time	Relative amount of stimulator (% of maximum)	[Acid-revealed Fru 6-P] (nmol/ml of reaction mixture)	[Fru 1,6-bis-phosphate] (nmol/ml of reaction mixture)
	0 min	0	0	10
0°C	5 min	62	59	10
25°C	1 min	100	102	30
	10 min	79	109	150
	60 min	85	114	680
20°C	6 h.			1550[a]
	24 h.			3100[a]

[a] Determinations made on samples of the neutralized mixture, without chromatographic purification

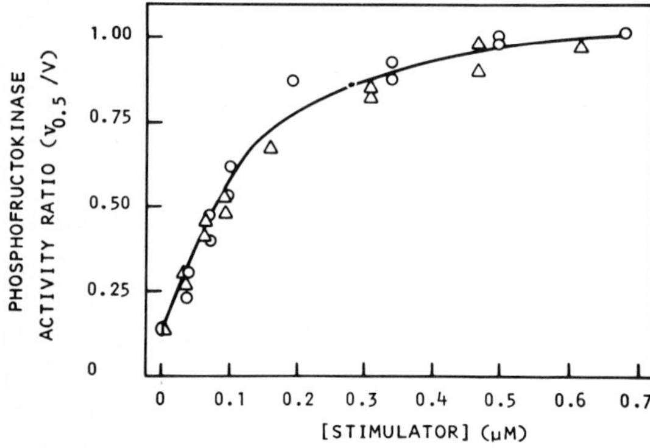

Fig. 8. Dose-response curve of liver phosphofructokinase to the natural (○) and to the synthetic (△) stimulator. The concentration of stimulator was determined by the amount of fructose 6-phosphate liberated by acid hydrolysis. [From Van Schaftingen and Hers (1980)]

7 Conclusion

From this work two important conclusions can be drawn: (1) The large effects of glucose and glucagon on the recycling of metabolites between fructose 6-phosphate and triose phosphates can be explained by the variations in the concentration of fructose 2,6-bisphosphate, which are induced by the same effectors; (2) The control of gluconeogenesis by glucose and by glucagon appears to occur in great part by a change in the rate of glycolysis.

Acknowledgements. This work was supported by the Belgian Fonds de la Recherche Scientifique Médicale and by the U.S. Public Health Service (Grant AM 9235). E.V.S. is Aspirant and L.H. is Chercheur Qualifié of the Fonds National de la Recherche Scientifique.

Note Added in Proof: It has now been established that fru-2,6-P_2 is a strong inhibitor of fructose 1,6-bisphosphatase.

References

Castano JG, Nieto A, Feliu JE (1979) J Biol Chem 254:5576-5579
Hofmann E (1976) Rev Physiol Biochem Pharmacol 75:1-68
Kagimoto T, Uyeda K (1979) J Biol Chem 254:5584-5587
Pilkis S, Schlumpf J, Pilkis J, Claus TH (1979) Biochem Biophys Res Commun 88:960-967
Rognstadt R, Katz J (1976) Arch Biochem Biophys 177:337-345
Van Schaftingen E, Hers HG (1980) Biochem Biophys Res Commun 96:1524-1531
Van Schaftingen E, Hue L, Hers HG (1980a) Biochem J 192:263-271
Van Schaftingen E, Hue L, Hers HG (1980b) Biochem J 192:887-895
Van Schaftingen E, Hue L, Hers HG (1980c) Biochem J 192:897-901

Discussion[2]

The questions were related to the structure and function of Fru-2,6-P_2. Dr. Hers stressed that the information given in the discussion is based on preliminary experiments. Accordingly, Fru-2,6-P_2 neither inhibits FBPase nor activates pyruvate kinase, but rather is a potent positive effector which like AMP and Fru-1,6-P_2 increases the affinity of the enzyme for F-6 P and counteracts the inhibition by ATP, however, at a much lower concentration than the other known positive effectors.

---

2 Rapporteur: Elmar A. Siess

The Relationship Between Latent Phosphorylase Phosphatases and the Precursors of the Heat-stable Inhibitor in a Liver High-speed Supernatant

M.F. Jett and H.G. Hers[1]

The high-speed supernatant fraction, which was obtained from a fed rat liver homogenate prepared in 5 mM-Imidazole, 200 mM-sucrose, 45 mM-benzamidine and 0.5 mM-dithiotreitol, contained only a small amount of spontaneously active phosphorylase phosphatase(s) (about 10% of the total activity of the fresh homogenate). This activity can be increased 25-fold by treatment with ammonium sulfate/ethanol as described by Brandt et al. (1974) and 10-fold by incubation with trypsin (Laloux et al. 1978). When the same high-speed supernatant fraction was incubated at 90°C for 5 min, homogenized, centrifuged, filtered on Sephadex G-25 (fine) and assayed in the presence of enzymes activated by either trypsin or ammonium sulfate/ethanol, an inhibitory component was detected. Under identical assay conditions, the enzyme(s) activated by trypsin were about 25-fold more sensitive to this heat-stable inhibitor than the enzyme(s) activated by ethanol. Furthermore, although the precursor of the heat-stable inhibitor was destroyed during the activation by trypsin, it could be recovered (about 80%) along with the activated enzyme in the final step of the ammonium sulfate/ethanol activation process.

Fractionation of the high-speed supernatant by cellulose phosphate chromatography yielded two main fractions (Fig. 1). The flow-through fraction (Fraction 1) contained approximately 70% of the latent enzyme which is activated by ammonium sulfate/

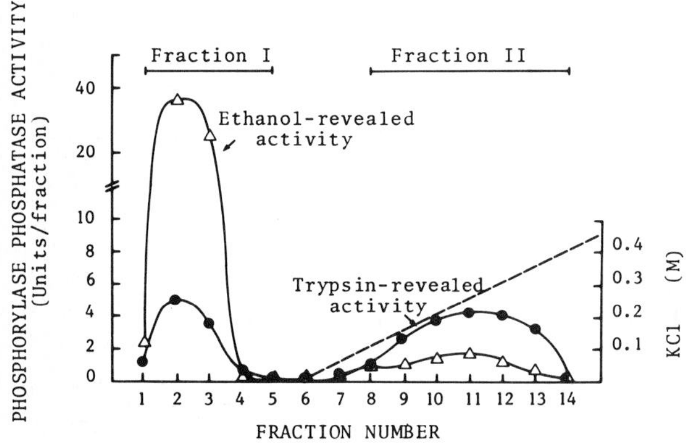

Fig. 1. Cellulose phosphate chromatography of a liver high-speed supernatant. A 10 ml aliquot of a filtered (Sephadex G-25) high-speed fraction was subjected to chromatography on cellulose phosphate at pH 6.8. Fractions (5.7 ml) were collected, dialyzed, treated with either trypsin (●—●) or ammonium sulfate/ethanol (△—△), and assayed for phosphorylase-phosphatase activity

1 Laboratoire de Chimie Physiologique, Université de Louvain, UCL 75 39, Faculté de Médecine, 75 Avenue Hippocrate, 1200 Brussels, Belgium

ethanol. Although enzyme(s) in this fraction could be partially activated by trypsin, ammonium sulfate/ethanol treatment was far more effective than trypsin in revealing the latent phosphorylase phosphatase activity. Fraction I also contained a protein which after heating was inhibitory to the trypsin-activated enzyme(s), but was without major effect on the ammonium sulfate/ethanol-activated enzyme in this fraction. When, however, the latter enzyme was subjected to limited trypsin digestion, between 70% and 80% of the original activity was lost, and the enzyme was sensitized to the heat-stable inhibitor to about the same degree as the trypsin-activated enzyme from the same fraction.

Another fraction (Fraction II) was eluted from the cellulose phosphate column at 0.3 M-KCl. This fraction contained enzyme(s) which were preferentially activated by trypsin and to a lesser degree by ammonium sulfate/ethanol. It also contained a precursor of the heat-stable inhibitor. When the latent enzyme(s) in this fraction were activated by either method, the resulting enzymes were readily inhibited by the heat-stable inhibitor obtained from the same fraction. When fraction II was further purified by DEAE-cellulose chromatography (Fig. 2), the latent phosphorylase phosphatase could be partially separated from the precursor of the inhibitor, suggesting that activation by trypsin is not solely the result of the proteolytic digestion of the precursor of the inhibitor.

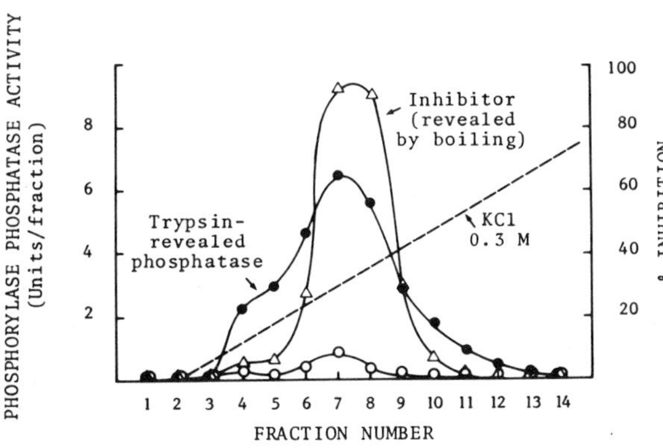

Fig. 2. DEAE-cellulose chromatography of Fraction II: The enzyme obtained from Fraction II was subjected to chromatography on DEAE-cellulose at pH 7.4. The fractions (3.4 ml) were dialyzed and assayed directly (○—○), or following treatment with trypsin (●—●). The heat-stable inhibitor (△—△) was treated at 90°C and was assayed with the trypsin-activated enzyme

When the peak fraction of the DEAE-cellulose chromatography was used both as the source of phosphorylase phosphatase (after trypsin treatment) and the inhibitor (after heating), a marked inhibition was observed (see Table 1). No inhibition, however, was detected when the heating step was omitted.

From these observations, we conclude that: (1) there are at least two latent phosphorylase phosphatases in the high-speed supernatant fraction; (2) activation of the latent phosphatases by either trypsin or ammonium sulfate/ethanol treatment is not necessarily related to the dissociation of the heat-stable inhibitor from the catalytic moiety; (3) the sensitivity to the heat-stable inhibitor can be induced by limited proteolysis; (4) heating is an essential step in the conversion of the precursor of the heat-stable inhibitor to its inhibitory form.

Table 1. Heat dependency of the inhibitor

Enzyme	Inhibitor	Phosphorylase phosphatase activity	
+	–	1.49	0.164
–	+	0.15	
+	+	1.58	
+	+ (Boiled)	0.54	

The same preparation (Fraction II on phospho-cellulose, peak fraction on DEAE-cellulose) was used as a source of the enzyme (after trypsin treatment) and the inhibitor

Acknowledgements. This work was supported by the Belgian Fonds de la Recherche Scientifique Médicale and by the U.S. Public Health Service (Grant AM 9235). M.F. Jett is a recipient of a National Research Service Award AM. 05843.

References

Brandt H, Killilea D, Lee YC (1974) Biochem Biophys Res Commun 61:598-604
Laloux M, Stalmans W, Hers HG (1978) Eur J Biochem 92:15-24

Multimodulation of Phosphofructokinases in Metabolic Regulation[1]

A. Sols, J.G. Castaño, J.J. Aragón, C. Domenech, P.A. Lazo, and A. Nieto[2]

1 Introduction

Enzyme physiology has proceeded within less than three decades from mere catalysis (no matter how efficient) to specific regulatory mechanisms, and to multimodulation of enzyme activity. Multimodulation makes possible the integration of a variety of signals by certain enzymes (Sols 1979, 1981). The importance of multimodulation has been obscured by uncritical claims piling up and reaching in some cases a bewildering multiplicity. For proper understanding of metabolic regulation it is necessary to sort out facts from artefacts, and biologically relevant mechanisms from in vitro curiosities. An outstanding case in this respect is phosphofructokinase, certainly a key enzyme in metabolic regulation (Sols 1976) and one long subjected to inflationary pressure with respect to mechanisms for the regulation of its activity, particularly in higher animals. The recent finding of metabolic interconversion of this enzyme in certain animal tissues makes a detailed discussion of the multimodulation of phosphofructokinase of particular interest in the context of this Symposium.

2 Regulation of Enzyme Activity: Mechanisms and Principles

To discuss multimodulation of enzyme activity it is important to determine a framework of definitions and general rules (Sols 1981):

1. Allosteric regulation of enzyme activity involves regulatory sites, catalytic interacting sites, or both.

2. Any compound can be an allosteric effector for an enzyme, no matter how unrelated to substrate and product, also even if it is a substrate or product. This definition includes different sites for the same physiological effector.

3. Allosteric effectors can be either positive or negative. They can act on the molecular activity, the affinity for a substrate or cofactor, the degree of cooperatively, the ratio of affinities for substrate and product, the relative specificity, or any combination of these.

4. Cooperative effects can be either positive or negative, and the more common positive ones can have either an oligomeric or a mnemonical basis.

---

1 This paper is dedicated to Dr. Helmut Holzer on the occasion of his 60th birthday

2 Instituto de Enzimología y Patología Molecular del CSIC, Facultad de Medicina, Universidad Autónoma, Arzobispo Morcillo 4 – Madrid 34, Spain

5. If there is more than one kind of allosteric effect on an enzyme:
   a. The enzyme must integrate virtually simultaneous signals.
   b. Physiological activation frequently involves counteraction of an inhibition (oligomeric decrease in affinity for a substrate or allosteric effector).
   c. Multiple effects may range from essentially independent (contributing algebraically) to strongly concerted.

6. Covalent modification of an enzyme can affect any of the above parameters, either catalytic or allosteric, positively or negatively.

7. Metabolic interconversion of enzymes by reversible covalent modification is usually regulated by some allosteric effect on either the metabolic enzyme or on a converter enzyme.

The farsighted concept of chemical transduction proposed by Monod, borrowing from Kosland's induced fit concept, crystallized with the term allosteric, a catchy word that soon became a mixed blessing (Sols 1973) because of confusion between two frequently linked but essentially independent concepts: specifically regulatory sites and multiplicity of interacting equal sites in oligomeric proteins. The common mixture of these two concepts can be exemplified by a recent description by Perutz (1978): „Monod et al. proposed the term allostery for *enzymes that possess two, or at least two, stereospecifically different, nonoverlaping receptor sites* (1963). One of these, the active site, binds the substrate, while the other, the allosteric site, binds the effector. *Such enzymes contain more than one site of each kind, and these act cooperatively*" (italics added).

Curiously, the first identification of a specific regulatory site in an enzyme was for a primary product: a site in animal hexokinase for inhibition by glucose-6-P, reported in 1954 on the basis of clearcut differential specificity (Crane and Sols 1954), and now confirmed by the demonstration that the monomeric enzyme can bind two molecules of glucose-6-P (Lazo et al. 1980). Allosteric effects involving specific regulatory sites occur in some phosphofructokinases for one of the substrates and one of the products (Sols 1973).

The generalization that many cases of allosteric physiological activation of an enzyme involve the counteraction of a prior inhibition, easier to accomplish than primary activation, stems not only from certain antagonistic, frequently called "deinhibitory" allosteric effects, but also from the basic fact that the so-called positive cooperativity for substrate binding in oligomeric proteins involves an initial hindrance by quaternary constraint on the binding of the substrate in the oligomer, as compared with binding to the corresponding monomer. In these proteins the affinity for the substrate can rise steeply, but in fact it is never greater than that of the monomer at any given concentration of substrate.

It is important to emphasize that "allosteric enzyme", in the initial sense introduced by Monod et al. (1963), and the restricted one used here are not equivalent, nor necessarily linked, to cooperativity in oligomeric proteins.

Multimodulation of enzyme activity arises from the accumulation in a given enzyme of several regulatory mechanisms, whether of different types (cooperative, allosteric, covalent modification), of the same type (multiple allosteric effects), or any combination of these.

Multiple phosphorylation has recently been observed in several enzymes, including phosphofructokinase (Pilkis et al. 1980). In some cases it can be an additional mechanism for finer metabolic control; but it also can be mere noise due to incomplete specificity of a converter enzyme, since it is almost certain that the number of protein phosphorylations observable after labeling in vivo with high specific activity [^{32}P] is greater than that of modulatory mechanisms potentially involved in metabolic regulation.

In summary, with the simplified nomenclature used in this article there are allosteric enzymes, cooperative enzymes, and convertible enzymes, as well as nonmodulated enzymes. Many regulatory enzymes are allosteric and cooperative; many convertible enzymes are also allosteric; and some regulatory enzymes are allosteric, cooperative, and convertible.

3 Multimodulation of Phosphofructokinase

Phosphofructokinase has long been known as a difficult enzyme. When I first encountered it, in Cori's laboratory in 1951, I heard comments about its being an "unreliable enzyme". Since then, much time has been wasted by many people because of the propensity of this enzyme to give disconcerting and frequently irreproducible results. This is related to the fact that this enzyme is sensitive to many factors, and in a perverse way (for the experimenter), since the animal enzyme is inhibited by a substrate (ATP) and a by-product (H^+), activated by a product (fructose-1,6-P_2) and indirectly by the other (AMP made out of ADP ba the ubiquitous adenylate kinase); moreover, as a ultimate trick for the unsuspecting biochemist, the kinetic behavior of the enzyme is markedly affected by its own concentration! Phosphofructokinase is the prototype of multimodulated enzyme, ever since Passonneau and Lowry published a short paper in 1962 in which under the title "Phosphofructokinase and the Pasteur effect" they reported that the activity of a partially purified preparation of the muscle enzyme could be affected by physiological concentrations of half a dozen metabolites, including inhibition by a substrate and activation by the two products. By 1966 Lowry and Passonneau gave kinetic evidence for multiple binding sites in the brain enzyme concluding that there ought to be at least seven and "a possible dozen!"

Time and work in many laboratories have increased the number of presumptive effectors. Up to 23 different effectors, some of them acting in more than one way, have been listed in a recent review (Table 1). Out of this jungle a critical selection is made

Table 1. Effectors of phosphofructokinase as listed by Tejwani (1978)

Inhibitors	Activators	Deinhibitors of ATP citrate, or Mg^{2+}
ATP	NH_4^+	cAMP
Citrate		AMP
Mg^{2+}	K^+	ADP
Ca^{2+}	Pi	Fru-6-P
P-creatine		Pi
Glycerate-3-P	AMP	
P-enolpyruvate	cAMP	Fru-1,6-P_2
Glycerate-2-P	ADP	Glc-1,6-P_2
Glycerate-2,3-P_2	Fru-1,6-P_2	Man-1,6-P_2
Oleate or palmitate	Peptide stabilizing factor	
Fructosebisphosphate		
3',5'-cyclic GMP		

below of the known modulatory mechanisms that seem to be really involved in the physiological regulation of phosphofructokinase from eukaryotic organisms. No attempt has been made to criticise every claim that is not accepted as likely. The burden of proof should lie on the shoulders of the claimants; in a field as overcrowded as that of the regulation of phosphofructokinase this has become one of the more important rules in practice.

3.1 Animal Phosphofructokinase

Phosphofructokinase has been most extensively studied in higher animal tissues. Tissue differences in kinetic behavior indicate the occurrence of isozymes that seem to be based on three types of genetically determined subunits called M (muscle type), L (liver type), and F (fibroblast type) (Kahn et al. 1980). Most regulatory properties seem common to all of the mammalian enzymes, regardless of organisms and tissue, although probably with quantitative differences (Meldolesi and Laccetti 1979).

Cooperativity for fructose-6-P is a widespread fact in native phosphofructokinase of most origins, including prokaryotes. It is appropriate for an oligomeric enzyme that catalyzes the first irreversible step in the glycolytic pathway below the glucose-6-P crossroad. Changes in the oligomeric state of animal (but not yeast) phosphofructokinase markedly affect its kinetic behavior and sensitivity to allosteric effectors (Hofer 1971; Hulme and Tipton 1971); the dissociation constant is in the micromolar range and is markedly affected by the pH (Hofer and Krystek 1975).

A large excess of ATP over the MgATP available inhibits the generality of kinases by isosteric competition with the true substrate. In addition to this possibility, phosphofructokinases of many origins are inhibited by ATP in the physiological range, as first reported by Passonneau and Lowry (1962), and shown to be an allosteric effect corresponding to end product inhibition (Viñuela et al. 1963). Kinetic results with eukaryotic phosphofructokinases of various origins suggested that the inhibition by ATP was related to the concentration of Mg^{2+} as if free ATP were a better inhibitor than MgATP, in contrast to the requirements for phosphoryl transfer at the active site. Pettigrew and Frieden (1979) have shown that free ATP binds to the inhibitory site about tenfold more tightly than MgATP. This specificity can be physiologically significant, since because of the cellular constancy of total Mg higher levels of ATP within the physiological range will involve greater relative abundance of free ATP. The complication that ATP is both a substrate and an inhibitor was compounded when Ramaiah et al. (1964) found that the inhibition by ATP could be countered by AMP. By affinity labeling with AMP derivatives evidence has been obtained that animal phosphofructokinase has a site for activation by AMP different from the allosteric site for inhibition by ATP (Mansour and Martensen 1978; Pettigrew and Frieden 1978), as was previously shown to be the case by trypsin treatment of the yeast enzyme (Salas et al. 1968). The three different kinds of sites for nucleotides have markedly different specificities as shown in Table 2 for the Ehrlich ascites tumor enzyme. The complexity regarding the number of different sites for adenylnucleotides is increased by the report of cAMP as an activator of animal phosphofructokinases, first made by Mansour and Mansour as early as 1962 (Mansour 1979). Nevertheless, cAMP, at least in higher animals, is merely an analog of AMP acceptable in vitro but

Table 2. Phosphofructokinase of ascites tumor

Differential specificity of the three types of adenylnucleotide sites		
NTPs	as substrates:	ATP ≃ GTP > CTP ≃ UTP ≃ ITP
NTPs	as inhibitors:	ATP ⋙ GTP ≃ ITP > CTP, UTP
NMPs	as activators:	AMP ⋙ IMP, GMP, CMP, UMP.

NTP and NMP stand for nucleoside tri- and monophosphates respectively (Castaño et al., unpublished results)

never able to act in vivo, because it can operate only in the concentration range physiological for AMP (Pinilla and Luque 1977), which cAMP never approaches because of thermodynamic limitations. For this reason there never has been selective evolutionary pressure to have an AMP site discriminating against cAMP. Exactly the opposite should be expected for a cAMP regulatory site, which needs both high affinity for cAMP and sharp discrimination with respect to the much more abundant AMP.

The combination of an inhibitory site for ATP and an activatory site for AMP appears to make phosphofructokinase precisely fitted to buffer the energy charge of the cell (Atkinson 1968), adjusting overall energy mobilization to energy expenditure, maintaining a stable supply of appropriately charged nucleotides (with the contribution of adenylate kinase and nucleosidediphosphate kinase) to perhaps hundreds of reactions. Phosphofructokinase would thus be a major factor in energy charge homeostasis, which in turn was probably a major acquisition for metabolic regulation.

Phosphofructokinases are also feedback inhibited by a carbon end product, citrate in most eukaryotic organisms (Parmeggiani and Bowman 1963; Salas et al. 1965). Covalent modification by pyridoxal-P blocks this regulatory site (Colombo and Kemp 1976). Glycerate-3-P and P-enolpyruvate can inhibit the muscle enzyme, but apparently they do so by binding at the citrate inhibitory site (Colombo et al. 1975). It is not clear whether glycerate-2,3-P_2, with a Ki of 1.4 mM (Otto et al. 1977) also acts at the citrate site or involves an additional site. The low specificity citrate site could be involved even in the very strong inhibition by decavanadate (Ki 4 nM; Choate and Mansour 1979), an unlikely candidate for metabolic regulation. Decreased sensitivity to inhibition by citrate has been reported for the enzyme in lipomas (Atkinson et al. 1974).

Animal phosphofructokinase was found by Trivedi and Danforth (1966) to be extremely sensitive to small changes of pH in the physiological range, with dramatic inhibition at the lower levels. Bock and Frieden (1976) have found with the muscle enzyme that one of the two substrates, fructose-6-P, binds preferentially to unprotonated forms of the enzyme, while inhibitory ATP binds preferentially to the protonated forms at two ionisable groups (Pettigrew and Frieden 1979; Wolfman and Hammes 1979). With the ascites tumor phosphofructokinase we have also found that binding of ATP to the inhibitory site is markedly enhanced by protonation of the enzyme. The inhibition by a decreasing pH presumably allows for a protective feedback slowdown of glycolysis in anaerobic tissues; it is conspicuously absent in the yeast enzyme, an organism in which glycolysis does not produce H^+.

NH_4^+ ions can activate phosphofructokinases within the physiological concentration range even in the presence of physiological K^+ concentrations (about 0.1 M) (Sols and

Salas 1966; Otto et al. 1976), allowing a high rate of glycolysis when there is demand for carbon skeletons derived from glycolysis. Pi, closely related to energy metabolism, is a strong activator of phosphofructokinase, synergistically with AMP (Bañuelos et al. 1977) and NH_4^+.

Fructose-1,6-P_2 in the micromolar range is a strong activator of the phosphofructokinases from muscle (El-Bawdry et al. 1973; Hood and Hollaway 1976) (synergistically with AMP; Tornheim and Lowenstein 1976), liver and testis, but not that of the Ehrlich ascites tumor (unpublished results). Both glucose-1,6-P_2 and mannose-1,6-P_2 can also activate the enzyme, although at higher concentrations (Rose and Warms 1974). Very recently Hers and co-workers have found in liver extracts a powerful activator of the liver enzyme that seems to be fructose-2,6-P_2, with a Ka of ca. 0.1 μM (Hers, this volume). It would be interesting to ascertain whether more than one site is involved in this activation and, in case of a single site, which of the two fructose bisphosphates is the main physiological effector.

The reported inhibition of muscle phosphofructokinase by creatine phosphate has proved to be due to a contaminant, as yet unidentified (Fitch et al. 1979). The possibility of spurious effects due to impurities should be considered particularly when rather high concentrations of a compound are required for an observable effect, regardless of the physiological concentration of the presumptive effector.

Among other presumptive physiological factors reported to be able to affect phosphofructokinase activity in certain animal tissues, but as yet of doubtful significance, are a liver peptide stabilizing factor (Dunaway and Segal 1976; Dunaway et al. 1979), a muscle protein inhibitor (Wohnlich 1975), and fructosebisphosphatase (Uyeda and Luby 1974).

Information on ligand bindings has been recently summarized by Goldhammer and Paradies (1979); some of the bindings are synergistic, in parallel with kinetic observations with effectors of the same sign.

The kinetic behavior of phosphofructokinases is so complex and variable that Mansour ended a review in 1972 saying that "the reader may well come to the conclusion that phosphofructokinase is beyond control" and "that future advances in our knowledge of the regulation of phosphofructokinase will have to await studies on the nature of the enzyme in the resting cell and its response to different physiological conditions." This goal has been partly accomplished through the in situ approach for the kinetic study of intracellular enzymes in permeabilized cells, with the confirmation that at least most of the allosteric effectors studied in vitro do operate in cells, with kinetic parameters similar or better suited for physiological regulation of metabolism than those obtained in vitro. In erythrocytes (Aragón et al. 1980) and in ascites tumors (unpublished results) there are marked quantitative differences in several kinetic and allosteric parameters, as illustrated for the ATP inhibition at different pH values (Fig. 1). Another kind of attempt to study the allosteric properties of a mammalian phosphofructokinase in near-physiological conditions (pH 7.0, 120 mM KCl, and 3 mM MgATP) has been made by Reinhart and Lardy (1980a, b, c) with the rat liver enzyme. The behavior of the enzyme in these conditions is markedly different from that observed at unphysiologically high pH, with respect to the various potential effectors. The major limitation of this approach was the unphysiologically low concentration of the enzyme for technical reasons.

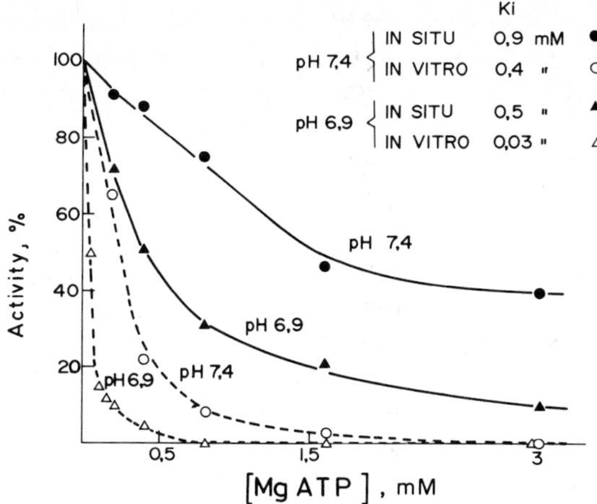

Fig. 1. Inhibition by ATP within the physiological pH range of rat erythrocyte phosphofructokinase in situ and in vitro. (Aragón and Sols, unpublished results)

The phosphofructokinases of both liver and muscle have been found to be phosphorylable in vivo (Brand and Söling 1975; Hofer and Furst 1976), but for years no physiologically useful correlation between phosphorylation and metabolic regulation could be found. In 1979, three laboratories independently found a glucagon-induced, cyclic AMP-mediated reversible inactivation of phosphofructokinase in hepatocytes when assayed in suboptimal, near physiological conditions (Castaño et al. 1979; Pilkis et al. 1979; Kagimoto and Uyeda 1979). And this inactivation seems to be accompanied by increased phosphorylation of the enzyme (Kagimoto and Uyeda 1979, 1980; Claus et al. 1980; Pilkis et al. 1980; Castaño and Nieto, unpublished results) which in vitro seems to depend on at least two phosphorylable sites per tetramer susceptible to the cyclic AMP-dependent protein kinase. The glucagon-dependent inactivation in vivo does not disappear (although it may decrease) upon several 100-fold purification of the enzyme. However, the observation that the inactivation found in crude extracts can be markedly decreased by gel filtration has led Hers (this volume) and Söling (this volume) to conclude that it was due merely to change in some effector metabolite, tentatively identified by Hers as the powerful new activator fructose-2,6-P_2. Nevertheless, the fact that some inactivation withstands extensive purification of the enzyme supports the conclusion that inactivating phosphorylation is also involved.

The multimodulation of animal phosphofructokinases is summarized in Table 3: there are at least nine regulatory mechanisms, involving at least nine distinct sites per subunit (Table 4).

Table 3. Multimodulation of animal phosphofructokinases

Cooperativity for Fru-6-P			
⊖	ATP	⊕	AMP
⊖	Citrate	⊕	Pi
⊖	H^+	⊕	NH_4^+
		⊕	Fru-1,6-P_2
[o ⇌ m] phosphorylation, inactivating?			

Table 4. Different specific sites in animal phosphofructokinase

Active site (2 subsites: Fru-6-P and nucleosidetriphosphate)				
ATP	inhibitory	site	AMP	activatory site
H^+	"	subsite	Pi	" "
Citrate	"	site	NH_4^+	" "
			Fru-1,6-P_2	" "
Phosphorylable site, inactivating?				
Oligomerization site(s)				

3.2 Yeast Phosphofructokinase

Allosteric modulation of yeast phosphofructokinase has been extensively studied particularly in our own laboratory (Viñuela et al. 1963; Salas et al. 1965; Sols and Salas 1966; Salas et al. 1968; Bañuelos et al. 1977) and also in Los Angeles, Freiburg, and Leipzig. These and other studies have led to the characterization of this enzyme as truly multimodulated, as summarized in Table 5. A minimum of six different regulatory sites (including the oligomerization site(s) on which the cooperativity depends) are likely to occur in the *Saccharomyce cerevisiae* enzyme. Moreover, some of these effects have been confirmed in situ, studying the enzyme in permeabilized cells (Bañuelos and Gancedo 1978). It is to be noted that in this case, in contrast to the enzyme in animal tissues (Sect. 3.1), there are no marked differences in the kinetic behavior of the enzyme in situ in respect to in vitro. This parallelism is related to the fact that the yeast enzyme, in contrast to those in animal tissues, does not undergo changes in the degree of polymerization leading to concentration-dependent behavior (Laurent et al. 1979).

Table 5. Multimodulation of yeast phosphofructokinase

Cooperativity for fructose-6-P	
⊖ ATP ⊖ citrate	synergistic
⊕ AMP ⊕ NH_4 ⊕ Pi	synergistic

3.3 Bacterial Phosphofructokinases

Bacterial phosphofructokinases in general seem to be markedly different from those of eukaryotic organisms. Feedback inhibition of most of them is produced by P-enolpyruvate instead of by ATP, they are generally activated by ADP instead of AMP (although the fructosebisphosphatase of *E. coli* is inhibited by AMP), and their subunits are much smaller.

The most extensive kinetic characterization among bacterial phosphofructokinases has been carried out in *E. coli*, studied by Atkinson and Walton in 1965 and by Blangy et al. in 1968. Later, Reeves and Sols (1973) studied the enzyme in more physiological conditions including the in situ approach and obtained kinetic and allosteric parameters well suited for metabolic regulation in vivo. Afterwards Fraenkel et al. (1973) identified the occurrence in *E. coli* of an apparently nonallosteric isozyme (see also Doelle and McIvor 1980). Nevertheless, when studied by us in more physiological conditions in strains containing only one or the other of the two isozymes, it become evident that both are allosteric, although qualitatively different. Isozyme II is inhibited by ATP instead of by P-enolpyruvate (Fig. 2 and Table 6). This result has already been confirmed by D. Kotlarz and H. Buc (pers. comm.).

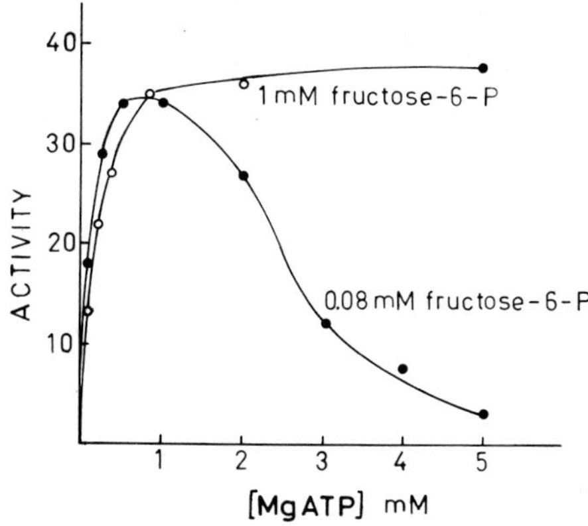

Fig. 2. Inhibition by ATP of *E. coli* phosphofructokinase II. (Domenech and Sols, unpublished results)

Table 6. Modulation of *E. coli* phosphofructokinases in situ and in vitro (pH 7.5, 0.1 M KCl) (Domenech and Sols, unpublished results)

	I	II
Cooperativity for Fru-6-P	Yes	
$S_{0.5}$ with 5 mM ATP	ca 0.2 mM	
nH	ca 1.5	
Inhibition by P-enolpyruvate	Yes	No
Inhibition by ATP	No	Yes
Ki with 0.08 mM Fru-6-P		2 mM
Inhibition by citrate	Yes	
(synergistic with P-enolpyruvate)		
Activation by ADP	Yes	
Activation by NH_4	Yes	

Robinson and Fraenkel (1978) reported that strains with only isozyme II can grow almost as efficiently as those with only the I isozyme. Aside from the consideration that a "little" effect in artificial laboratory conditions would be severely crippling in the harsh competition in nature, the fact that each of the two isozymes of *E. coli* phosphofructokinase does have a mechanism for allosteric feedback regulation changes the potential significance of the observation with respect to the in vivo operation and overall biological value of the allosteric modulation of phosphofructokinases.

4 Conclusions and Perspectives

Phosphofructokinases in general are truly multimodulated enzymes, although there is much noise in the literature. Complexity of the multimodulation seems to increase from prokaryotic organisms to yeast to higher animals, although the differences might be due in part to the fact that the latter enzymes have been much more extensively studied. Most phosphofructokinases are highly sophisticated integrators of a rich variety of metabolic signals through a complex series of regulatory mechanisms.

The precise involvement of phosphofructokinases in metabolic regulation could be clarified through convergence of several different approaches:

1. Kinetic and binding studies carried out for the different ligands in assay conditions similar to those prevailing in the cells of origin of each enzyme studied, including as far as possible the concentration of the enzyme itself. It would be of considerable interest to carry out systematic studies of each of the major isozymes in standard physiological conditions as basic reference for the identification of new possible varieties and for the particular conditions in individual tissues.

2. Determination of the primary structure, with identification of the residues involved in the active and regulatory sites. Advances in this direction include a study on the AMP site of heart phosphofructokinase (Weng et al. 1980), the complete primary structure of the *Bacillus stearothermophilus* enzyme (Kolb et al. 1980), and the partial structure of the muscle enzyme (Walker et al. 1976). X-ray analyses and other techniques will allow the exact quantification of sites for the various ligands, their topology on the enzyme molecule, and the influence of their respective specific bindings on the tertiary and quaternary structures.

3. Several complementary approaches by noninvasive techniques could contribute to ascertain the precise physiological significance of the in vitro studies to the in vivo regulation, with the in situ approach partly bridging the gap between in vivo and in vitro.

4. Characterization of the molecular pathology of abnormal phosphofructokinases could contribute to the ascertaining of the in vivo consequences of the loss or change of different modulatory mechanisms. To find regulatory mutants of mammalian phosphofructokinases, careful kinetic exploration of the enzyme in biopsies (or necropsies) should be carried out in cases indicating abnormalities of glycolysis in vivo, particularly when not accountable by obvious quantitative change in the level of some enzyme.

References

Aragón JJ, Feliu JE, Frenkel RA, Sols A (1980) Permeabilization of animal cells for kinetic studies of intracellular enzymes: In situ behavior of the glycolytic enzymes in erythrocytes. Proc Natl Acad Sci USA 77:6324-6328

Atkinson DE (1968) The energy charge of the adenylate pool as a regulatory parameter. Interaction with feed-back modifiers. Biochemistry 7:4030-4034

Atkinson DE, Walton GM (1965) Kinetics of regulatory enzymes. *Escherichia coli* phosphofructokinase. J Biol Chem 240:757-763

Atkinson JNC, Galton JJ, Gilbert C (1974) Regulatory defect of glycolysis in human lipoma. Br Med J 1:101-102

Bañuelos M, Gancedo C (1978) In situ study of the glycolytic pathway in *Saccharomyces cerevisiae*. Arch Microbiol 117:197-201

Bañuelos M, Gancedo C, Gancedo JM (1977) Activation by phosphate of yeast phosphofructokinase. J Biol Chem 252:6394-6398

Blangy D, Buc H, Monod J (1968) Kinetics of the allosteric interactions of phosphofructokinase from *Escherichia coli*. J Mol Biol 31:13-35

Bock PE, Frieden C (1976) Phosphofructokinase II. Role of ligands in pH-dependent structural changes of the rabbit muscle enzyme. J Biol Chem 251:5637-5643

Brand IA, Söling HD (1975) Activation and inactivation of rat liver phosphofructokinase by phosphorylation-dephosphorylation. FEBS Lett 57:163-168

Castaño JG, Nieto A, Feliu JE (1979) Inactivation of phosphofructokinase by glucagon in rat hepatocytes. J Biol Chem 254:5576-5579

Choate G, Mansour TE (1979) Studies on heart phosphofructokinase. Decavanadate as a potent allosteric inhibitor at alkaline and acidic pH. J Biol Chem 254:11457-11462

Claus TH, El-Maghrabi R, Riou JP, Pilkis SJ (1980) Modulation of hepatic pyruvatekinase and phosphofructokinase activity. Fed Proc 39:1670

Colombo G, Kemp RG (1976) Specific modification of an effector binding site of phosphofructokinase by pyridoxal phosphate. Biochemistry 15:1774-1780

Colombo G, Tate PW, Girotti AW, Kemp RG (1975) Interactions of inhibitors with muscle phosphofructokinase. J Biol Chem 250:9404-9412

Crane RK, Sols A (1954) The non-competitive inhibition of brain hexokinase by glucose-6-phosphate and related compounds. J Biol Chem 210:597-606

Doelle HW, McIvor S (1980) The Pasteur effect in *Escherichia coli* K-12 mutants lacking the allosteric form of phosphofructokinase. FEMS Microbiol Lett 7:337-340

Domenech C, Sols A (1979) Allosteric modulation of the isozymes of phosphofructokinase studied by the in situ approach. Arch Biol M 12:266

Dunaway GA, Segal HL (1976) Purification and physiological role of a peptide stabilizing factor of rat liver phosphofructokinase. J Biol Chem 251:2323-2329

Dunaway GA, Lanham R, Kruep D (1979) Composition of the rat liver phosphofructokinase regulatory factor. Fed Proc 38:413

El-Bawdry AM, Otani A, Mansour TE (1973) Studies on heart phosphofructokinase. Role of fructose 1,6-diphosphate in enzyme activity. J Biol Chem 248:557-563

Fitch CD, Chevli R, Jellinek M (1979) Phosphocreatine does not inhibit rabbit muscle phosphofructokinase or pyruvate kinase. J Biol Chem 254:11357-11359

Fraenkel DG, Kotlarz D, Buc H (1973) Two fructose 6-phosphate kinase activities in *Escherichia coli*. J Biol Chem 248:4865-4866

Goldhammer AR, Paradies HH (1979) Phosphofructokinase: structure and function. In: Horecker BL, Stadtman ER (eds) Current topics in cellular regulation, vol 15. Academic Press, London New York, pp 109-141

Hers HG (1980) This volume

Hofer HW (1971) Influence of enzyme concentrations on the kinetic behaviour of rabbit muscle phosphofructokinase. Hoppe-Seyler's Z Physiol Chem 352:997-1004

Hofer HW, Furst M (1976) Isolation of a phosphorylated form of phosphofructokinase from skeletal muscle. FEBS Lett 62:118-122

Hofer HW, Krystek E (1975) Determination of the kinetic parameters of phosphofructokinase dissociation. FEBS Lett 53:217-220

Hood K, Hollaway MR (1976) Significant role of fructose-1,6-diphosphate in the regulatory kinetics of phosphofructokinase. FEBS Lett 68:8-14

Hulme EC, Tipton KE (1971) The dependence of phosphofructokinase kinetics upon protein concentration. FEBS Lett 12:197-200

Kagimoto T, Uyeda K (1979) Hormone-stimulated phosphorylation of liver phosphofructokinase in vivo. J Biol Chem 254:5584-5587

Kagimoto T, Uyeda K (1980) Regulation of rat liver phosphofructokinase by glucagon-induced phosphorylation. Arch Biochem Biophys 203:792-799

Kahn A, Cottreau D, Meienhofer MC (1980) Purification of F_4 phosphofructokinase from human platelets and comparison with the other phosphofructokinase forms. Biochim Biophys Acta 611:114-126

Kolb E, Hudson PJ, Harris JI (1980) Phosphofructokinase: Complete amino-acid sequence of the enzyme from *Bacillus stearothermophilus.* Eur J Biochem 108:587-598

Laurent M, Seydoux FJ, Dessen P (1979) Allosteric regulation of yeast phosphofructokinase. Correlation between equilibrium binding, spectroscopic and kinetic data. J Biol Chem 254:7515-7520

Lazo PA, Sols A, Wilson JE (1980) Brain hexokinase has two spatially discrete sites for binding of glucose-6-phosphate. J Biol Chem 255:7548-7551

Lowry OH, Passonneau JV (1966) Kinetic evidence for multiple binding sites on phosphofructokinase. J Biol Chem 241:2268-2279

Mansour TE (1972) Phosphofructokinase. In: Horecker BL, Stadtman ER (eds) Current topics in cellular regulation, vol 5. Academic Press, London New York, pp 1-46

Mansour TE (1979) Chemotherapy of parasitic worms: New biochemical strategies. Science 205: 462-469

Mansour TE, Martensen TM (1978) Heart phosphofructokinase. Allosteric kinetics following affinity labeling modification of the enzyme by 8-[m-(m-fluorosulfonylbenzamide)benzylthio] adenine. J Biol Chem 253:3628-3634

Meldolesi MF, Laccetti P (1979) Rat thyroid phosphofructokinase. Comparison of the regulatory and molecular properties with those of rat muscle enzyme. J Biol Chem 254:9200-9203

Monod J, Changeux JP, Jacob J (1963) Allosteric proteins and cellular control systems. J Mol Biol 6:306-329

Otto M, Jacobash G, Rapoport S (1976) A comparison of the influence of potassium and ammonium ions on the phosphofructokinases from rabbit muscle and rat erythrocytes. Eur J Biochem 65: 201-206

Otto M, Heinrich R, Jacobash G, Rapoport S (1977) A mathematical model for the influence of anionic effectors on the phosphofructokinase from rat erythrocytes. Eur J Biochem 74:413-420

Parmeggiani A, Bowman RH (1963) Regulation of phosphofructokinase activity by citrate in normal and diabetic muscle. Biochem Biophys Res Commun 12:268-273

Passonneau JV, Lowry OH (1962) Phosphofructokinase and the Pasteur effect. Biochem Biophys Res Commun 7:10-15

Perutz MF (1978) Electrostatic effects in proteins. Science 201:1187-1191

Pettigrew DW, Frieden C (1978) Rabbit muscle phosphofructokinase. Modification of molecular and regulatory kinetic properties with the affinity label 5'-p-(fluorosulfonyl)benzoyl adenosine. J Biol Chem 253:3623-3627

Pettigrew DW, Frieden C (1979) Binding of regulatory ligands to rabbit muscle phosphofructokinase. A model for nucleotide binding as a function of temperature and pH. J Biol Chem 254:1887-1895

Pilkis S, Schlumpf J, Pilkis J, Claus TH (1979) Regulation of phosphofructokinase activity by glucagon in isolated rat hepatocytes. Biochem Biophys Res Commun 88:960-967

Pilkis S, Schlumpf J, Claus TH, El-Maghrabi R (1980) Studies on phosphorylation state of rat hepatic phosphofructokinase. Fed Proc 39:1940

Pinilla M, Luque J (1977) Activation of rat erythrocyte phosphofructokinase by AMP and by nonphysiological concentrations of cyclic AMP. Mol Cell Biochem 15:219-223

Ramaiah A, Hathaway JA, Atkinson DE (1964) Adenylate as a metabolic regulator. Effect on yeast phosphofructokinase kinetics. J Biol Chem 239:3619-3622

Reeves RE, Sols A (1973) Regulation of *Escherichia coli* phosphofructokinase in situ. Biochem Biophys Res Commun 50:459-466

Reinhart GD, Lardy HA (1980a) Rat liver phosphofructokinase: kinetic activity under near-physiological conditions. Biochemistry 19:1477-1483

Reinhart GD, Lardy HA (1980b) Rat liver phosphofructokinase: use of fluorescence polarization to study aggregation at low protein concentrations. Biochemistry 19:1484-1490

Reinhart GD, Lardy HA (1980c) Rat liver phosphofructokinase: kinetic and physiological ramifications of the aggregation behavior. Biochemistry 19:1491-1495

Robinson JP, Fraenkel DG (1978) Allosteric and nonallosteric *E. coli* phosphofructokinase: effects on growth. Biochem Biophys Res Commun 81:858-863

Rose IA, Warms JVB (1974) Glucose- and mannose-1,6-P_2 as activators of phosphofructokinase in red blood cells. Biochem Biophys Res Commun 59:1333-1340

Salas ML, Viñuela E, Salas M, Sols A (1965) Citrate inhibition of phosphofructokinase and the Pasteur effect. Biochem Biophys Res Commun 19:371-376

Salas ML, Salas J, Sols A (1968) Desensitization of yeast phosphofructokinase to ATP inhibition by treatment with trypsin. Biochem Biophys Res Commun 31:461-466

Söling HD (1980) This volume

Sols A (1973) Allosteric effects by products and substrates involving specifically regulatory sites. In: Mechanisms and control properties of phosphotransferases. Akademie-Verlag, Berlin, pp 239-251

Sols A (1976) The Pasteur effect in the allosteric era. In: Kornberg A, Horecker BL, Cornudella L, Oró J (eds) Reflections on biochemistry. Pergamon Press, Belfast, pp 199-206

Sols A (1979) Multimodulation of enzyme activity. Physiological significance and evolutionary origin. In: Atkinson DE, Fox F (eds) Modulation of protein function, vol XIII. Academic Press, London New York, pp 27-45

Sols A (1981) Multimodulation of enzyme activity. In: Horecker BL, Stadtman ER (eds) Current topics in cellular regulation. Academic Press, London New York to be published

Sols A, Salas ML (1966) Phosphofructokinase Yeast. In: Wood WA (ed) Methods in enzymology, vol 9. Academic Press, London New York, pp 436-442

Tejwani GA (1978) The role of phosphofructokinase in the Pasteur effect. TIBS 3:30-33

Tornheim K, Lowenstein JM (1976) Control of phosphofructokinase from rat skeletal muscle. Effects of fructose diphosphate, AMP, ATP, and citrate. J Biol Chem 251:7322-7328

Trivedi B, Danforth WH (1966) Effect of pH on the kinetics of frog muscle phosphofructokinase. J Biol Chem 241:4110-4112

Uyeda K (1979) Phosphofructokinase. In: Meister A (ed) Advances in enzymology, vol 48. John Wiley & Sons, New York, pp 193-244

Uyeda K, Luby LJ (1974) Studies on the effect of fructose diphosphatase on phosphofructokinase. J Biol Chem 249:4562-4570

Viñuela E, Salas M, Sols A (1963) End-product inhibition of yeast phosphofructokinase by ATP. Biochem Biophys Res Commun 12:140-145

Walker ID, Harris JI, Runswick MJ, Hudson P (1976) The subunits of rabbit muscle phosphofructokinase. A search for sequence repetition. Eur J Biochem 68:255-269

Weng L, Heinrikson RL, Mansour TE (1980) Amino acid sequence at the allosteric site of sheep heart phosphofructokinase. J Biol Chem 255:1492-1496

Wohnlich J (1975) Purification and some properties of rabbit skeletal muscle fructose 6-phosphate kinase inhibitor. Biochimie 57:683-694

Wolfman NM, Hammes GG (1979) A calorimetric study of the interaction of ATP with rabbit muscle phosphofructokinase. J Biol Chem 254:12289-12290

Discussion[3]

In a comment to Dr. Sols' presentation Dr. Wieland pointed to studies on metabolic compartmentation in glucagon stimulated hepatocytes (Siess EA, Brocks DG, Lattke HK, Wieland OH (1977) Biochem J 166:225-235) which seem to exclude ATP and citrate as potential effectors in PFK regulation.

---

3 Rapporteur: Elmar A. Siess

Structure, Function, and Regulation of Mammalian Pyruvate Dehydrogenase Complex

L.J. Reed, F.H. Pettit, D.M. Bleile, and T-L. Wu[1]

1 Introduction

The mammalian pyruvate dehydrogenase complex consists of three catalytic components – pyruvate dehydrogenase (E_1), dihydrolipoyl transacetylase (E_2), and dihydrolipoyl dehydrogenase (E_3). These three enzymes, acting in sequence, catalyze the reactions shown in Fig. 1 (Reed 1974). E_1 catalyzes both the decarboxylation of pyruvate (reac-

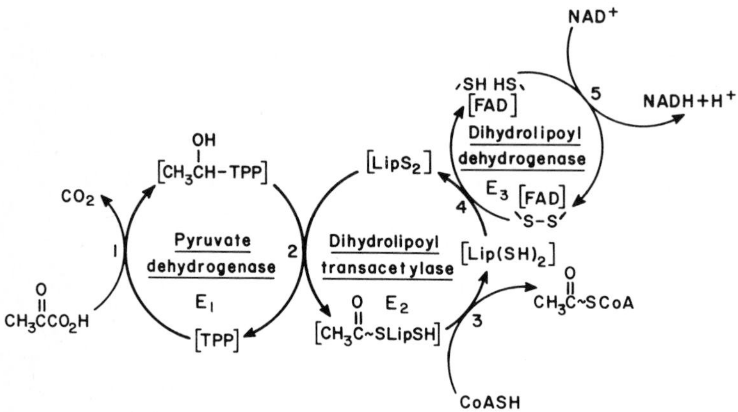

Fig. 1. Reaction sequence in pyruvate oxidation. *TPP*, thiamin pyrophosphate; *LipS*$_2$ and *Lip(SH)*$_2$, lipoyl moiety and its reduced form; *CoASH*, coenzyme A; *FAD*, flavin adenine dinucleotide; *NAD*$^+$ and *NADH*, nicotinamide adenine dinucleotide and its reduced form

tion 1) and the subsequent reductive acetylation of the lipoyl moiety (reaction 2) which is covalently bound to E_2. E_2 catalyzes the transacetylation step (reaction 3), and E_3 catalyzes the reoxidation of the dihydrolipoyl moiety NAD^+ as the ultimate electron acceptor (reactions 4 and 5). The mammalian complex also contains small amounts of two regulatory enzymes, a kinase and a phosphatase, which modulate the activity of E_1 by phosphorylation and dephosphorylation, respectively (Linn et al. 1969). This paper

1 Clayton Foundation Biochemical Institute and Department of Chemistry, The University of Texas at Austin, Austin, TX 78712, USA

discusses some aspects of the structural organization of the mammalian pyruvate dehydrogenase complex and regulation of its activity by a phosphorylation-dephosphorylation cycle.

2 Subunit Composition and Structure of Pyruvate Dehydrogenase Complex

The pyruvate dehydrogenase complex is organized about a core, consisting of the oligomeric E_2, to which multiple copies of E_1 and E_3 are bound by noncovalent bonds. The pyruvate dehydrogenase complexes isolated from bovine kidney and heart mitochondria have molecular weights (M_r) of about seven million and 8.5 million, respectively. The component enzymes of the two complexes are very similar, if not identical (Barrera et al. 1972). E_1 has a M_r of about 154,000 and possesses the subunit composition $\alpha_2\beta_2$ (Table 1). The M_r of the two subunits are about 41,000 and 36,000, respectively. The

Table 1. Subunit composition of bovine heart pyruvate dehydrogenase complex

Enzyme	M_r	Subunits No.	Subunits M_r	Subunits per molecule of complex
Native Complex	8,500,000			
E_1 [a]	154,000	2	41,000	60
		2	36,000	60
E_2 [b]	3,100,000	60	52,000	60
E_3 [c]	110,000	2	55,000	12
Kinase	∿100,000	1	48,000	
		1	45,000	
Phosphatase	150,000	1	97,000	
		1	50,000	

[a] E_1, pyruvate dehydrogenase; [b] E_2, dihydrolipoyl transacetylase; [c] E_3, dihydrolipoyl dehydrogenase

core enzyme, E_2, has a M_r of about 3.1 million and consists of 60 apparently identical polypeptide chains of M_r about 52,000 (Barrera et al. 1972; Kresze et al. 1980). Each E_2 chain contains one covalently bound lipoyl moiety (White et al. 1980). E_3 has a M_r of about 110,000 and contains two apparently identical polypeptide chains and two molecules of FAD. The bovine kidney pyruvate dehydrogenase complex contains about 20 E_1 tetramers ($\alpha_2\beta_2$) and about 6 E_3 dimers, whereas the heart complex contains about 30 E_1 tetramers and about 6 E_3 dimers. The kidney complex can bind about ten additional E_1 tetramers, but neither complex can bind additional E_3 dimers. The kinase is tightly bound to E_2. The amount of endogenous kinase in the bovine kidney and heart pyruvate dehydrogenase complexes is small, about three molecules per molecule of complex in the kidney complex and less in the heart complex. The phosphatase also binds to E_2, and this attachment requires the presence of Ca^{2+} ions (Pettit et al. 1972). Based on activity measurements of mitochondrial extracts and highly purified preparations of phosphatase and pyruvate dehydrogenase complex, there appear to be about five molecules of phosphatase per molecule of complex in bovine kidney mitochondria and somewhat more in bovine heart mitochondria (Pettit and Reed, unpublished data). The kinase and the phosphatase have been isolated in homogeneous form. The kinase consists of two subunits of M_r about 48,000 and 45,000, as estimated by sodium dodecyl sulfate-polyacrylamide gel electrophoresis (Pettit and Reed, unpublished data). The phosphatase has a M_r of about 150,000 and consists of two subunits of M_r about 97,000 and 50,000 (Teague, Pettit, Yeaman, and Reed, unpublished data). Siess and Wieland (1972) have reported that pyruvate dehydrogenase phosphatase from pig heart consists of a single polypeptide chain of M_r 92,000 to 95,000.

The appearance of E_2 in the electron microscope is that of a pentagonal dodecahedron, and its design appears to be based on icosahedral (532) symmetry (Reed and Oliver 1968). The E_1 tetramers appear to be located on the 30 edges and the E_3 dimers in the 12 faces of the pentagonal dodecahedron.

2.1 Subunit Structure of Dihydrolipoyl Transacetylase

Studies of limited proteolysis with trypsin have revealed unique architectural features of the mammalian dihydrolipoyl transacetylase (Bleile et al. 1981). The bovine heart transacetylase subunit consists of two different domains: a compact domain of M_r about 26,000, designated the subunit binding domain, to which is attached a flexible extension of M_r about 28,600 bearing the covalently bound lipoyl moiety. These M_r were determined by sedimentation equilibrium analysis. The subunit binding domain contains the intersubunit binding sites of E_2 and the catalytic site for transacetylation. The subunit binding domain confers quaternary structure on the transacetylase, i.e., the assemblage of subunit binding domains is responsible for the pentagonal dodecahedron-like appearance of E_2 in the electron microscope. The lipoyl domain is acidic, and it contains a high proportion of prolyl residues. It exhibits anomalous migration in sodium dodecyl sulfate-polyacrylamide gel electrophoresis (apparent M_r about 37,000). Presumably there is a trypsin-sensitive hinge region between the two domains (Fig. 2). The lipoyl moiety is bound in amide linkage to the ϵ-amino group of a lysyl residue in the extended lipoyl domain. Movement of the lipoyl domain with its attached lipoyl moiety conceivably permits interaction of the latter with successive active sites on the complex.

These architectural features of the bovine heart dihydrolipoyl transacetylase subunit are very similar to those found in *Escherichia coli* dihydrolipoyl transacetylase (Bleile et al. 1979) and in *E. coli* dihydrolipoyl transsuccinylase (McRorie, Bleile and Reed, unpublished data). All three subunits consists of a compact domain that confers quaternary structure on these transacylases and contains the catalytic site for transacylation; to this domain is attached a large extension containing the lipoyl moiety.

Kresze and Steber (1979) reported that the major initial products of limited proteolysis of bovine kidney dihydrolipoyl transacetylase by a lysosomal thiol protease were two fragments with apparent M_r of about 29,000 and 36,000 as estimated by sodium dodecyl sulfate-gel electrophoresis. Although these two fragments were not characterized further, it appears that they may correspond to the subunit binding domain and the lipoyl domain, respectively, of the bovine heart transacetylase subunit. In contrast

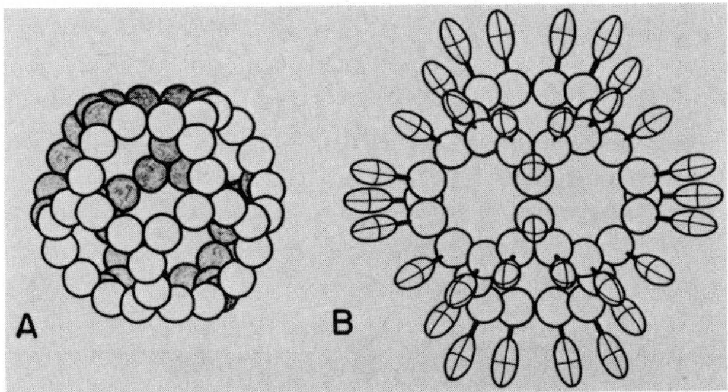

Fig. 2 A, B. Interpretative models of the quaternary structure of mammalian dihydrolipoyl transacetylase. **A** Model of the transacetylase having 60 spherical subunits placed in groups of three about the 20 vertices of a pentagonal dodecahedron. **B** Model of the transacetylase illustrating its proposed domain and subunit structure. Each of the 60 transacetylase subunits is represented by one *sphere* and its attached *ellipsoid*. The *spheres* represent the assemblage of subunit binding domains, and the *ellipsoids* represent the extended lipoyl domains. The figure is viewed down a twofold axis of symmetry

to the findings of Bleile et al. (1981) and Kresze and Steber (1979), Machicao and Wieland (1980), using limited tryptic digestion, have reported that the bovine kidney transacetylase subunit is composed of two homologous if not identical domains, each containing protein-bound lipoic acid. They also suggested that the native enzyme may be comprised of 30 subunits whose 60 domains are arranged about the vertices of a pentagonal dodecahedron.

The anomalous properties of the lipoyl domain may be responsible, at least in part, for conflicting data in the literature concerning the M_r of the mammalian E_2 subunit and the subunit composition of the pyruvate dehydrogenase complex. Hamada et al. (1975) claim that the E_2 component of the pig heart pyruvate dehydrogenase complex has a M_r of 1.8 million and consists of 24 subunits of M_r about 74,000. This stoichiometry was also proposed by Sugden and Randle (1978). However, the morphology of pig heart dihydrolipoyl transacetylase as seen in the electron microscope (Junger and Reinauer 1972) is strikingly similar to that of bovine heart and kidney dihydrolipoyl transacetylase (Reed and Oliver 1968), which has the appearance of a pentagonal dodecahedron and is probably composed of 20 morphological units (E_2 trimers) with 532 symmetry. It seems unlikely that 24 (or 30) subunits are compatible with a pentagonal dodecahedron-like structure. It is interesting to note that the modified transacetylase produced by limited tryptic digestion of the native bovine heart transacetylase has a M_r (1,580,000) similar to the M_r of pig heart dihydrolipoyl transacetylase reported by Hamada et al. (1975). In view of the sensitivity of the mammalian transacetylase to proteolysis, it seems possible that the transacetylase studied by the latter investigators may have been nicked by an endogenous protease, releasing the lipoyl domains. In any case, the mass of evidence argues that the mammalian dihydrolipoyl transacetylase is composed of 60 subunits of M_r about 52,000 organized into a pentagonal dodecahedron.

Crystalline E_1 catalyzes a rapid reductive acetylation of the lipoyl moiety on the lipoyl domain in the presence of [2-^{14}C]pyruvate and thiamin pyrophosphate (Bleile et al. 1981). The apparent K_m and V_{max} values obtained with the isolated lipoyl domain as substrate were 18.4 μM and 3.3 μmol · min^{-1} · mg^{-1}, respectively. This finding supports the conclusion (Barrera et al. 1972) that E_1 catalyzes both the decarboxylation of pyruvate and the subsequent reductive acetylation of the lipoyl moiety covalently bound to E_2 (reactions 1 and 2, Fig. 1). No incorporation of radioactive acetyl groups into lipoamide was observed, indicating that the lipoyl domain possesses structural features necessary for interaction with E_1.

3 Regulation of Mammalian Pyruvate Dehydrogenase Complex

In eukaryotic cells the pyruvate dehydrogenase complex is located in mitochondria, within the inner membrane-matrix compartment. The activity of the mammalian pyruvate dehydrogenase complex is regulated by two different mechanisms: end product inhibition by acetyl-CoA and NADH (Garland and Randle 1964; Tsai et al. 1973) and a phosphorylation-dephosphorylation cycle (Linn et al. 1969). Phosphorylation and concomitant inactivation of the complex is catalyzed by an ATP-specific, Mg^{2+}-dependent kinase, and dephosphorylation and concomitant reactivation is catalyzed by a Mg^{2+}-dependent and Ca^{2+}-stimulated phosphatase. The site of this covalent regulation is the pyruvate dehydrogenase (E_1) component of the complex (Barrera et al. 1972). Phosphorylation occurs at three serine residues in the α subunit of bovine kidney and heart pyruvate dehydrogenase (Yeaman et al. 1978). Tryptic digestion of [^{32}P]-labeled pyruvate dehydrogenase complex yielded three phosphopeptides, a mono-(site 1) and a di-(sites 1 and 2) phosphorylated tetradecapeptide and a monophosphorylated nonapeptide (site 3) (Fig. 3). These findings have been confirmed with pyruvate dehydrogenase from pig heart, except that the tryptic tetradecapeptide obtained from the pig heart enzyme contains Asp instead of Asn at residue 8 (Sugden et al. 1979).

Site I (over Ser(P)); Site 2 (over second Ser(P))

Tyr–His–Gly–His–Ser(P)–Met–Ser–Asn–Pro–Gly–Val–Ser(P)–Tyr–Arg

Site 3 (over Ser(P))

Tyr–Gly–Met–Gly–Thr–Ser(P)–Val–Glu–Arg

Fig. 3. Phosphorylation sites on pyruvate dehydrogenase from bovine kidney and heart

Studies with uncomplexed E_1 from bovine kidney and heart and highly purified pyruvate dehydrogenase kinase showed that site 1 is phosphorylated much faster than sites 2 and 3 and that phosphorylation at site 1 correlates closely with inactivation of E_1 (Yeaman et al. 1978). When E_1 and its kinase are both bound to the transacetylase (E_2), the rate of phosphorylation is increased markedly, particularly at sites 2 and 3. In the early stages of phosphorylation of E_1 complexed with E_2, inactivation of E_1 correlates closely with phosphorylation at site 1. However, appreciable phosphorylation at sites 2 and 3 accompanies inactivation of the final 25%-30% of the activity of the pyruvate dehydrogenase complex (Yeaman et al. 1978).

Earlier studies of the relationship between the degree of phosphorylation at site 1 and the extent of inactivation of E_1 indicated that this tetrameric enzyme may exhibit half-site reactivity, i.e., phosphorylation at site 1 on one α subunit may be sufficient to inactivate the tetramer (Yeaman et al. 1978; Sugden and Randle 1978). However, the evidence is not conclusive because of uncertainty in accurately determining the subunit composition of individual preparations of the pyruvate dehydrogenase complex and the lack, until recently, of a suitable method of separating α chains with different degrees of phosphorylation. Using two dimensional gel electrophoresis in conjunction with determination of site occupancy by analysis of tryptic phosphopeptides, Wahl and Utter (1980) have shown that both α chains of uncomplexed E_1 tetramers are phosphorylated at site 1 in the inactive enzyme. These techniques are now being applied to E_1 complexed with E_2, i.e., the pyruvate dehydrogenase complex.

3.1 Role of Multisite Phosphorylation

To study the role of multisite phosphorylation, Teague et al. (1979) used [γ-^{32}P]ATP and 5'[γ-thio]triphosphate (ATPγS). A [^{32}P]-labeled phosphoryl group was inserted mainly into site 1 and thiophosphoryl groups into sites 2 and 3 of bovine kidney pyruvate dehydrogenase [$PDH(OP)_1(SP)_{2,3}$]. Dephosphorylation of this enzyme with pyruvate dehydrogenase phosphatase removed the phosphoryl group at site 1, whereas the phosphatase-resistant thiophosphoryl groups remained intact at sites 2 and 3. This modified enzyme was virtually inactive. In a similar manner, [^{32}P]-labeled phosphoryl groups were inserted mainly into sites 1 and 2, and a thiophosphoryl group was inserted mainly into site 3 [$PDH(OP)_{1,2}(SP)_3$]. The phosphatase released the phosphoryl groups from sites 1 and 2, and the monothiophosphorylated enzyme (site 3) was active (Reed et al. 1980). Comparable findings have been made with the pig heart pyruvate dehydrogenase complex (Radcliffe et al. 1980). These results indicate that phosphorylation sites 1 and 2, but not site 3, are inactivating sites. Since site 1 is phosphorylated at a faster rate than sites 2 and 3, and it is dephosphorylated more slowly than the others, site 2 may serve as a safeguard for the function of site 1 (Chock et al. 1980).

Randle and co-workers (Sugden et al. 1978; Kerbey and Randle 1979) reported evidence indicating that phosphorylation at sites 2 and 3, in addition to site 1, markedly inhibited the rate of reactivation of the phosphorylated pyruvate dehydrogenase by its phosphatase. On the other hand, Teague et al. (1979) observed that the presence of phosphoryl groups at sites 2 and 3 on bovine kidney pyruvate dehydrogenase did not significantly affect the rate of its reactivation by highly purified phosphatase. These investigators showed that dephosphorylation at sites 1, 2, and 3 proceeds via a random mechanism and that the fractional phosphorylation of each site exerts no effect on the dephosphorylation rate constants of its neighboring sites. Since dephosphorylation at site 1 is the slowest step and the same phosphatase catalyzes dephosphorylation at all three sites, the presence of phosphoryl groups at all three sites would be expected to result in a slower rate of reactivation by a factor of about 3 when the phosphatase concentration is significantly lower than the dehydrogenase concentration (Chock et al. 1980). The basis of the inconsistency in the results of the two laboratories is not apparent. It is possible that different molar ratios of phosphatase and its protein substrate were used in the two laboratories. Since Randle and co-workers used a partially purified phosphatase preparation of unspecified activity, it is not possible to compare the amounts of phosphatase used in the two laboratories. In the experiments of Teague et al. (1979) the concentration of highly purified phosphatase was approximately 10 nM and the concentration of E_1 was varied between 0.71 and 28.6 μM. These values correspond to molar ratios of phosphatase:phosphorylated pyruvate dehydrogenase complex of about 1:3.6 to 1:143. In bovine kidney mitochondria this ratio is estimated to be about 5:1. Since this ratio is considerably higher than that used in the in vitro experiments, uncertainty remains as to the physiological significance of multisite phosphorylation of pyruvate dehydrogenase.

Phosphorylation of pyruvate dehydrogenase does not have a significant effect on its affinity for E_2. The dissociation constant (K_d) of dephosphorylated E_1 and monophosphorylated E_1 (site 1) is about 13 nM, and the K_d of heavily phosphorylated E_1 (sites 1, 2, and 3) is about 40 nM (Wu and Reed, unpublished data). These observations are inconsistent with the report of Pratt et al. (1979) that dephosphorylated E_1 exhibits at least a ninefold greater affinity than phosphorylated E_1 for their binding sites on E_2, as estimated from kinetic studies. In view of the high sensitivity of E_1 α subunit and of E_2 to proteolysis, it seems likely that the preparations of E_1 and E_2 used by Pratt et al. (1979) may have been nicked by an endogenous protease.

4 Control of Kinase and Phosphatase Activities

The pyruvate dehydrogenase system is well designed for fine regulation of its activity. Pyruvate dehydrogenase and its two converter enzymes, the kinase and the phosphatase, comprise a monocyclic interconvertible enzyme cascade (Stadtman and Chock 1977). Interconversion of the active and inactive, phosphorylated forms of pyruvate dehydrogenase is a dynamic process that leads to steady state. The system is endowed with an unusual capacity to be regulated by multiple metabolites, and it is remarkably flexible with respect to increasing concentrations of individual effectors. Figure 4 summarizes

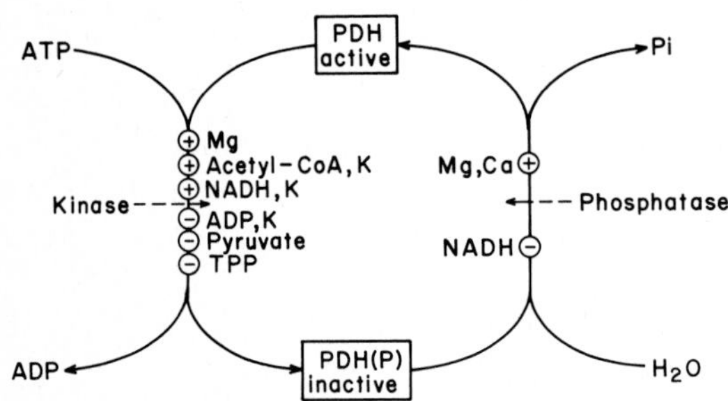

Fig. 4. Schematic representation of the covalent modification of pyruvate dehydrogenase and its control by various effectors

the control of kinase and phosphatase activities by metal ions, cofactors and metabolites. Mg^{2+} is required by both the kinase and the phosphatase. However, the apparent K_m of the phosphatase for Mg^{2+} ($\sim$2 mM) is about 100 times the apparent K_m of the kinase for Mg^{2+}. Ca^{2+} markedly stimulates phosphatase activity (about tenfold) in the presence, but not in the absence, of the transacetylase (Pettit et al. 1972; Denton et al. 1972). Ca^{2+} lowers the apparent K_m of the phosphatase for phosphorylated E_1 about 20-fold (from about 58 μM to about 2.9 μM), with little change, if any, in V_{max}. The phosphatase binds to E_2 in the presence, but not in the absence of Ca^{2+} (Pettit et al. 1972). The highly purified phosphatase binds one mol of Ca^{2+} per mol of enzyme with a $K_d \sim 35$ μM (Wu and Reed, unpublished data). There are no high affinity sites for Ca^{2+} on E_2. These observations suggest that Ca^{2+} alters the conformation of the phosphatase, permitting specific binding of this converter enzyme to E_2. Favorable topographical positioning of the phosphatase and phosphorylated E_1 on E_2 apparently facilitates the Mg^{2+}-dependent dephosphorylation.

NADH inhibits phosphatase activity and stimulates kinase activity. Kinase activity is also stimulated by acetyl-CoA and is inhibited by ADP and pyruvate. The effects of NADH and acetyl-CoA on kinase activity require the presence of K^+ or NH_4^+ ions. ADP is competitive with respect to ATP, and this inhibition apparently requires the presence of monovalent cation. The coenzyme thiamin diphosphate inhibits kinase activity, apparently as a result of binding at the catalytic site of E_1 and thereby altering the conformation about phosphorylation site 1 so that the serine hydroxyl group is less accessible to the kinase (Butler et al. 1977).

In the presence of ATP, Mg^{2+}, one or more allosteric effectors and all the proteins involved in the monocyclic cascade, a steady state is rapidly established for the fractional phosphorylation of pyruvate dehydrogenase that determines the specific activity of the enzyme. Moreover, the dehydrogenase activity can be varied progressively over a wide range by varying the concentration or molar ratios of the different effectors (Roche and Reed 1974; Pettit et al. 1975).

It is apparent that E_1, the kinase and the phosphatase can be separate targets for one or more of the effectors. It is also possible that regulatory effects could be transmitted via the transacetylase (E_2) to which E_1 and its two converter enzymes are bound. We have used peptide substrates to distinguish between direct and indirect effects of modulators of kinase and phosphatase activities. The phosphorylated tetradecapeptide (Fig. 3) produced by tryptic digestion of [^{32}P]-labeled E_1 is an effective substrate for the phosphatase, and the dephosphotetradecapeptide can serve as a substrate for the kinase (Davis et al. 1977). Reed et al. (1980) showed that dephosphorylation of the [^{32}P]-labeled tetradecapeptide by highly purified phosphatase was inhibited by NADH, and this inhibition was reversed by NAD^+. The rate of phosphorylation of the dephosphotetradecapeptide by highly purified kinase was stimulated by acetyl-CoA and NADH and inhibited by ADP, pyruvate and dichloroacetate. These results strongly indicate that the effectors act directly on the kinase and the phosphatase.

The activity of the pyruvate dehydrogenase complex may also be regulated by hormone action on tissues (Denton and Hughes 1978). Of particular interest is the effect of insulin, which increases the fraction of the active, nonphosphorylated form of pyruvate dehydrogenase in adipose tissue. It has been suggested that an oligopeptide may be the mediator of the insulin effect and that this mediator stimulates pyruvate dehydrogenase phosphatase activity (Larner et al. 1979; Popp et al. 1980).

Acknowledgments. We thank Robert Oliver, Dr. Stephen Ernst and Dr. Marvin Hackert for producing the transacetylase models. This work was supported in part by Grant GM06590 from the U.S. Public Health Service.

References

Barrera CR, Namihira G, Hamilton L, Munk P, Eley MH, Linn TC, Reed LJ (1972) Studies on the subunit structure of the pyruvate dehydrogenase complexes from bovine kidney and heart. Arch Biochem Biophys 148:343-358

Bleile DM, Munk P, Oliver RM, Reed LJ (1979) Subunit structure of dihydrolipoyl transacetylase component of pyruvate dehydrogenase complex from *Escherichia coli.* Proc Natl Acad Sci USA 76:4385-4389

Bleile DM, Hackert ML, Pettit FH, Reed LJ (1981) Subunit structure of dihydrolipoyl transacetylase component of pyruvate dehydrogenase complex from bovine heart. J Biol Chem 256:514-519

Butler JR, Pettit FH, Davis PF, Reed LJ (1977) Binding of thiamin thiazolone pyrophosphate to mammalian pyruvate dehydrogenase and its effects on kinase and phosphatase activities. Biochem Biophys Res Commun 74:1667-1674

Chock PB, Rhee SG, Stadtman ER (1980) Interconvertible enzyme cascades in cellular regulation. Annu Rev Biochem 49:813-843

Davis PF, Pettit FH, Reed LJ (1977) Peptides derived from pyruvate dehydrogenase as substrates for pyruvate dehydrogenase kinase and phosphatase. Biochem Biophys Res Commun 75:541-549

Denton RM, Hughes WA (1978) Pyruvate dehydrogenase and the hormonal regulation of fat synthesis in mammalian tissues. Int J Biochem 9:545-552

Denton RM, Randle PJ, Martin BR (1972) Stimulation by calcium ions of pyruvate dehydrogenase phosphate phosphatase. Biochem J 128:161-163

Garland PB, Randle PJ (1964) Control of pyruvate dehydrogenase in perfused rat heart by concentration of acetyl CoA. Biochem J 91:6c

Hamada M, Otsuka K-I, Tanaka N, Ogasahara K, Koike K, Hiraoka T, Koike M (1975) Purification, properties, and subunit composition of pig heart lipoate acetyltransferase. J Biochem (Tokyo) 78:187-197

Junger E, Reinauer H (1972) Untersuchungen zur Struktur der Pyruvatdehydrogenase aus Schweineherzmuskel. Biochim Biophys Acta 250:478-490

Kerbey AL, Randle PJ (1979) Role of multi-site phosphorylation in regulation of pig heart pyruvate dehydrogenase phosphatase. FEBS Lett 108:485-488

Kresze G-B, Steber L (1979) Inactivation and disassembly of the pyruvate dehydrogenase multienzyme complex from bovine kidney by limited proteolysis with an enzyme from rat liver. Eur J Biochem 95:569-578

Kresze G-B, Dietl B, Ronft H (1980) Mammalian acetyltransferase: Molecular weight determination by gel filtration in the presence of guanidinium chloride. FEBS Lett 112:48-50

Larner J, Galasko G, Cheng K, DePaoli-Roach AA, Huang L, Daggy P, Kellog J (1979) Generation by insulin of a chemical mediator that controls protein phosphorylation and dephosphorylation. Science 206:1408-1410

Linn TC, Pettit FH, Reed LJ (1969) Regulation of the activity of the pyruvate dehydrogenase complex from beef kidney mitochondria by phosphorylation and dephosphorylation. Proc Natl Acad Sci USA 62:234-241

Machicao F, Wieland OH (1980) Evidence for two-domain subunit structure of kidney lipoate acetyltransferase. FEBS Lett 115:156-158

Pettit FH, Roche TE, Reed LJ (1972) Function of calcium ions in pyruvate dehydrogenase phosphatase activity. Biochem Biophys Res Commun 49:563-571

Pettit FH, Pelley JW, Reed LJ (1975) Regulation of pyruvate dehydrogenase kinase and phosphatase by acetyl-CoA/CoA and $NADH/NAD^+$ ratios. Biochem Biophys Res Commun 65:575-582

Popp DA, Kiechle FL, Kotagal N, Jarett L (1980) Insulin stimulation of pyruvate dehydrogenase in an isolated plasma membrane-mitochondrial mixture occurs by activation of pyruvate dehydrogenase phosphatase. J Biol Chem 255:7540-7543

Pratt ML, Roche TE, Dyer DW, Cate RL (1979) Enhanced dissociation of pyruvate dehydrogenase from the pyruvate dehydrogenase complex following phosphorylation and regulatory implications. Biochem Biophys Res Commun 91:289-296

Radcliffe PM, Kerbey AL, Randle PJ (1980) Inactivation of pig heart pyruvate dehydrogenase complex by adenosine-5'-*O*(3-thiotriphosphate). FEBS Lett 111:47-50

Reed LJ (1974) Multienzyme complexes. Acc Chem Res 7:40-46

Reed LJ, Oliver RM (1968) The multienzyme α-keto acid dehydrogenase complexes. Brookhaven Symp Biol 21:397-411

Reed LJ, Pettit FH, Yeaman SJ, Teague WM, Bleile DM (1980) Structure, function and regulation of the mammalian pyruvate dehydrogenase complex. In: Mildner P, Ries B (eds) Enzyme regulation and mechanism of action. Pergamon Press, Oxford New York, pp 47-56

Roche TE, Reed LJ (1974) Monovalent cation requirement for ADP inhibition of pyruvate dehydrogenase kinase. Biochem Biophys Res Commun 59:1341-1348

Siess EA, Wieland OH (1972) Purification and characterization of pyruvate dehydrogenase phosphatase from pig heart muscle. Eur J Biochem 26:96-105

Stadtman ER, Chock PB (1977) Superiority of interconvertible enzyme cascades in metabolic regulation: Analysis of monocyclic systems. Proc Natl Acad Sci USA 74:2761-2765

Sugden PH, Randle PJ (1978) Regulation of pig heart pyruvate dehydrogenase by phosphorylation. Studies on the subunit and phosphorylation stoichiometries. Biochem J 173:659-668

Sugden PH, Hutson NJ, Kerbey AL, Randle PJ (1978) Phosphorylation of additional sites on pyruvate dehydrogenase inhibits its reactivation by pyruvate dehydrogenase phosphate phosphatase. Biochem J 169:433-435

Sugden PH, Kerbey AL, Randle PJ, Waller CA, Reid KBM (1979) Amino acid sequences around the sites of phosphorylation in the pig heart pyruvate dehydrogenase complex. Biochem J 181:419-426

Teague WM, Pettit FH, Yeaman SJ, Reed LJ (1979) Function of phosphorylation sites on pyruvate dehydrogenase. Biochem Biophys Res Commun 87:244-252

Tsai CS, Burgett MW, Reed LJ (1973) A kinetic study of the pyruvate dehydrogenase complex from bovine kidney. J Biol Chem 248:8348-8352

Wahl M, Utter MF (1980) Resolution of the phosphorylated forms of pyruvate dehydrogenase. Fed Proc 39:1975

White RH, Bleile DM, Reed LJ (1980) Lipoic acid content of dihydrolipoyl transacylases determined by isotope dilution analysis. Biochem Biophys Res Commun 94:78-84

Yeaman SJ, Hutcheson ET, Roche TE, Pettit FH, Brown JR, Reed LJ, Watson DC, Dixon GH (1978) Sites of phosphorylation on pyruvate dehydrogenase from bovine kidney and heart. Biochemistry 17:2364-2370

Discussion[2]

In a comment Dr. Cohen pointed out that the clustering of phosphate groups seems to be a common phenomenon in multiply phosphorylated proteins, e.g., glycogen synthase, spectrin, rhodopsin, and casein. Since two of the serine residues in pyruvate dehydrogenase are only six residues apart in the primary sequence it appears possible that the third site is also very close in the primary structure. Dr. Reed answered that a peptide containing all three sites has not been obtained yet. Furthermore, with respect to a question on Ca^{2+} binding to PDH-phosphatase it remains to be established which of the two subunits of this enzyme interacts with this ligand.

---

2 Rapporteur: Elmar A. Siess

Regulation of Adipose Tissue Pyruvate Dehydrogenase by Insulin: Possible Messenger Role of Peroxide(s)

O.H. Wieland and I. Paetzke-Brunner[1]

It is now well established that the pyruvate dehydrogenase complex (PDC) of adipose tissue is susceptible to hormonal control by insulin [for literature see Ref. 1]. The fact that incubation of rat epididymal fat pad or isolated fat cells with insulin leads to a several-fold increase in the proportion of the active form (PDH_a) of the complex has provided a unique model for studying the crucial problem of how the hormone's signal from interaction with the plasma membrane receptors is transmitted to the mitochondrial matrix space where the PDH complex is situated.

According to recent reports insulin can stimulate PDC activity even in a cell-free system consisting of a mixture of isolated plasma membranes and mitochondria. It has been proposed that interaction of insulin with the isolated plasma membrane results in the generation of a peptide factor which than triggers the activation of the PDH complex [2-5]. Although these studies open an interesting possibility, the observed increases in enzyme activity are rather modest and further evidence in support of this mechanism must be awaited.

Earlier studies from this laboratory were focused on the possibility that regulation of PDC interconversion in adipose tissue by insulin might result from changes of metabolites known to influence the activity of PDH_a kinase [1, 6-8]. Compounds of physiological interest are listed in Table 1. These are mainly the three couples: ATP/ADP,

Table 1. Physiological compounds acting on PDH interconversion

Kinase reaction		Phosphatase reaction	
+	–	+	–
ATP, Mg	ADP	Mg (or Mn)	NADH
NADH, K	NAD	Ca	
Acetyl-CoA, K	CoASH		
Pyruvate (low conc.)	Pyruvate		
Dihydrolipoamide			

NAD^+/NADH, and acetyl-CoA/CoASH which exert control over the kinase, whereby an increase of the proportion of these couples stimulates, and a decrease lowers, the enzyme's activity. Easiest to understand is the mechanism for ATP and ADP, from the fact

1 Klinisch-chemisches Institut und Forschergruppe Diabetes, Städt. Krankenhaus München-Schwabing, Kölner Platz 1, 8000 München 40, FRG

that ADP acts as a competitor for ATP, the substrate of kinase [9]. High ratios of ATP/ADP will therefore favor phosphorylation, i.e., inactivation of PDC, and vice versa. In the case of NAD^+/NADH and acetyl-CoA/CoA, which are no reactants of the kinase, the interrelationships are more complicated. Evidence has been provided that the oxidation-reduction (acetylation) state of the lipoyl residues of lipoate acetyltransferase plays a regulatory role. Thus, an increase in kinase activity was observed when the lipoyl residues were reduced and acetylated by NADH or dihydrolipoamide, and by acetyl-CoA or small concentrations of pyruvate, respectively [10-12]. This would explain why an increase in NADH and acetyl-CoA causes phosphorylation and inactivation of the PDH-complex, in vitro [10, 13, 14]. So far, most of these results were obtained using purified preparations of PDH complexes, while little is known about the physiological significance of the described effectors for the regulation of PDC interconversion, in vivo. Compartmentation studies on isolated liver cells in our laboratory brought first clear evidence that the phosphorylation state of the PDC system was related with the ATP/ADP ratio within the mitochondrial matrix in a similar manner, as observable on the isolated enzyme, in vitro [15].

Concerning the phosphatase, Ca^{2+} at micromolar concentrations is essential as cofactor [16-18]. In addition, enzyme activity depends on high Mg^{2+} concentration in the millimolar range [19]. NADH may act as inhibitor of the phosphatase [13].

With regard to the mechanism of hormonal regulation of PDC interconversion in adipose tissue we had proposed the possibility that insulin, by influencing the phosphorylation state of the mitochondrial adenine nucleotides, might lead to a lowering of the ATP/ADP ratio and hence to an increase in PDH_a activity [6, 7]. In order to get more direct information on the levels of the mitochondrial adenine nucleotides of intact fat cells we have adopted the digitonin fractionation procedure originally developed for the study of metabolite compartmentation in isolated liver cells [20]. Figure 1 represents a

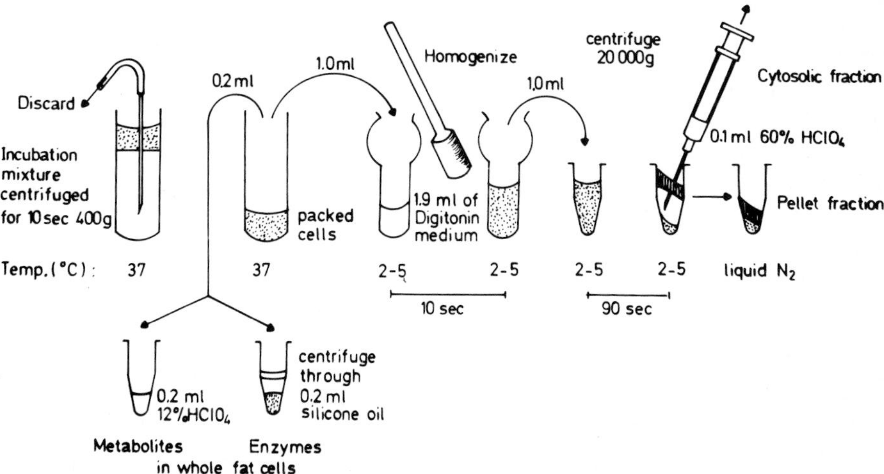

Fig. 1. Digitonin fractionation of isolated fat cells. The procedure developed for metabolite compartmentation studies of adipocytes as described in detail elsewhere [8] is schematically illustrated here

schematic view on the procedure of digitonin fractionation of isolated adipocytes. Data on the validity of this method for determination of the adenine nucleotides in the cytosol and the mitochondrial pellet of isolated fat cells have been reported elsewhere [8].

With regard to the action of insulin it is clear from the results of Table 2 that, against what we had expected, there was no change of the mitochondrial ATP/ADP ratio of fat

Table 2. Effect of insulin (1 mU/ml) on ATP/ADP ratios and PDC activities of fat cells and mitochondria after digitonin fractionation

Compartment	Additions	ATP/ADP	PDH_a	PDH_{total}	$\frac{PDH_a \times 100}{PDH_{total}}$
			(mU/g fresh cells)		
Whole cells	None	4.3 ± 0.6	63 ± 8	220 ± 26	29
	Insulin	4.9 ± 0.7	116 ± 12	227 ± 28	51*
	CFCCP[a)]	4.1 ± 0.7	112 ± 21	209 ± 26	53*
Pellet fraction	None	2.1 ± 0.1	68 ± 12	226 ± 34	30
	Insulin	2.2 ± 0.2	116 ± 16	229 ± 34	51*
	CFCCP[a)]	0.7 ± 0.1*	131 ± 32	208 ± 38	63*

Digitonin fractionation of isolated adipocytes was performed as outlined in Fig. 1. Data are taken from [8]. a) 2 μM; * $P < 0.01$

cells incubated with insulin although PDC did respond to the hormone by an about two-fold activation. A similar activation was obtained in the uncoupler experiments in Table 2 where the predictable drop of the mitochondrial ATP/ADP ratio became clearly apparent. This assures the validity of our compartmentation method in assessing changes in mitochondrial adenine nucleotides relevant for PDC regulation. In view of these results a regulatory role of the phosphorylation state of the mitochondrial adenine nucleotides as a mechanism for PDC activation by insulin has become rather unlikely.

An interesting new aspect came from the studies of May and de Haën [21], and Mukherjee et al. [22], which showed that hydrogen peroxide, similarly to insulin, leads to activation of PDC in isolated fat cells. From these observations a second messenger role was ascribed to hydrogen peroxide responsible for the communication of the peripheral insulin signal to the mitochondria. In the following experiments we have investigated these interrelationships and their physiological significance in more detail. As shown in Table 3, a nearly threefold increase in PDH_a was obtained upon incubation of isolated fat cells with H_2O_2, confirming the results of the above mentioned authors.

In order to exclude the possibility that this effect was merely due to a disturbance of mitochondrial ATP production by H_2O_2, we have measured in parallel the mitochondrial ATP levels by digitonin fractionation. From the data included in Table 3 it appears that neither in whole cells nor in the mitochondrial pellet was there a significant change of ATP in the presence of H_2O_2.

With regard to the possible role of H_2O_2 as a second messenger of insulin at the mitochondrial level it was of interest to study if H_2O_2 has an effect on the PDH complex also when added directly to isolated mitochondria. As indicated in Table 4, isolated fat cell mitochondria not only did respond to H_2O_2 but moreover displayed much higher

Table 3. Effect of hydrogen peroxide on PDH_a activity and ATP levels of isolated fat cells and their mitochondrial fraction

	PDH_a (in % of total)	ATP (nmol/g fresh cells)	
Additions	Whole cells	Whole cells	Mitochondrial fraction
None	13.4 ± 3.1	120.6 ± 9.7	7.5 ± 1.0
	$p < 0.005$	n.s.	n.s.
H_2O_2 (5 mM)	29.6 ± 4.0	113.2 ± 1.8	6.9 ± 0.9

Rat epididymal fat cells were incubated for 15 min at 37°C and further treated for determination of PDH activities and of subcellular distribution of ATP and ADP as described in [26]. Mean values ± SEM of four experiments are given. Total PDC activity was 289 ± 29 mU/g fresh cells

Table 4. Effect of H_2O_2 on PDH_a activity and ATP/ADP ratio of isolated fat cell mitochondria

Addition	PDH_a (in % of total)	$\frac{ATP}{ADP}$
None	10.5 ± 2.0	0.54 ± 0.06
H_2O_2, 0.1 mM	33.1 ± 5.6*	0.53 ± 0.07
H_2O_2, 1.0 mM	70.3 ± 7.8*	0.31 ± 0.04*
ATP – trap	72.1 ± 5.4*	0.25 ± 0.04*

Isolated fat cell mitochondria were incubated for 9 min at 25°C and further treated for determination of PDC activities and adenine nucleotides as described in [8]. ATP-trap: 20 μM ADP, 10 mM glucose, 250 μg/ml hexokinase. Mean values ± SEM of five experiments are given. Asterisks indicate significant difference from control with $p < 0.01$

sensitivity, as indicated by a threefold increase in the level of PDH_a at already 0.1 mM H_2O_2. It should be noted in Table 4 that again there appeared no lack in ATP supply under these conditions as indicated by the unchanged ATP/ADP ratios. Only at higher H_2O_2 concentrations (1 mM) was there a drop in mitochondrial ATP/ADP, which could explain the large further increase of PDH_a up to 70% of total PDC activity. For comparison it was shown that a similar lowering of ATP/ADP induced by ATP trapping led to the same elevation of the fraction of PDH_a as was seen with H_2O_2. Thus in the following experiments we have kept peroxide concentrations within a range that did not influence the phosphorylation state of the mitochondrial adenine nucleotides.

Further studies revealed that isolated fat cell mitochondria are liable to PDC activation not only by H_2O_2 but also by an organic peroxide tert. butyl hydroperoxide. As illustrated by the dose-response curve of Fig. 2, tert. butyl hydroperoxide is an even more potent stimulator causing maximal (2.4-fold) increase in PDH_a activity at already 10 μM concentration, while about ten times as much of H_2O_2 is required to achieve the same effect. This difference may be due to catalase activity present in the mitochondrial preparations, which degrades H_2O_2 but does not react with organic peroxides [23]. On the other hand, tert. butyl hydroperoxide and other organic peroxides are good substrates

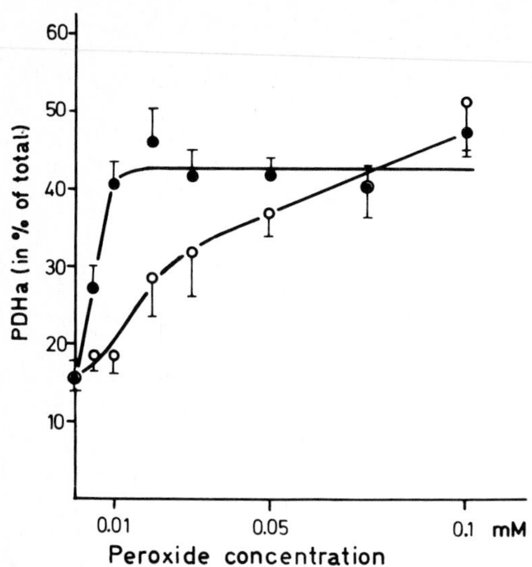

Fig. 2. Activation of PDH complex of fat cell mitochondria by H_2O_2 and tert. butyl hydroperoxide. Isolated fat cell mitochondria were incubated for 9 min at 25°C at the peroxide concentrations indicated on the abscissa. ○—○, H_2O_2; ●—●, tert. butyl hydroperoxide. Each *point* represents the mean value ± SEM of five experiments. [Data from Ref. 26]

for the enzyme(s) glutathione peroxidase which occurs in the cytoplasma and the mitochondria of mammalian cells [23].

Thus it seems possible that (a) mitochondrial glutathione peroxidase is specifically involved in the peroxide induced conversion of PDH_b to PDH_a. In this context it should be noted that tert. butanol, the enzymatic product of tert. butyl hydroperoxide reduction did not change PDC activities when added to mitochondria at corresponding concentrations (data not shown).

The time course of PDC activation by peroxides is illustrated in Fig. 3. It can be seen that the rate of PDH_a formation proceeds by an initial lag period of several minutes, reaching its maximum after about 15 min. There appears no difference in the activation kinetics between H_2O_2 and tert. butyl hydroperoxide. It seems noteworthy in this context that a similar time dependency was observed for PDC activation by insulin in fat cells (Fig. 4). In Fig. 3 there is some increase in PDH_a activity also in the peroxide-free controls, which becomes faster after 10 min of incubation. As the mitochondria in these experiments were incubated without exogenous substrate, deficient ATP generation is likely to be responsible for this change.

In experiments illustrated in Fig. 5 it was demonstrated that the peroxide effect on the PDH system is reversible. As can be seen, the elevated PDH_a levels after inital incubation of mitochondria with H_2O_2 returned toward normal after addition of a small amount of catalase. Thereafter, a second dephosphorylation cycle was released by addition of tert. butyl hydroperoxide. In these experiments there was essentially no change in PDH_a activity in the control mitochondria due to the addition of succinate as energy-yielding substrate.

The potential role of H_2O_2 as a physiological messenger of insulin's action on PDC interconversion has gained support from the finding that insulin stimulates H_2O_2 production in adipocytes as judged from oxidation of ^{14}C-formate [24, 25]. This is confirmed by experiments documented in Table 5 showing a twofold stimulation of ^{14}C-formate

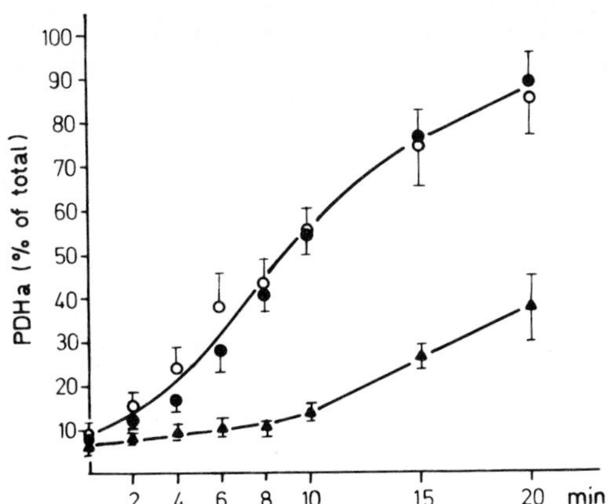

Fig. 3. Time course of PDC activation in isolated fat cell mitochondria by peroxides. Mitochondria were incubated for the times indicated on the abscissa in the presence of 0.01 mM H_2O_2 (●—●, n = 8) or 0.01 mM tert. butyl hydroperoxide (○—○, n = 3), and further treated as described in [26]; (▲—▲) controls (n = 5). Mean values ± SEM are given

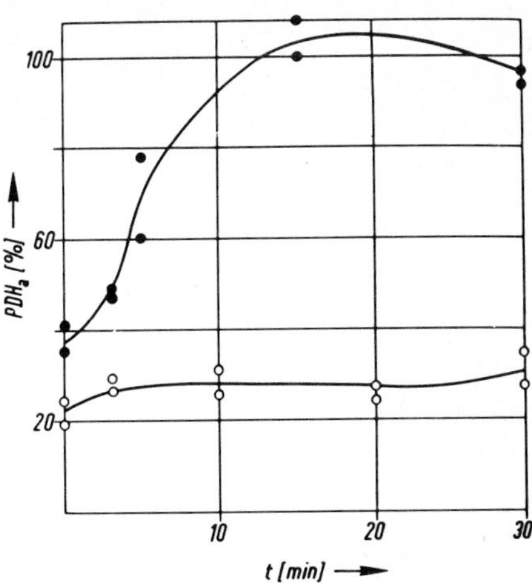

Fig. 4. Time course of PDC activation in isolated fat cells by insulin. The figure is taken from Ref. [7] by permission of Walter de Gruyter Co. Berlin-New York. ●—●, insulin, 0.2 mU/ml; ○—○, Controls. Data from two separate experiments

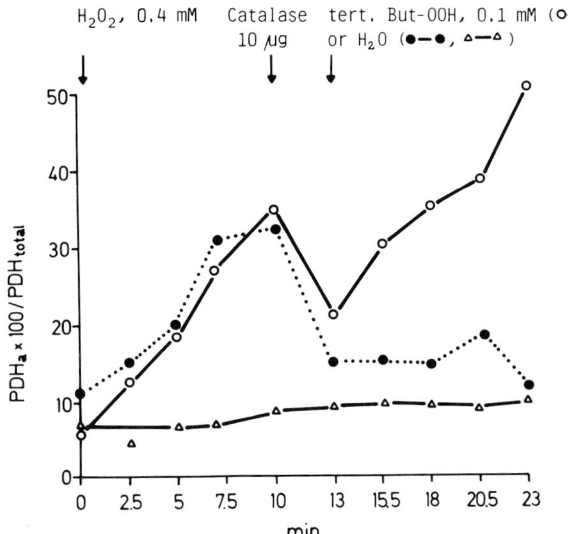

Fig. 5. Reversibility of the peroxide effect on PDC interconversion in isolated fat cell mitochondria. Isolation and incubation medium of fat cell mitochondria was as described in [26] except inclusion of 2 mM succinate in the medium. After 10 min of incubation with H_2O_2, catalse was added. After 3 min tert. butyl hydroperoxide (ButOOH) (○—○) or H_2O (●– –●) was added and incubations continued as indicated. △—△, Controls without peroxides

oxidation by insulin. Concomitantly there was a threefold increase in PDH_a activity. In agreement with earlier results from this laboratory [7] Table 5 further indicates that the effect of insulin on PDC activity requires the presence of glucose in the incubation me-

Table 5. Effect of insulin on PDC activity and formate oxidation in isolated fat cells

Experimental condition	Glucose added	PDH_a (in % of total)	^{14}C-formate oxidation % of the basal rate
Control	None	9.6 ± 1.9	
Insulin 2 mU/ml	None	10.1 ± 2.4	107 ± 10
Control	0.5 mM	12.2 ± 1.8	
Insulin 2 mU/ml	0.5 mM	36.1 ± 2.8	211 ± 25

Fat cells were incubated for 30 min at 37°C with [^{14}C]-formate and further treated as described in [26]. Mean values ± SEM from five experiments are given. Total PDC activity was 23.2 ± 2.8 mU/assay

dium. This is shown to hold also for the stimulation of H_2O_2 production, at variance with other reports where increased H_2O_2 production by insulin was found even in the absence of glucose in the medium [22]. The reason for this discrepancy is not clear. From our results it would appear that glucose has to serve as substrate for the insulin-stimulated process(es) of intracellular peroxide generation, and consequently activation of the PDC system. In line with this interpretation we have observed that no glucose is required for stimulation of PDC activity if H_2O_2 is used as activator [26].

In conclusion, our studies have confirmed that insulin in isolated adipocytes stimulates H_2O_2 production, and that H_2O_2 activates the PDH complex by conversion of PDH_b to PDH_a. Moreover we could show that isolated fat cell mitochondria are also susceptible to PDC activation by H_2O_2, and that tert. butyl hydroperoxide is effective at much lower concentrations than H_2O_2. Further we have established that, like in the case of insulin, a decrease in mitochondrial ATP can not explain the peroxide-dependent dephosphorylation of PDH_b.

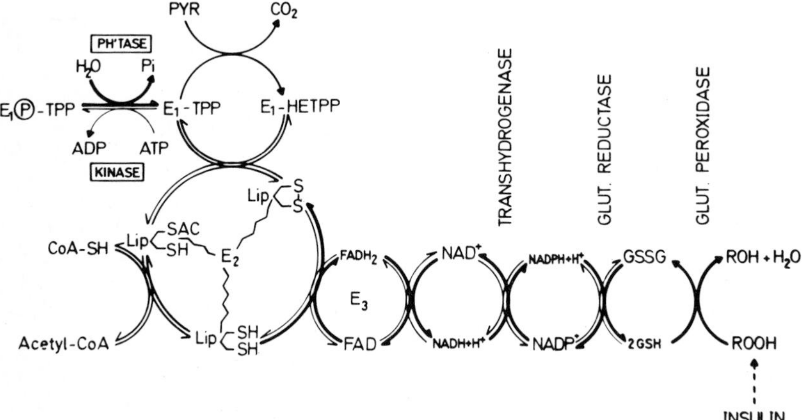

Fig. 6. Tentative scheme for the mechanisms involved in PDC activation by peroxide in adipose tissue. E_1, E_2, E_3 indicate the component enzymes of the PDH-complex: Pyruvate dehydrogenase, dihydrolipoate acetyltransferase, lipoate dehydrogenase, respectively. *TPP*, thiamine pyrophosphate; *HETPP*, hydroxyethylthiamine pyrophosphate. For further explanation see text

A tentative explanation of the sequence of reactions involved in the interaction of peroxide with the phosphorylation state of the PDC system is schematically outlined in Fig. 6. This mechanism is based on reported experimental evidence already mentioned before [10-12], indicating that the activity of PDH_a kinase (which leads to inactivation) is stimulated through the reduction and acetylation of the lipoyl moieties of the acetyltransferase core of PDC. Accordingly, an increase in the proportion of oxidized lipoyl residues would lower the activity of the kinase and thus lead to PDC activation. We have observed that tert. butyl hydroperoxide up to 15 mM does not change kinase activity when added to highly purified kidney PDC (Lynen, unpublished observations). This excludes direct oxidation of SH-groups by peroxide and suggests the participation of (an) enzyme catalyzed process(es). In our scheme (Fig. 6) oxidation of dihydrolipoamide (and S-acetyldihydrolipoamide) residues would involve coupling of the mitochondrial enzymes glutathione peroxidase, glutathione reductase and $NAD(P)^+$-transhydrogenase with the flavoprotein lipoate dehydrogenase of the PDH complex. However, the existence of another still unknown peroxidative pathway for regulation of the redox state of the lipoyl system of the PDH complex can not be excluded at present.

Our studies with tert. butyl hydroperoxide point to the possibility that not H_2O_2 itself but an intracellularly generated organic peroxide may be the postulated mitochondrial messenger of insulin. In conjunction with present available evidence our results lend substantial support to further research on the nature of this hypothetical compound and the underlying mechanisms of regulation of the activity of the PDH complex in adipose tissue.

Acknowledgements. This work was supported by the Deutsche Forschungsgemeinschaft, Bad Godesberg, FRG. The excellent technical assistance of Mrs. H. Schön and Miss G. Feil is much appreciated. Our thanks are also due to Dr. E.A. Siess for valuable discussions and suggestions during this work.

References

1. Löffler G, Bard S, Wieland OH (1975) FEBS Lett 60:269-274
2. Seals JR, McDonald JM, Jarrett L (1979) J Biol Chem 254:6991-6996
3. Seals JR, McDonald JM, Jarrett L (1979) J Biol Chem 254:6997-7001
4. Seals JR, Jarrett L (1979) Science 206:1407-1408
5. Seals JR, Czech MP (1980) J Biol Chem 255:6529-6531
6. Wieland OH, Weiss L, Löffler G, Brunner I, Bard S (1973) In: Fischer EH, Krebs EG, Neurath H, Stadtman ER (eds) Metabolic interconversion of enzymes. Springer, Berlin Heidelberg, New York, pp 117-129
7. Weiss L, Löffler G, Wieland OH (1974) Hoppe-Seyler's Z Physiol Chem 355:363-377
8. Paetzke-Brunner I, Schön H, Wieland OH (1978) FEBS Lett 93:307-311
9. Hucho F, Randall DD, Roche TE, Burgett MW, Pelley JW, Reed LJ (1972) Arch Biochem Biophys 151:328-340
10. Roche TE, Cate RL (1976) Biochem Biophys Res Commun 72:1375-1383
11. Cate RL, Roche TE (1978) J Biol Chem 253:496-503
12. Cate RL, Roche TE (1979) J Biol Chem 254:1659-1665
13. Pettit FH, Pelley JW, Reed LJ (1975) Biochem Biophys Res Commun 65:575-582
14. Kerbey AL, Randle PJ, Cooper RH, Whitehouse S, Pask HT, Denton RM (1976) Biochem J 154:327-348
15. Siess EA, Wieland OH (1976) Biochem J 156:91-102
16. Pettit FH, Roche TE, Reed LJ (1972) Biochem Biophys Res Commun 49:563-571

17. Siess EA, Wieland OH (1972) Eur J Biochem 26:96-105
18. Denton RM, Randle PJ, Martin BR (1972) Biochem J 128:161-163
19. Linn TC, Pettit FH, Reed LJ (1969) Proc Natl Acad Sci USA 62:234-241
20. Zuurendonk PF, Tager JM (1974) Biochim Biophys Acta 333:393-399
21. May JM, de Haën C (1979) J Biol Chem 254:9017-9021
22. Mukherjee SP, Mukherjee Ch, Lynn WS (1980) Biochem Biophys Res Commun 93:36-41
23. Chance B, Sies H, Boveris A (1979) Physiol Rev 59:527-605
24. Mukherjee SP, Lane RH, Lynn WS (1978) Biochem Pharmacol 27:2589-2594
25. May JM, de Haën C (1979) J Biol Chem 254:2241-2220
26. Paetzke-Brunner I, Wieland OH (1980) FEBS Lett 122:29-32

Discussion[2]

Answering the questions put forward Dr. Wieland pointed out that the effect of insulin on H_2O_2 production is linked to glucose metabolism rather than to its transport. As to the mechanism of action a distinction whether H_2O_2 itself or subsequently arising GSSG affects (in)directly the PDH-system is not yet possible. Lipid peroxidation in the surrounding of the PDH complex by highly reactive oxygen derivatives also has to be considered. Dr. Switzer commented that millimolar concentrations of H_2O_2 and NAD(P)H form a rather powerful reagent for the nonspecific inactivation of enzymes.

2 Rapporteur: Elmar A. Siess

Proteolytic Modification

Mast Cell Proteases

R.G. Woodbury[1] and H. Neurath[2]

1 Introduction

Although proteases are widely distributed in mammalian tissues and biological fluids, much of the early research in this area has concentrated on the study of the proteases of the digestive tract, notably the pancreas. Studies on these serine proteases provided information which added significantly to our basic understanding of the structure-function relationships of enzymes (Neurath and Bradshaw 1970). Also, comparison of the amino acid sequences of a variety of serine proteases from many diverse organisms provided data that had a major influence on formulating the fundamental concepts of the evolution of proteins in general (e.g., Dayhoff 1972; de Haën et al. 1975).

In recent years, it has become apparent that serine proteases are not restricted to the digestive tract. They are found in abundance in plasma as well as intracellularly in many types of tissues. Limited or extensive proteolysis due to serine proteases plays an important role in many biological processes including blood coagulation (Davie and Fujikawa 1975), complement-dependent immunity (Hatcher et al. 1977), fertilization (Morton 1972), and intracellular protein turnover (Schimke and Doyle 1970). In addition, it is apparent that imbalances in tissue proteolytic activities may result in diseases such as emphysema, muscular dystrophy, and rheumatoid arthritis (Barrett 1977). Thus, proteolysis is widely utilized in control mechanisms of many diverse biological functions and frequently is mediated by the relatively "simple" serine proteases (Neurath and Walsh 1976).

We have studied several proteases originating from the mast cells of various rat tissues. These intracellular, but probably secretory, enzymes include two serine proteases and a carboxypeptidase. The data obtained from studies of the mast cell serine proteases not only add to our understanding of the structure-function relationships of serine proteases, but they also give evidence that the mammalian serine proteases have evolved into several distinct groups including the digestive tract proteases originating from the pancreas, the coagulation enzymes produced in the liver, and the tissue proteases contained in granulocytes such as neutrophils and mast cells.

2 Mast Cell Proteases

The purification of two similar but distinct chymotrypsin-like serine proteases from rat tissues have been described previously (Woodbury and Neurath 1978; Everitt and Neu-

1 Institute for Cell and Tumor Biology, German Cancer Research Center, 6900 Heidelberg, FRG
2 Department of Biochemistry, University of Washington, Seattle, WA 98195, USA

rath 1979). These proteases, one isolated from the small intestine and the other from skeletal muscle, differ from each other mostly in their relative esterase and protease activities and in the conditions necessary for extracting them from tissues (Katunuma et al. 1975).

Studies of the localization of the protease from the small intestine by immunofluorescent and histological staining indicated that this enzyme was contained within cells distributed throughout the connective tissue of the intestinal mucosa (Woodbury et al. 1978a) and led to the identification of the protease-containing cells as the "atypical" mast cells, also known as the mucosal mast cells.

Immunological cross-reactivity and amino-terminal sequence identity (Woodbury et al. 1978b) provided proof that the serine protease obtained from skeletal muscle is in fact identical to the enzyme known as chymase obtained from peritoneal mast cells (Lagunoff and Benditt 1963). Both of these mast cell proteases had been isolated previously by several groups from a variety of rat tissues (Katunuma et al. 1975; Fräki and Hopsu-Haru 1975; Haas and Heinrich 1978). In most cases, the cellular locations of the enzymes were unknown, and as a result, the nomenclature for these two proteases is multifarious and confusing. We have chosen simply to refer to the protease from peritoneal mast cells as rat mast cell protease I (RMCP I) and to that from atypical mast cells as rat mast cell protease II (RMCP II). It will become apparent, however, that even these designations may be inadequate descriptions.

Because these two mast cell serine proteases have similar properties and originate from related cell types, one might be inclined to conclude that they carry out similar functions and that any differences in their properties reflect tissue specificity. If this is true, then a comparative analysis of the subtle differences between the structure and/or function of each protease may provide valuable information about the mechanism of their action. Such information should be particularly useful in identifying regions of the enzymes that are important in conferring unique protease specificity. Before such comparisons are seriously attempted it is obviously necessary to establish the physiological roles of the two mast cell enzymes. Although the functions of the enzymes are not known, it may be informative to compare properties of the two types of rat "mast" cells in order to establish how they relate to each other. A comparison of some of the known properties of these cells (Table 1) suggests that, in fact, these cells are not as closely related to each other as previously thought. Both cells contain similar serine proteases in basophilic secretory granules. On the basis of this criterion alone, however, the basophilic leucocyte of blood could also be called an "atypical" mast cell. Hence, it would seem inappropriate to refer to the basophilic granulocyte of mucosal tissues as an atypical or mucosal mast cell. Since it is not always possible to ascertain in retrospect precisely what kind of cell or type of protease have been investigated, comparison of the data obtained in different laboratories may lead to erroneous conclusions regarding the nature of the protease being studied. A comparison of the known properties of the two mast cell proteases (Table 2) shows that the enzymes share several physical and chemical properties with many other serine proteases, e.g., similar molecular weights and similar sensitivities to the inhibitor diisopropyl phosphorofluoridate (Katunuma et al. 1975). They also have similar chymotrypsin-like specificity toward ester substrates. There are, however, significant differences in the properties of these enzymes, most notably in their relative activities toward protein substrates and in the conditions needed for their ex-

Table 1. Comparison of properties of rat normal and atypical mast cells

Parameter	Normal cell	Atypical Cell
Location	Connective tissue [a] Peritoneum	Mucosal tissue [b]
Origin	Connective tissue [c]	Bone marrow [c]
Thymus dependence	No [d]	Yes [d]
Basophilic granules	Large, insoluble in H_2O [e]	Small, soluble in H_2O [e]
Histamine	High level [f]	Low level [f]
Serotonin	High level [f]	Low level [f]
Heparin	High level [f]	None [f]
Degranulating agents (Compound 48/80 etc.)	Effective [g]	No effect [g]
Proteases	RMCP I [h] Carboxypeptidase A [i]	RMCP II [j]

[a] Padawer 1978; [b] Miller and Jarrett 1971; [c] Kitamura et al. 1977; [d] Ruitenberg and Elgersma 1976; [e] Enerbäck 1966; [f] Whur and Gracie 1967; [g] Veilleux 1973; [h] Woodbury et al. 1978b; [i] Everitt and Neurath 1980; [j] Woodbury et al. 1978a

Table 2. Comparison of properties of RMCP I and RMCP II

Property	RMCP I	RMCP II
Esterase specificity	Chymotrypsin-like [a]	Chymotrypsin-like [a]
Specific activity BzTyrOEt – (units/mg)	58.3 [b]	8.5 [c]
pH optimum	8 – 9 [a]	8 – 9 [a]
Relative protease activity	Fair – Good [a]	Very poor [a]
Inactivated by DFP	Yes [a]	Yes [a]
Inactivated by TPCK	Yes, slowly [b]	No [c]
Inactivated by α_1-protease inhibitor	Yes [b]	Yes [c]
Conditions for extraction	1 M KCl [a]	0.15 M KCl [a]
Molecular weight	26,000 [b]	24,655 [d]
Inactive zymogen form	No [a,b]	No [a,c]

[a] Katunuma et al. 1975; [b] Everitt and Neurath 1979; [c] Woodbury and Neurath 1978; [d] Woodbury et al. 1978c

traction from tissues (Katunuma et al. 1975). This latter difference may prove to be related to the function of each protease. Since it is necessary to use 1 M KCl to extract significant amounts of RMCP I from the mast cell granules, it is likely that under conditions of physiologic ionic strength the protease remains tightly associated with secretory granules and therefore its action might be restricted to a relatively small region surrounding the affected mast cell. In contrast, RMCP II is readily extracted with water and may therefore freely diffuse away from secreted granules and function in tissues remote from the degranulated cell.

An enzyme which appears similar to bovine carboxypeptidase A has been recently isolated from peritoneal mast cell granules and partially characterized (Everitt and Neurath 1980). It appears to be present in the granules in fully active form and probably lacks a zymogen form. It is not known yet whether the mucosal basophilic granulocyte also contains a similar carboxypeptidase. Everitt (1980), however, has failed to detect positive immunofluorescent staining in the mucosal basophilic granulocytes using antiserum directed toward peritoneal mast cell carboxypeptidase.

3 Structure

When the amino acid compositions of the two "mast" cell serine proteases were compared (Woodbury et al. 1978b), it became clear that the major differences between them were the presence of two additional half-cystine residues in RMCP I and nearly twice as much lysine in RMCP I as compared with RMCP II. The high lysine content of RMCP I may explain why this protease associates so strongly with the highly sulfated acid mucopolysaccharide, heparin, that is present in mast cell granules (Lagunoff and Pritzl 1976). In contrast, the granules of the mucosal basophilic leukocyte do not contain heparin (Whur and Gracie 1967) and apparently this reflects the lower lysine level of RMCP II and the easy extractability of this protease.

The complete covalent structure of RMCP II is known (Woodbury et al. 1978c). There is clearly a homologous relationship between this enzyme and the pancreatic serine proteases (approximately 35% sequence identity relative to the bovine proteases). Various features of the structure of RMCP II that are either similar or different from the other mammalian serine proteases have been discussed in detail previously (Woodbury et al. 1978c; Woodbury and Neurath 1980). The most striking structural feature of the enzyme is the lack of a disulfide bond (cys 191-cys 220 in chymotrypsin) which is present in all other known serine proteases including those of bacterial origin (Olsen et al. 1970; Johnson and Smillie 1974; Jurásek et al. 1974). The potential importance of this structural alteration may be appreciated if one examines the regions of the protease molecule that are involved in enzymatic function. The sequence data presented in Fig. 1 illustrate that in other serine proteases this disulfide bond occurs in the areas of both the primary and the extended substrate binding regions. Thus, the absence of the disulfide bond in RMCP II may well influence the substrate specificity of the enzyme.

The sequence of the first 52 residues of the amino terminus of RMCP I is known (Woodbury et al. 1978b). When this sequence is compared to the same region in RMCP II, it is found that approximately 75% of the amino acid residues are identical and the substitutions which occur are extremely conservative. In contrast, only about 40% of the amino acid residues of bovine chymotrypsin are in identical positions in RMCP I or RMCP II.

Recently, the sequence of the first 23 amino acid residues of the amino terminus of cathepsin G from human neutrophils was determined (Travis et al. 1979). Twenty of these residues are identical to those of RMCP I (Fig. 2). Considering the unrelatedness

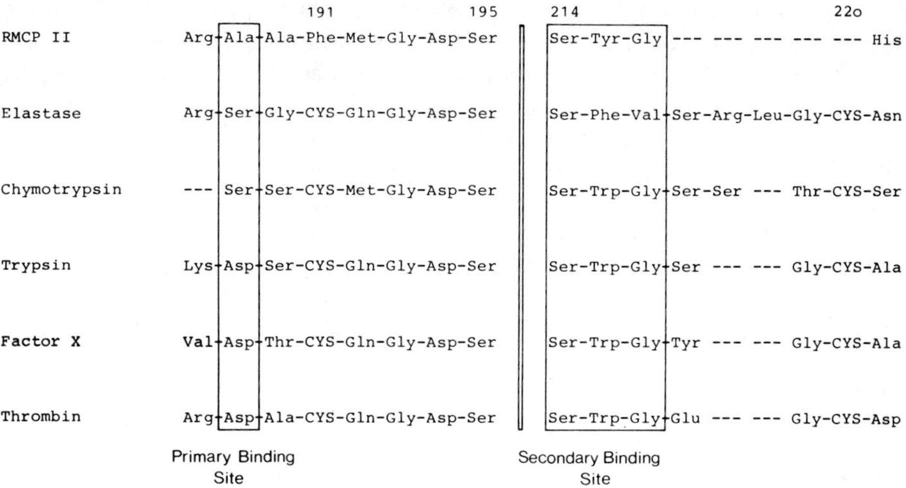

Fig. 1. Structures of several serine proteases in the vicinity of a disulfide bond (Cys 191 – Cys 220 in chymotrypsin) common to all known serine proteases except RMCP II. Half-cystine residues are in *capital letters* and the regions (primary and secondary) believed to participate in substrate binding are indicated by the *boxed areas. Gaps* (—) in the sequences have been inserted in order to optimize the alignment of residues among the proteases. The *numbers* designate the residues in chymotrypsin

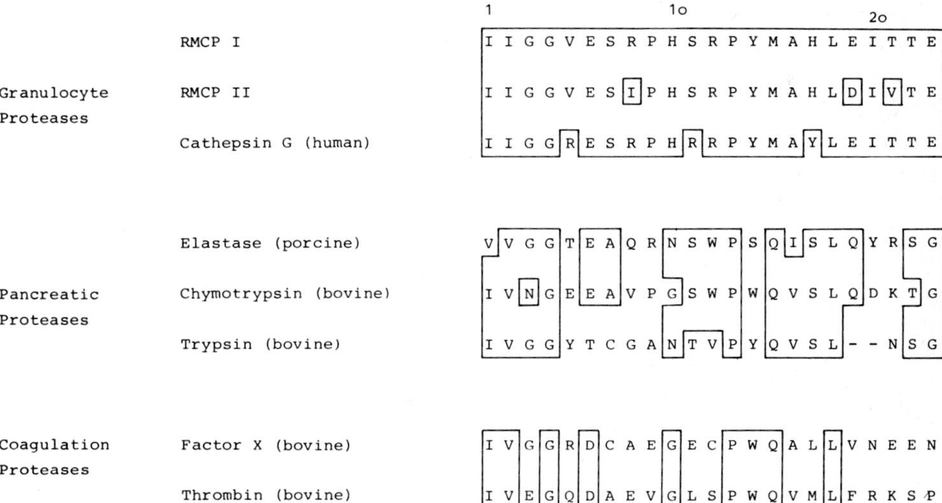

Fig. 2. Comparison of the amino-terminal sequences of a representative panel of mammalian serine proteases. On the basis of sequence similarities the proteases can be placed in three hypothetical groups of enzymes: the granulocyte proteases, represented by RMCP I, RMCP II and human cathepsin G; the pancreatic proteases, represented by porcine elastase and bovine chymotrypsin and trypsin; and the blood coagulation proteases represented by bovine factor X and thrombin. Sequence identity among the proteases is indicated by the *boxed* areas

of the species involved, this represents a remarkable degree of sequence identity. This sequence similarity would, of course, be expected to occur if cathepsin G and RMCP I were, in fact, the same enzyme synthesized by different granulocytes. Starkey (1977) suggested that this was the case because the proteases share many physical and chemical properties. A detailed analysis of the substrate specificities of the two proteases (Yoshida et al. 1980), however, clearly indicates that cathepsin G and RMCP I are different enzymes.

Except for the structures of RMCP I and RMCP II, the sequence information for human cathepsin G represents the only existing structural data for the intracellular serine proteases of granulocytes. Together these structural data suggest that intracellular secretory enzymes of this type may have followed a common evolutionary pathway which has led to a distinct group of serine proteases in the sense that the pancreatic serine proteases comprise a group that is distinctively different from the blood coagulation enzymes. Comparison of the amino-terminal sequences of a representative group of serine proteases is consistent with this hypothesis (Fig. 2). The extent of amino-terminal sequence identity among the first 23 residues of the group of the three granulocyte proteases, RMCP I, RMCP II, and cathepsin G, is about 85%. For the pancreatic enzymes, procine elastase, bovine chymotrypsin, and trypsin, the sequence identity is greater than 50%. The blood coagulation proteases, bovine factor X and thrombin, possess almost 40% sequence identity in this region.

The amino terminal region of the pancreatic serine proteases is apparently related to the specificity of the enzyme. Thus, as one progresses from lower to higher animals the sequences in this region of the chymotrypsins are well conserved and those of the trypsins also change very little. For example, frog chymotrypsin and bovine chymotrypsin have over 90% sequence identity in this region (Pies et al. 1980). The high degree of sequence identity of the three granulocyte proteases may be explained, in part, on the basis of their common chymotrypsin-like specificity. However, an intergroup comparison of the sequences of human cathepsin G and bovine pancreatic chymotrypsin reveals only 30% sequence identity, suggesting that the granulocyte proteases are distinctly different from the pancreatic enzymes. Similarly, cathepsin G and bovine thrombin possess less than 20% sequence identity in this region.

Other lines of evidence support the conclusion that the mammalian pancreatic, blood coagulation, and granulocyte serine proteases each form a distinct group of enzymes. The comparison of a number of properties of these three hypothetical groupings of mammalian serine proteases (Table 3) indicates that when several parameters are taken together the granulocyte proteases may be clearly distinguished from pancreatic or blood coagulation serine proteases. Such a distinction cannot be based on one or two properties alone. Thus, a knowledge of molecular weight and/or esterase specificity is not sufficient to distinguish a granulocyte protease from a pancreatic one. However, the inability to identify a zymogen form of a protease may indicate that the enzyme is of granulocyte origin since as far as is known the serine proteases of these cells are fully active in vivo (Lagunoff and Pritzl 1976; Starkey 1977). Similarly, other properties related to questions such as how restricted is the enzymatic activity (i.e., does the protease have a unique substrate or can it degrade many native proteins), how extensive is the proteolysis, how large is the activation peptide of a protease possessing a zymogen form, together with structural data may indicate in which group a newly discovered serine protease should be placed.

Table 3. Characteristic properties of three distinct classes of mammalian serine proteases

Property	Protease Pancreatic	 Blood Coagulation	 Granulocyte
Origin	Pancreas	Liver	Bone marrow [a] Connective tissue
Enzyme specificity	Low	High [b]	Intermediate/High [c,d]
Proteolytic effectiveness	Extensive	Limited [b]	Limited [c]
Molecular weight (approx.)	25,000 [e]	> 50,000 [b]	25,000-30,000 [f,g,h]
Zymogen form	Yes [i]	Yes [i]	No [c,f,g,h]
Size of activation peptide(s)	Small [i]	Large [i]	– –
Amino-terminal sequence identity (intragroup)	50%	40%	85%
Amino terminal sequence identity (intergroup)	30% (Chymotrypsin/ Factor X)	20% (Cathepsin G/ Thrombin)	30% (Cathepsin G/ Chymotrypsin)
Function	Digestion	Coagulation	Inflammation?

[a] Zucker-Franklin 1978; [b] Davies and Fujikawa 1975; [c] Katunuma et al. 1975; [d] Yoshida et al. 1980; [e] de Haën et al. 1975; [f] Woodbury and Neurath 1978; [g] Everitt and Neurath 1979; [h] Starkey 1977; [i] Neurath and Walsh 1976

Obviously, this list (Table 3) is not complete. For example, within the cell granulocyte serine proteases are bound to a variety of acid mucopolysaccharides (Davies et al. 1971; Yurt and Austen 1977) perhaps as a protective measure to confine the active enzymes. The protease-mucopolysaccharide interaction can be strong, often requiring concentrated salt solutions to extract the enzyme from the complex (Lagunoff and Pritzl 1976; Starkey 1977).

Similarity in function probably is an important reason why the pancreatic proteases or the blood coagulation enzymes as a group have many physical and chemical properties in common. Perhaps the similarities in the properties of the granulocyte serine proteases analogously indicate that these enzymes have related functions. Since granulocytes, along with monocytes, function at one level or another in promoting inflammation (Bloom 1965; Miller and Jarrett 1971; Starkey 1977; Zucker-Franklin 1978), it is conceivable that the proteases of these cells serve similar physiological purposes.

4 Substrate Specificity

Recently, the substrate specificities of RMCP I and RMCP II were studied using several different polypeptide derivatives of 4-nitroanilide (Yoshida et al. 1980). Although these proteases and human cathepsin G have similar chymotrypsin-like esterase activities, the proteolytic specificities of the rat enzymes are, in some respects, quite unlike those of either cathepsin G or bovine chymotrypsin.

The best substrates for RMCP I and RMCP II that have been tested (Table 4) have a hydrophobic residue at the P_1 subsite, as designated by Schechter and Berger (1967). The enzymes also prefer hydrophobic residues at subsites P_2 and P_3, indicating that these proteases require for optimal activity substrates with rather extended regions of hydrophobic structure. This requirement may explain why both RMCP I and RMCP II have limited effectiveness in degrading many native proteins (Katunuma et al. 1975). The substrate structure favored by these proteases is likely to occur in the interior of the native protein molecule, and thus may be inaccessible to the enzymes.

Clearly, as shown by the data in Table 5, both RMCP I and RMCP II show a marked preference for hydrophobic residues at the extended subsites of substrate. Substitution of a Gly for a Phe at subsite P_3 results in a significant loss in activity of both proteases which, in the case of RMCP II, is due to a 600-fold increase in K_m. Apparently the primary and secondary binding regions of RMCP I and RMCP II are hydrophobic and relatively extended. This may explain, in part, another aspect of the substrate specificity of these enzymes that appears to distinguish them uniquely from the other serine proteases. Yoshida et al. (1980) observed that peptide 4-nitroanilides containing proline at the P_3 position were good substrates for both RMCP I and RMCP II. For example, with the substrate Suc-Phe-Pro-Thr-Phe-4-nitroanilide for RMCP I, k_{cat}/Km was 73000 $M^{-1}s^{-1}$. In contrast, it was previously thought (Segal 1972; Powers and Tuhy 1973) that peptides of this type, containing a P_3 proline, could not be accepted by serine proteases as substrates because the prolyl residue distorted the configuration of the peptide backbone of the substrate in a way that would preclude favorable interaction at both the primary and secondary binding sites of the protease. This has been observed for chymotrypsin, elastase, and cathepsin G (Segal 1972; Powers and Tuhy 1973; Yoshida et al. 1980). The same structural feature that allows RMCP I and RMCP II to interact with substrates containing several hydrophobic and relatively large residues may also allow peptides containing a P_3 proline to interact favorably with the appropriate catalytic binding sites of the proteases.

With the best substrates (Table 4), the specificity constants, k_{cat}/Km, for RMCP I are about 50-100 times greater than those for RMCP II. As already noted, RMCP II lacks a disulfide bond in the vicinity of both the primary and secondary binding sites, which characteristically has been observed in serine proteases, and this structural feature may be related to the low level of activity of RMCP II toward protein substrates (Katunuma et al. 1975). RMCP I contains two additional half-cystines per molecule compared with RMCP II (Woodbury et al. 1978b). If RMCP I were to contain the disulfide bond lacking in RMCP II, then this would perhaps explain the significant difference in the proteolytic activities between the two enzymes. Knights and Light (1976) observed that trypsin, generated from trypsinogen containing a reduced and alkylated disulfide bond at this position, had a much lower activity than did trypsin generated from native tryp-

Table 4. Substrate specificities of RMCP I, RMCP II and human cathepsin G toward 4-nitroanilides

	RMCP I [a]			RMCP II [a]			Cathepsin G [b]		
Substrate	K_m (mM)	k_{cat} (s^{-1})	k_{cat}/K_m ($M^{-1}s^{-1}$)	K_m (mM)	k_{cat} (s^{-1})	k_{cat}/K_m ($M^{-1}s^{-1}$)	K_m (mM)	k_{cat} (s^{-1})	k_{cat}/K_m ($M^{-1}s^{-1}$)
Suc-Phe-Leu-Phe-NA [c]	0.023	19	830,000	0.57	10	18,000	0.62	0.95	1,500
Suc-Phe-Pro-Phe-NA	0.053	24	450,000	2.5	19	7,600	1.5	5.3	3,500
Suc-Phe-Pro-Thr-Phe-NA	0.15	11	73,000	13	5.4	420	1.0	0.047	47
MeOSuc-Ala-Ala-Pro-Phe-NA	0.42	10	24,000	1.4	0.53	380	–	–	–

[a] Reactions done in 50 mM phosphate buffer, pH 8.0, containing 10% DMSO at 25°C
[b] Reactions done in 0.1 M Hepes buffer, pH 7.5, containing 10% DMSO and 0.5 M NaCl at 25°C
[c] NA, 4-nitroanilide

Table 5. Substrate specificity of RMCP I and RMCP II – Effect of a hydrophobic residue at subsite P_3

Substrate	RMCP I	RMCP II		
		K_m	k_{cat}	k_{cat}/K_m
P_5 P_4 P_3 P_2 P_1	Specific activity (mol/min/mg)	(mM)	(s^{-1})	$(M^{-1}s^{-1})$
Suc-Ala-Phe-Leu-Phe-NA [a]	97	0.74	0.85	1800
Suc-Ala-Gly-Leu-Phe-NA [a]	2	270	2.80	10

[a] NA, 4-nitroanilide

sinogen. The Km values for the substrates of the modified trypsin were approximately 1000 times greater than those of native trypsin, yet the k_{cat} values were nearly the same for both enzymes. It may be significant, therefore, that RMCP II is a much less efficient protease than RMCP I, primarily because the Km values are much higher for the former enzyme compared with those of the latter (Table 4).

In summary, one would have to conclude that the general substrate specificities of RMCP I and RMCP II are the same. The specificities are significantly different from those of chymotrypsin and cathepsin G. The major difference in the proteolytic action of RMCP I and RMCP II is that the former is much more efficient than the latter. This may simply mean, however, that so far suboptimal substrates have been used to study the specificity of RMCP II.

5 Carboxypeptidase

It has been reported that rat peritoneal mast cells contain a carboxypeptidase activity which is associated with the secretory granules (Haas et al. 1979). Recently, Everitt and Neurath (1980) successfully purified this enzyme and reported its partial characterization. The enzyme has a specificity similar to that of bovine pancreatic carboxypeptidase A. Both enzymes also have similar molecular weights and amino acid compositions (Table 6). Unlike pancreatic carboxypeptidase A, the mast cell carboxypeptidase has an uneven number of half-cystine residues per molecule and therefore may have a free sulfhydryl group. Also, the mast cell carboxypeptidase has a substantially higher content of lysine than that of pancreatic origin. This may explain the requirement of concentrated salt solutions to extract the enzyme from the mast cell granule (Everitt and Neurath 1980) and suggests that the carboxypeptidase, as well as RMCP I, is bound tightly to heparin within the granules. The carboxypeptidase is fully active, and no procarboxypeptidase form has been detected.

6 Physiological Role

Because RMCP II is a recently discovered tissue protease, so far only a small effort has been made to determine its physiological function. It may play an important role in the

Table 6. Amino acid composition of mast cell carboxypeptidase and bovine pancreatic carboxypeptidase A

Amino acid	Mol Amino acid / Mast cell enzyme	Mol protein Carboxypeptidase A
Alanine	17	21
Arginine	16	11
Aspartic acid/NH_2	30	29
Glutamic acid/NH_2	18	25
Glycine	19	23
Half-cystine [a]	5	2
Histidine	9	8
Isoleucine	20	21
Leucine [b]	21	23
Lysine	27	15
Methionine	7	3
Phenylalanine	13	16
Proline	15	10
Serine [c]	28	32
Threonine [c]	19	26
Tryptophan [d]	9	7
Tyrosine	14	19
Valine	15	16
Total residues	306	307

[a] Determined as cysteic acid after performic acid oxidation. [b] Value after 96 h hydrolysis. [c] Value after extrapolation to zero time of hydrolysis.
[d] Determined after alkaline hydrolysis

defense against parasites which invade the intestinal and respiratory tracts since the mucosal basophilic granulocytes has been implicated in this immune process (Miller and Jarrett 1971).

During the two decades that RMCP I (chymase) has been studied several functions of the protease were proposed in order to explain, in part, the role of mast cells in inflammation (Fig. 3). In addition to the long-held belief that the secreted enzyme(s) promotes widespread proteolysis of tissue matrix proteins, there is evidence indicating that mast cell proteases may selectively degrade vascular basement membrane Type IV collagen (Sage et al. 1979), activate and/or inactivate vasoactive peptides (Seppä 1980; Everitt 1980), selectively degrade connective tissue proteolycans (Seppä et al. 1979), function in the granule secretory process (Lichtenstein and Osler 1966), and generate complement-dependent chemotactic peptides for other cells involved in inflammation (Hatcher et al. 1977) (although there is yet no proof that the enzyme responsible for this last observation originates from mast cells). However, none of these hypothetical roles has been shown to be an actual physiological function of mast cell protease.

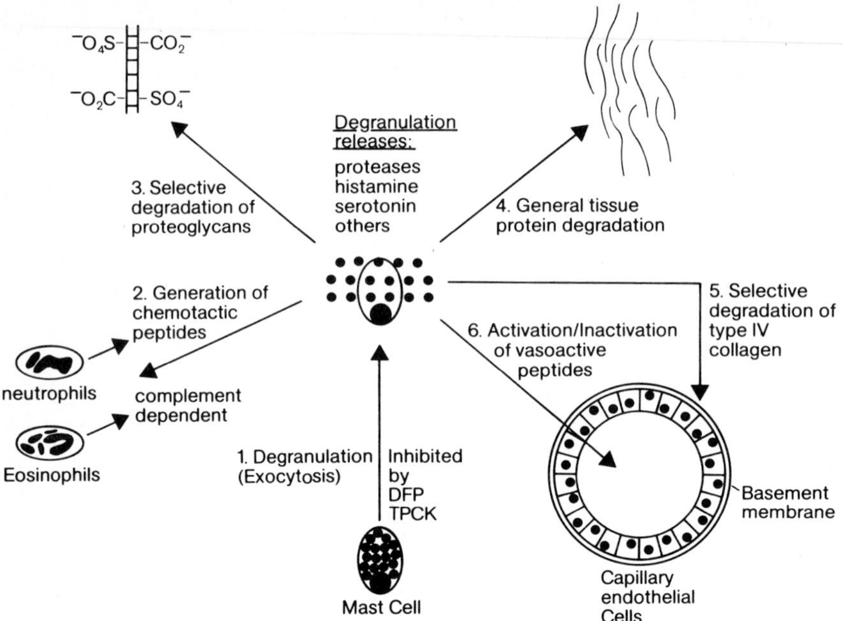

Fig. 3. Illustration of some of the proposed physiological functions of mast cell protease(s). References: degranulation: Lichtenstein and Osler 1966; chemotaxis: Hatcher et al. 1977; degradation of proteoglycans: Seppä et al. 1979; degradation of Type IV collagen: Sage et al. 1979; activation and/or inactivation of vasoactive peptides: Seppä 1980; Everitt 1980

Since RMCP I, as well as the carboxypeptidase, is fully active, not only as an esterase but also as a protease (Yurt and Austen 1977) within the mast cell granule, one must consider the possibility that the enzyme normally functions intracellularly. Mast cell degranulation is a pathological condition, and therefore any effects due to the action of the secreted proteases could be viewed as being abnormal. Austen (1978) has argued persuasively that, although the mast cell may function in the defense against parasites, an equally important role of these cells could be to help in the maintenance of tissue homeostasis of the microenvironment, particularly near small blood vessels.

The mast cell is frequently used as a model to study the process of exocytosis, yet it is not generally appreciated, that under physiological conditions the cell exhibits to a considerable degree phagocytosis and pinocytosis (Padawer 1978). Indeed, it has been observed by electron microscopy (Padawer 1978) that the granules of normal mast cells are often connected to the outside of the cell by a network of channels formed by the fusion of the plasma and perigranular membranes. The extensive network of channels connecting the granules to the outside of the cell could allow the passage of extracellular materials through the cell and over the granules where they may be acted upon by the protease and possibly by other enzymes so that the products could then diffuse out of the cell through the channels (Padawer 1978). In this manner the mast cell would serve as a biological filter or processor that aids the maintenance of tissue homeostasis.

Whether this hypothesis has any validity remains to be tested. It is important, however, that in attempts to establish the function(s) of the mast cell proteases, or of proteases of other granulocytes, we remember that their effects in a pathological state may be secondary to their roles in the normal one.

References

Austen KE (1978) J Immunol 121:793-805

Barrett AJ (ed) (1977) Proteinases in mammalian cells and tissues. Elsevier/North-Holland, Amsterdam New York

Bloom GD (1965) In: Zweifach BW, Grant L, McCluskey RT (eds) The inflammatory process. Academic Press, London New York, pp 355-388

Davie EW, Fujikawa K (1975) Annu Rev Biochem 44:799-829

Davies P, Rita GA, Krakauer K, Weissmann G (1971) Biochem J 123:559-569

Dayhoff MD (1972) Atlas of protein sequence and structure, vol 5. National Biomedical Research Foundation, Washington DC

de Haën C, Neurath H, Teller DC (1975) J Mol Biol 92:225-259

Enerbäck L (1966) Acta Pathol Microbiol Scand 66:289-302

Everitt MT (1980) Doctoral Dissertation. University of Washington, Seattle, Washington

Everitt MT, Neurath H (1979) Biochemie 61:653-662

Everitt MT, Neurath H (1980) FEBS Lett 110:292-296

Fräki JE, Hopsu-Haru VK (1975) Arch Derm Forsch 253:261-276

Haas R, Heinrich PC (1978) Eur J Biochem 91:171-176

Haas R, Heinrich PC, Sasse D (1979) FEBS Lett 103:168-171

Hatcher VB, Lazarus GS, Levine N, Burke PG, Yost FJ Jr (1977) Biochem Biophys Acta 483:160-171

Iodice AA, Leong V, Weinstock IM (1966) Arch Biochem Biophys 117:477-486

Johnson P, Smillie LB (1974) FEBS Lett 47:1-6

Jurásek L, Carpenter MR, Smillie LB, Gertler A, Levy S, Ericsson LH (1974) Biochem Biophys Res Commun 61:1095-1100

Katunuma N, Kominami E, Kobayshi K, Banno Y, Suzuki K, Chichibu K, Hamaguchi Y, Katsunuma T (1975) Eur J Biochem 52:37-50

Kitamura Y, Shimada M, Hatanaka K, Miyano Y (1977) Nature(London) 268:442-443

Knights RI, Light A (1976) J Biol Chem 251:222-228

Lagunoff D, Benditt EP (1963) Ann NY Acad Sci 103:185-197

Lagunoff D, Pritzl P (1976) Arch Biochem Biophys 173:554-563

Lichtenstein LM, Osler AG (1966) J Immunol 96:159-168

Miller HRP, Jarrett WFH (1971) Immunology 20:277-287

Morton DB (1972) In: Barrett AJ (ed) Proteinases in mammalian cells and tissues. Elsevier/North-Holland, Amsterdam New York, pp 445-500

Neurath H, Bradshaw RA (1970) Acc Chem Res 3:249-257

Neurath H, Walsh KE (1976) Proc Natl Acad Sci USA 73:3825-3832

Olsen MOJ, Nagabhushan N, Dzwiniel M, Smillie LB, Whitaker DR (1970) Nature (London) 228: 438-442

Padawer J (1978) In: Bach MK (ed) Immediate hypersensitivity, vol 7. Marcel Dekker Inc, New York, pp 301-307

Pies W, Zwilling R, Woodbury RG, Neurath H (1980) FEBS Lett 109:45-49

Powers JC, Tuhy PM (1973) Biochemistry 12:4767-4774

Ruitenberg EJ, Elgersma A (1976) Nature (London) 764:258-260

Sage H, Woodbury RG, Bornstein P (1979) J Biol Chem 254:9893-9900

Schechter I, Berger A (1967) Biochem Biophys Res Commun 27:157-162

Schimke RT, Doyle D (1970) Annu Rev Biochem 39:929-976

Segal DM (1972) Biochemistry 11:349-356
Seppä H (1980) Inflammatory 4:1-8
Seppä H, Väänänen K, Korhonen K (1979) Acta Histochem 64:64-70
Starkey PM (1977) In: Barrett AJ (ed) Proteinases in mammalian cells and tissues. Elsevier/North-Holland, Amsterdam New York, pp 57-89
Travis J, Giles PJ, Porcelli L, Reilly CF, Baugh R, Powers J (1979) CIBA Symp
Veilleux R (1973) Histochemie 34:157-161
Whur P, Gracie M (1967) Experientia 23:655-661
Woodbury RG, Neurath H (1978) Biochemistry 17:4298-4304
Woodbury RG, Neurath H (1980) FEBS Lett 114:189-196
Woodbury RG, Gruzenski GM, Lagunoff D (1978a) Proc Natl. Acad Sci USA 75:2785-2789
Woodbury RG, Everitt M, Sanada Y, Katunuma N, Lagunoff D, Neurath H (1978b) Proc Natl Acad Sci USA 75:5311-5313
Woodbury RG, Katunuma N, Kobayashi K, Titani K, Neurath H (1978c) Biochemistry 17:811-819
Yoshida N, Everitt MT, Woodbury RG, Neurath H, Powers JC (1980) Biochemistry 19:799-804
Yurt RW, Austen KF (1977) J Exp Med 146:1405-1419
Zucker-Franklin D (1978) In: Bach MK (ed) Immediate Hypersensitivity. Marcel Dekker Inc, New York, pp 407-430

Discussion[3]

Shaltiel and Fischer suggested looking for a possible involvement of these proteinases in anaphylaxis and related pathogenic mechanisms in humans.

Katunuma reported on own findings suggesting the involvement of RMCP I in the degranulation of mast cell granules. (The antihistaminic drug oxatoamide, which prevents degranulation and thereby histamine release by compound 48/80, inhibits the activity of RMCP I in vitro).

Söling argued that a direct participation of these proteinases in the exocytosis of mast cell granules would suggest that a similar type of proteinase would be present in other secretory cells.

Neurath replied that mast cells were known not only to undergo exocytosis but also pinocytosis and phagocytosis. Hence they could conceivably take up proteins from the external environment which bypass the granule-associated RMCP I, diffuse out into the medium and thus exert a homeostatic action. In this manner mast cells could be viewed as a biological filter.

Ballard asked for a possible area in the typical mast cell proteinase where basic amino acids would be concentrated.

Neurath answered that he had not yet the complete sequence of the typical mast cell proteinase, but that he would expect to have a domain which would be largely basic in character.

Heinrich pointed out that these proteinases were easily co-purified as contaminants since they would be associated with granules and therefore sediment along with other organelles. Moreover, they could not simply be inhibited by PMSF, since they would remain active as long as they were located within the granule.

Neurath – asked by Shaltiel – mentioned that histamine had a very small inhibitory effect, probably without physiological significance.

---

3 Rapporteur: Matthias Müller

Chemical Steps in the Selective Inactivation and Degradation of Glutamine Phosphoribosylpyrophosphate Amidotransferase in *Bacillus subtilis*

R.L. Switzer, D.A. Bernlohr, M.E. Ruppen, and J.Y. Wong[1]

1 Introduction

1.1 A Statement of the Problem

The mechanism and regulation of selective intracellular proteolysis present major unsolved problems to the field of biochemical regulation of enzyme activity. A number of useful experimental approaches may be taken toward solving these problems. One such approach is to characterize the proteolytic apparatus of the cell by isolation and biochemical characterization of individual proteolytic enzymes and their inhibitors. This approach is strengthened in studies with microorganisms by use of genetic and physiological analysis of protease-deficient mutants. A second means of studying proteolysis consists in labeling the proteins of the cell with a radioactive amino acid and determining the physiological variables that influence the release of the amino acid from proteins during their degradation in the presence of a large excess of the nonradioactive amino acid. A third approach, which we have chosen, is to focus on specific substrates of selective proteolytic events and to attempt to dissect the inactivation and degradation of a given protein into discrete biochemical steps. Ultimately, the objective is to reconstruct the degradation process from purified components in a manner that faithfully reproduces its regulation and physiological requirements.

1.2 The Experimental System: Sporulating Bacillus subtilis

A large number of instances of selective inactivation of enzymes in microorganisms have been described (Switzer 1977). Many of these involve proteolysis. We have chosen to study the inactivation and degradation of certain biosynthetic enzymes in *Bacillus subtilis*. As these bacteria enter stationary phase as a result of exhaustion of glucose (or the nitrogen source) from the growth medium, many of the enzymes of nucleotide and amino acid biosynthesis are selectively inactivated. Presumably this inactivation serves to commit the sporulating cell to synthesis of new proteins and RNA from amino acids and nucleotides derived from degradation of previously synthesized macromolecules. The inactivation of some of these enzymes is much more rapid than the turnover of bulk protein at the same phase of growth and precedes the extensive turnover of protein that occurs later during sporulation (Maurizi et al. 1978). Where it has been examined, the enzymes in question are stable in growing cells and are first inactivated at the end of ex-

1 Department of Biochemistry, University of Illinois, Urbana, IL 61801, USA

ponential growth. Thus, these processes provide instances of selective, physiologically regulated inactivation and – as will be seen – degradation.

1.3 The "Substrate": Glutamine PRPP Amidotransferase

This paper will concentrate exclusively on studies of the inactivation and degradation of the first enzyme of purine nucleotide biosynthesis de novo, glutamine phosphoribosylpyrophosphate (PRPP) amidotransferase. This enzyme catalyzes the following reaction:

$$\text{glutamine} + \text{PRPP} + H_2O \rightarrow \text{phosphoribosylamine} + \text{glutamate} + \text{PPi}.$$

Phosphoribosylamine is transformed in a series of steps to the purine nucleotide IMP, which is converted to the major end products of the pathway AMP and GMP. As would be expected, this enzyme is subject to tight metabolic regulation in *Bacillus subtilis*. Synthesis of the enzyme is repressed by addition of purines – most effectively by a mixture of adenosine plus guanosine – to the growth medium (Nishikawa et al. 1967). The enzyme displays a complex pattern of feedback inhibition by purine nucleotides (Meyer and Switzer 1979). Most important in that pattern are (1) very potent, cooperative inhibition by AMP and appreciably weaker inhibition by GMP, GDP, and ADP, (2) antagonism of inhibition by PRPP, and (3) strongly synergistic inhibition by several pairs of purine nucleotides. Like enzymes of de novo pyrimidine biosynthesis (Maurizi et al. 1978; Paulus and Switzer 1979), amidotransferase is subject to inactivation in the early stationary phase of growth (Turnbough and Switzer 1975a). It is this inactivation and subsequent degradation that is the focus of this paper.

2 The Oxygen-dependent Inactivation of Glutamine PRPP Amidotransferase

2.1 Inactivation of Amidotransferase in Vivo

The inactivation of amidotransferase in stationary phase *B. subtilis* cells was discovered and characterized by Turnbough and Switzer (1975a). It was shown that deprivation of the culture of oxygen interrupted inactivation of the enzyme, but a variety of inhibitors of electron transport or energy-yielding metabolism did not. Thus, oxygen itself appears to be required for inactivation. Inactivation proceeds normally in the presence of rifampin or chloramphenicol (antibiotic inhibitors of RNA and protein synthesis, respectively) and in protease-deficient mutant strains. Amidotransferase is probably stable in growing cells, because addition of adenosine to repress further synthesis did not result in a decline in the amidotransferase content of the cells until the cells ceased growing. This result implies that the inactivation is subject to physiological regulation.

2.2 Inactivation of Amidotransferase in Vitro Involves Oxidation of an Iron-Sulfur Center

The discovery by Turnbough that amidotransferase is inactivated by exposure to oxygen in cell extracts (Turnbough and Switzer 1975b) pointed the way to a detailed biochemical examination of the process. Since purified amidotransferases from avian liver had been shown to contain nonheme iron (Hartman 1963; Rowe and Wyngaarden 1968) and the crude *Bacillus* enzyme was sensitive to inhibition by iron chelators, it was suggested that oxidation of enzyme-bound iron might account for the inactivation in vitro (Turnbough and Switzer 1975b).

Development of efficient procedures for purification of *B. subtilis* amidotransferase to homogeneity (Wong et al. 1977, and Biochemistry, 1981, in press) allowed us to demonstrate that this suggestion was, to a major extent, correct. Purified amidotransferase is a brown-colored, oxygen-labile enzyme. Exposure of the enzyme to oxygen leads to loss of catalytic activity which is correlated with partial bleaching of the visible spectrum (Switzer et al. 1979). No additional oxygen-dependent "inactivation enzyme" is required for inactivation. The nature of the oxygen-sensitive site was clarified by the discovery that the enzyme contains not only iron, but inorganic sulfide (Wong et al. 1977). Identification of the structure of the iron-sulfur center was complicated by the difficulty of demonstrating conventional oxidation-reduction behavior of the center and the fact that the isolated enzyme generally contains iron and sulfur in somewhat less than the theoretical stoichiometry. However, it has been recently shown by iron-sulfur cluster displacement techniques and Mössbauer spectroscopy that amidotransferase contains [4Fe-4S] centers in the diamagnetic state (Averill et al. 1980). Such centers are formally analogous to those found in the oxidized forms of bacterial ferredoxins or the reduced forms of the high potential iron proteins (HiPIP) of photosynthetic bacteria (Yoch and Carithers 1979). The discovery of such a center in an enzyme that does not catalyze an obvious redox reaction was quite unanticipated. Subsequently, Ruzicka and Beinert (1978) reported that aconitase from mammalian mitochondria contains an iron-sulfur center. These discoveries open a new vista in the biochemistry of iron-sulfur proteins.

2.3 Chemical and Physical Changes Accompanying Oxidative Inactivation of Amidotransferase

The [4Fe-4S] center in amidotransferase is essential for catalytic activity (Wong et al. 1977), but it is not known whether it participates directly in catalysis. When the enzyme is exposed to oxygen, loss of catalytic activity, bleaching of the visible chromophore, and oxidation of enzyme-bound sulfide are coincident (Switzer et al. 1979). The following description of the reaction of the enzyme with oxygen was obtained from a careful study using purified enzyme and standardized reaction conditions (Bernlohr and Switzer, Biochemistry, 1981, in press):

All of the iron is converted to the high spin ferric form, which remains bound in a rather heterogeneous state to the protein.

All of the sulfide is oxidized. About one-third of the sulfide is found in the enzyme at the oxidation state of S^0 bound in the form of thiocystine. The remainder is converted to unidentified products, possibly SO_3^{2-} or SO_4^{2-}.

The three free sulfhydryl groups in the native enzyme are not altered. Of the four sulfhydryl groups originally liganded to the iron-sulfur center, two become free sulfhydryl groups and the other two form the thiocystine residue.

No other changes in the amino acid composition of the enzyme were detected upon standard amino acid analysis.

The visible circular dichroic spectrum of amidotransferase disappears on oxidation, which confirms that the iron-sulfur center has been drastically altered. The ellipticity in the ultraviolet region is also sharply reduced, suggesting changes in the polypeptide folding corresponding to loss of α-helix content.

The oxidized amidotransferase is much less soluble than the native enzyme and aggregates very readily, especially in the presence of Mg^{2+} ions.

Experiments with specific trapping enzymes or reagents indicated that the true oxidant is O_2, rather than peroxide, superoxide, hydroxyl radical or singlet oxygen. Inactivation proceeds normally in the dark and is not inhibited by thiols. No means of chemical reactivation of the oxidized enzyme has been found.

Altogether, the results indicate that reaction of amidotransferase results in a total destruction of the [4Fe-4S] center and major changes in the tertiary and quaternary structure of the enzyme. It is likely that the oxidized enzyme is more susceptible to proteolysis; preliminary experiments indicate that this is the case.

2.4 Regulation of the Oxidative Inactivation of Amidotransferase by Ligands

If, as was suggested in Sect. 2.1, the inactivation of amidotransferase is subject to physiological regulation, how is this regulation brought about? The early experiments of Turnbough and Switzer (1975b) on the inactivation of amidotransferase in crude extracts provided a possible solution. It was noted that some allosteric inhibitors, notably AMP, *stabilized* the enzyme against inactivation by oxygen while others, such as GMP, *destabilized* the enzyme. Availability of the purified enzyme has enabled a thorough study of the effects of ligands on the rate of inactivation. The decay in activity was exponential, so the results can be conveniently expressed as half-lives ($t_{1/2}$) of the enzyme under carefully standardized conditions (O_2-saturated Tris/HCl buffer, pH 7.9, 10 mM $MgCl_2$, 37°). The following results deserve emphasis:

Of a wide range of nucleotides tested only three markedly influenced the rate of inactivation at physiological levels. These were AMP, GMP, and GDP. A fourth nucleotide, ADP, was not effective alone, but was active as a partner in synergistic effects with GMP and GDP.

The only nucleotide to protect against inactivation was AMP, which increased the $t_{1/2}$ of the enzyme from 26 min to 170 min at 0.3 mM. Stabilization by AMP was very specific and was antagonized by the substrate PRPP.

GMP and GDP increased the rate of inactivation, reducing the $t_{1/2}$ values to 12 and 16 min, respectively.

Three pairs of nucleotides (ADP plus GMP, ADP plus GDP, and GMP plus GDP), which had been previously identified as very effective synergistic inhibitors of amidotransferase (Meyer and Switzer 1979), also decreased the $t_{1/2}$ values of amidotransferase to 6 and 12 min. The concentration of nucleotides required for maximal effects was 0.5 mM or less.

The effects of the stabilizing nucleotide AMP were antagonized by the destabilizing nucleotides. Thus, with appropriate mixtures of nucleotides the $t_{1/2}$ of amidotransferase could be manipulated between 6 min and 180 min under standard conditions.

While all of the effectors of the rate of oxidative inactivation of amidotransferase were also allosteric inhibitors of the enzyme, there was no simple correlation between inhibition and stabilization or destabilization. It should be stressed that the concentrations of nucleotides needed for half-maximal effects on the rates of inactivation of amidotransferase were from 10- to 40-fold lower than the concentration needed to inhibit the enzyme by 50%. Thus, these effectors are capable of regulating the stability of a catalytically active – as opposed to an inhibited – form of the enzyme, as would be expected if such regulation is physiologically significant.

The effects of substrates and products of amidotransferase on the rate of inactivation by oxygen were relatively small. Glutamine alone had no effects; glutamine plus PRPP showed stabilization to a $t_{1/2}$ of 42 min. PRPP (but not glutamine) had the ability to modulate nucleotide effects, e.g., by antagonizing stabilization by AMP and both antagonizing and enhancing destabilization by various other nucleotides.

We suggest that the stability of amidotransferase to inactivation in vivo by oxygen is regulated by some combination of these effects, primarily the effects of purine nucleotides. This suggestion is certainly not proven, however. Experiments attempting to correlate the stability of amidotransferase in vivo with the pools of the proposed effectors of stability are in progress.

3 The Degradation of Glutamine PRPP Amidotransferase in Vivo

3.1 Immunochemical Evidence that Inactivation of Amidotransferase is Accompanied by Degradation (Ruppen, Wong, and Switzer, unpublished experiments)

The catalytic activity of amidotransferase is lost in an oxygen-dependent process both in vivo and in vitro. We wish now to pose the following question: Does an inactive, oxidized form of the enzyme accumulate inside the cells in vivo? Alternatively, is the enzyme degraded? We have attempted to answer this question by preparing monospecific antibodies to both native and oxidized amidotransferase using the purified native and in vitro oxidized enzyme. These antibodies have been used to determine whether immunochemically cross-reactive forms of amidotransferase remain in cells that have lost enzymatic activity. A detailed description of this approach can be found in the studies of Maurizi et al. (1978). The antibodies against native and oxidized amidotransferase were shown to cross-react with each type of antigen. The antibodies were also demonstrated to be capable of detecting enzyme that had been partially degraded by trypsin treatment. The loss of cross-reactive protein from cells in which the inactivation of amidotransferase was proceeding was followed by three methods: (1) microcomplement fixation; (2) direct immunoprecipitation of antigen from cells labeled with [^{3}H]leucine, followed by analysis of the washed immunoprecipitate by SDS polyacrylamide gel electrophoresis, and (3) a procedure as described in (2) except that double precipitation with goat anti-rabbit IgG antibody was used to detect possible soluble complexes. In all cases the loss of cross-reactive protein from the cells was coincident or only very slightly slower than the loss of amidotransferase activity. No immunoprecipitable species smaller than the normal subunit was detected on SDS gel electrophoresis of immunoprecipitates. Given the experiments already quoted concerning the specificity and versatility of the antibody preparations used, the results provide strong presumptive evidence for the concomitant inactivation and degradation of amidotransferase to an immunochemically unrecognizable form in vivo.

3.2 Physiological Characteristics of the Degradation of Amidotransferase in Vivo
(Ruppen and Switzer, unpublished experiments)

Using the methods described in the preceding paragraph, it is possible to determine whether the physiological properties of the degradation of amidotransferase differ from those already described (Sect. 2.1) for the inactivation of amidotransferase. These studies are still in a very early stage, but three significant differences have already been found.

The disappearance of cross-reactive material is significantly slowed by addition of fluoroacetate, an inhibitor of the tricarboxylic acid cycle, which is the major source of metabolic energy in these bacteria at this phase of growth. This was the case even when fluoroacetate was added during the inactivation. The loss of amidotransferase activity proceeded normally under these conditions, as previously reported (Turnbough and Switzer 1975b).

The rate of loss of cross-reactive protein was also decreased, but clearly not blocked, in a protease-deficient mutant strain, S-87 (Hageman and Carlton 1973). This strain is deficient in the major known intracellular serine protease of *B. subtilis*, which is produced in large amounts in stationary phase cells and appears to play a role in the turnover of bulk protein in this phase (Maurizi and Switzer 1980).

The disappearance of protein that is cross-reactive with amidotransferase was strongly inhibited by the addition of chloramphenicol to the culture 0.5 h before the inactivation began. Inactivation of amidotransferase activity was unaffected, as previously shown (Turnbough and Switzer 1975a). Addition of chloramphenicol at a later time when the amidotransferase activity had declined to 50% of its maximal value did not block the loss of cross-reactive protein. Most interestingly, preliminary experiments indicate that the inactive immunoprecipitate protein has a molecular weight substantially less than the native protein.

These experiments, while very preliminary, indicate that amidotransferase is indeed degraded in vivo, that the metabolic requirements for inactivation and degradation are different, and that dissection of the degradation process into individual steps may be possible. Such experiments should provide important criteria for evaluating the physiological significance of any in vitro reconstruction of the degradation that may be developed.

4 Discussion

A very generalized scheme for the degradation of proteins is postulated below.

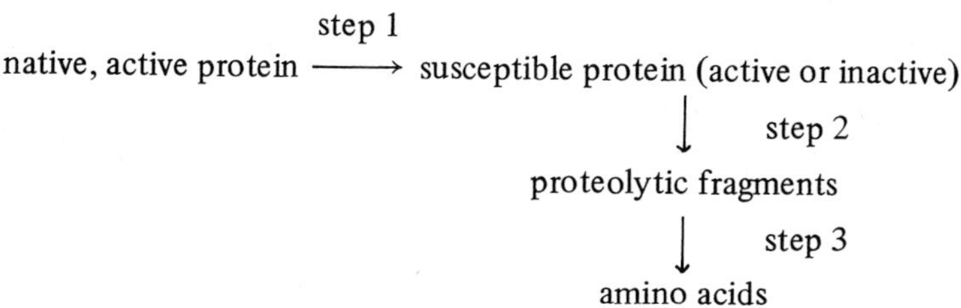

It is suggested that step 1, conversion of the active protein to a form that is much more susceptible to proteolysis, may proceed by one of several mechanisms. It could involve a conformational change or an alteration in the quaternary structure. In other cases a covalent modification, such as a phosphorylation or attachment of another polypeptide (Hershko et al. 1980) may occur. In the case of glutamine PRPP amidotransferase from *B. subtilis* we propose that step 1 involves the irreversible oxidation of an iron-sulfur prosthetic group by molecular oxygen. In all cases it is to be expected that step 1 should be regulated by appropriate metabolites. These metabolites could obviously act by regulating the activity of an enzyme that catalyzes step 1, but in many cases they may act by binding to substrate, allosteric, or special sites on the target protein itself. We have suggested here that the allosteric effectors and possibly PRPP play this latter role in regulating the inactivation of amidotransferase. At the present time this suggestion is not supported by adequate evidence, but it provides a useful model for further experimentation.

Very little is known about steps 2 and 3 in any system, except that they generally seem to be appreciably faster than step 1. It must be anticipated that a number of endoproteases and peptidases will be involved. The initial proteolytic steps provide a basis for additional physiological control of proteolysis, such as the regulation of bulk protein degradation by ppGpp in *E. coli* proposed by Voellmy and Goldberg and their associates (1980). A number of specific proteases with defined target proteins or groups of target proteins, such as the spore-specific protease of *B. megaterium* described by Dignam and Setlow (1980), probably exist, but it seems unlikely that specific proteases are elaborated for each target protein that is subject to selective degradation. Very little is known about the peptidases involved in step 3, but it is likely that these show little specificity other than for the residues cleaved. The genetic experiments of Miller (1975) have provided the best available picture of the peptidase composition of a microbial system.

Our studies with amidotransferase have as yet done little to characterize steps 2 and 3 for this system, except to show that they can apparently be separated from step 1 and may be analyzed into a sequence of steps by the use of inhibitors of metabolism and use of mutant strains. It should be made clear that it has not yet been demonstrated conclusively that the inactive, oxidized form of amidotransferase as prepared in the laboratory is actually formed transiently in vivo. Neither has it been shown that fragments of amidotransferase can be accumulated in chloramphenicol-treated cells, although this seems likely. An intense study of the degradation of amidotransferase in vitro, using purified components, will be required to complete our understanding of this particular instance of selective proteolysis. It is to be hoped that the results will provide insight into the general problem of selective proteolysis in all cell types.

Acknowledgments. This research has been supported by Grant No. AI11121 from the National Institute of Allergy and Infectious Diseases, U.S. Public Health Service.

References

Averill BA, Dwivedi A, Debrunner P, Vollmer SJ, Wong JY, Switzer RL (1980) Evidence for a tetranuclear iron-sulfur center in glutamine phosphoribosylpyrophosphate amidotransferase from *Bacillus subtilis.* J Biol Chem 255:6007

Dignam SS, Setlow P (1980) *Bacillus megaterium* spore protease. J Biol Chem 255:8408

Hageman JH, Carlton BC (1973) Effects of mutational loss of specific intracellular proteases on the sporulation of *Bacillus subtilis.* J Bacteriol 114:612

Hartman SC (1963) Phosphoribosylpyrophosphate amidotransferase. J Biol Chem 238:3024

Hershko A, Ciechanover A, Heller H, Haas AL, Rose IA (1980) Proposed role of ATP in protein breakdown. Proc Natl Acad Sci USA 77:1783

Maurizi MR, Switzer RL (1980) Proteolysis in bacterial sporulation. Curr Top Cell Regul 16:163

Maurizi MR, Brabson JS, Switzer RL (1978) Immunochemical studies of the inactivation of aspartate transcarbamylase by stationary phase *Bacillus subtilis* cells. J Biol Chem 253:5585

Meyer E, Switzer RL (1979) Regulation of *Bacillus subtilis* glutamine phosphoribosylpyrophosphate amidotransferase activity by end products. J Biol Chem 254:5397

Miller CG (1975) Peptidases and proteases of *Escherichia coli* and *Salmonella typhimurium.* Annu Rev Microbiol 29:485

Nishikawa H, Momose H, Shiio I (1967) Regulation of purine nucleotide synthesis in *Bacillus subtilis.* J Biochem (Tokyo) 62:92

Paulus TJ, Switzer RL (1979) Synthesis and inactivation of the carbamyl phosphate synthetase isozymes of *Bacillus subtilis* during growth and sporulation. J Bacteriol 140:769

Rowe PB, Wyngaarden JB (1968) Glutamine phosphoribosylpyrophosphate amidotransferase. J Biol Chem 243:6373

Ruzicka FJ, Beinert H (1978) The soluble "high potential" type iron-sulfur protein from mitochondria is aconitase. J Biol Chem 253:2514

Switzer RL (1977) The inactivation of microbial enzymes in vivo. Annu Rev Microbiol 31:135

Switzer RL, Maurizi MR, Wong JY, Flom KJ (1979) Selective inactivation and degradation of enzymes in sporulating bacteria. In: Atkinson DE, Fox CF (eds) Modulation of protein function. Academic Press, London New York, p 65

Turnbough CL, Switzer RL (1975a) Oxygen-dependent inactivation of glutamine phosphoribosylpyrophosphate amidotransferase in stationary-phase cultures of *Bacillus subtilis.* J Bacteriol 121:108

Turnbough CL, Switzer RL (1975b) Oxygen-dependent inactivation of glutamine phosphoribosylpyrophosphate amidotransferase in vitro. J Bacteriol 121:115

Voellmy R, Goldberg AL (1980) Guanosine-5'-diphosphate-3-diphosphate (ppGpp) and the regulation of protein breakdown in *Escherichia coli.* J Biol Chem 255:1008

Wong JY, Meyer E, Switzer RL (1977) Glutamine phosphoribosylpyrophosphate amidotransferase from *Bacillus subtilis.* J Biol Chem 252:7424

Yoch DC, Carithers RP (1979) Bacterial iron-sulfur proteins. Microbiol Rev 43:384

Discussion[2]

Stadtman asked whether the effect of oxidation could also be due to a radical-mediated modification of some amino acids.

Switzer pointed out that they had not found any altered amino acid in the oxidized form of the protein and that the addition of hydroxyl radical traps, superoxide-dismutase, catalase and had been without effect on the rate of inactivation of the enzyme.

Katunuma reported on his own finding that oxidation of the Fe^{2+} in the Gln-PRPP-amidotransferase from hepatoma was paralleled by a loss of activity.

Switzer replied that the *Bacillus subtilis*-amidotransferase definitely contains a [4Fe-4S] center rather than Fe^{2+} only. The situation with the human and avian enzymes is still unclear (Itakura M, Holmes EW (1979) J Biol Chem 254:333-338; Wyngaarden JB (1972) Curr Top Cell Regul 5:135-176).

Holzer asked for the mechanism of action of chloramphenicol with respect to the accumulation of intermediates in the inactivation.

Switzer speculated that chloramphenicol either blocked the derepression of the responsible proteinase in the stationary phase or altered the accumulation of guanosine-tetraphosphate as had been shown by Goldberg and others (Goldberg AL, St John AC (1976) Ann Rev Biochem 45:747-803). Guanosine-tetraphosphate could play an important role in the regulation of a pre-existing proteinase.

Switzer – when asked by Snell – reported that they could remove the Fe-S center by treating with o-phenanthroline which is accompanied by inactivation. Only very small amounts of activity were regenerated by treating with iron, inorganic sulfide, and thiols. No evidence exists that the enzyme is a bifunctional one.

Ballard asked whether cross-reacting material would be conserved in highly oxygenated cultures of a proteinase-deficient mutant.

Switzer answered that they could end up with a condition where one has about 60% protein present as cross-reacting material with no remaining activity.

Severin asked for the mechanism of AMP binding.

Switzer answered that there is a very specific allosteric site for AMP at which not even closely related structural analogues would bind.

2 Rapporteur: Matthias Müller

Inactivation and Turnover of Fructose-1,6-Bisphosphatase from *Saccharomyces cerevisiae*

M.J. Mazón, J.M. Gancedo, and C. Gancedo[1]

1 Introduction

The commonest way to regulate the amount of an enzyme in microorganisms is to adjust the rate of synthesis to the particular metabolic situation prevailing in the environment. Nevertheless, in some cases the amount of enzyme is regulated by inactivating the existent protein. The term inactivation will be used in this article to mean the disappearance of an enzymatic activity more rapidly than repression of its synthesis and dilution by growth could account for. This concept is illustrated in Fig. 1. In principle, this definition does not make any assumption on the reversible or irreversible nature of the process.

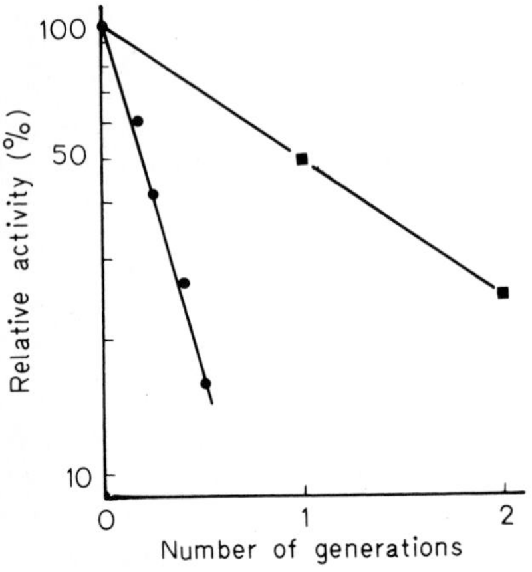

Fig. 1. Inactivation of fructose-1,6-bisphosphatase by glucose. ●—●, observed values; ■—■, calculated on the basis of repression and dilution by growth

In yeast many cases of inactivation have been observed when glucose is added to cultures growing on noncarbohydrate carbon sources (Table 1). By analogy with the existing term of catabolite repression, these inactivation phenomena caused by glucose are designated as catabolite inactivation (Holzer 1976).

In this report we present some of our results obtained during the study of the catabolite inactivation of the fructose-1,6-bisphosphatase from the yeast *Saccharomyces cerevisiae.*

1 Instituto de Enzimologia del C.S.I.C., Facultad de Medicina de la Universidad Autónoma, Madrid, Spain

Table 1. Some cases of enzyme inactivation in yeasts

Enzyme	Inactivated by	Reference
"Galactozymase"	Glucose	Spiegelman and Rainer (1947)
Malate dehydrogenase	Glucose	Duntze et al. (1968)
Fructose bisphosphatase	Glucose	Gancedo (1971)
Phosphoenol pyruvate carboxykinase	Glucose	Gancedo and Schwerzmann (1976)

2 Results and Discussion

Fructose-1,6-bisphosphatase is the terminal enzyme of the gluconeogenic pathway in microorganisms. In yeast the synthesis of the enzyme is repressed by glucose (Gancedo and Gancedo 1971) and, in addition, when glucose (or the related sugars fructose or mannose) is added to a culture growing on gluconeogenic precursors the existing enzyme is rapidly inactivated (Gancedo 1971). The kinetic of this inactivation in a culture growing on a *minimal salts* medium is shown in Fig. 2.

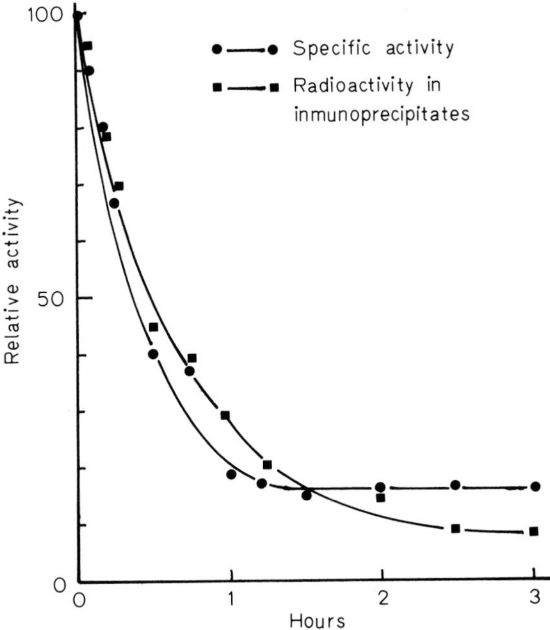

Fig. 2. Time course of the inactivation of fructose-1,6-bisphosphatase. At zero time 1% glucose was added to the culture and samples withdrawn at the indicated times

The inactivation seemed irreversible by a variety of indirect approaches (Gancedo 1971), however direct evidence of the irreversibility could be only provided when antibodies against the purified enzyme were available (Funayama et al. 1979). Figure 3 shows the result of an experiment where radioactively labeled fructose-1,6-bisphosphatase was inactivated and transferred to unlabeled medium with ethanol as carbon source. The reappearance of enzyme activity and of cross-reacting material was followed in parallel. No labeled cross-reacting material was found, thus indicating irreversibility of the process.

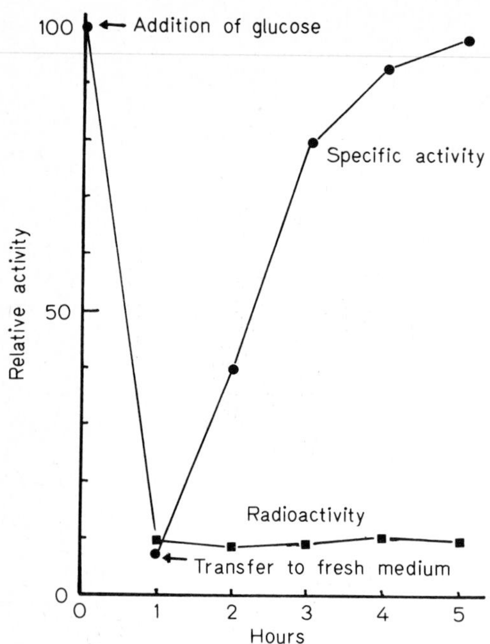

Fig. 3. Irreversibility of the inactivation of fructose-1,6-bisphosphatase. Glucose added as in Fig. 2

The fate of the protein was followed using antibodies also during the course of inactivation to try to gain an insight into the mechanism of the process. As it may be seen in Fig. 2, the cross-reacting material disappears almost in parallel with the enzymatic activity even if samples are taken at a very short time after the addition of glucose. This fact together with the irreversibility of the process makes proteolysis a likely mechanism to explain the observations made. Indeed, the concentration of an enzyme is always a balance between its synthesis and its degradation. If the rate of degradation of fructose-1,6-bisphosphatase were always high, that is if the enzyme has a high turnover rate, the rapid disappearance of activity upon glucose addition could be explained simply by an immediate turning-off of the synthesis without changing the rate of degradation. To examine this possibility, the half-life of the enzyme in different metabolic situations was calculated (Table 2). As it can be seen, the enzyme is highly stable during growth so that continuous turnover cannot explain its behavior.

Table 2. Rate constants of degradation for yeast FbPase

Metabolic conditions	Rate constant of degradation (h^{-1})	Half-life (hours)
Growing on ethanol	0.008	87
Inactivation by glucose	1.6	0.5
Resynthesis after inactivation	0.6	1.2

At this point Lenz and Holzer (1980) reported an interesting observation on the problem. When yeast was grown on *rich medium* with glucose and then left to utilize the ethanol produced in the fermentation of the sugar, the characteristics of the inactivation of fructose-1,6-bisphosphatase by glucose were different (Fig. 4). A very rapid inactivation was

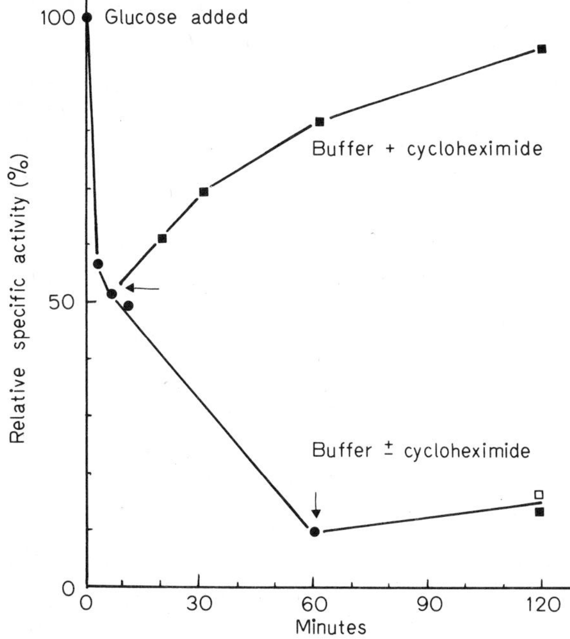

Fig. 4. Reversible and irreversible inactivation of fructose-1,6-bisphosphatase. Conditions as Lenz and Holzer (1980). At the times shown by the *arrows* the yeasts were resuspended in new media as indicated

observed during the first 5 min after the addition of glucose, and this inactivation seemed to be reversible since the activity could be recovered in the presence of cycloheximide, an inhibitor of protein synthesis in yeast. If the time of action of the glucose is prolonged, the inactivation becomes irreversible. To study the nature of the reversible part of the inactivation we measured the incorporation of [^{32}P] in the fraction precipitated by antibodies against fructose-1,6-bisphosphatase. As it can be seen in Table 3, an incorporation was found that paralleled the observed inactivation. We cannot at present ascertain the nature of this incorporation although we are tempted to speculate that it consists in a

Table 3. Incorporation of [^{32}P] in FbPase during reversible inactivation

Time (min)	Activity (%)	[^{32}P] incorporation (cpm in FbPase)
0	100	< 200
1	55	6100
3	61	5200
6	61	4200
9	59	6000

Table 4. Growth on glucose of a mutant that does not show inactivation of fructose-1,6-bisphosphatase by glucose

Transfer form	to	Growth	FbPase mU/mgp
Ethanol	Glucose	Negative [a]	62
Galactose	Glucose	Negative [a]	< 1

[a] Generation time of parental strain in these conditions approx. 90 min

phosphorylation. Preliminary evidence shows that reactivation is accompanied by a loss of the bound radioactivity.

It was difficult to visualize how a stable protein could be rapidly degraded without undergoing a previous change in structure. The possible existence of a phosphorylated form of the enzyme makes the whole process easier to understand. First the protein could be modified and then the modified form would be proteolyzed. It should be mentioned however that attractive as this sequence may appear we have not yet definite proof of its working in vivo. Indeed evidence is lacking to establish if the inactivation observed in the two different metabolic conditions (minimal ethanol vs rich glucose) occurs by the same mechanism or is due to different processes. Further work would (we hope!) help to clarify these points.

The physiological significance of the inactivation seems to be to avoid futile cycling between fructose-1,6-bisphosphatase and phosphofructokinase. (Why a simple allosteric inhibition of fructose-1,6-bisphosphatase by a metabolite that is low in gluconeogenesis and high in glycolysis, e.g., fructose-1,6-bisphosphate, has not been favored by evolution is an open question.) However, the only information on the possible influence of a working cycle at this level in yeast was derived from a mutant of *S. carlsbergensis* isolated some years ago (Van de Poll et al. 1974). This mutant does not grow on glucose and its fructose-1,6-bisphosphatase is not inactivated by addition of this sugar to the culture. The lack of growth was therefore related to the operation of a futile cycle phosphofructokinase-fructose-1,6-bisphosphatase. To test this hypothesis, we devised conditions in which the cycle could be avoided, and tested the growth on glucose in these conditions. It can be seen in Table 4 that even in conditions where the cycle cannot operate, due to the absence of fructose-1,6-bisphosphatase, the growth on glucose is negative. This result disqualifies in our opinion this mutant as supporter of the futile cycle hypothesis (that may nevertheless be true).

Work is in progress toward a deeper understanding of the complex regulation of this enzyme in *S. cerevisiae.*

References

Duntze W, Neumann D, Holzer H (1968) Glucose induced inactivation of malate dehydrogenase in intact yeast cells. Eur J Biochem 1:21-25

Funayama S, Molano J, Gancedo C (1979) Purification and properties of a D-fructose 1,6 bisphosphatase from *S. cerevisiae*. Arch Biochem Biophys 197:170-177

Gancedo C (1971) Inactivation of fructose 1,6 bisphosphatase by glucose in yeast. J Bacteriol 107: 401-405

Gancedo JM, Gancedo C (1971) Fructose 1,6 diphosphatase, phosphofructokinase and glucose 6 phosphate dehydrogenase from fermenting and non-fermenting yeast. Arch Microbiol 76:132-138

Gancedo C, Schwerzmann K (1976) Inactivation by glucose of phosphoenolpyruvate carboxykinase from *S. cerevisiae*. Arch Microbiol 109:221-225

Holzer H (1976) Catabolite inactivation in yeast. Trends Biochem Sci 1:178-181

Lenz AG, Holzer H (1980) Rapid reversible inactivation of fructose 1,6 bisphosphatase in *S. cerevisiae* by glucose. FEBS Lett 109:271-274

Spiegelman S, Rainer JM (1947) The formation and stabilization of an adaptive enzyme in the absence of its substrate. J Gen Physiol 31:175-193

Van de Poll KW, Kerkenaar A, Schamhart DHJ (1974) Isolation of a regulatory mutant of fructose 1,6 diphosphatase in *S. carlsbergensis*. J Bacteriol 117:965-970

Catabolite Inactivation of Phosphoenolpyruvate Carboxykinase and Aminopeptidase I in Yeast

M. Müller and J. Frey[1]

1 Catabolite Inactivation of Enzymes in Yeast

Catabolite inactivation of yeast enzymes provides a typical example for the control of enzyme activity by selective inactivation. This inactivation has been defined by Holzer (1976) as the rapid irreversible loss of catalytic activity following the addition of glucose or metabolically related sugars to yeast cells, which had been starved for carbon or grown on a nonsugar carbon source.

Phosphoenolpyruvate carboxykinase (PEPCK) is subject to this catabolite inactivation (Haarasilta and Oura 1975; Gancedo and Schwerzmann 1976) like other enzymes which are involved in the formation of glucose either by gluconeogenesis or by the uptake of sugars (for a recent review on catabolite inactivation see Müller and Holzer 1981).

Aminopeptidase I is a vacuolar enzyme. It has a molecular weight of 640 000, and consists of 12 subunits. The dodecameric enzyme sediments at 22S (for details see Frey and Röhm 1978). Like most of the proteinases from yeast which have so far been characterized, aminopeptidase I is repressed by glucose (Frey and Röhm 1978). Frey and Röhm (1978) reported that 1% glucose was the critical limit, below which aminopeptidase I activity began to rise. A similar sensitive response towards the nitrogen source (Frey and Röhm 1978) distinguishes aminopeptidase I from the other proteinases (Hansen et al. 1977). The most outstanding feature of aminopeptidase I is, however, the rapid loss of activity following the addition of glucose to glucose-starved cells. From these findings one may conclude that aminopeptidase I has a primarily catabolic function, i.e., to provide amino acids when the cells are limited for carbon or nitrogen.

2 Immunochemical Studies on the Catabolite Inactivation of Phosphoenolpyruvate Carboxykinase and Aminopeptidase I

In principle, inactivation of an enzyme could be the result of either a modification of the protein without change of the primary structure or the result of a partial or complete degradation. The use of immunochemical techniques for the determination of cross-reacting enzyme should permit distinction between these two possibilities.

Direct immunoprecipitates of radioactively labeled PEPCK in crude extracts prepared during catabolite inactivation is depicted in Fig. 1. At the times indicated aliquots

1 Biochemisches Institut der Universität Freiburg, Hermann-Herder-Str. 7, 7800 Freiburg, FRG

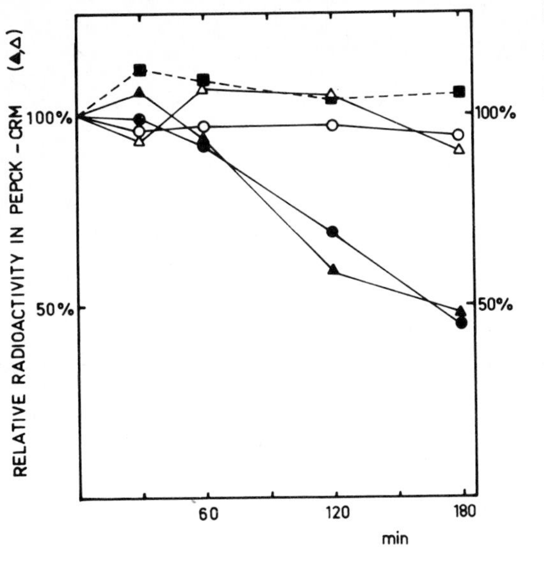

Fig. 1. Immunoprecipitation of phosphoenolpyruvate carboxykinase during catabolite inactivation. At zero time L-[4,5-^{3}H]leucine-labeled cells were transferred to a glucose-containing (*closed symbols*) or to a glucose-free medium (*open symbols*). Nonradioactive leucine had been added to both incubations. At the times indicated enzymatic activity and immunoprecipitable radioactivity in equal amounts of crude extract-protein were determined. Quantitative precipitation was verified by the complete elimination of enzymatic activity. ▲, △, radioactivity recovered from the phosphoenolpyruvate carboxykinase-band (PEPCK) of immunoprecipitates separated on SDS-polyacrylamide gels; ●, ○, enzymatic activity of phosphoenolpyruvate carboxykinase (PEPCK); ■, enzymatic activity of glucose 6-P dehydrogenase (G-6-PDH)

were withdrawn, cells homogenized and PEPCK immunoprecipitated from equal amounts of crude extract-protein using monospecific, affinity-purified antibodies. The immunoprecipitates were separated on SDS gels and each gel analyzed for radioactivity. When the radioactivity recovered from each PEPCK band of the SDS gels is plotted versus time after the beginning of the glucose-induced inactivation and compared to the catabolic activity, a parallel rate of degradation is observed. This indicates that no inactive material accumulated that could be detected by direct immunoprecipitation. From the loss of radioactivity, the apparent half-lives of the enzyme were calculated to be 34.5 h in the absence, and 2.5 h in the presence of glucose. Thus, glucose leads to a 14-fold increase in the rate of degradation. The very small loss of radioactivity from cross-reacting material in the control incubation without glucose, clearly demonstrates that catabolite inactivation is not the result of a normal high rate of degradation of PEPCK together with a repression of enzyme synthesis.

In the experiment described above, cross-reacting enzyme was found exclusively in the same band as the native enzyme on SDS gels. Thus no intermediate of the inactivation – either a modified enzyme or a cleavage product – with a different electrophoretic mobility was detected. The failure to detect such intermediates could be overcome by use of a more sensitive method. In Fig. 2 the fate of aminopeptidase I was followed by the microcomplement fixation technique. The microcomplement fixation technique is based upon the fact that complement is bound to immunocomplexes in an irreversible manner, thereby losing its hemolytic activity. By this immunological method it was shown that in 100 000 g supernatants of cell homogenates the loss of aminopeptidase I activity is accompanied by a loss of immunologically detectable protein.

Immunological evidence for an intermediate in aminopeptidase I during catabolite inactivation was provided by experiments where cell extracts were analyzed after density gradient centrifugation (Fig. 3). At zero time all the cross-reacting material sedimented

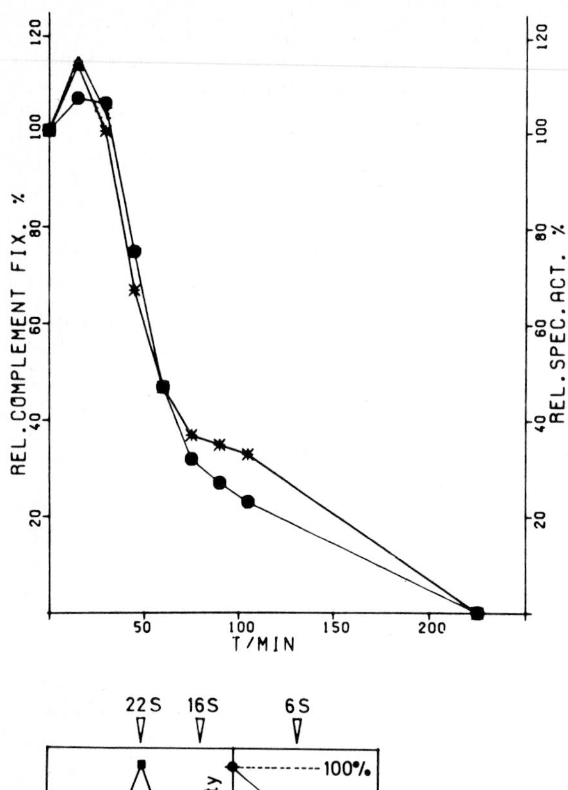

Fig. 2. Response of intracellular level of aminopeptidase I to addition of glucose to starved cells. Cell extracts were prepared and enzyme activities measured as described (Frey and Röhm 1979). Enzyme activities are expressed as % of the specific activity at t = 0, the values for immunoreactive aminopeptidase I are % of the amount present at zero time (0.18 μg/mg protein). Shown is the specific activity of aminopeptidase I (•) as a function of time elapsed after transfer of cells to glucose. In addition, the amounts of aminopeptidase I protein (*) detected by microcomplement fixation are given

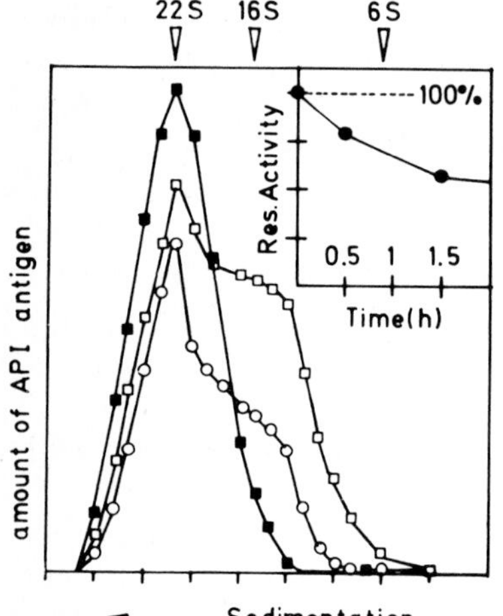

Fig. 3. Immunological evidence for an intermediate in aminopeptidase I (AP I) inactivation. Cells were homogenized a) immediately (▪), b) 0.5 h (□), and c) 1.5 h (○) after transfer to glucose. Samples of "10^5 g supernatants" were separated by sucrose density gradient centrifugation. The resulting fractions were assayed for immunological reactive aminopeptidase I protein. The complement fixation data are expressed in arbitrary units. The absolute amounts of aminopeptidase I were in the range of 1-50 ng. A 16 S aminopeptidase I species is formed from the active 22 S enzyme during inactivation. In the experiment shown the inactivation process had a half-time of about 40 min and led to a residual activity of 50%

at 22 S – like the native, dodecameric enzyme –, whereas during the course of catabolite inactivation an additional peak of cross-reacting aminopeptidase I appeared at 16 S. This species was observed only in the presence of glucose and exhibited a sedimentation behavior different from the hexameric form of aminopeptidase I (12 S) (Frey 1979). Therefore, the glucose-induced appearance of the cross-reacting 16 S form of aminopeptidase I seems to be related to the accumulation of an intermediate in the inactivation. Using the less sensitive direct immunoprecipitation, a similar finding could not be observed for PEPCK.

The concomitant loss of both catalytic activity and antigenicity of an enzyme in a cell homogenate supernatant could also be due to disappearance of the enzyme into the particulate cell fraction. However, the small amount of cross-reacting aminopeptidase I which could be immunologically detected in membrane preparations remained constant during catabolite inactivation (Frey 1979). Similarly, the loss of immunoreactive PEPCK was the same, whether determined in total cell homogenates or in high-speed supernatants.

If proteolysis is involved in catabolite inactivation one could expect a delay in the degradation of the enzyme in a mutant lacking considerable amounts or proteolytic activity. A mutant lacking the activities of proteinase B, carboxypeptidase Y, and carboxypeptidase S, has been recently constructed (Achstetter et al. 1981). This mutant exhibits catabolite inactivation of PEPCK similar to that in wild type (Wolf and Ehmann 1981). Quantitation of cross-reacting PEPCK by rocket immunoelectrophoresis during catabolite inactivation showed a decrease of immunoreactive antigen parallel to the loss of catalytic activity (Fig. 4). Proteinase B and the carboxypeptidases Y and S thus do not seem to be involved in the process of catabolite inactivation of PEPCK.

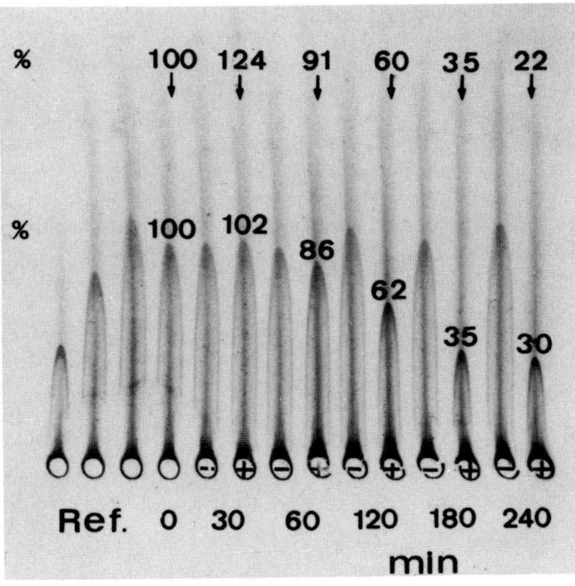

Fig. 4. Immunoprecipitation of PEPCK during glucose-induced inactivation in a yeast mutant lacking activities of three proteolytic enzymes. Rocket immunoelectrophoresis of crude extracts was carried out on agarose containing 2% (v/v) affinity-purified antibodies. Ref. = dilutions of the crude extract prepared at zero time for standardization. In order to calculate the relative amount of cross-reacting PEPCK, the rocket height at t = 0 min was set = 100%. *Numbers of the upper row* indicate specific activity, *numbers on top of the rockets* indicate relative amount of immunoreactive PEPCK. (+), incubation with glucose; (–) incubation without glucose

3 Summary

Direct immunoprecipitation of phosphoenolpyruvate carboxykinase (PEPCK) and microcomplement fixation with aminopeptidase I during the glucose-induced catabolite inactivation of these enzymes in yeast cells revealed a concomitant loss of catalytic activity and cross-reacting material, respectively. This indicates that both enzyme molecules are extensively altered during the course of catabolite inactivation rendering them unrecognizable to their antibodies. Only when cell homogenates are fractionated on sucrose density gradients and the amount of immunologically cross-reacting aminopeptidase I was measured by microcomplement fixation, putative intermediates in the inactivation process were detected. This is probably due to the fact that microcomplement fixation does not require precipitating antigen – antibody complexes. No decrease in the rate of PEPCK loss in terms of cross-reacting material was observed in a mutant lacking the activities of proteinase B, carboxypeptidase Y and S.

References

Achstetter T, Ehmann C, Wolf DH (1981) New proteolytic enzymes in yeast. Arch Biophys Biochem 207:445-454

Frey J (1979) Untersuchungen zur intrazellulären Lokalisierung und Regulation von Hefepeptidasen. Doctoral thesis, University Marburg

Frey J, Röhm K-H (1978) Subcellular localization and levels of aminopeptidases and dipeptidase in *Saccharomyces cerevisiae*. Biochim Biophys Acta 527:31-41

Frey J, Röhm K-H (1979) The glucose-induced inactivation of aminopeptidase I in *Saccharomyces cerevisiae*. FEBS Lett 100:261-264

Gancedo C, Schwerzmann K (1976) Inactivation by glucose of phosphoenolpyruvate carboxykinase from *Saccharomyces cerevisiae*. Arch Microbiol 109:221-225

Haarasilta S, Oura E (1975) On the activity and regulation of anaplerotic and gluconeogenetic enzymes during the growth process of baker's yeast. The biphasic growth. Eur J Biochem 52:1-7

Hansen RJ, Switzer RL, Hinze H, Holzer H (1977) Effects of glucose and nitrogen source on the levels of proteinases peptidases and proteinase inhibitors in yeast. Biochim Biophys Acta 496: 103-114

Holzer H (1976) Catabolite inactivation in yeast. Trends Biochem Sci (TIBS) 1:178-181

Müller M, Holzer H (1981) Proteolysis and catabolite inactivation in yeast. In: Turk V, Vitale L (eds) Proteinases and their inhibitors. Structure function and applied aspects. pp 57-65. Mladinska knjiga – Pergamon Press. Ljubljana Oxford

Wolf DH, Ehmann C (1981) Carboxypeptidase S – and Carboxypeptidase Y – deficient mutants of *Saccharomyces cerevisiae*. J Bacteriol 147 (in press)

Metabolic Interconversion of Yeast Fructose-1,6-Bisphosphatase

H. Holzer[1,2], P. Tortora[1], M. Birtel[1], and A.-G. Lenz[2]

After the addition of glucose to acetate- or ethanol-grown yeast cells a small group of enzymes is rapidly inactivated. This phenomenon has been called "catabolite inactivation" (Holzer 1976). Among other enzymes participating in gluconeogenesis, fructose-1,6-bisphosphatase is inactivated during catabolite inactivation (Harris and Ferguson 1967; Gancedo 1971). As shown in Fig. 1, the simultaneous presence of phosphofructokinase

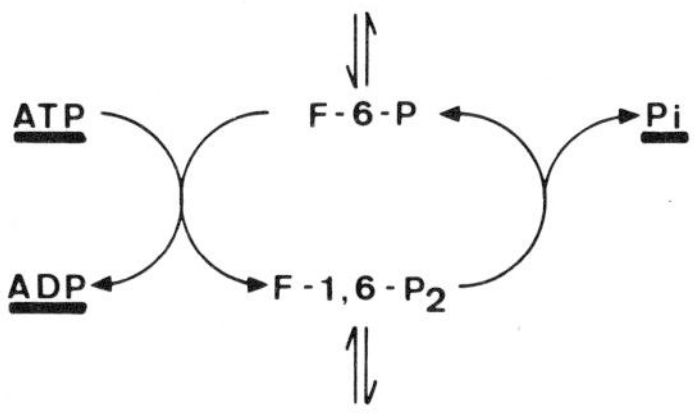

Fig. 1. Phosphofructokinase and fructose-1,6-bisphosphatase form an ATP-splitting "futile cycle"

and fructose-1,6-bisphosphatase would lead after addition of glucose to a "futile cycle" which continously splits ATP to ADP and inorganic phosphate. The rapid inactivation of fructose-1,6-bisphosphatase after addition of glucose therefore protects the cells from ATP depletion. Experiments with specific antibodies have shown that catabolite inactivation of cytoplasmic malate dehydrogenase (Neeff et al. 1978), aminopeptidase I (Frey and Röhm 1979), fructose-1,6-bisphosphatase (Funayama et al. 1980), and phosphoenolpyruvate carboxykinase (Müller et al. 1981), is the result of proteolytic degradation of the respective enzymes.

The kinetics of inactivation of fructose-1,6-bisphosphatase as measured by Anke-Gabriele Lenz in *Saccharomyces cerevisiae* M1 are shown in Fig. 2 (Lenz 1980). The very rapid inactivation of the enzyme in the first 3 min after addition of glucose indicates that not proteolysis but another, more rapid process causes the "short-term" inactivation. Further evidence for a nonproteolytic short-term inactivation results from experiments with chloroquine, an inhibitor of lysosomal proteolytic processes in mammalian tissues (Wibo and Poole 1974). It is shown in Table 1 that inactivation of fructose-1,6-bisphosphatase is relatively insensitive to chloroquine: 10 mM chloroquine inhibits catabolite inactivation of phosphoenolpyruvate carboxykinase and of cytoplasmic malate dehydrogenase almost completely, whereas inactivation of fructose-1,6-bisphosphatase is not inhibited. As shown in Table 2, the reactivation of fructose-1,6-bisphosphatase

1 Biochemisches Institut der Universität Freiburg, Hermann-Herder-Str. 7, D-7800 Freiburg, FRG
2 GSF-Abteilung für Enzymchemie, D-8042 Neuherberg/bei München, FRG

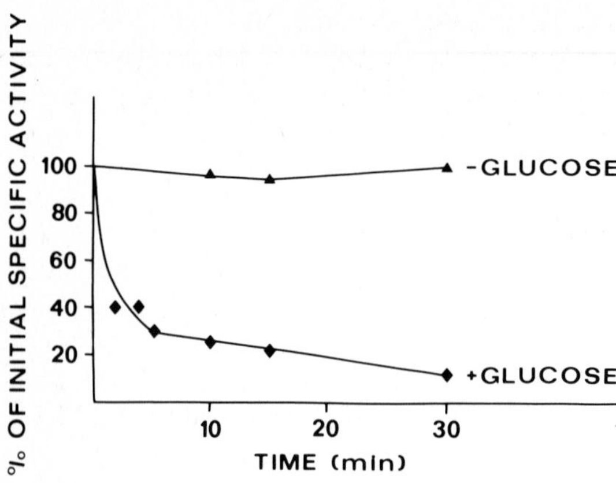

Fig. 2. Glucose-initiated inactivation of fructose-1,6-bisphosphatase (Lenz 1980)

Table 1. Influence of chloroquine on catabolite inactivation (Lenz 1980)

	Concentration of chloroquine (mM)					
	0	5	10	20	50	100
Phosphoenolpyruvate carboxykinase	81	78	2	13	1	1
Malate dehydrogenase	37	42	16	17	1	1
Fructose-1,6-bisphosphatase	96	96	92	69	47	35

Values represent % activity lost after 2 h of incubation with glucose

Table 2. Effect of inhibitors on reactivation of fructose-1,6-bisphosphatase (Lenz and Holzer 1980)

Addition to reactivation medium	Activity (%) [a]		Reactivation in % of the value without addition of inhibitor
	After 3 min inactivation with glucose	After 120 min reactivation	
–	37	91	100
100 µg Cycloheximide/ml	41	94	98
20 mM Fluoride	37	65	52
10 mM Cyanide	38	48	19
0.01 mM Carbonylcyanide-m-chlorophenylhydrazon (CCCP)	42	42	0

[a] Fructose-1,6-bisphosphatase activity prior to addition of glucose was set 100%

after short-term inactivation, is insensitive to cycloheximide and therefore independent on de novo protein synthesis. It is, however, dependent on energy because cyanide, as well as the uncoupling reagent carbonylcyanide-m-chlorophenylhydrazon (CCCP) prevent reactivation almost completely. These findings favor the idea of an ATP-dependent covalent interconversion such as phosphorylation/dephosphorylation, adenylation/deadenylation, ADPribosylation/deADPribosylation, being the mechanism for the rapid inactivation/reactivation of fructose-1,6-bisphosphatase. Further evidence for an interconversion of two different stable forms of the enzyme may be seen from the data shown in Table 3. The ratio of the activities of fructose-1,6-bisphosphatase with Mg^{2+} and Mn^{2+},

Table 3. Effect of Mg^{2+} and Mn^{2+} on the activity of fructose-1,6-bisphosphatase (Lenz and Holzer 1980)

	FBPase activity in crude extract (units/ml)	
	Without incubation with glucose	3 min after incubation with glucose
10 mM Mg^{2+}	0.46	0.24
2 mM Mn^{2+}	0.27	0.24
$\frac{\text{Activity with } Mg^{2+}}{\text{Activity with } Mn^{2+}}$	1.7	1.0

respectively, in crude extracts from yeast shows a significant change during the rapid short-term inactivation. A similar change in the dependency of the activity of glutamine synthetase from *Escherichia coli* upon Mg^{2+} and Mn^{2+} was shown some years ago to be characteristic for the inactivation of glutamine synthetase by ammonia-induced enzymic interconversion (Mecke et al. 1966a, 1966b).

Definite proof that proteolysis is not the mechanism for the rapid inactivation of fructose-1,6-bisphosphatase was obtained in experiments with specific antibodies against the enzyme. The proteins in glucose-derepressed yeast cells were labeled with [^{3}H]-leucine. Before and after addition of glucose samples of cells were rapidly collected, extracted, and treated with specific antiserum against fructose-1,6-bisphosphatase. After gel electrophoresis the radioactivity of the cross-reacting material was counted. It is shown in Fig. 3 that the specific activity of fructose-1,6-bisphosphatase decreased in the first 3 min to about 40% of the initial value, whereas the cross-reacting material showed in the first minute only an insignificant change. Only after a lag period of about 10 min did the cross-reacting material decrease rapidly to about 40% of the initial value. This result disproves proteolysis and points, in agreement with the other data discussed above, to covalent conversion being the mechanism of the short-term inactivation that occurs during the first 3 min after addition of glucose. The covalent conversion is then followed by proteolytic degradation. This conception is summarized in Fig. 4. We assume that the conversion of active fructose-1,6-bisphosphatase to a less active form of the enzyme initiates subsequent proteolytic degradation of the latter form. Glucose and/or its catabolites, which initiate inactivation and finally the degradation of fructose-1,6-bisphosphatase,

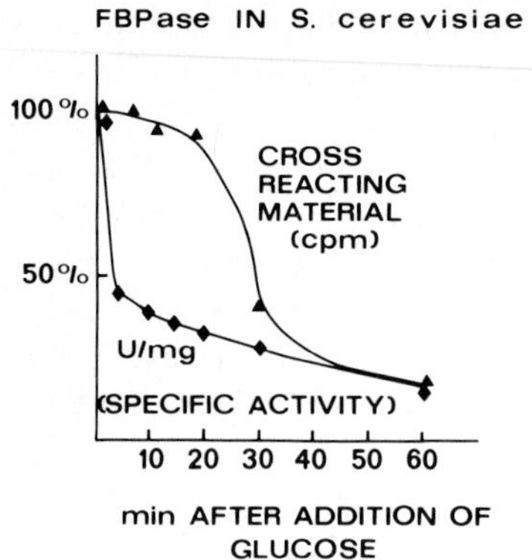

Fig. 3. Immunological cross-reacting material and specific catalytic activity of fructose-1,6-bisphosphatase after addition of glucose to derepressed yeast cells

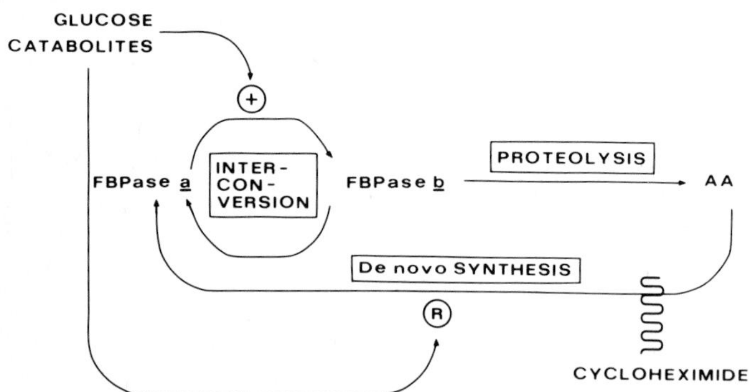

Fig. 4. Hypothetical scheme of catabolite inactivation of fructose-1,6-bisphosphatase by interconversion-initiated proteolysis. + , positive allosteric effect (activation); R , catabolite repression, FBPase a , active fructose-1,6-bisphosphatase; FBPase b , less active form of fructose-1,6-bisphosphatase

could act as allosteric effectors of the inactivating, i.e., converting enzyme. As discussed in a recent review article (Holzer and Heinrich 1980), selectivity of the glucose-induced proteolytic degradation of fructose-1,6-bisphosphatase could be explained by the effector-stimulated conversion of fructose-1,6-bisphosphatase to a less active form which is susceptible to proteolysis. "Interconversion-initiated proteolysis" might probably be a more general mechanism of "selective proteolysis".

Conclusions

The glucose-induced "catabolite inactivation" of fructose-1,6-bisphosphatase (FBPase) in acetate- or ethanol-grown yeast cells shows at first a rapid phase of ca. 3 min in which about 60% of the enzyme's activity disappear. In a subsequent slower phase of inactivation the activity disappears almost completely in about 1 h. The observations listed below indicate that metabolic conversion, i.e., covalent modification, is the mechanism of the "short-term inactivation": (1) The enzyme from derepressed cells shows a ratio of catalytic activity with 10 mM Mg^{2+} to 2 mM Mn^{2+} of 1.7 compared to the enzyme from short-term inactivated cells which exhibits a ratio of 1.0; the pH optimum of catalytic activity of fully active enzyme is 8.5 as compared to an optimum of 9.0 for the 3 min inactivated enzyme. (2) In contrast to catabolite inactivation of cytoplasmic malate dehydrogenase and phosphoenolpyruvate carboxykinase, the short-term inactivation of FBPase in intact yeast cells is relatively insensitive to chloroquine, an inhibitor of intracellular proteolysis. (3) Reactivation of FBPase in a glucose-free medium is insensitive to cycloheximide, i.e., it requires no protein synthesis. (4) the reactivation is prevented by CCCP, an uncoupler of oxidative phosphorylation, i.e., it is dependent on energy-rich phosphate bonds.

With specific antibodies against FBPase it is shown that the cross-reacting material from yeast cells with [^{3}H]-leucine-labeled protein does not decrease in the short-term inactivation period. In the following hour of incubation with glucose, however, the cross-reacting material decreases to a few percent of the initial value. This result clearly indicates that the initial short-term inactivation is not a consequence of proteolysis but very probably represents a covalent modification, i.e., metabolic interconversion, of the enzyme. It is postulated that metabolic interconversion renders the enzyme susceptible to proteolysis and thereby initiates proteolysis. The well-known selectivity of metabolic interconversion (cf. Holzer and Duntze 1971: Metabolic regulation by chemical modification of enzymes. Ann Rev Biochem 40:345-374) would then explain how selective proteolysis of an enzyme might be initiated.

Acknowledgements. The authors are indebted to Drs. Carlos Gancedo and Maria Jesus Mazón for a gift of antiserum against yeast fructose-1,6-bisphosphatase and for discussion of the manuscript.

References

Frey J, Röhm K-H (1979) The glucose-induced inactivation of aminopeptidase I in *Saccharomyces cerevisiae*. FEBS Lett 100:261-264

Funayama S, Gancedo JM, Gancedo C (1980) Turnover of yeast fructose-1,6-bisphosphatase in different metabolic interconversion. Eur J Biochem 109:61-66

Gancedo C (1971) Inactivation of fructose-1,6-bisphosphatase by glucose in yeast. J Bacteriol 107: 401-405

Harris W, Ferguson JJ Jr (1967) Inactivation of yeast fructose diphosphatase in the course of catabolite repression. Fed Proc 26:678

Holzer H (1976) Catabolite inactivation in yeast. Trends Biochem Sci (TIBS) 1:178-181

Holzer H, Heinrich PC (1980) Control of proteolysis. Annu Rev Biochem 49:63-91

Lenz A-G (1980) Studien zum Mechanismus der Katabolit-Inaktivierung in Hefe. Doctoral thesis, University Freiburg

Lenz A-G, Holzer H (1980) Rapid reversible inactivation of fructose-1,6-bisphosphatase in *Saccharomyces cerevisiae* by glucose. FEBS Lett 109:271-274

Mecke D, Wulff K, Liess K, Holzer H (1966a) Characterization of a glutamine synthetase inactivating enzyme from *Escherichia coli*. Biochem Biophys Res Commun 24:452-458

Mecke D, Wulff K, Holzer H (1966b) Metabolit-induzierte Inaktivierung von Glutaminsynthetase aus *Escherichia coli* im zellfreien System. Biochim Biophys Acta 128:559-567

Müller M, Müller H, Holzer H (1981) Immunochemical studies on catabolite inactivation of phosphoenolpyruvate carboxykinase in *Saccharomyces cerevisiae*. J Biol Chem (in press)

Neeff J, Hägele E, Nauhaus J, Heer U, Mecke D (1978) Evidence for catabolite degradation in the glucose-dependent inactivation of yeast cytoplasmic malate dehydrogenase. Eur J Biochem 87: 489-95

Wibo M, Poole B (1974) Protein degradation in cultured cells. II. The uptake of chloroquine by rat fibroblasts and the inhibition of cellular protein degradation and cathepsin B_1. J Cell Biol 63: 430-440

Discussion[3]

after Gancedo, Müller, and Holzer

Gancedo agreed with Shaltiel that it would be worthwile to use antibodies against the denatured form of the enzyme.

Cohen argued that there could be the possibility of an initial removal of a small peptide from fructose-1,6-bisphosphatase.

Holzer answered that the immunoprecipitated protein at early time points would exhibit no change in the molecular weight on SDS-gels. Reactivation without protein-synthesis, however, would be hard to explain after removal of such a peptide.

Gancedo – asked by Wolf – pointed out that the phosphate label would be lost from the enzyme-band on SDS-gels during reactivation.

Hers asked for measurements of glucose metabolites during the very rapid inactivation of fructose-1,6-bisphosphatase.

Gancedo replied that fructose-1,6-bisphosphate would be accumulated very quickly after the shift from ethanol to glucose. Furthermore, in a phosphoglucose-isomerase-deficient mutant the inactivation would be decreased with glucose or fructose alone, whereas it would be essentially unchanged after addition of both sugars together. A phosphofructokinase-deficient mutant, however, would exhibit an effective inactivation.

Holzer added that ATP would rapidly disappear after addition of glucose reaching a minimum by one minute and a steady-state concentration after about 10 min. Furthermore, in *E. coli* the activity of glutamine synthetase would disappear within seconds after the addition of NH_4^+ by action of glutamine, which would be rapidly formed and cause the inactivation.

Switzer and Gancedo explained that the different rates of inactivation of fructose-1,6-bisphosphatase observed in rich medium versus minimal medium could be due to the amino acids present in the rich medium as well as to the carbon source, which could influence the activity of proteinases involved in the degradation process. (i.e., change of the ratio of the activities of the converting system to the proteolytic system according to a comment by Holzer.)

Ballard pointed out that the phosphoenolpyruvate carboxykinase in mammalian liver would be controlled by the level of mRNA of the enzyme.

Gancedo replied that the rate of synthesis had not been determined in yeast grown on a rich medium.

Further information given by Holzer and Gancedo upon questions of Betz and Wieland:

3 Rapporteur: Matthias Müller

Glucose itself does not induce the inactivation (inactivation is observed with fructose and in a hexokinase-deficient mutant).

The inducer should be easily detected after purification of the converting enzyme.

The present idea is that phosphorylation or related modifications do not only make the protein susceptible to proteolysis but also to the uptake into vacuoles or lysosomes.

Fructose-1,6-Bisphosphatase Converting Enzymes in Rabbit Liver Lysosomes: Purification and Effect of Fasting

S. Pontremoli[1], E. Melloni[1], and B.L. Horecker[2]

All vertebrate fructose-1,6-bisphosphatases (Fru-P_2ase) analyzed, including the purified enzymes from rabbit liver [1] and muscle [2], rat liver [2] and muscle [2], chicken muscle [2], snake muscle [2], and pig kidney [3], were found to differ significantly in primary structure except for a highly conservative region including amino acid residues 40 to 75 from the amino terminal end.

In Fig. 1 are reported, for example, the similarities and the differences between the amino acid sequences of rabbit liver and pig kidney Fru-P_2ases. The differences in the primary structure of the two enzymes tend to decrease approaching that portion of the

```
                                                     10                                      20
Rabbit Liver   AcAla-Asp-Lys-Ala-Pro-Phe-Asp-Thr-Asp-Ile-Ser-Thr-Met-Thr-Arg-Phe-Val-Met-Glu-Glu
Pig Kidney     AcThr     Glx     Ala             Asn     Val     Leu
                                                     30                                      40
                 Gly-Arg-Lys-Ala-Gly-Gly-Thr-Gly-Glu-Met-Thr-Glu-Leu-Leu-Asn-Ser-Leu-Cys-Thr-Ala
                                 Arg
                                                     50                                      60
                 Val-Lys-Ala-Ile-Ser-Thr-Ala-Val-Arg-Lys-Ala-Gly-Ile-Ala-His-Leu-Tyr-Gly-Ile-Ala
```

Taken from Botelho et al. [1] and Marcus et al. [3]

Fig. 1. Amino-terminal sequences of rabbit liver and pig kidney Fru-P_2ase

molecule having a complete conservation of structure. In all Fru-P_2ases examined, within this structure, an exposed peptide region between residues 57 and 67 is susceptible to limited proteolysis by subtilisin. In this region the primary sites of hydrolysis are the peptide bonds Ala^{60}-Gly^{61} and Thr^{63}-Asn^{64} [4, 5, 6].

Digestion with subtilisin does not alter the molecular weight of the enzyme protein measured under nondissociating conditions, however in SDS slab gel electrophoresis it has been found [5], in accordance with the limited proteolysis described above, that subtilisin produces the progressive and complete conversion of the native 36.000 molecular weight subunits into shorter chains having molecular weights of 29.000 and 7.000 respectively (Fig. 2). As shown in Fig. 2, during these changes in molecular properties Fru-P_2ase activity increases three- to fourfold at pH 9.2, while its sensitivity to AMP inhibition significantly decreases.

Abbreviations: Fru-P_2ase, fructose 1,6-bisphosphatase; SDS, sodium dodecyl sulphate

1 Department of Biochemistry, University of Genoa, Genoa, Italy
2 Roche Institute of Molecular Biology, Nutley, NJ 07 110, USA

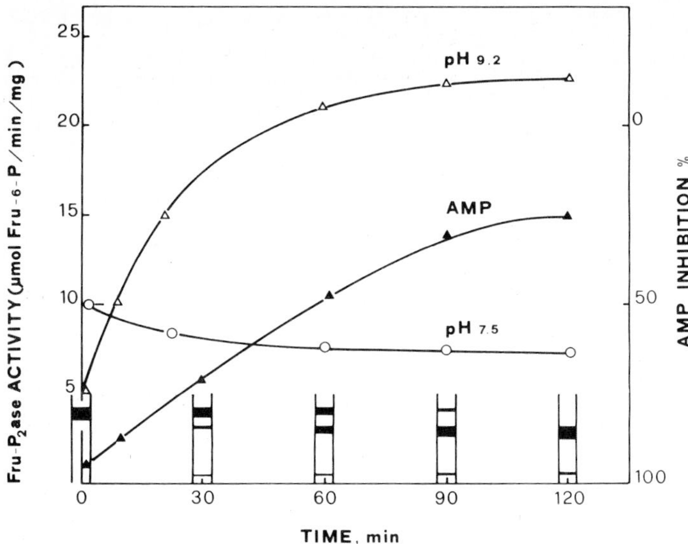

Fig. 2. Changes in catalytic, regulatory, and molecular properties of rabbit liver Fru-P_2ase during digestion by subtilisin. For details, see Pontremoli et al. [4]

The occurrence of similar changes in structural, regulatory, and catalytic properties in all Fru-P_2ases so far tested, and the existence of sequence identity in the protease-sensitive region of all of these enzymes, are consistent with the hypothesis that limited proteolysis plays a role in the physiological regulatory process involving all Fru-P_2ases.

The physiological significance of this phenomenon was further supported by our previous observations indicating that rabbit [7] and rat [8] liver lysosomes contain a proteolytic system which produces in purified Fru-P_2ase structural and catalytic modifications similar to those observed with subtilisin, and that this proteolytic activity can be expressed, although to different extent, by intact lysosomes [9], isolated and washed membranes [10], and by soluble intralysosomal enzymes [10, 11].

In the present report we will describe the further characterization of this proteolytic system designated by us as Fru-P_2ase converting enzyme activities, in terms of the number and types of these proteolytic activities and their distribution within the lysosomal fractions. The experiments reported here have been performed using the rabbit liver as the source of lysosomes and of purified Fru-P_2ase.

As shown in Table 1, the total Fru-P_2ase converting enzyme activity present in the rabbit liver lysosomes and assayed by following the increase in Fru-P_2ase activity at pH 9.2 is distributed as follows: 86% in a soluble form, 14% as a membrane-bound activity, a large portion of which (50%) is expressed by isolated membranes or intact lysosomes.

We will discuss this point later but it is important to emphasize that Fru-P_2ase digestion by intact lysosomes is not due to leakage of soluble lysosomal enzymes since, as shown later, no release of such proteolytic enzymes occurs during the time of incubation of intact lysosomes with Fru-P_2ase.

Table 1. Content and distribution of Fru-P_2ase converting enzymes in rabbit liver lysosomes

Fraction	Converting enzyme activity [a] (Units/g fresh liver)	%
Disrupted lysosomes [b]	1.24	100
Soluble content	1.07	86
Solubilized membranes [c]	0.17	14
Intact membranes [d]	0.085	7
Intact lysosomes [d]	0.078	7

[a] Evaluated as described by Melloni et al. [11]
[b] Assayed in the presence of 0.5% triton X-100
[c] Solubilized with 1 M NaCl
[d] Assayed in 0.25 M sucrose containing 0.05 M sodium acetate, pH 6.5

The soluble Fru-P_2ase converting enzyme activities in rabbit liver were characterized as follows: the heavy particle fraction containing lysosomes was disrupted by freezing and thawing, and the membrane fraction by removal through centrifugation.

The soluble fraction was then chromatographed at pH 6.0 on Ultrogel AcA34 and yielded three distinct peaks with converting enzyme activity (Fig. 3), fraction I corresponding to a molecular weight of approximately 150.000 and being completely free of any of the known cathepsin, while fraction II with an approximate molecular weight of

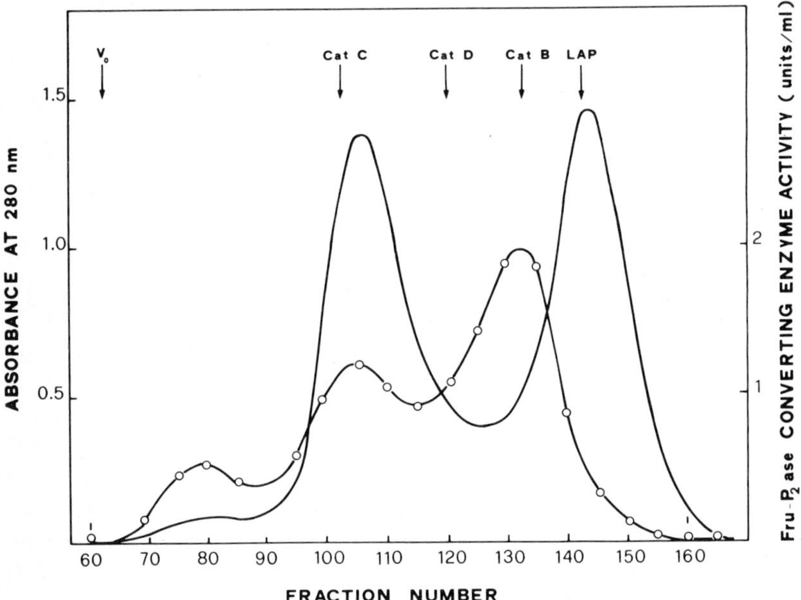

Fig. 3. Chromatography on Ultrogel AcA34 of the soluble Fru-P_2ase converting enzyme activities from rabbit liver lysosomes. For details, see Melloni et al. [11]. *LAP* refers to leucine amino-peptidase

75.000 emerged immediately after cathepsin C and before cathepsin D; fraction III with approximate molecular weight of 30.000 was eluted together with cathepsin B. The three fractions were designated CE-I, CE-II, and CE-III, respectively. The same pattern of converting enzyme activities was obtained from lysosomes that were purified from the heavy particle fraction by centrifugation through a linear sucrose gradient.

As shown in Fig. 4, the converting enzyme activity eluted from the membrane fraction with NaCl and chromatographed on the same column was identified on the basis of its elution volume as CE-II. No activity appeared in the positions corresponding to CE-I or CE-III.

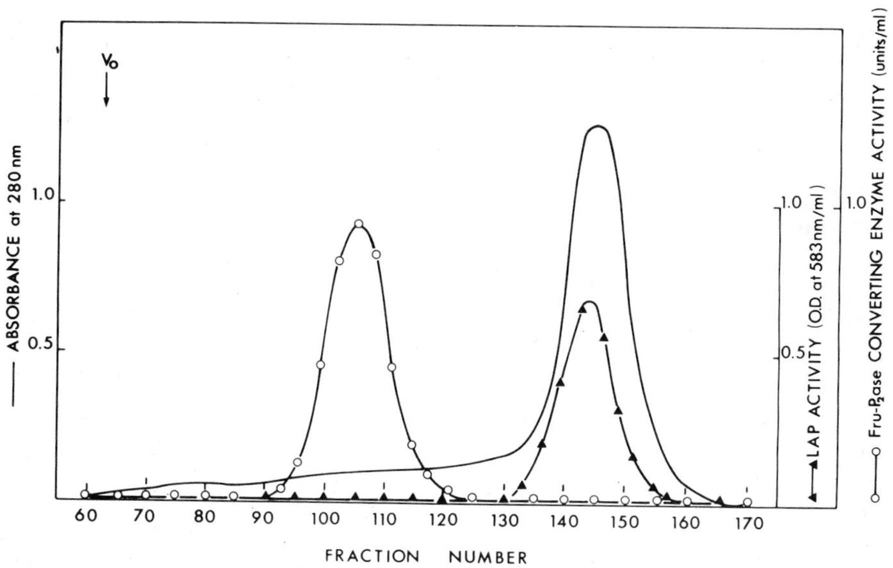

Fig. 4. Chromatography on Ultrogel AcA34 of the membrane-bound Fru-P$_2$ase converting enzyme activities from rabbit liver lysosomes. For details, see Melloni et al. [11]

Each of the three fractions from Ultrogel AcA34 and corresponding to CE-I, CE-II, and CE-III were then concentrated by ultrafiltration and purified by adsorption at pH 6.0 on a DEAE cellulose column and elution with a linear gradient of NaCl from 0 to 0.2 M. Ce-I again emerged free of known cathepsin, CE-II was recovered in a single peak free of cathepsin D, C, or B activity; while CE-III again emerged in coincidence with the BANA hydrolyzing activity. CE-III, further chromatographed at pH 5.0 on CM-cellulose, yielded two peakes both containing Fru-P$_2$ase converting activity and BANA hydrolyzing activity. These two enzyme activities designated as CE-III A and B, correspond presumably to two different isozymes of cathepsin B. Identical chromatography on DEAE cellulose was used to purify the CE-II solubilized from isolated membranes. The properties of all the purified enzymes are listed in Table 2.

Converting enzymes I and II appear to be new cathepsins distinguishable from each other by their molecular weights of 150.000 and 75.000, but indistinguishable in a num-

Table 2. Effect of activators and inhibitors on the activity of purified Fru-P_2ase converting enzymes

Effector	Converting enzyme activity (% of original activity)						
	Ce I	CE II_S	CE II_M	CE III_A		CE III_B	
				Fru-P_2ase	BANA	Fru-P_2ase	BANA
None	100	100	100	100	100	100	100
Cysteine (5 mM)	260	255	250	380	380	320	350
NaCl (0.5 M)	90	95	95	45	47	41	44
Leupeptin (1 μg/ml)	96	94	90	7	10	11	15
Pepstatin (2.5 μg/ml)	100	100	100	80	90	80	95

CE II_S: CE II Purified from lysosomal soluble fraction
CE II_M: CE II Purified from lysosomal membranes
CE III_A and CE III_B: obtained by chromatography on CM cellulose
Data taken from Melloni et al. [11]

ber of common properties such as activation by thiol reagents and insensitivity to NaCl, leupeptin, and pepstatin, typical inhibitors of lysosomal endopeptidase.

Converting enzyme III was found to be associated with two cathepsin B isozymes characterized by different activities with protein and BANA as substrates; both being NaCl- and leupeptin-sensitive but pepstatin-insensitive.

On the basis of these results we may summarize the distribution of the lysosomal Fru-P_2ase converting enzymes as reported in Table 3 as follows: (1) of the total converting enzyme activity present in the rabbit liver lysosomes, CE-III is the predominant form;

Table 3. Levels and distribution of the various Fru-P_2ase converting enzyme activities in rabbit liver lysosomes

Lysosomal fraction	Converting enzyme activity (units/g fresh liver)			
	CE I	CE II	CE III_A	CE III_B
Soluble	0.16 (15%)	0.27 (25%)	0.32 (30%)	0.33 (30%)
Membrane	–	0.17 (100%)	–	–

Data taken from Figs. 3 and 4

(2) a significant fraction of the total CE-II activity (30%-40%) is associated with the membrane fraction. The bound and soluble converting enzymes exhibit differences with respect to their pH optima and stability at slightly alakaline pH. The purified converting enzymes, including CE-II solubilized and purified from the membrane fraction, show maximum activity at pH 5.0; only 10%-15% being expressed at neutral pH values (Fig. 5A). As shown in Fig. 5B, all of the soluble converting enzymes become unstable above pH 6.0. At pH 7.5 only 30% of the original activity is retained in 30 min at room tem-

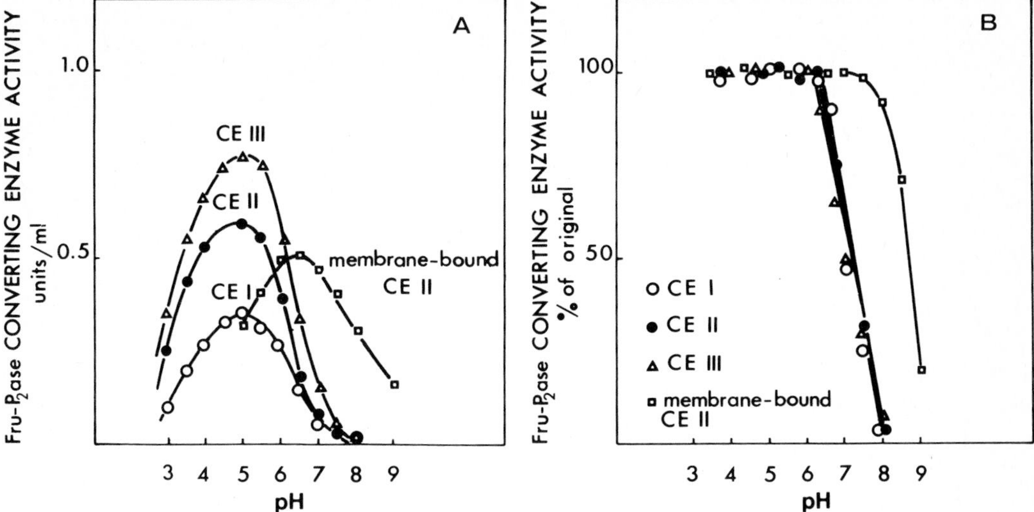

Fig. 5A, B. Effect of pH on the activity (**A**) and stability (**B**) of Fru-P_2ase converting enzyme purified from lysosomes or still bound to lysosomal membranes. **A** The incubation mixtures contained, in a final volume of 1 ml, 0.01-0.015 mg of proteolytic enzyme, 0.7 mg of Fru-P_2ase, and 0.05 M buffer at the indicated pH. For the assay of the anzyme activity still bound to the membrane, 1,2 mg of membrane proteins were used in each incubation mixture. Between pH 2.5 and 3.5 the buffer used was sodium citrate; between 4.0 and 6.5 sodium acetate; and between 7.0 and 9.0 sodium borate. **B** One ml of solution containing 0.02-0.04 mg of purified converting enzyme or 2 mg lysosomal membrane was adjusted to the indicated pH. The samples were incubated for 30 min at room temperature and 0.5 ml of each was taken for the assay of Fru-P_2ase converting enzyme. Melloni et al. [11]

perature. On the other hand, as shown in Fig. 5A, the bound form of CE-II shows maximum activity at pH 6.5, rather than at pH 5.0 as is the case for the solubilized form of this enzyme. In addition (Fig. 5B), the membrane-bound CE-III is more stable above neutral pH; instability is first detected above pH 7.5-8.0.

The physiological relevance of these properties of the membrane-bound CE-II will become further evident below, where we will present results of Fru-P_2ase digestion by intact lysosomes.

The changes in catalytic properties of Fru-P_2ase as a result of the limited proteolysis with the purified converting enzymes are very similar with all three enzymes.

In each case, as shown in Fig. 6, the activity measured at pH 9.2 increases three- to fourfold, while the activity at pH 7.5 remains unchanged. At the same time the sensitivity to AMP inhibition decreases by more than 50%. These changes resemble those observed with subtilisin digestion. The changes in subunit structure are somewhat different. With purified CE-III the pattern of digestion is identical to that previously reported for the digestion with subtilisin: the native subunit molecular weight 36.000 is replaced by two peptides corresponding to molecular weights of 29.000 and 7.000, respectively (Fig. 7A). With purified CE-I and CE-II from the soluble lysosomal fraction and with the corresponding CE-II eluted from the membrane fraction, in addition to the predominant fragments of molecular weights 29.000 and 7.000, fragments corresponding to

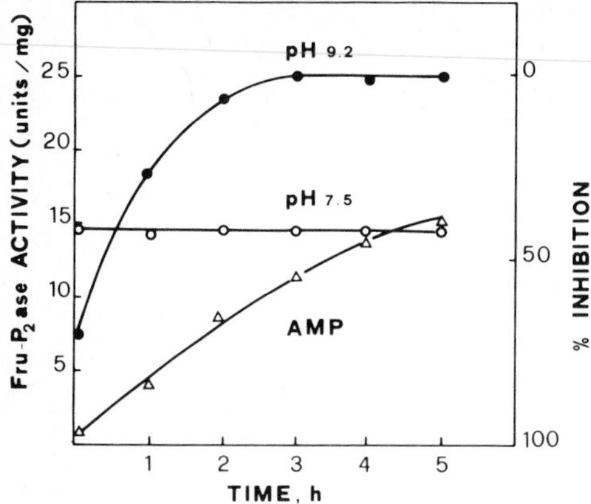

Fig. 6. Changes in catalytic and regulatory properties of Fru-P_2ase digested with purified converting enzymes from rabbit liver lysosomes. The incubation mixture (1 ml) contained 15 μg of purified converting enzyme II, 1 mg of Fru-P_2ase and 0.1 M sodium acetate, pH 5.0. At the times indicated, aliquots (5 μl) were removed and assayed for Fru-P_2ase activity at pH 9.2 and pH 7.5. Inhibition by *AMP* was determined at pH 7.5 using 0.1 mM inhibitor. Similar results were obtained with converting enzyme I and III and with converting enzyme II purified on DEAE-cellulose after elution from the membranes with M NaCl

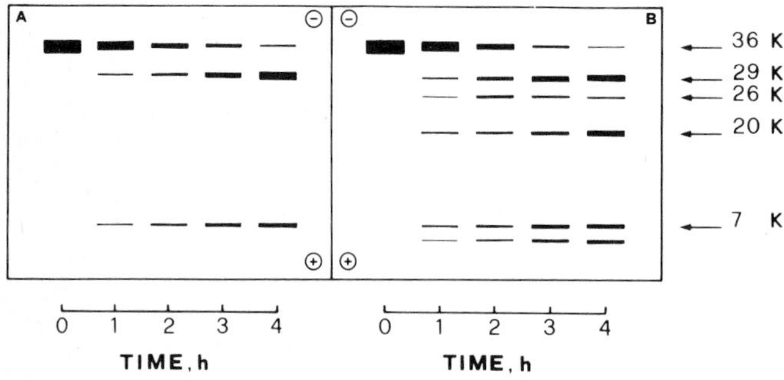

Fig. 7A, B. Changes in molecular properties of Fru-P_2ase digested with purified converting enzyme II (**A**) and converting enzyme III (**B**). For the analysis of the digestion products by slab gel polyacrylamide electrophoresis, 50 μl of the incubation mixtures, prepared as described in Fig. 6, were removed and acidified with 10 μl of 3 M trichloroacetic acid. For each analysis 20-25 μg of Fru-P_2ase was used

molecular weights 26.000 and 22.000 and a smaller peptide were also detected in the SDS polyacrylamide gel (Fig. 7B). In the case of the lysosomal rat liver CE [10], when the 29.000 and 7.000 dalton peptides were isolated and sequenced at the NH_2 and COOH terminal regions respectively, it was found that the hydrolysis occurred at the ASN^{64}-VAL^{65} bond adjacent to one of the subtilisin-sensitive bonds.

Based on this observation and on the size of the fragments produced during digestion of Fru-P_2ase, we presume that the limited proteolysis observed with the rabbit liver converting enzymes described here occurs in the highly conserved proteinase-sensitive region at a peptide bond(s) close to one of the bounds hydrolyzed by subtilisin.

In order to explore the possibility that limited proteolysis of Fru-P_2ase produced by the lysosomal converting enzymes is related to a physiological regulatory mechanism, we have studied the levels and distribution of the lysosomal converting enzymes in a condition of enhanced gluconeogenesis such as that induced by fasting. The heavy particle fraction containing lysosomes was prepared from livers of fed and 60 h fasted rabbits. Lysosomes were disrupted and the soluble and membrane-bound enzyme fraction applied to an Ultrogel AcA34 column as previously described [12].

As shown in Fig. 8A, the activities of all three converting enzymes from the soluble lysosomal fraction are increased during fasting, with the greatest increase in the activity of CE-II.

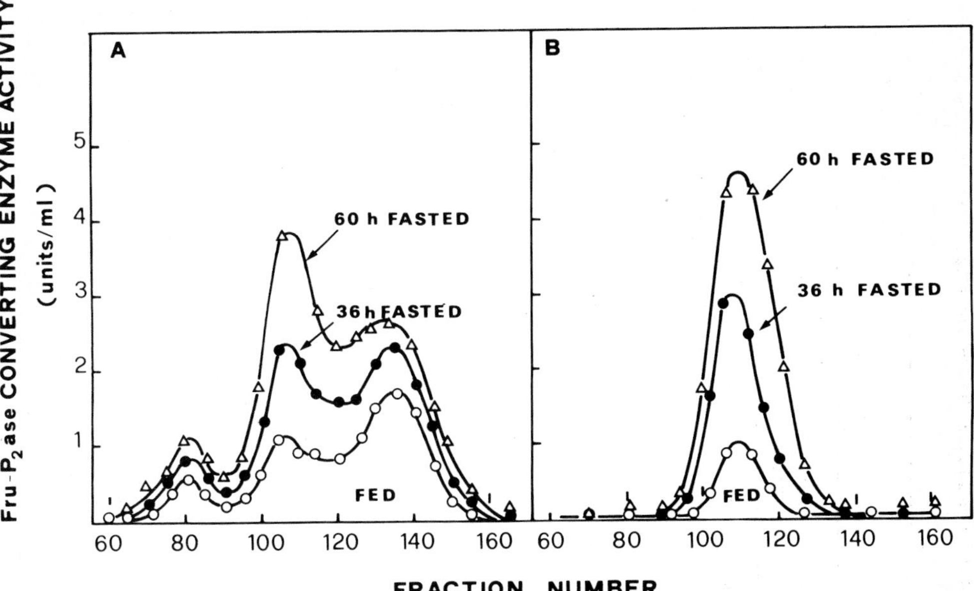

Fig. 8A, B. Chromatography on Ultrogel AcA34 of soluble **(A)** and membrane-bound **(B)** lysosomal converting enzyme activities obtained from liver lysosomes of fed and fasted rabbits. The membrane-bound enzyme was solubilized with 1 M NaCl before chromatography. For details see Melloni et al. [12]

The most stricking increase, however, was observed in the fraction of CE-II associated with the lysosomal membrane fraction (Fig. 8B). The levels of the three Fru-P_2ase converting enzymes and marker cathepsin in fed and fasted conditions are reported in Table 4.

In the soluble fraction the activity of CE-II increases, after 60 h fasting, by a factor of more than 5 compared with increases of less than twofold in the activities of CE-I and CE-III as well as of that of cathepsin B, C and leucylaminopeptidase. It must be emphasized that as a result of these changes in the converting enzyme levels, CE-II replaces cathepsin B (CE-III) as the predominant form in lysosomes prepared from fasted rabbits.

Particularly interesting is the change in the level of CE-II associated with membranes which after 60 h fasting increases 11-fold. There is also an increase in the amount of membrane-bound leucylaminopeptidase, but this increase is much smaller than that observed for CE-II. It must be considered that the increase in activity of CE-II cannot be attributed to the general increase in number and activity of rabbit liver lysosomes since its increase is by far greater than that of other converting enzymes or cathepsins.

Table 4. Levels of Fru-P_2ase converting enzymes and cathepsins in the soluble and membrane-bound fractions of lysosomes isolated from fed and fasted rabbit livers

Proteolytic enzyme activity	Soluble fraction					Membrane-bound fraction				
	Fed (A)		Fasted (B)		In-crease	Fed (A)		Fasted (B)		In-crease
	U/g liver	U/Mg protein	U/g liver	U/Mg protein	B/A	U/g liver	U/Mg protein	U/g liver	U/Mg protein	B/A
Converting enzyme I	0.16	0.07	0.30	0.11	1.6	ND	–	ND	–	
Converting enzyme II	0.27	0.11	1.40	0.53	5.2	0.17	0.42	2.1	4.2	11
Converting enzyme III	0.65	0.27	0.94	0.35	1.4	ND	–	ND	–	
Cathepsin B	0.33	0.14	0.52	0.22	1.4	ND	–	ND	–	
Cathepsin C	1.02	0.42	2.12	0.80	1.8	ND	–	ND	–	
Leu-aminopeptidase	0.11	0.046	0.24	0.1	2.0	0.012	0.03	0.05	0.1	4

U = units
ND = not detectable
For details, see Melloni et al. [11, 12]

Comparison of the levels and distribution of the various CE in the lysosomes from livers of fed and fasted animals indicates that in fasting the sum of the activities of the soluble and membrane-bound CE-II accounts for more than 70% of the total converting enzyme activity compared with less than 40% in the fed livers, where CE-III (cathepsin B) is the major form (Table 5).

Table 5. Distribution of Fru-P_2ase converting enzymes in lysosomes isolated from livers of fed and fasted rabbits

Lysosomal fraction	Converting enzyme activity (units/g of fresh liver)							
	CE I		CE II		CE III_A		CE III_B	
	Fed	Fasted	Fed	Fasted	Fed	Fasted	Fed	Fasted
Soluble	0.16	0.30 (1.8x)	0.27	1.40 (5.2x)	0.32	0.44 (1.5x)	0.33	0.50 (1.5x)
Membrane	–	–	0.17	2.10 (11x)	–	–	–	–

Data taken from Fig. 8A and B

As shown in Fig. 9A and 9B, isolated washed membranes or unbroken lysosomes prepared from livers of fed rabbits and incubated at pH 6.5 in isotonic sucrose solution express Fru-P_2ase converting enzyme activity. The increase in activity of membrane-bound

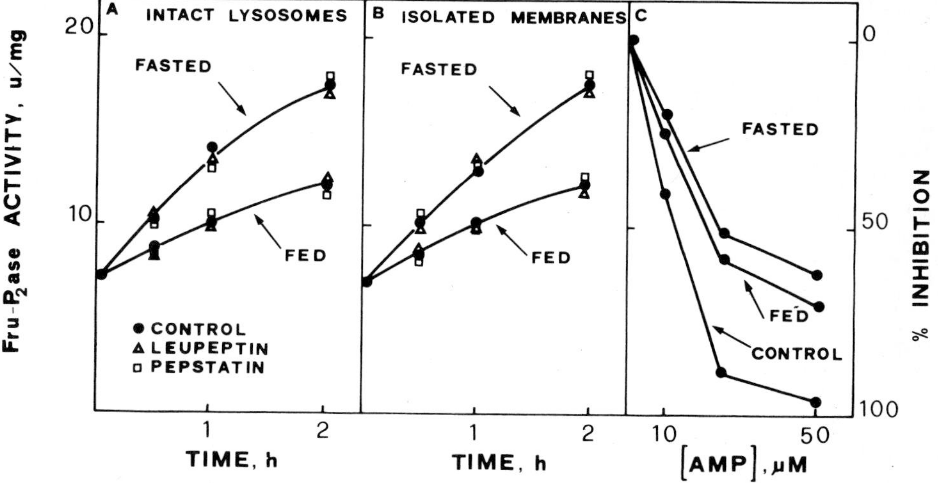

Fig. 9A-C. Changes in the catalytic activity and sensitivity to AMP inhibition of Fru-P_2ase digested with lysosomal membranes and with intact lysosomes. The amount of intact lysosomes or lysosomal membranes obtained from 5 g of fresh liver was incubated with 1 mg of Fru-P_2ase in 0.25 M sucrose containing 0.05 M sodium acetate, pH 6.5. At the times indicated 5 μl of the incubation mixtures were removed and assayed for Fru-P_2ase activity at pH 7.5 and 9.2. In separate experiments, leupeptin (1 μM) or pepstatin (10 μM) was added to the incubation mixtures. After 2 h of incubation, sensitivity of Fru-P_2ase to AMP inhibition was determined at pH 7.5 using 0.1 mM AMP in the assay system

CE-II during fasting was found to be associated with an increase in the rate of modification of Fru-P_2ase by isolated membranes or intact lysosomes (Fig. 9A and 9B). The activity expressed by intact lysosomes is almost identical to that observed with the membranes isolated from the same lysosomes. Furthermore, the rate of Fru-P_2ase conversion is unaffected by the presence of leupeptin or pepstatin, thus confirming that the activity expressed by intact lysosomes has properties identical to those of CE-II present in membranes and differing from those of known intralysosomal cathepsin. As with subtilisin or with purified soluble lysosomal converting enzymes, limited proteolysis by intact lysosomes produces not only increase in catalytic activity at pH 9.2 but also a significant desensitization to AMP inhibition (Fig. 9C).

In order to exclude the possibility that the activity observed with intact lysosomes was due to leakage of lysosomal proteases, we measured the activity of several lysosomal hydrolases on aliquots of the lysosomal suspension prepared from fed and fasted rabbits during the period of incubation with Fru-P_2ase (Fig. 10). The data obtained showed that negligible amounts of each enzyme marker was released during the 2 h of incubation.

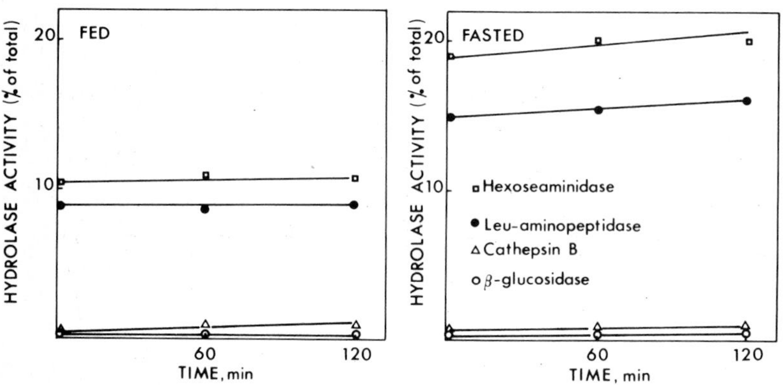

Fig. 10. Hydrolase activities expressed by intact lysosomes at 37°C in isotonic sucrose solution. The intact lysosomes obtained from 5 g of fresh tissue were diluted in 1 ml of 0.25 M sucrose solution and incubated at 37°C for 2 h at pH 6.5. At the time indicated, aliquots (0.02 ml) of the suspension were removed and used for the assay of the various hydrolases. All assay systems were in 0.25 M sucrose containing 0.05 M sodium acetate, pH 6.5 and 1 mM substrate. Hexosaminidase activity was evaluated using 4-methylumbelliferyl-N-acetyl-β-D-glucosaminide; β-glucosidase using 4-methylumbelliferyl-β-D-glucoside; cathepsin B using benzoyl-arginine-β-naphthylamide and leucylaminopeptidase using leucine-β-naphthylamide. For all enzymes, the activity was evaluated by monitoring the increase in fluorescence intensity during 2 min of incubation at 25°C

Intact lysosomes prepared from 1 g of fresh liver from fed rabbit express 7.7% of the total converting enzyme activity in the membrane and soluble fractions (Table 6). Intact lysosomes from an equal weight of fresh liver from fasted rabbits express 15% of the total converting enzymes activity, due to the enrichment of the membrane-bound CE-II. In addition, these data show that the intact lysosomes from fasted rabbits exhibit almost seven times more CE activity than intact lysosomes from fed animals, and indeed the activity they express is equal to 50% of the total converting activity present in lyso-

Table 6. Converting enzyme activity expressed by intact lysosomes and by isolated membranes from livers of fed and fasted rabbits

Condition	Converting enzyme activity (units/g of fresh liver)			
	Total	Intact lysosomes	Intact membranes	Increase
Fed	1.2	0.084 (7%)	0.092 (7.7%)	–
Fasted	4.3	0.62 (14.4%)	0.70 (16%)	7x

Data taken from Fig. 9A and B

somes from fed animals. From these figures it is evident that lysosomes during fasting become very efficient in catalyzing the limited proteolysis of Fru-P_2ase, which may somehow contribute to the molecular adaptation of this enzyme to prolonged gluconeogenetic conditions.

Several considerations may apply to the results presented in this report. To begin with, (1) the region susceptible to limited proteolysis is highly conserved in both liver and muscle Fru-P_2ases; (2) the fact that all Fru-P_2ases are modified and show similar increases in the activity at alkaline pH value and, in the case of the liver enzymes, similar decreased sensitivity to the allosteric inhibitor AMP; (3) the fact that under conditions of enhanced gluconeogenesis a significant and specific enrichment of Fru-P_2ase converting enzyme II activity occurs in the membrane fraction and that a large part of this activity is expressed by intact lysosomes, and finally (4) the recent indications according to which limited proteolysis of Fru-P_2ase induces dissociation of the Fru-P_2ase/aldolase complex [13], which may be a very important molecular enzyme arrangement in the regulation of both glycolysis and gluconeogenesis.

Acknowledgement. The Institute of Biological Chemistry, University of Genoa, Italy, acknowledges support from the Italian Consiglio Nazionale delle Ricerche.

References

1. Botelho LH, El-Dorry HA, Crivellaro O, Chu DK, Pontremoli S, Horecker BL (1977) Digestion of rabbit liver fructose 1,6-bisphosphatase with subtilisin: sites of cleavage and activity of the modified enzyme. Arch Biochem Biophys 184:535
2. MacGregor JS (1981) Fructose 1,6-bisphosphatase and fructose 1,6-bisphosphatase aldolase: Structure, catalytic properties and their role in gluconeogenesis. Ph D thesis, Cornell University Graduate School, New York
3. Marcus F, Edelstein L, Keim PS, Heinieckson RL (1979) Sequence of the amino-terminal region of pig kidney fructose 1,6-bisphosphatase. Fed Proc 38:507
4. Pontremoli S, Melloni E, De Flora A, Horecker BL (1973a) Conversion of neutral to alkaline liver Fructose 1,6-Bisphosphatase: Chenges in molecular properties of the enzyme. Proc Natl Acad Sci USA 70:661
5. Dzugaj A, Chu DK, El-Dorry HA, Horecker BL, Pontremoli S (1976) Isolation of the S-peptide formed on digestion of Fructose 1,6-bisphosphatase with subtilisin and its non-covalent association with the enzyme protein. Biochem Biophys Res Commun 70:638
6. Horecker BL, MacGregor JS, Singh VN, Tsolas O, Sun SC, Crivellaro O, Pontremoli S (1979) Partial amino acid sequence of rabbit liver Fructose 1,6-Bisphosphatase and sites of cleavage by proteinases. FEBS "Special Meeting on Enzyme", Dubrovnik-Cavtat-Iugoslavia, April 17-21

7. Pontremoli S, Melloni E, De Flora A, Accorsi A, Balestrero F, Tsolas O, Horecker BL, Poole B (1976) Evidence for the selective release of lysosomal proteinases in fasted rabbits. Biochimie 58:149
8. Lazo PS, Tsolas O, Sun SC, Pontremoli S, Horecker BL (1978) Modification of Fructose 1,6-Bisphosphatase by a proteolytic enzyme from rat liver lysosomes. Arch Biochem Biophys 188:308
9. Pontremoli S, Melloni E, Balestrero F, Franzi AT, De Flora A, Horecker BL (1973b) Fructose 1,6-Bisphosphatase: The role of lysosomal enzyme in the modification of catalytic and structural properties. Proc Natl Acad Sci USA 70:303
10. Pontremoli S, Accorsi A, Melloni E, Schiavo E, De Flora A, Horecker BL (1974) Transformation of neutral to alkaline Fructose 1,6-Bisphosphatase. Arch Biochem Biophys 164:716
11. Melloni E, Pontremoli S, Salamino F, Sparatore B, Michetti M, Horecker BL (1980a) Characterization of three rabbit liver lysosomal proteases with Fructose 1,6-Bisphosphatase converting enzyme activity. Arch Biochem Biophys (in press)
12. Melloni E, Pontremoli S, Salamino F, Sparatore B, Michetti M, Horecker BL (1980) Fructose 1,6-bisphosphatase converting enzyme in rabbit liver lysosomes: Effect of fasting on the total and membrane-bound activities and on the activity expressed by intact lysosomes. Proc Natl Acad Sci USA (in press)
13. MacGregor JS, Snight VN, Davoust S, Melloni E, Pontremoli S, Horecker BL (1980) Evidence for the formation of a rabbit liver aldolase: rabbit liver Fructose 1,6-Bisphosphatase complex. Proc Natl Acad Sci USA 77:3889

Discussion[3]

Answer by Dr. Pontremoli to a question by Dr. Hers:
Under fasting conditions FBPase activity in the rat does not change but in the rabbit activity and total protein amount increase. In terms of physiology the most relevant effect of the limited proteolytic cleavage is certainly the desensitation to AMP, an allosteric inhibitor of the enzyme. Furthermore, activity at pH 9.2 is increased about 3-4-fold in the converted FBPase. Phosphorylation of FBPase occurs only with the rat liver enzyme, and in this case there seems to be no effect on the catalytic properties. It may be that phosphorylation enhances proteolysis of the enzyme.
Comment by Dr. Cohen:
The reason that only the rat liver FBPase can be phosphorylated by cyclic AMP-dependent protein kinases, whereas the enzyme from mouse, pig, and bovine liver cannot is that the rat enzyme is slightly bigger (MW = 40K) than the other mammalian FBPase (MW = 36K). Therefore, the rat liver enzyme appears to contain an extra 30-40 residues which contain the site of phosphorylation. This suggests that either there is a real difference in the structure and regulation of the rat liver enzyme, or that the other mammalian FBPases have lost a small peptide from either the N- or C-terminus of the polypeptide chain.
Answer by Dr. Pontremoli to the question by Dr. Hers whether FBPase is a regulated enzyme:
The rabbit liver enzyme certainly shows a response to a number of modulators. The purified enzyme is partially inactive because it contains zinc at the high affinity inhibitory sites. In order to assay the enzyme you have to add EDTA or histidine, whose liver concentration fluctuates in parallel to gluconeogenic conditions. There may be a fast regulatory control mechanism by modulators and a slower mechanism by limited proteolysis which mediates adaptation to gluconeogenic conditions.

3 Rapporteur: Peter Bünning

Release and Modulation of Membrane-bound γ-Glutamyltranspeptidase by Limited Proteolysis

N. Katunuma, Y. Matsuda, A. Tsuji, and T. Miura[1]

γ-Glutamyltranspeptidase (γ-GTP) is known to be widely distributed in cells of organisms, from microorganisms to mammals. It is closely associated with cell membranes, such as bile canaliculi of liver, the brush border of the small intestine and kidney, and the choroid plexus, but it has also been found in the serum, bile, urine, and colostrum. This enzyme catalyzes the transfer of the γ-glutamyl moiety from γ-glutamyl derivatives to certain acceptors, and hydrolysis of glutamine in the presence of maleic acid. The level of serum γ-GTP is known to increase in certain hepatobiliary diseases, and it is accepted that serum γ-GTP is derived from the liver. In this paper we present the structure of membrane-bound γ-GTP, the hydrophobic form, and discuss the possible mechanism of leakage of this enzyme into the serum from the membranes.

1 Structure of Rat Kidney γ-Glutamyltranspeptidase

γ-GTP was solubilized from the membranes of rat kidney with Triton X-100 or papain and purified to a homogeneous state. Triton-solubilized γ-GTP (T-γ-GTP) was a hydrophobic protein, binding to Octyl-Sepharose and soluble in buffer containing detergent, whereas papain-solubilized γ-GTP (P-γ-GTP) was hydrophilic, and freely soluble in aqueous buffer. The specific activities of purified T-γ-GTP and P-γ-GTP were the same and Papain treatment did not affect the catalytic activity of γ-GTP. T-γ-GTP and P-γ-GTP were both composed of two nonidentical subunits. The molecular weights of the subunits of T-γ-GTP were estimated as 50,000 and 23,000, and those of P-γ-GTP as 46,000 and 23,000 by SDS-polyacrylamide gel electrophoresis. The amino acid compositions of the subunits of T-γ-GTP and P-γ-GTP are shown in Table 1. The heavy subunit of T-γ-GTP has 52 amino acid residues more than that of P-γ-GTP, but the light subunits of T-γ-GTP and P-γ-GTP are very similar. In addition, when T-γ-GTP was treated with papain and applied on Sephadex G-200 column, hydrophilic γ-GTP (PT-γ-GTP) was eluted as same position as P-γ-GTP and small product (peptide mixture) was separated. Amino acid composition of PT-γ-GTP was very similar to that of P-γ-GTP and that of small product was very similar to the difference (A-B) of amino acid compositions of heavy subunits of T-γ-GTP and P-γ-GTP, as shown in Table 1.

Presumably, papain treatment removed part of the heavy subunit of T-γ-GTP but did not modify the light subunit. The end groups of the subunits were analyzed to con-

1 Department of Enzyme Chemistry, Institute for Enzyme Research, School of Medicine, Tokushima University, Tokushima, Japan

Table 1. Amino acid compositions of subunits of γ-glutamyltranspeptidase. Values are averages of those after 24, 48, and 72 h of hydrolysis

Amino acid	Calculated number of residues/mol protein					
	T-γ-GTP		P-γ-GTP			
	Heavy subunit (A)	Light subunit	Heavy subunit (B)	Light subunit	A-B	Small product
Aspartic acid	34	20	31	22	3	4
Threonine [a]	29	17	25	17	4	4
Serine [a]	32	20	27	20	5	5
Proline	22	12	19	12	3	2
Glutamic acid	48	18	43	18	5	3
Glycine	40	20	35	20	5	5
Alanine	44	18	39	18	5	5
Valine	28	21	23	20	5	4
Methionine [a]	9	2	8	2	1	1
Isoleucine	27	11	25	12	2	3
Leucine	43	15	35	14	8	6
Tyrosine	7	2	6	1	1	1
Phenylalanine	16	9	14	9	2	2
Histidine	7	4	7	5	0	1
Lysine	20	10	20	11	0	2
Arginine	24	8	21	7	3	1
Total	430	207	378	208	52	49

[a] Value obtained by extrapolation to zero h hydrolysis

firm this. The amino terminal residues of the heavy and light subunits of T-γ-GTP were methionine and threonine, respectively, while those of P-γ-GTP were glycine and threonine, respectively. Thus it is evident that the amino terminal of the heavy subunit of T-γ-GTP was modified by papain treatment, whereas the light subunits were not modified at the amino terminus, as summarized in Table 2. Next, the carboxyl terminals were analyzed with carboxypeptidase. The patterns of the amino acids released from T-γ-GTP and P-γ-GTP with carboxypeptidase Y in the presence of sodium dodecyl sulfate were very similar, as shown in Table 3, suggesting that the carboxyl terminal sequences of T-γ-GTP and P-γ-GTP are identical.

Therefore, the amino terminal portion of γ-GTP contains a hydrophobic domain that anchors the enzyme to the membrane (Tsuji et al. 1980a). The active site of the enzyme is reported to be located in the light subunit, and to be exposed on the surface of the cell (Inoue et al. 1977; Tate and Ross 1977).

This enzyme is reported to have many sugar residues and to have multiple forms (Tate and Meister 1976).

Therefore, we next separated the multiple forms by DEAE cellulose column chromatography. Four fractions with different isoelectric points were obtained. Examination of the properties of these fractions showed that the multiple forms of the enzyme did not differ in protein structure, but in their contents of bound carbohydrates, espe-

Table 2. Molecular weights and amino terminal amino acids of T-γ-glutamyltranspeptidase and P-γ-glutamyltranspeptidase

	Subunit	T-γ-GTP	P-γ-GTP
Molecular weight	Heavy	50,000	46,000
	Light	23,000	23,000
Amino terminus	Heavy	Methionine	Glycine
	Light	Threonine	Threonine

The molecular weights of the subunits were determined by SDS-polyacrylamide gel electrophoresis. Amino terminal amino acids were determined by dansylation

Table 3. Amino acids released from Triton-solubilized and papain-solubilized γ-glutamyltranspeptidase by treatment with carboxypeptidase Y in the presence of SDS

		mol/mol protein		
	Amino acid	15 min	30 min	60 min
T-γGTP	Tyrosine	0.36	0.48	0.73
	Phenylalanine	0.16	0.27	0.39
	Leucine	0.20	0.28	0.38
P-γGTP	Tyrpsine	0.23	0.42	0.56
	Phenylalanine	0.16	0.20	0.26
	Leucine	0.09	0.18	0.23

Samples (2 mg protein) were incubated with carboxypeptidase Y (36 μg) in 100 mM pyridine-acetate buffer, pH 5.5, in the presence of 1.0% of SDS at 25°C. The reaction was terminated by addition of 4 vol of acetone and the supernatants obtained by centrifugation were dried in a rotary evaporator. Then the free amino acids liberated were analyzed with an amino acid analyzer

cially sialic acid (Matsuda et al. 1980). Figure 1 shows the relation of the isoelectric points and contents of sialic acid of the multiple forms. The peptides released from T-γ-GTP with papain also contain some carbohydrate. Presumably, some carbohydrate residues bind close to the amino terminal of the enzyme.

The structure of the membrane-bound γ-GTP proposed from our data is shown in Fig. 2.

A hydrophobic portion with methionine at the amino terminus is buried in the phospholipid bilayer of the plasma membrane. When membrane-bound γ-GTP is treated with papain, a new amino terminal, glycine, appears and the enzyme becomes hydrophilic. Thus, hydrophilic γ-GTP (P-γ-GTP) results from removal of a hydrophobic portion from membrane-bound γ-GTP (T-γ-GTP) with protease, without change in catalytic activity.

The immunochemical properties of P-γ-GTP and T-γ-GTP were also examined (Miura et al. 1980). The single precipitin lines between T-γ-GTP and P-γ-GTP, respectively, against antiserum obtained with immunization with the purified P-γ-GTP, fused completely, as shown in Fig. 3.

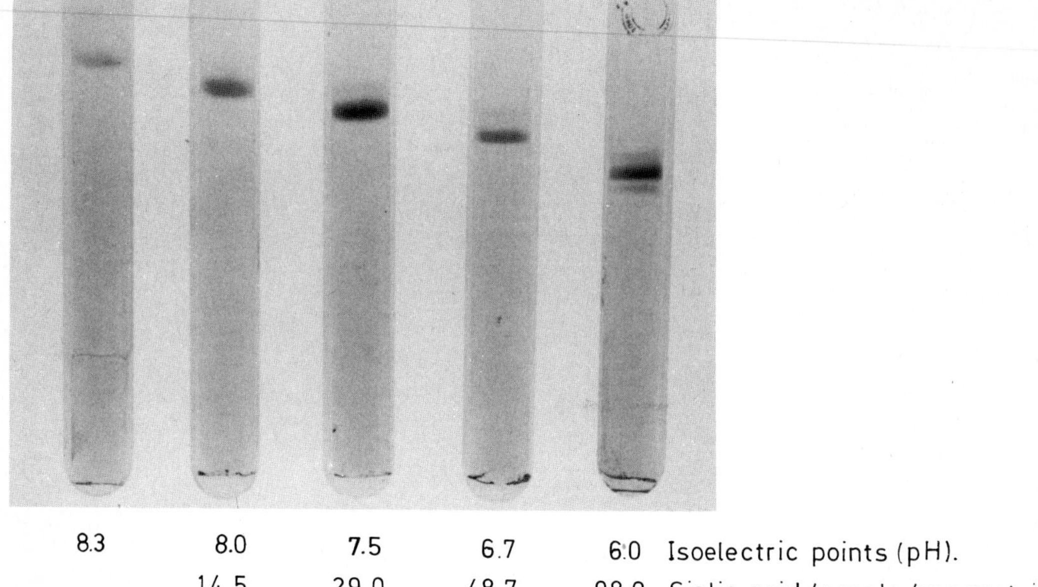

Fig. 1. Isoelectric focusing gel electrophoresis and contents of sialic acid in multiple forms of γ-Glutamyl-transpeptidase

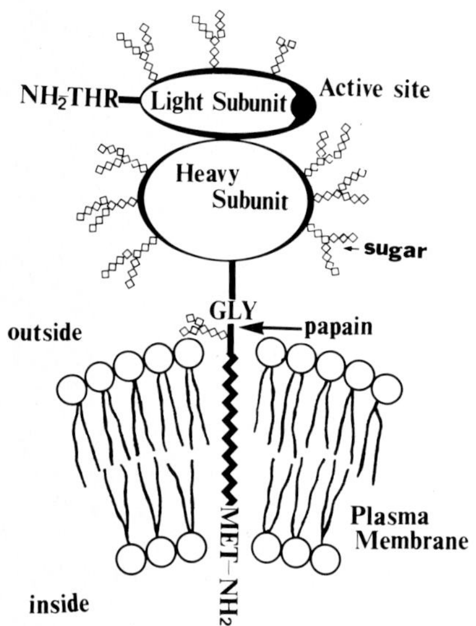

Fig. 2. Structure of membrane-bound γ-Glutamyl-transpeptidase

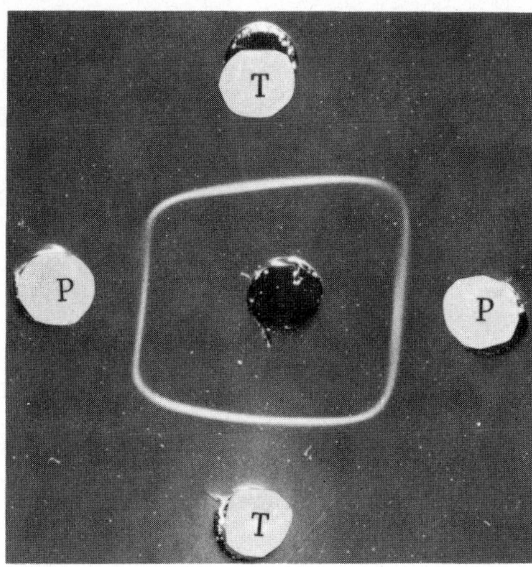

Fig. 3. Double immunodiffusion of Triton-solubilized and papain-solubilized γ-glutamyltranspeptidase against antiserum. The *center well* contains antiserum obtained with immunization wit papain solubilized γ-GTP from rat kidney. *P* contains purified papain-solubilized γ-GTP and *T* contains purified Triton-solubilized γ-GTP

Moreover 1 μl of antiserum precipitated 0.55 U of T-γ-GTP and 0.45 U of P-γ-GTP. Therefore, P-γ-GTP and T-γ-GTP are immunologically identical and the portion that binds the enzyme to the membrane does not contain antigenic determinants.

2 Proteolysis of Triton-solubilized γ-Glutamyltranspeptidase

Proteases can convert hydrophobic T-γ-GTP to hydrophilic P-γ-GTP. We next examined the activities of various proteases in this conversion. Papain, bromelin, trypsin, mast cell serine protease, cathepsin L, and cathepsin B converted T-γ-GTP to a hydrophilic γ-GTP and corresponding peptides. The hydrophilic γ-GTP formed with these endoproteases were very similar in size, mobility on polyacrylamide gel electrophoresis, and behavior on a Sephadex column, and the catalytic activity of γ-GTP was not affected by treatment with any of the proteases. The fact that various proteases all form a very similar hydrophilic P-γ-GTP could be explained by supposing that in the tertial structure of T-γ-GTP a certain portion that is susceptible to proteases is exposed, while the rest of the molecule is very resistant to proteases. Figure 4 shows the elution patterns on Sephadex G-200 of P-γ-GTPs treated with various proteases. Among the exopeptidase, amino peptidases, such as leucine amino peptidase, converted T-γ-GTP to a hydrophilic γ-GTP, whereas carboxypeptidase A, carboxypeptidase B, and carboxypeptidase Y did not release any amino acids or peptides from T-γ-GTP. After treatment with leucine amino peptidase, γ-GTP was adsorbed to concanavalin A (Con A) Sepharose and eluted with 20 mM α-methylmannoside without 1% Triton X-100, as shown in Fig. 5. Thus, the amino terminal portion, or a portion near the amino terminus, of T-γ-GTP is susceptible to proteolytic attack, but the carboxyl terminals are very resistant to proteases in the absence of a denaturant.

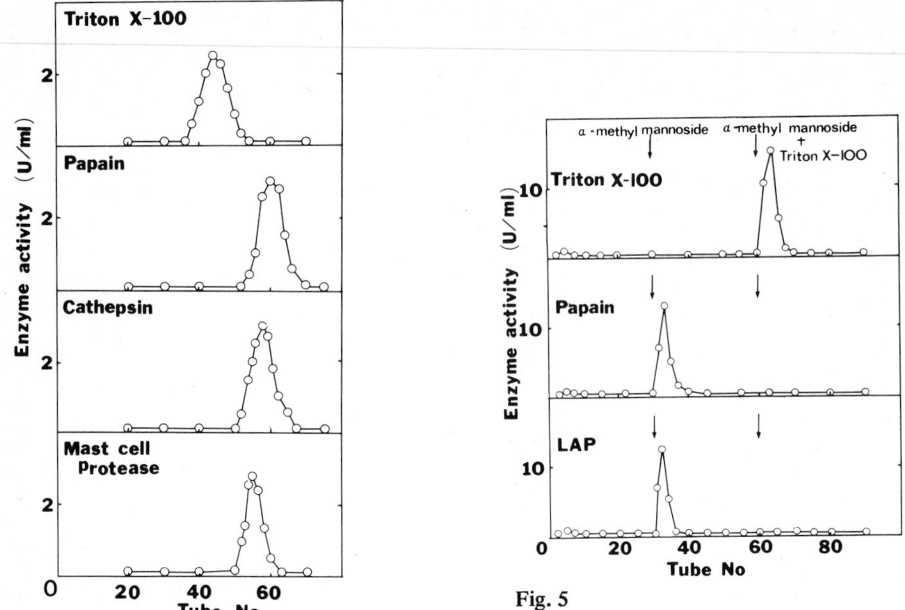

Fig. 5

Fig. 4. Elution profile of P-γ-glutamyltranspaptidase formed with various proteases on Sephadex G-200 column. Crude plasma membrane fraction of rat kidney was incubated with purified proteases (papain, or mast cell serine protease) in 50 mM sodium phosphate buffer pH 7.2, or with cathepsin B_1 and H in 50 mM acetate buffer pH 5.5 at 37°C for 6 h. After centrifugation at 105,000 g for 60 min, the supernatant was applied on Sephadex G-200 column equilibrated with 50 mM imidazole buffer pH 7.2 and eluted with the same buffer. Triton X-100-solubilized enzyme was eluted with the same buffer containing 1% Triton X-100. *Vertical* represents the activity of γ-GTP released from membranes

Fig. 5. Elution profiles from Con A Sepharose of γ-Glutamyltranspeptidases treated with leucine amino peptidase. Samples (2-4 mg T-γ-GTP) were incubated in 100 mM Tris-HCl buffer pH 8.0 containing 25 mM $MgCl_2$ and 1% Triton X-100 at 37°C for 20 h with leucine amino peptidase or papain, and then incubated with Con A Sepharose at 37°C for 3 h with shaking and then packed into a column and washed with 50 mM imidazole buffer pH 7.2. γ-GTP was eluted with buffer containing 100 mM α-methylmannoside and with same buffer containing 100 mM α-methylmannoside and 1% Triton X-100

3 Properties of Activity Releasing γ-Glutamyltranspeptidase from the Membrane

As mentioned above, membrane-bound γ-GTP was released by addition of purified proteases. Next we examined the location of the activity releasing γ-GTP from the membrane. Basal lateral membrane and brush border membrane were isolated according to the method of Wilfong and Neville (1970). As shown in Table 4, brush border membrane fraction showed high specific activities, both of γ-GTP and the activity releasing γ-GTP from the membrane, whereas basal lateral membrane fraction had both γ-GTP and releasing activity, but showed low specific activities. The optimum pH for release of γ-GTP was examined using crude plasma membrane, and a broad range of pH 9 to 11 was obtained, as shown in Fig. 6.

Table 4. Localization of γ-glutamyltranspeptidase and the activity releasing γ-glutamyltranspeptidase in rat kidney

	Protein	γ-GTP		Releasing activity	
		Total	Sp.Act.	γ-GTP released	Sp.Act.
	(mg)	(U)	(U/mg)	(U)	(mU/ng)
Basal lateral membrane	16.8	12	0.70	0.184	11.0
Brush border membrane	2.5	12	4.73	0.088	35.2

Membrane fraction was incubated in 50 mM Tris-HCl buffer pH 9.0 at 37°C for 2 h, cooled to 0°C and centrifuged at 105,000 g for 30 min. The activity of γ-GTP in the supernatant was determined. Unit of releasing activity was expressed as unit of γ-GTP released in the supernatant at 37°C for 2 h

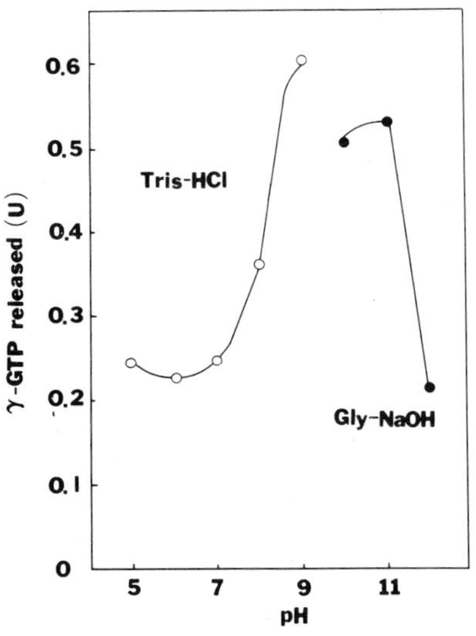

Fig. 6. Optimum pH for release of γ-Glutamyltranspeptidase from membrane. Crude plasma membrane fraction of rat kidney was incubated in the buffer with indicated pH at 37°C for 6 h and centrifuged at 105,000 g for 60 min. Activity of γ-GTP released in the supernatant was determined

Then various protease inhibitors were tested to characterize the activity releasing γ-GTP from the membrane.

As shown in Table 5, phenylmethylsulfonylfluoride was the most effective (82% inhibition), followed by chymostatin (25% inhibition). Therefore, the release seems to be due to a chymostatin-sensitive serine protease with an alkaline optimum pH.

4 Serum γ-Glutamyltranspeptidase and Bile γ-Glutamyltranspeptidase

γ-GTP is tightly bound to the cell membrane, and protease or detergent is required to solubilize it. Less than 1% of the total γ-GTP is found in the cytosol of kidney. The membrane-bound γ-GTP seems to be very similar to purified T-γ-GTP, containing a

Table 5. Effects of inhibitors on release of γ-glutamyltranspeptidase from membranes

Inhibitor	Concentration	Activity of γ-GTP released	
		Units	% Activity
	(mM)		
None		0.844	100
E-64	0.5	0.795	94
PMSF [a]	1.0	0.152	18
Pepstatin	60	0.714	85
Chymostatin	0.1	0.630	75
Elastatinal	0.02	0.930	111
EDTA	1.0	0.907	107
EGTA	1.0	0.777	92
2-Mercaptoethanol	5.0	0.793	94
Soybean trypsin inhibitor	0.01	0.940	111

[a] Phenylmethanesulfonylfluoride

Crude plasma membrane fraction was incubated in the 50 mM Tris-HCl buffer pH 9.0 with indicated inhibitors at 37°C for 2 h and the mixture was centrifuged at 105,000 g for 30 min and γ-GTP activity in the supernatant was determined

Table 6. Solubilities of detergent-solubilized, papain-solubilized, bile, and serum γ-glutamyltranspeptidase

	Triton-solubilized γ-GTP		Bile γ-GTP	
	Total activity (mU)	Recovery (%)	Total activity (mU)	Recovery (%)
Before acetone treatment	3,011.0	(100)	1,364.0	(100)
Sup A	152.1	(5.1)	365.4	(26.7)
Sup B	2,425.9	(80.5)	984.0	(72.1)
Insoluble	100.5	(3.3)	0	(0)
	Papain-solubilized γ-GTP		**Serum γ-GTP**	
	Total activity (mU)	Recovery (%)	Total activity (mU)	Recovery (%)
Before acetone treatment	2,784.0	(100)	1,165.0	(100)
Sup A	2,755.2	(99.7)	983.8	(84.4)
Sup B	48.0	(1.7)	34.9	(3.0)
Insoluble	0	(0)	16.3	(1.4)

Samples were treated with 10 vol of chilled acetone at −2°C for 20 min, and insoluble material was collected by centrifugation at 10,000 g for 20 min and extracted overnight with 50 mmol/l imidazole, pH 7.2, at 4°C. Then the preparation was centrifuged at 105,000 g for 60 min. The resulting supernatant is referred to as Sup A. The precipitate was extracted overnight with 50 mmol/l imidazole, pH 7.2, containing 1.0% of Triton X-100 at 4°C, and then centrifuged at 105,000 g for 60 min. The supernatant is referred to as Sup B, and the precipitate as the insoluble fraction

hydrophobic segment that anchors it to the membrane. Serum γ-GTP and bile γ-GTP were partially purified by a procedure not causing proteolytic modification. On acetone treatment to remove bound detergent, serum γ-GTP become hydrophilic while bile γ-GTP become hydrophobic, as shown in Table 6.

P-γ-GTP was adsorbed to Con A Sepharose and eluted with 100 mM α-methylmannoside. The elution profile of serum γ-GTP adsorbed to Con A Sepharose was the same as that of P-γ-GTP, whereas bile γ-GTP behaved like T-γ-GTP, as shown in Fig. 5.

Bile-γ-GTP also bound to Octyl-Sepharose, a hydrophobic resin, and was eluted with Tween 20, but, serum γ-GTP, like P-γ-GTP, did not bind to Octyl-Sepharose. Thus, bile γ-GTP seems to be solubilized with detergent from the cell membrane and still to contain a hydrophobic domain, whereas serum γ-GTP is a hydrophilic protein formed by proteolysis that removes the hydrophobic domain (Tsuji et al. 1980b).

Acknowledgements. This work was supported by a Grant-in-Aid for Scientific Research from the Ministry of Education, Science and Culture of Japan. We thank Mrs. E. Inai for help in preparing the manuscript.

References

Inoue M, Horiuchi S, Morino Y (1977) Eur J Biochem 73:335-342
Matsuda Y, Tsuji A, Katunuma N (1980) J Biochem 87:1243-1248
Miura T, Matsuda Y, Tsuji A, Katunuma N (1981) J Biochem 89:217-222
Tate SS, Meister A (1976) Proc Natl Acad Sci USA 73:2599-2603
Tate SS, Ross ME (1977) J Biol Chem 252:6042-6045
Tsuji A, Matsuda Y, Katunuma N (1980a) J Biochem 87:1567-1571
Tsuji A, Matsuda Y, Katunuma N (1980b) Clin Chim Acta 104:361-366
Wilfong RF, Neville Jr DM (1970) J Biol Chem 245:6106

Discussion[2]

Comment by Dr. Blobel:
A comparison of the T-γ-GTP with the P-γ-GTP shows that the amino terminal tail which is cleaved off by papain consists of about 50 amino acid residues. However, only 25 amino acid residues would be required to span the plasma membrane. The amino acid composition of the amino terminal tail indicates that it is not entirely hydrophobic and, therefore, it is possible that it contains a hydrophilic portion which sticks into the cytoplasm.

2 Rapporteur: Peter Bünning

Inactivation of Cytosol Enzymes by a Liver Membrane Fraction

G.L. Francis and F.J. Ballard[1]

1 Introduction

In experiments involving a range of cultured mammalian cells derived from several different tissues we have shown that lysosomotropic agents such as proteolytic inhibitors or weak bases are only able to reduce intracellular protein breakdown by about 30% (Knowles and Ballard 1976; Ballard 1977; Hopgood and Ballard 1980; Livesey et al. 1980). A similar maximal extent of inhibition is observed with growth factors and protein synthesis inhibitors, agents that are also thought to act on lysosomal proteolysis (Ballard 1977; Ballard et al. 1980). Furthermore, combinations of these different compounds are no more effective than any one effective agent added at an optimal concentration. Consequently, we have proposed that lysosomal protein breakdown only accounts for about a third of the total rate of intracellular protein breakdown. This value is almost certainly an upper limit produced by carrying out the experiments in serum-free medium, because lysosomotropic agents are usually without effect on protein breakdown when tested on cells incubated in normal growth medium containing 10% of fetal calf serum (unpublished data). A second difficulty with a lysosomal pathway of intracellular proteolysis is explaining how it could be compatible with the widely different half-lives of cell proteins (Goldberg and St. John 1976). Some type of selective uptake of proteins having short half-lives or selective exclusion of proteins with long half-lives must be proposed.

Accordingly, we have investigated nonlysosomal processes which might satisfy the restrictions listed above. We have emphasized initial reactions in protein breakdown which may be rate-limiting for the total degradative pathway, using as a model the loss of catalytic activity of liver enzymes.

2 Distribution and Characterization of the Inactivation Reaction

Initial experiments showed that when liver homogenates were incubated at 37°C several enzymes present in the cytosol fraction lost activity roughly in the same order as their degradation rates in vivo (Hopgood and Ballard 1974). This finding encouraged us to investigate the intracellular distribution of the factor which accounted for enzyme inactivation and to attempt its purification. The factor was present in all liver membranes that were examined (Francis and Ballard 1980a). Especially high activities were observed

1 CSIRO Division of Human Nutrition, Kintore Avenue, Adelaide, SA 5000, Australia

in plasma membranes with somewhat less in microsomes and the lowest in lysosomal membranes. Purification of plasma membranes from collagenase-isolated hepatocytes and the demonstration that such plasma membranes were as active as those obtained from total liver eliminated Kuppfer and other endothelial cells as the major source of the inactivation capacity (Francis and Ballard 1980a).

The inactivation factor could not be extracted from microsomes by washing with solutions of high ionic strength and even extraction with 1% Triton X100 resulted in only a modest yield of the factor. However, the inclusion of trypsin (1 mg per 10 mg of microsomal protein) with Triton X100 in the extraction mixture resulted in substantial solubilization of the factor from microsomes incubated at 37° for 90 min. Subsequent purification using, in sequence, ammonium sulfate precipitation, Sepharose CL-6B, DEAE-cellulose and hydroxyapatite chromatography resulted in approximately 100-fold purification. Triton X100 was required at all stages during the purification. Inactivation capacity was localized in a single region when the purified material was electrophoresed on nondenaturing gels. The factor appeared about 30% pure (Francis and Ballard 1980a).

The factor inactivated pure, rat liver, glucose-6-phosphate dehydrogenase (G6PD) without other additions and retained a selective action on enzymes with short half-lives in vivo (Table 1).

Table 1. Inactivation of liver cytosol enzymes by purified inactivation factor

Enzyme	Half-life in vivo [a]	Half-time for inactivation in vitro [b] (min)
Tyrosine aminotransferase	2 h	45
Phosphoenolpyruvate carboxykinase	6 h	34
Glucose-6-phosphate dehydrogenase	6 h	24
Fructose bisphosphatase	2 days	85
Aldolase	3 days	100
Alanine aminotransferase	3.5 days	120
Lactate dehydrogenase	6 days	stable
Aspartate aminotransferase	11 days	stable

[a] Half-lives are taken from Hopgood and Ballard (1974); Litwack and Rosenfield (1973); Peavy and Hansen (1979)

[b] Measured by using the standard inactivation assay described by Francis and Ballard (1980a)

L-cystine was included at a concentration of 2 mM in the experiment described in Table 1 since this agent had been shown earlier to accelerate the inactivation rate. Cystine alone did not cause loss of catalytic activity except at much higher concentrations.

We noted that inactivation of the three pyridoxal-containing enzymes, tyrosine aminotransferase, alanine aminotransferase and aspartate aminotransferase, was not only in the same order as their half-lives in vivo, but was in proportion to the relative dissociation rates for pyridoxal phosphate (Litwack and Rosenfield 1973). It is possible that inactivation occurs after loss of the cofactor, although we have no evidence on that point.

Liver was chosen for most studies because much more information was available for this tissue concerning the half-lives of enzymes in vivo. However, crude microsomal fractions from several rat tissues were shown to inactivate G6PD in the absence of cystine and at greater rates when 2 mM cystine was included (Table 2).

Table 2. Inactivation of liver glucose-6-phosphate dehydrogenase by crude membranes from different rat tissues

Source of membranes[b] (0.25 mg protein)	Percent G6PD activity lost in 30 min [a]	
	No cystine	+ 2 mM cystine
None (assay control)	0	0
Small intestine	60	86
Spleen	33	81
Lung	41	79
Thymus	41	79
Brain	31	69
Liver	13	64
Heart	28	63
Stomach	16	60
Pancreas	9	41
Kidney	0	40
Skeletal muscle	13	34
Testis	0	0
Reticulocytes	0	0
Erythrocytes	0	0

[a] Measured using the standard assay described by Francis and Ballard (1980a)
[b] Prepared by centrifuging tissue homogenates at 65000 g for 100 min and extracted in 1% Triton X100

Indeed membranes from small intestine, spleen, lung, and thymus had substantially higher activities than liver membranes. No activity could be detected in membranes from testis, reticulocytes, or erythrocytes.

3 Properties of the Inactivation Reaction

The inactivation reaction does not involve proteolytic cleavage. A wide range of proteolytic inhibitors including inhibitors of aspartate, serine and cysteine proteinases, as well as the unspecific inhibitors in serum, were without effect on the inactivation of phosphoenolpyruvate carboxykinase (PEPCK) by microsomes (Ballard and Hopgood 1976). Also when [^{3}H]-labeled PEPCK was incubated with microsomes, catalytic activity was lost before loss of reactivity toward a specific antibody and well before any evidence of proteolytic cleavage (Ballard et al. 1974). By including an internal marker of [^{14}C]-labeled PEPCK in the sodium dodecyl sulfate gels used to examine the breakdown products of [^{3}H]-labeled PEPCK, it was possible to detect any small changes in molecular weight as well as any internal cleavage in the PEPCK molecule. However, all catalytic activity was lost before any change in the [^{3}H] to [^{14}C] radioactivity ratio was seen across the PEPCK peak (Ballard et al. 1974).

A further line of evidence against a proteolytic involvement in the inactivation reaction is the ability to reactivate some of the inactive enzyme. This property was first shown with purified G6PD (Francis et al. 1980), but can also be demonstrated using cytosol as a source of the enzymes (Fig. 1). At early stages during the inactivation of a rel-

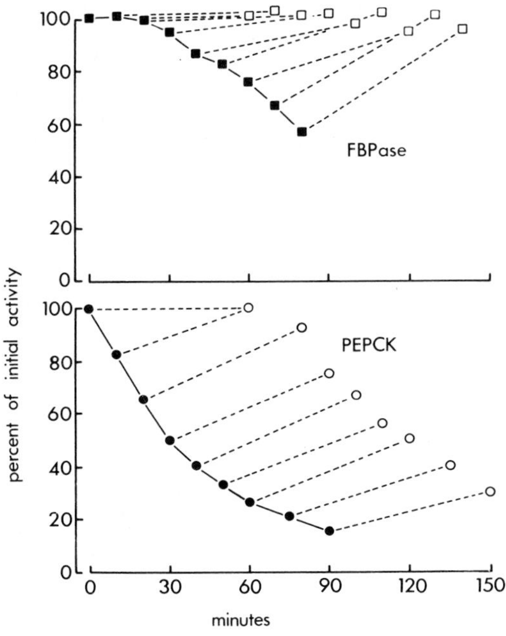

Fig. 1. Inactivation of phosphoenolpyruvate carboxykinase (*PEPCK*) and fructose 1,6-bisphosphatase (*FBPase*) by purified inactivation factor and reactivation with dithiothreitol. A Cytosol fraction was incubated with 2 mM cystine and purified inactivation factor at 37°C for various times (■, ●). Subsequently samples were taken for assay of *PEPCK* or *FBPase* and also for incubation with 25 mM dithiothreitol for a further 60 min at 37°C prior to the assay of *PEPCK* or *FBPase* (□, ○). Values are expressed as percentages of the initial activity and are corrected for slight losses in activity of control incubations incubated without the inactivation factor

atively unstable enzyme (PEPCK) and following 50% inactivation of a relatively stable enzyme (FBPase), complete recovery of catalytic activity occurred when dithiothreitol was added to the reaction mixture. Similar patterns of reactivation were observed with aldolase, alanine aminotransferase, and tyrosine aminotransferase. This reversibility argues against a proteolytic cleavage and also against destruction of the enzymes by oxygen radicals. Reversibility of inactivation is lost with increasing incubation of enzymes with the factor (Fig. 1). We consider that this long term change is due to denaturation of the modified enzymes.

Inactivation is stimulated not only by L-cystine but also by D-cystine, cystamine, and oxidised glutathione. Protection against inactivation is observed with L-cysteine, reduced glutathione and dithiothreitol (Ballard and Hopgood 1976).

4 Proposed Mechanism for the Inactivation Reaction

High concentrations of L-cystine (10 mM or greater) can inactivate PEPCK or G6PD without the inactivation factor being present. This occurs especially if purified enzymes are used in place of cytosol in the inactivation assay. A likely explanation for this finding is that surface thiols in the enzyme substrates are oxidized by disulfides in the inactivation factor, as illustrated in Fig. 2 (see also Ballard et al. 1977; Francis and Ballard 1980b).

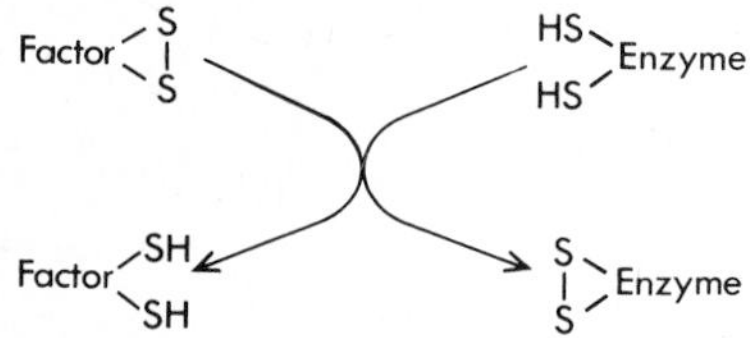

Fig. 2. A proposed model for the inactivation reaction. Details are explained in the text

One prediction of this mechanism is that disulfides would be essential in the membrane factor. In order to test this prediction we reduced membrane disulfides and subsequently tested the reduced factor in the inactivation assay. Plasma membranes from rat liver were incubated with 50 mM dithiothreitol, after which the reducing agent was removed by pelleting and washing the membranes. This treatment completely destroyed the inactivation capacity of the plasma membranes (Table 3). However, if these "reduced" membranes were left at 0°C for several hours, the inactivating capacity was partially restored, presumably due to oxidation of the thiols. A second incubation with 100 mM iodoacetamide immediately after dithiothreitol treatment prevented the restoration of inactivation capacity. We interpret this experiment as reaction of iodoacetamide with the membrane thiols produced by the dithiothreitol treatment so that no oxidation during the subsequent incubation could occur. Control experiments showed that iodoacetamide had no action on the membranes unless a prior reduction step was involved.

Table 3. Effect of Dithiothreitol (DTT) treatment of plasma membranes on the ability to inactivate glucose 6-phosphate dehydrogenase (G6PD)

First treatment [a]	Second treatment[a]	% G6PD inactivated in 120 min [b]
None	None	97
50 mM DTT	None	0
None	100 mM IAM [c]	97
50 mM DTT	100 mM IAM	0
50 mM DTT	Incubation in air	67
50 mM DTT	100 mM IAM, incubation in air	0

[a] Details are given in the text
[b] Measured as described by Francis and Ballard (1980b)
[c] IAM, iodoacetamide

A second prediction from the mechanism in Fig. 2 is that inactivation of enzymes should lead to a loss of titrable thiols. Reaction of native G6PD with radioactive iodoacetamide indicated 0.84 mol of available thiol per mol of enzyme. However, following 80% inactivation of G6PD there was a diminution of available thiols from 0.84 to 0.37

mol/mol G6PD (Table 4). An intermediate thiol content was observed if 1 mM NADP was included in the reaction mixture. Under these conditions inactivation of G6PD was partially prevented. Loss of available thiols in inactivated G6PD and the other enzymes

Table 4. Loss of thiols in glucose-6-phosphate dehydrogenase (G6PD) accompanies inactivation [a]

Incubation conditions	% G6PD inactivated in 3 h	Iodoacetamide bound (mol/mol G6PD)
No factor	24	0.84
Complete	80	0.37
Complete + 1 mM NADP	60	0.52

[a] Details of these experiments are given by Francis and Ballard (1980b)

tested would also account for the ability of dithiothreitol to reactivate the enzymes (Fig. 1). We have not included reversibility in the mechanism proposed in Fig. 2, but it would be expected to occur as a result of the high intracellular concentrations of reduced glutathione in tissues. These GSH concentrations as well as the high GSH to GSSG ratio would also protect enzymes against inactivation by the membrane factor.

Ligands may stabilize enzymes in a conformation that is less susceptible to inactivation. We have shown that NADP, a substrate for G6PD, can prevent inactivation of the enzyme if the nucleotide is added at high concentrations (see Table 4). Similarly, Mn^{2+} or Co^{2+}, which are required in the PEPCK reaction, can prevent the inactivation of that enzyme (Ballard and Hopgood 1976). Co^{2+} or Mn^{2+} did not protect PEPCK against "inactivation" catalyzed by trypsin.

What is the fate of the inactivated enzyme? Earlier experiments with [^{3}H]-labeled PEPCK showed that inactivation of the enzyme with microsomes led to the binding of label to the membranes (Ballard and Hopgood 1976). Whether the enzyme was attached to the microsomes via a mixed disulfide was not determined, but such a binding would be consistent with the proposed reaction mechanism. With G6PD, on the other hand, only a small proportion of the inactivated enzyme was bound to plasma membranes, suggesting that any disulfides formed were either intramolecular or between G6PD molecules. The latter possibility was excluded by showing that the molecular weight of G6PD did not change upon inactivation.

Enzyme inactivated by the factor might be a substrate for lysosomal proteolysis. The uptake of denatured proteins into lysosomes in vitro in the presence of ATP has been described in an abstract by Yumoto and Mori (1976). We have attempted similar experiments with G6PD inactivated by the purified factor but have been unable to show any significant binding or uptake of the protein into lysosomes. However, inactivated G6PD is a better substrate for proteolytic digestion than the native enzyme (Table 5). This was shown by preparing [^{125}I]-labeled G6PD, inactivating a portion of it with the purified factor and treating both enzyme forms with different amounts of trypsin. Subsequently the proteins were separated by sodium dodecyl sulfate electrophoresis and the radioactivity measured that was associated with the G6PD monomer. Low temperatures were required to prevent further inactivation prior to the proteolysis step. It can

Table 5. Tryptic degradation of native glucose-6-phosphate dehydrogenase (G6PD) and G6PD inactivated by the membrane factor

Ratio of trypsin to G6PD	Native G6PD	Inactivated G6PD
0	100 [a]	100
0.01	98	84
0.05	78	56
0.1	71	44
0.5	42	14

[a] Values are the percent of radioactivity coinciding with the mobility of G6PD on SDS gels after incubation of [^{125}I]-labeled G6PD for 10 min at 0°C

be seen from Table 5 that native G6PD is more resistant to proteolytic degradation than the modified enzyme. Similar results were obtained when trypsin was replaced by chymotrypsin or by a lysosomal extract from liver.

We have no direct evidence that the inactivation factor acts in a catalytic way, although incubation of a tenfold excess of G6PD with the purified factor certainly leads to complete inactivation of the enzyme. This could occur by an inactivation cycle producing thiols in the factor followed by oxidation of these thiols. The oxidation might occur in vivo via the microsomal mixed function oxidases.

Disulfide exchange reactions have been proposed to be involved in a number of cellular events, including the inactivation of tyrosine aminotransferase (Beneking et al. 1978), the reduction of insulin and other extracellular proteins having disulfide bridges (glutathione insulin transhydrogenase; Ansorge et al. 1973) and during protein synthesis (Freedman and Hawkins 1977). We are unsure about any relationship between the inactivation factor and these various reactions, although we could not measure any glutathione insulin transhydrogenase activity in the purified factor.

5 Relevance of the Inactivation Reaction in Intracellular Protein Breakdown

In an attempt to determine whether all labile proteins are susceptible to the inactivation factor, we prepared PEPCK containing azetidine-2-carboxylic acid in place of proline. PEPCK was induced in hepatoma cells in the presence of the analog and was shown to be more labile both in the intact cells and in extracts (Knowles and Ballard 1978). However, PEPCK containing azetidine-2-carboxylic acid was not a better substrate for the inactivation reaction than native enzyme. This was true whether the inactivation factor was obtained from liver or from the hepatoma (Francis et al. 1980).

We suggest that reaction with the inactivation factor represents just one way in which an intracellular protein can enter the degradation sequence. Another would be the inherent susceptibility to unfolding produced by errors in the primary sequence of a protein, as with the azetidine-2-carboxylic acid experiments. Ligands can be viewed as "locking" a protein conformation in a less susceptible form. Perhaps the common factor in all mechanisms that target a protein for degradation is the susceptibility of proteins to denaturation-like processes. These can be a result of surface hydrophobic regions

(Bohley 1968) or the exposure of normally buried regions that may be reactive to disulfides, to proteolytic attack, or to denaturation.

No direct experiments have been carried out to test whether the inactivation factor has a role in protein breakdown. Certainly the correlation between inactivation susceptibility and half-lives in vivo is relevant, but it can be explained by a number of trivial processes. Experiments designed to alter the intracellular ratio of thiols to disulfides have not been successful, perhaps because the manipulations are too toxic for the cells. However, a number of approaches are now available by which the inactivation factor in one cell type can be introduced into a second cell which does not contain the factor. Such cell fusion or microinjection experiments are crucial if we are to understand the initial, rate-limiting reactions in protein breakdown.

References

Ansorge S, Bohley P, Kirshke H, Langer J, Marquardt I, Hanson H (1973) The identity of the insulin degrading thiol-protein disulfide oxidoreductase (glutathione-insulin transhydrogenase) with the sulfhydryl-disulfide interchange enzyme. FEBS Lett 37:238-240

Ballard FJ (1977) Intracellular protein degradation. Essays Biochem 13:1-37

Ballard FJ, Hopgood MF (1976) Inactivation of phosphoenolpyruvate carboxykinase (GTP) by liver extracts. Biochem J 154:717-724

Ballard FJ, Hopgood MF, Reshef L, Hanson RW (1974) Degradation of phosphoenolpyruvate carboxykinase (GTP) in vivo and in vitro. Biochem J 140:531-538

Ballard FJ, Hopgood MF, Knowles SE, Francis GL (1977) Attempts to relate enzyme inactivation to degradation in vivo. Acta Biol Med Germ 36:1805-1813

Ballard FJ, Knowles SE, Wong SSC, Bodner JB, Wood CM, Gunn JM (1980) Inhibition of protein breakdown in cultured cells is a consistent response to growth factors. FEBS Lett 114:209-212

Beneking M, Schmidt H, Weiss G (1978) Subcellular distribution of a factor inactivating tyrosine aminotransferase. Study of its mechanism and relationship to different forms of the enzyme. Eur J Biochem 82:235-243

Bohley P (1968) Intrazelluläre Proteolyse. Naturwissenschaften 55:211-217

Francis GL, Ballard FJ (1980a) Distribution and partial purification of a liver membrane protein capable of inactivating cytosol enzymes. Biochem J 186:571-579

Francis GL, Ballard FJ (1980b) Enzyme inactivation via disulphide-thiol exchange as catalyzed by a rat liver membrane protein. Biochem J 186:581-590

Francis GL, Knowles SE, Ballard FJ (1980) Inactivation of cytosol enzymes by a liver membrane protein. In: Protein degradation in health and disease. Ciba Foundation Symposium 75. Excerpta Medica, Amsterdam, pp 123-137

Freedman RB, Hawkins HC (1977) Enzyme-catalyzed disulphide interchange and protein biosynthesis. Biochem Soc Trans 5:348-357

Goldberg AL, St John AC (1976) Intracellular protein degradation in mammalian and bacterial cells, part 2. Annu Rev Biochem 45:747-803

Hopgood MF, Ballard FJ (1974) The relative stability of liver cytosol enzymes incubated in vitro. Biochem J 144:371-376

Hopgood MF, Ballard FJ (1980) Regulation of protein breakdown in hepatocyte monolayers. In: Protein degradation in health and disease. Ciba Foundation Symposium 75. Excerpta Medica, Amsterdam, pp 205-218

Knowles SE, Ballard FJ (1976) Selective control of the degradation of normal and aberrant proteins in Reuber H35 hepatoma cells. Biochem J 156:609-617

Knowles SE, Ballard FJ (1978) Effects of amino acid analogues on protein synthesis and degradation in isolated cells. Br J Nutr 40:275-287

Litwack G, Rosenfield S (1973) Coenzyme dissociation, a possible determinant of short half-life of inducible enzymes in mammalian liver. Biochem Biophys Res Commun 52:181-188

Livesey G, Williams KE, Knowles SE, Ballard FJ (1980) Effects of weak bases on the degradation of endogenous and exogenous proteins by rat yolk sacs. Biochem J 188:895-903

Peavy DE, Hansen RJ (1979) Influence of diet on the in vivo turnover of glucose-6-phosphate dehydrogenase in rat liver. Biochim Biophys Acta 586:22-30

Yumoto S, Mori S (1976) Energy-dependent proteolysis by isolated rat liver lysosomes. J Cell Biol 70:2809

Intracellular Protein Topogenesis

G. Blobel[1]

1 Introduction

A cell contains millions of protein molecules. These are steadily being synthesized and degraded. At homeostasis, a given species of protein is represented by a characteristic number of molecules that is kept constant within a narrow range. Very little is known about the cell's accounting procedures, i.e., how it balances and controls biosynthesis and biodegradation.

An important aspect of biosynthesis [1, 2] as well as biodegradation [3] is the intracellular topology of proteins. Many protein species spend their entire life in the same compartment in which they are synthesized, others have to be translocated across the hydrophobic barrier of one or two distinct cellular membranes in order to reach the intracellular compartment or extracellular site where they exert their function. Numerous protein species have to be integrated asymmetrically into distinct cellular membranes. For many proteins this requires partial translocation, i.e., selective transfer of one or several distinct hydrophilic or charged segments of the polypeptide chain across the hydrophobic barrier of one or two intracellular membranes. Following complete or partial translocation across a translocation-competent membrane(s), subpopulations may undergo further "posttranslocational" traffic. Soluble or membrane proteins may be shipped in bulk or by receptor-mediated processes from a translocation-competent donor compartment to a translocation-incompetent receiver compartment. This posttranslocational traffic may be unidirectional (in which case the protein ends up as a permanent resident of a particular cellular membrane) or may follow a cyclic pattern between distinct cellular membranes (e.g., recycling of receptors).

The collective term "topogenesis" has been introduced [2] to encompass protein translocation (partial or complete) across membranes as well as subsequent posttranslocational protein traffic. Not included in these processes that define topogenesis are distinct traffic patterns that may be required for protein degradation. Theoretical considerations on the topology of protein degradation have been presented elsewhere [3] and will not be dealt with here: in essence, these considerations argue for the existence of three (animal cells) or even four (plant cells) separate compartments for protein degradation, each containing a distinct set of proteases. Detailed proposals have been made also for protein topogenesis [2]. The essence of these proposals is that the information for intracellular protein topogenesis resides in discrete "topogenic" sequences that constitute a permanent or transient part of the polypeptide chain. The repertoire of distinct topo-

1 Laboratory of Cell Biology, The Rockefeller University, New York, NY 10021, USA

genic sequences was predicted to be relatively small because many different proteins would be topologically equivalent, i.e., targeted to the same intracellular address. The information content of topogenic sequences would be decoded, each by a distinct effector. Four types of topogenic sequences were distinguished: (1) *Signal* sequences initiate translocation of proteins across specific membranes. They would be decoded and processed by protein translocators that, by virtue of their signal sequence-specific domain and their unique location in distinct cellular membranes, effect unidirectional translocation of proteins across specific cellular membranes. (2) *Stop-transfer* sequences interrupt the translocation process that was previously initiated by a signal sequence and, by excluding a distinct segment of the polypeptide chain from translocation, yield asymmetric integration of proteins into translocation-competent membranes. (3) *Sorting* sequences would act as determinants for posttranslocational traffic of subpopulations of proteins, originating in translocation-competent donor membranes (and compartments) and going to translocation-incompetent receiver membranes (and compartments). A specific example for sorting, namely the separation of secretory from lysosomal proteins is discussed. (4) *Insertion* sequences initiate unilateral integration into the lipid bilayer without the mediation of a distinct protein effector.

An attempt is made here to amplify some of these previous proposals [1, 2] and to discuss some of the recent experimental data are relevant to these proposals.

2 Signal Sequences

Table 1 lists the biological membranes or membrane pairs (see *h* and *i*) that have been proposed [2] to be endowed with a transport system (translocator) for unidirectional

Table 1. Cellular membranes proposed to be endowed with a transport system (translocator) for the unidirectional translocation of nascent or newly synthesized proteins [taken from ref. 2]

Mode of translocation	Membrane	Code
Cotranslational	a. Prokaryotic plasma membrane	PPM
	b. Inner mitochondrial membrane	IMM
	c. Thylakoid membrane	TKM
	d. Rough endoplasmic reticulum	RER
Posttranslational (across *one* membrane)	e. Outer mitochondrial membrane	OMM
	f. Outer chloroplast membrane	OCM
	g. Peroxisomal membrane	PXM
Posttranslational (across *two* membranes)	h. Mitochondrial envelope	MEN
	i. Chloroplast envelope	CEN

Each of the translocation-competant membranes [1] listed here (*a-i*) is proposed to contain only *one* distinct "translocator" (in multiple copies). Each translocator responds to *one* type of signal sequence. Translocation can proceed across a *single* membrane (*a-g*), or *two* membranes (*h-i*), cotranslationally (*a-d*), or posttranslationally (*e-i*). Suggested abbreviations for these translocation-competent membranes might serve as useful codes. For example, a signal sequence (Si) addressed to the rough endoplasmic reticulum (RER), to the chloroplast envelope (CEN), etc., might be designated Si(RER), Si(CEN), etc. Likewise, a particular signal receptor (SiR), or signal peptidase (SiP) could be classified as SiR(RER), SiR(CEN), or SiP(RER), SiP(CEN), etc.

translocation of nascent or completed polypeptide chains. The conjecture was made [2], based on evolutionary relationships between various cellular membranes (see Fig. 1), that the present translation-coupled (cotranslational) translocation systems (Table 1, *a-d*) are derived from a common ancestral system and that they might be highly conserved [2]. A high degree of conservation has indeed been demonstrated for the RER translocator within the animal and plant kingdoms [4]. Recently it has been demonstrated [5,6] that a

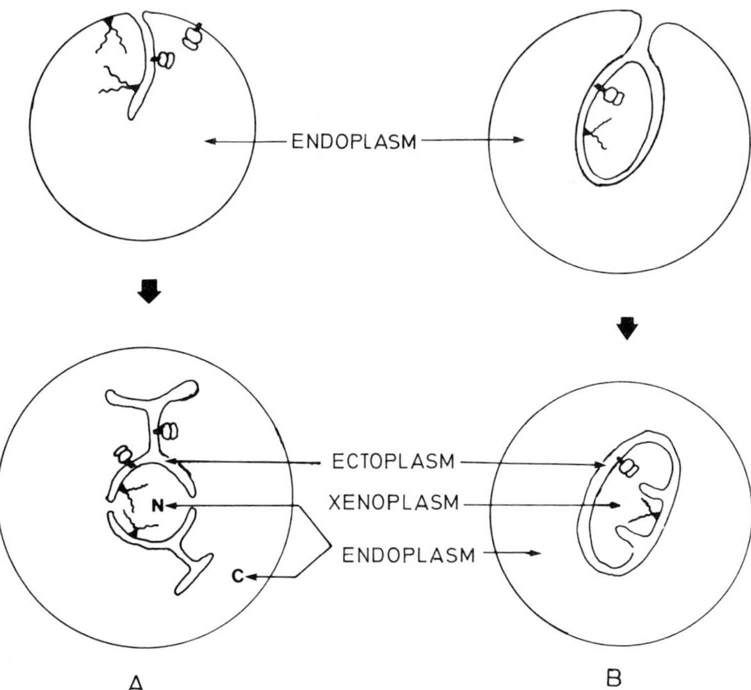

Fig. 1A, B. Schematic illustration of the evolution of intracellular membranes and compartments [taken from ref. 2]. **A** Aggregation of certain membrane functions in the plane of the pluripotent plasma membrane; nonrandom removal of these functions from the plasma membrane by invagination and fission resulting in the formation of a nuclear envelope (pore complexes omitted) continuous with the endoplasmic reticulum (rough and smooth) and generating an ectoplasmic compartment. The endoplasmic compartment is thereby subdivided into nucleoplasm (*N*) and cytoplasm (*C*). Note, however, that *N* and *C* remain connected via nuclear pores that do not present a membraneous barrier. Other intracellular membranes that are distinct from the endoplasmic reticulum, such as lysosomal, peroxisomal, and Golgi complex membranes, could have also developed by invagination from the plasma membrane or could be an outgrowth of the endoplasmic reticulum. **B** Symbiotic capture of another cell generating an additional xenoplasmic compartment. Green plant cells have two such xenoplasmic compartments (mitochondrial matrix and chloroplast stroma). Only the inner mitochondrial membrane and the inner chloroplast membrane (including derived thylakoid membrane) would be of xenoplasmic origin, whereas the outer mitochondrial and chloroplast membrane would be of orthoplasmic origin, like all other cellular membranes. The proposed terminology may be useful for describing the precise topology of IMP's (see Fig. 2). For example, monotopic IMP's of the thylakoid membrane may be exposed ectoplasmically (i.e., toward the intradisc space) or xenoplasmically (i.e., toward the stroma); bitopic IMP's of the outer mitochondrial membrane have an ectoplasmic and an endoplasmic domain, etc.

signal sequence addressed to the RER can be decoded by the putative signal receptor of the prokaryotic plasma membrane and be cleaved correctly by signal peptidase of the prokaryotic plasma membrane.

It should be emphasized that signal sequences only initiate the translocation process and that they do not guarantee its completion nor do they assure completion of subsequent posttranslocational traffic (see below).

In fact, chain translocation can be prematurely terminated by a distinct "stop-transfer" sequence (see below). It might also be prematurely terminated by other less specific sequence constellations which are not normally translocated [e.g., those occuring in "hybrid" proteins; for review see ref. 7].

It has been assumed that the boundary for an amino terminal signal sequence is more or less defined by its signal peptidase cleavage site. This, however, has recently been called into question in the case of the signal sequence addressed to the prokaryotic plasma membrane. It has been argued that the information necessary for translocation extends beyond the cleavage site into the mature portion of the protein [8] or may require additional information that is expressed only by the completely synthesized chain [9]. However, the experimental evidence is not definitive in either of these two studies. Evidence in the first case [8] rests on data with a hybrid protein for which the translocation process may have been properly initiated by a normal signal sequence but subsequently may have been aborted by some other unspecific sequence that is not permissive with the translocation process. Evidence in the second case [9] rests on in vivo experiments with UV-radiated bacteria. Thus, the conjecture that the signal peptide portion of the nascent chain contains all the information necessary to function as a signal sequence to initiate translocation has not been invalidated by these experiments. In fact, the question could be raised whether the amino terminal cleaved signal peptide contains more sequence information than is necessary to initiate chain translocation, and whether some of the signal peptide's amino terminal or carboxy terminal residues could be trimmed without affecting its interaction with the putative signal receptor to initiate chain translocation.

Elegant genetic studies [for review see ref. 7] on the signal sequence addressed to the prokaryotic plasma membrane have provided strong support for the concept that the information for initiating the translocation process does indeed reside in the signal peptide. Furthermore, these studies have also revealed [7] that the apparent degeneracy in the primary structure of the signal sequence has its limits: replacement of hydrophobic residues (that are clustered in the center region of the signal sequence) by charged residues prevents translocation; or, in the case of *E. coli* pre-lipoprotein, substitution of Gly by Asp in position 14 of the 20-residue-long signal peptide results in an abolition of cleavage but not of translocation [10].

There are now numerous examples of proteins that are translocated without cleavage of a signal sequence [see ref. 11 for recent example]. In the case of hen ovalbumin it has been shown that the uncleaved signal sequence might be located not at the amino terminus but in the central region of the molecule [12].

The primary structural features have been determined and exemplified by numerous proteins [see ref. 1 for a compilation] in the case of Si(RER) and Si(PPM). The primary structure for Si(CEN) has been elucidated so far only for one protein [13]. Primary structure information is not yet available for signal sequences addressed to IMM, TKM, OMM, OCM, PXM and MEN (see Table 1).

As already mentioned, the primary structure of the signal peptide for RER and PPM reveals common features, a finding that supports the evolutionary relationship between these two membranes (Fig. 1). Although there is variability in the length of these signal peptides (total number of amino acid residues varies between 15 and 30) their common feature appears to be a stretch of hydrophobic residues in the central region with charged or hydrophilic residues on either side of this hydrophobic core. Moreover, the penultimate residue, in all cases, is a small side-chain amino acid (Gly, Ala, Cys, Ser, Thr). As expected the primary structure of the signal sequence addressed to CEN [13] differs considerably from that addressed to RER or PPM. However, the primary structure of other examples needs to be established before the distinct features of Si(CEN) can be delineated.

Although detailed models for the mechanism of co- and posttranslocational translocation have been formulated [2], little is known about the various translocators (see Table 1). Attempts to isolate and characterize them have so far been reported only for the RER translocator. Salt extraction [14-16] as well as controlled proteolytic digestion [17-19] of isolated microsomal vesicles have been shown to abolish their protein translocation activity either partially or completely. Translocation activity can be fully restored to the salt- or protease-extracted vesicles upon readdition of a salt or protease extract [14-19]. The active components of the protease extract [19] as well as of the salt extract [16] have been purified. It remains to be investigated, however, how these components function in initiating the translocation process.

3 Stop-transfer Sequences

A "stop-transfer" sequence was proposed to contain the information required to interrupt the process of chain translocation that was previously initiated by a signal sequence [20-22]. Because translocation of the polypeptide chain [2] is likely to proceed sequentially and asymmetrically in both cotranslational and posttranslational translocation, stop-transfer sequences are effective means for asymmetric integration into the membrane of certain integral membrane proteins (IMPs) by either mode of translocation (see Table 1).

The sequence features that constitute the stop-transfer sequence remain to be defined. The stop-transfer sequence may not simply be that stretch of ~25 primarily hydrophobic residues which is typical of the transmembrane portion of bitopic IMP's and which might be envisioned to act as a stop-transfer sequence by virtue of being nonpermissive with the translocation process. There are, e.g., viral bitopic IMP's which possess stretches of at least 28 hydrophobic residues in their ectoplasmic domain [23, 24]. Since this domain is translocated it is clear that a long stretch of hydrophobic residue per se is not sufficient to stop the translocation process.

There could exist as many translocator-specific stop-transfer sequences as there are translocator-specific signal sequences. On the other hand, there could be only one stop-transfer sequence addressed to one component common to all translocators, e.g., phosphilipid molecules.

4 Insertion Sequences

It has been known for a long time that certain IMP's can insert into a lipid bilayer spontaneously, i.e., by a mechanism that is not receptor-mediated. The existence of specific insertion sequences has therefore been postulated [2]. These insertion sequences function to anchor proteins to the hydrophobic core of the lipid bilayer (see Fig. 2). This

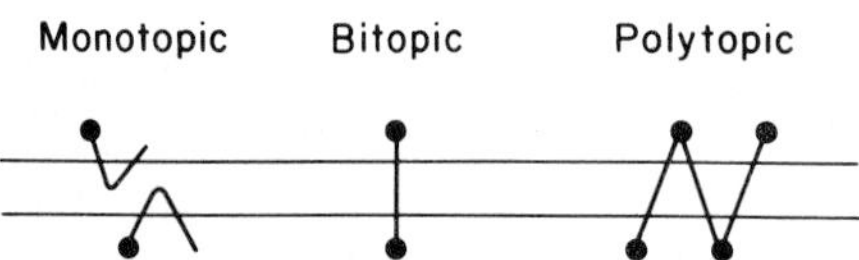

Fig. 2. Classification of integral membrane proteins (IMP's) as monotopic, bitopic, and polytopic [taken from ref. 2]. The hydrophobic boundary of the lipid bilayer is indicated by *two parallel lines. Dots* on polypeptide chains indicate major hydrophilic domains. The hydrophilic domain of a *monotopic* IMP is exposed only on one side of the lipid bilayer. A hydrophobic domain is indicated to anchor the polypeptide chain to the hydrophobic core of the lipid bilayer. A monotopic IMP may contain several hydrophilic and hydrophobic segments alternating with each other (not indicated here). All hydrophilic domains, however, are unilaterally exposed. The polypeptide chain of *bitopic* IMP's spans the lipid bilayer once and contains a hydrophilic domain on opposite sides of the membrane. In variants of bitopic IMP's (not indicated) the bilateral hydrophilic domains could be further subsegmented by interspersed hydrophobic domains that are capable of monotopic integration. The polypeptide chain of *polytopic* IMP's spans the membrane more than once and contains multiple hydrophilic domains on both sides of the membrane. The existence of polytopic IMP's remains to be demonstrated. Two structurally monotopic IMP's located on opposite sides of the membrane could interact via their hydrophobic anchorage domain and form a functionally bilateral ensemble

anchorage is not accompanied by the translocation across the membrane's lipid bilayer of charged domains of the polypeptide chain. The latter can be achieved presumably only by a signal sequence in a receptor-mediated process [2].

At this point it is necessary to comment on an alternative proposal [25] which argues that the cleaved amino terminal sequence extension of bacteriophage f1 (or M 13) coat protein (CP) functions not as a signal sequence, as proposed earlier [22], but functions essentially as an insertion sequence that "folds" this bitopic IMP into the lipid bilayer of PPM by a receptor-independent mechanism. Although the cleaved amino terminal sequence extension of the nascent CP is structurally analogous to all the other RER or PPM signal sequences, it will be possible to validate or to invalidate the receptor-independent mechanism of integration of CP into PPM in a definitive manner only after identification and characterization of the putative PPM translocator has advanced to the same analytical level as that achieved for its RER counterpart (see above).

As is the case for the stop-transfer sequence, the structural features of an insertion sequence remain to be defined. It is conceivable that there are several unique insertion sequences that can distinguish lipid composition and therefore insert only into specific membranes. On the other hand, the specificity of insertion into distinct membranes may be largely dictated by protein-protein interaction (i.e., by an affinity of a protein to be inserted to another IMP).

5 Sorting Sequences

Sorting sequences were defined as those discrete portions of the polypeptide chain that act as determinants for posttranslocational traffic and that are shared by proteins with an identical travel objective [2]. The structural features and the number of distinct sorting sequences remain to be defined. The cell's machinery which decodes sorting sequences and effects displacement remains to be identified. In eukaryotic cells, the RER is one of the most important origins for posttranslocational traffic of a large variety of soluble and integral membrane proteins. Let us consider here the case only of soluble proteins. Recent data [26, 27] strongly suggest that at least two subpopulations of soluble proteins are found mixed within the cisternae of the RER, namely lysosomal and secretory proteins. Thus, lysosomal enzymes have been shown [26, 27] to be synthesized with a signal sequence addressed to the RER that is structurally and functionally indistinguishable from that common to secretory proteins. Besides secretory and lysosomal proteins the intracisternal space of the RER may contain other soluble proteins that are neither lysosomal nor secretory but that are either permanent residents or in transit to be routed to other membrane-bounded compartments. Would each species of these groups of soluble proteins possess a common sorting sequence? And, would there be a corresponding number of sorting sequence-specific receptors in the RER that effects the distribution of these proteins?

In considering the problem of sorting of soluble protein from the cisternae of the RER into other compartments it might be useful to search for precedents and analogies. The plasma membrane, for example, is able to "sort" extracellular proteins into intracellular compartments by a variety of routes and mechanisms. There is unspecific uptake of soluble protein (pinocytosis), probably as part of an exchange of liquid quanta between the extracellular compartment and a distinct intracellular vesicular compartment. The only requirement for a protein to be transported this way might be for it to be water-soluble rather than to possess a sorting sequence. Liquid phase transport between these two compartments is probably accomplished by a set of one or several distinct IMP's which, by fission-fusion processes, shuttle between two distinct membranes, and which in the process of doing so may not mix (by lateral diffusion) with other IMP's of the two membranes.

Analogous bidirectional transport systems for water-soluble proteins may exist between the RER and specific cisternae of the Golgi, between specific cisternae of the Golgi and the extracellular space, or between other distinct intracellular compartments. Each of these transport systems would likely to be represented by a unique set of IMP's. If such a transport system would indeed exist between the cisternae of the RER and a specific "downstream" cisterna (which, ultrastructurally, could be part of the Golgi complex), then any water-soluble protein in the cisternae of the RER can potentially undergo downstream transport, unless it is immobilized by binding to a constitutive IMP (permanent resident) of the ER membrane. It is proposed here that secretory proteins might be displaced from the RER by such a liquid phase transport system.

How about lysosomal proteins? Some time ago, Neufeld [28, 29] proposed that lysosomal enzymes are endowed with a recognition marker that is common to all lysosomal enzymes and which functions in receptor-mediated sorting. Considerable evidence has since accumulated in support of this concept [for recent refs. see 30, 31]. The recognition marker appears to be phosphorylated mannose residues of the Asn-linked core sugars [32-34]. Because many secretory proteins also contain Asn-linked core sugars (but so far have not been detected to contain phosphorylated mannose residues), and because both secretory and lysosomal proteins are cosegregated [26, 27] in the cisternae of the RER, it is likely that the mannose modifying enzyme(s) recognize a protein sequence feature that is common to all (and unique for all) lysosomal enzymes. This sequence feature therefore would be an example of a sorting sequence. More recently Sly [31, 35] has proposed that receptor-mediated sorting of lysosomal proteins from secretory proteins occurs primarily upstream, at the level of the RER.

If secretory and lysosomal proteins are displaced from the RER, each into a separate and unique downstream compartment, secretory proteins by liquid phase transport and lysosomal proteins by receptor-mediated transport, one could conceive of a variety of possibilities for malfunctioning. For example, mutations in secretory proteins resulting in their having a lower solubility product might cause accumulation of the secretory protein within the cisternae of the RER; mutations in the sorting sequence of a lysosomal enzyme might result in its secretion. Alterations in the receptor might result in secretion of all lysosomal enzymes, etc.

How about other soluble proteins that may be permanently mobilized by binding to constitutive IMP's of the RER or of further downstream compartments? Immobilization may occur by protein-protein interaction, distinct in each case and therefore not mediated by a sorting sequence, at least not as defined here. On the other hand, it is conceivable that a permanently immobilized receptor could act as a ligand for several distinct species of proteins by virtue of a sequence feature that they share. This common sequence feature would constitute a sorting sequence.

6 Multiple Topogenic Sequences

The occurence of multiple topogenic sequences, such as a signal sequence and a sorting sequence for a given species or group of proteins has already been described above in the case of lysosomal enzymes (signal sequence plus sorting sequence) or in the case of bitopic IMP's (signal sequence and stop-transfer sequence). Figure 3 illustrates a further application of the concept of multiple topogenic sequences to the problem of how to achieve asymmetric integration of the polypeptide backbone of IMP's into the lipid bilayer. Although the precise orientation of the polypeptide backbone with respect to the lipid bilayer is unknown for most species of IMP's, the proposed [2] hypothetical schemes of multiple topogenic sequences can explain any one orientation by what essentially is a limited number of highly redundant mechanisms.

7 Pleiotopic Proteins

Pleiotopic proteins have been defined [2] as proteins which are similar in structure and function but differ in topology. The presence or absence of a topogenic sequence or the acquisition of another topogenic sequence would be an effective means to achieve several distinct cellular localizations of a given species of protein. Pleiotopic proteins may

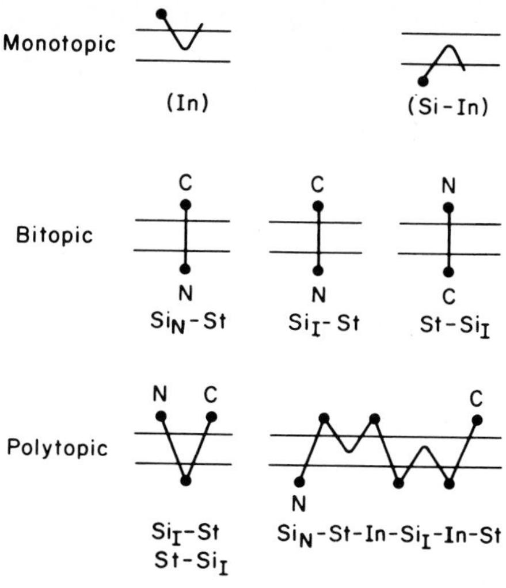

Fig. 3. Program of topogenic sequences for the asymmetric integration into membranes of some representative examples of monotopic, bitopic, and polytopic integral membrane proteins (IMP's) [taken from ref. 2]. Hydrophobic boundary of lipid bilayer is indicated by *two parallel lines*, with *upper line* facing the protein biosynthetic compartment. *Dots* represent major hydrophilic domains which, when indicated, contain amino (*N*) or carboxy (*C*) terminus of the polypeptide chain. Topogenic sequences are: insertion sequence (*In*), signal sequence (*Si*), and stop-transfer sequence (*St*). Si_N and Si_I indicate amino terminal and internal signal sequence, respectively. Examples given here (except for monotopic IMP on *upper left*) are for cotranslational integration into RER. Similar programs are conceivable also for cotranslational integration into PPM, IMM and TKM as well as for posttranslational integration into PXM, OMM, OCM, IMM (using Si(MEN)) and ICM/TKM (using Si(CEN)). An attempt has been made to list topogenic sequences in order of their location along the polypeptide chain starting from the amino terminus. The problems encountered in predicting the order relate to uncertainties as to the order of chain translocation. In particular in the case of an internal signal sequence (Si_I) there are several possibilities depending on the order of translocation [12]. The orientation of a polytopic IMP, such as indicated at the *lower right*, is entirely hypothetical and is illustrated here only to indicate how such a polypeptide chain could be integrated into the membrane by a program of multiple topogenic sequences

be represented by proteins that exist, for example, as a cytoplasmic form (no topogenic sequence) and a secreted form (+ Si(RER)); or as a secreted form (+ Si(RER)) and a membrane-bound form (several possibilities, see Fig. 3); or as a cytoplasmic form (no topogenic sequence) and a mitochondrial matrix form (+ Si(MEN)), and a peroxisomal form (+ Si(PXM)), etc., etc.

References

1. Blobel G, Walter P, Chang CN, Goldman BM, Erickson AH, Lingappa VR (1979) In: Hopkin CR, Duncan CJ (eds) Symposium of the Society of Experimental Biology (Great Britain), vol 33. Cambridge Univ Press, London, pp 9-36
2. Blobel G (1980) Proc Natl Acad Sci USA 77:1496-1500
3. Blobel G (1978) In: Cohen CN, Holzer H (eds) Limited proteolysis in microorganisms. US Department of Health, Education and Welfare, DHEW Publication No (NIH) 79-1591, pp 167-169

4. Shields D, Blobel G (1977) Proc Natl Acad Sci USA 74:2059-2063
5. Talmadge K, Stahl S, Gilbert W (1980) Proc Natl Acad Sci USA 77:3369-3373
6. Talmadge K, Kaufman J, Gilbert W (1980) Proc Natl Acad Sci USA 77:3988-3992
7. Emr SD, Hall MN, Silhavy TJ (1980) J Cell Biol 86:701-711
8. Moreno F, Fowler AV, Hall M, Silhavy TJ, Zabin I, Schwartz M (1980) Nature (London) 286: 356-359
9. Koshland D, Botstein D (1980) Cell 20:749-760
10. Lin JJC, Kanazawa H, Ozols J, Wu HC (1978) Proc Natl Acad Sci USA 75:4891-4895
11. Bonatti S, Blobel G (1979) J Biol Chem 254:12261-12264
12. Lingappa VR, Lingappa JR, Blobel G (1979) Nature (London) 281:117-121
13. Schmidt GW, Devillers-Thiery A, Desruisseaux H, Blobel G, Chua N-H (1979) J Cell Biol 83: 615-622
14. Warren G, Dobberstein B (1978) Nature (London) 273:569-571
15. Jackson RC, Walter P, Blobel G (1980) Nature (London) 286:174-176
16. Walter P, Blobel G (1980) Proc Natl Acad Sci USA 77:7112-7116
17. Walter P, Jackson RC, Marcus MM, Lingappa VR, Blobel G (1979) Proc Natl Acad Sci USA 76:1795-1799
18. Meyer DL, Dobberstein B (1980) J Cell Biol 87:498-502
19. Meyer DL, Dobberstein B (1980) J Cell Biol 87:503-508
20. Blobel G (1977) In: Brinkley BR, Porter KR (eds) International cell biology 1976-1977. Rockefeller University Press, New York, pp 318-325
21. Lingappa VR, Katz FN, Lodish HF, Blobel G (1978) J Biol Chem 253:8667-8670
22. Chang CN, Model P, Blobel G (1979) Proc Natl Acad Sci USA 76:1251-1255
23. Scheid A, Graves MC, Silver SM, Choppin PW (1978) In: Mahy BWJ, Barry RD (eds) Negative strand viruses and the host cell. Academic Press, London New York, pp 181-193
24. Gething MJ, White JM, Waterfield MD (1978) Proc Natl Acad Sci USA 75:2737-2740
25. Wickner W (1979) Annu Rev Biochem 48:23-45
26. Erickson AH, Blobel G (1979) J Biol Chem 254:11771-11774
27. Erickson AH, Conner G, Blobel G (1981) J Biol Chem (in press)
28. Hickman S, Neufeld EF (1972) Biochem Biophys Res Commun 49:992-999
29. Neufeld EF, Sando GN, Garvin AJ, Rome LH (1977) J Supramol Struct 6:95-101
30. Hasilik A, Neufeld EF (1980) J Biol Chem 255:4946-4950
31. Fischer DH, Gonzales-Noriega A, Sly WS, Morré JD (1980) J Biol Chem 255:9608-9615
32. Kaplan A, Achord DT, Sly WS (1977) Proc Natl Acad Sci USA 74:2026-2030
33. Sando GN, Neufeld EF (1977) Cell 12:619-627
34. Tabas I, Kornfeld (1980) J Biol Chem 255:6633-6639
35. Sly WS, Stahl P (1978) In: Silverstein S (ed) Transport of macromolecules in cellular systems. Life Sciences Research Report II. Dahlem Konferenzen, Berlin, pp 229-245

Discussion[2]

Helmreich raised the question on the mechanisms of insertion and assembly of multicomponent membrane systems, such as the acetylcholine receptor and the adenylate cyclase. Blobel answered that all four subunits of the acetylcholine receptor have recently been synthesized in a cell-free system supplemented with dog pancreas microsomal membranes and integrated into the microsomal membrane. The integration was coupled to translation. Furthermore, it was found that all four subunits of the acetylcholine receptor are transmembrane proteins, with hydrophilic domains on both sites of the membrane.

---

2 Rapporteur: Peter C. Heinrich

Since all lysosomal proteins tested so far are glycoproteins, a question of Afting aimed at the possibility that the sugar chain of these glycoproteins may play a role in protein sorting in the endoplasmic reticulum. There is a lysosomal disease known, in which an incorrect sugar chain results in a wrong compartmentation of the glycoprotein.

Blobel agreed that such a mechanism might exist, but it should be proven experimentally.

Precursor Processing in the Biosynthesis of Rat Liver Cytochrome c Oxidase

P.C. Heinrich, E. Schmelzer, and T. Nagasawa[1]

1 Introduction

Although chloroplasts and mitochondria are capable of protein synthesis [for a recent review see Buetow and Wood 1978], the majority of proteins of both organelles is synthesized on cytoplasmic ribosomes and imported into the mitochondria or chloroplasts. In the case of cytochrome c oxidase from *Neurospora crassa* (Sebald et al. 1972), *Saccharomyces cerevisiae* (Mason and Schatz 1973; Rubin and Tzagoloff 1973), and *Xenopus laevis* (Koch 1976) it has been clearly demonstrated that three subunits designated as I, II, and III are synthesized on mitochondrial ribosomes, and four subunits (IV to VII) are manufactured on cytoplasmic ribosomes. The biogenesis of this inner mitochondrial membrane enzyme raises the interesting question of how the subunits made in different cellular compartments under the direction of two physically separated genetic systems are assembled. Among the many questions about the biosynthesis of cytochrome c oxidase are (1) how does the cell regulate the synthesis of the seven subunits in the correct stoichiometry, and (2) how are the cytoplasmic subunits transported across the outer mitochondrial membrane? An intriguing mechanism to obtain the subunits IV to VII of yeast cytochrome c oxidase in the correct stoichiometry has been proposed by Poyton and McKemmie (1976, 1979). These authors have described the synthesis of a polyprotein precursor to all four cytoplasmically translated subunits (Fig. 1). They postulated that during or after its import into the inner mitochondrial membrane the polyprotein precursor is converted to the individual subunits IV, V, VI, and VII by a highly specific proteinase.

At that time, when Poyton was looking for a mitochondrial proteinase which could convert the polyprotein precursor of the yeast cytochrome c oxidase into the subunits IV to VII, we had isolated a highly specific proteinase from rat liver mitochondria (Jusic et al. 1976). The enzyme was originally found as a contaminant of chromatin (Heinrich et al. 1976). In subsequent studies we have subfractionated mitochondria into outer membrane, inner membrane, and matrix fractions. The proteinase was associated with the inner mitochondrial membrane fraction (Haas and Heinrich 1978). In a search for the physiological function of this proteinase it was our working hypothesis that, by analogy to the yeast cytochrome c oxidase, the four cytoplasmically synthesized cytochrome c oxidase subunits IV-VII of rat liver might also be synthesized as a polyprotein precursor and that this precursor might be cleaved by the proteinase in a highly specific manner into the four polypeptides IV-VII.

1 Biochemisches Institut der Universität Freiburg, Hermann-Herder-Str. 7, 7800 Freiburg, FRG

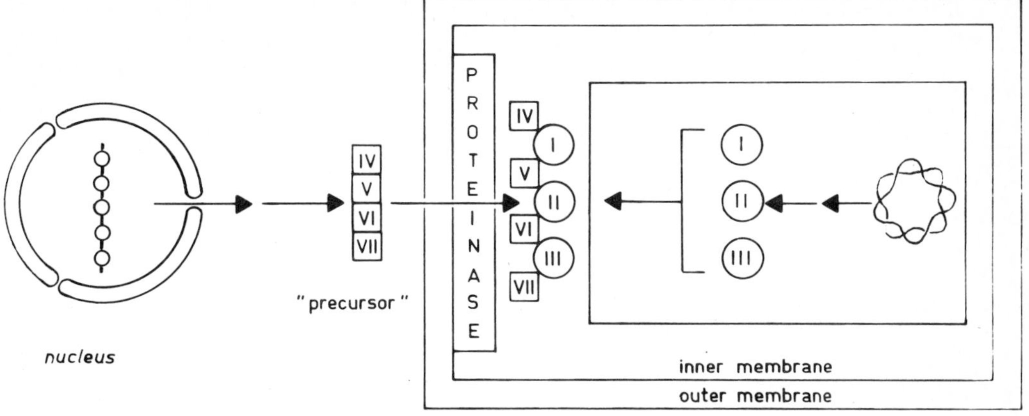

Fig. 1. Schematic representation of the assembly of cytochrome c oxidase. The postulated polyprotein precursor for the subunits *IV* to *VII* (Poyton and McKemmie 1976, 1979) is cleaved into the individual subunits by limited proteolysis

For the identification of a cytochrome c oxidase polyprotein precursor it was first necessary to isolate pure cytochrome c oxidase from rat liver mitochondria, and then to raise specific antibodies to the rat liver cytochrome c oxidase subunits.

2 Results

2.1 Purification of Rat Liver Cytochrome c Oxidase by Hydrophobic Chromatography (Nagasawa et al. 1979)

The purification of cytochrome c oxidase from rat liver mitochondria was particularly effective by the use of hydrophobic chromatography. Among several alkyl- and ω-aminoalkyl-agarose gels octyl- and hexyl-agarose showed a particularly strong interaction with cytochrome c oxidase (Fig. 2). Not shown in Fig. 2 is phenyl-Sepharose, which binds cytochrome c oxidase very strongly. We have examined the possible use of octyl- and ω-aminohexyl-agarose and phenyl-Sepharose for the purification of cytochrome c oxidase. As shown in Table 1, the highest heme a per mg protein content and essentially no reductase activities were obtained after phenyl-Sepharose chromatography. When the purified enzyme was subjected to dodecylsulfate-polyacrylamide gel electrophoresis, seven polypeptide constituents were distinguished (Fig. 3). For comparison the subunits of yeast and beef heart cytochrome c oxidase have also been separated on the same gel.

2.2 Preparation of Antibodies Against Rat Liver Cytochrome c Oxidase

The purified cytochrome c oxidase was used to raise antibodies in rabbits. As shown in Fig. 4, a single precipitin line was obtained. No cross-reaction with cytochrome c oxidase

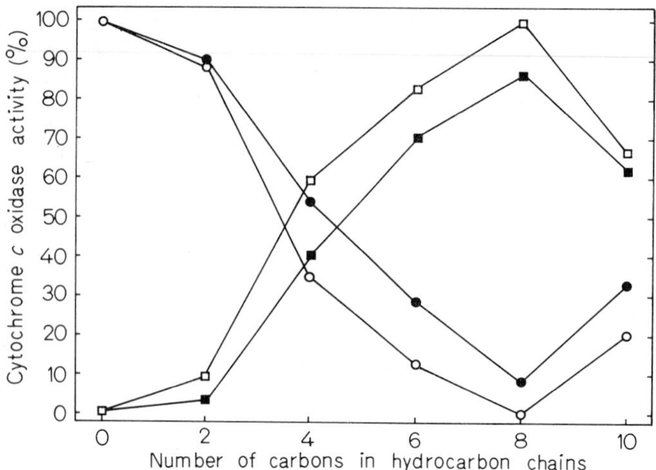

Fig. 2. Retention and elution behavior of cytochrome c oxidase on various alkyl and ω-aminoalkyl agarose gels. Rat liver cytochrome c oxidase (10.7 nmol heme a per mg of protein) in a volume of 200 μl was applied to the hydrophobic column (1 × 2.0 cm), equilibrated with 50 mM sodium phosphate buffer (pH 7.5) containing 1.5% sodium cholate. Each column was then washed with three column volumes of 1.5% sodium cholate in 50 mM sodium phosphate buffer (pH 7.5). Alkyl-agaroses and ω-aminoalkyl-agaroses were washed with three column volumes of 1.5% sodium cholate plus 3% Triton X-100 in 50 mM sodium phosphate buffer or 1.5% sodium cholate in 1.0 M NaCl and 50 mM sodium phosphate buffer, respectively. Fractions of 400-μl volume were collected and 10-μl aliquots of each fraction were assayed for cytochrome c oxidase activity. The total activity eluted with each buffer wash was calculated and is expressed as a percentage of the original activity applied. Cytochrome c oxidase which passed through the alkyl (○) and ω-aminoalkyl (●) agarose columns and enzyme which was bound to alkyl (□) and ω-aminoalkyl (■) agarose. The data are given as a percentage of the initial enzyme activity

Table 1. Effect of hydrophobic chromatography on the purification of rat liver cytochrome c oxidase. Seven fractions containing the highest cytochrome c oxidase activity eluted from the three kinds of hydrophobic columns were pooled (2.5 ml) and assayed. NADH-$K_3Fe(CN)_6$ reductase activities are given as μmol $K_3Fe(CN)_6$ reduced × min^{-1} × mg $protein^{-1}$

Preparation	Heme a/ protein	Heme a recovery	Reductase activities		
			NADH-nitroblue tetrazolium	NADPH-nitroblue tetrazolium	NADH-$K_3Fe(CN)_6$
	nmol/mg	%	ΔA_{510} × min^{-1} × mg^{-1}		μmol × min^{-1} × mg^{-1}
Deoxycholate-soluble extract	8.0	100	5.39	0.58	1.95
ω-Aminohexyl-agarose	10.3	67.3	0.50	0.08	0.07
Octyl-sepharose	10.2	42.9	0.95	0.28	0.41
Phenyl-sepharose	15.4	75.4	n.m. [a]	n.m.	n.m.

[a] Not measurable

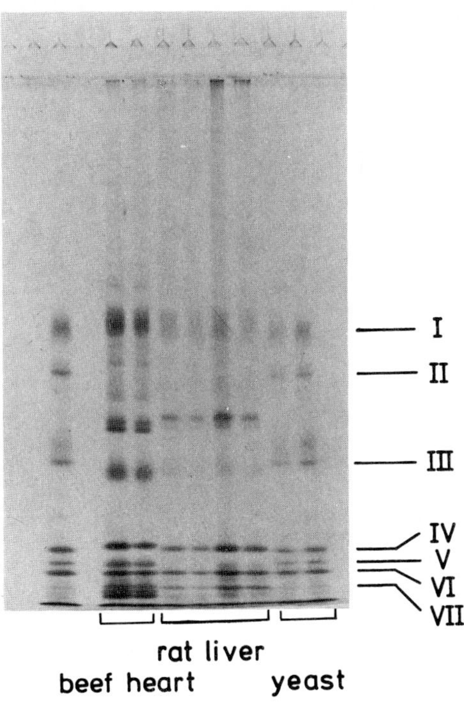

Fig. 3. Sodium dodecylsulfate-polyacrylamide gel electrophoretic separation of cytochrome c oxidase from rat liver, beef heart, and yeast mitochondria

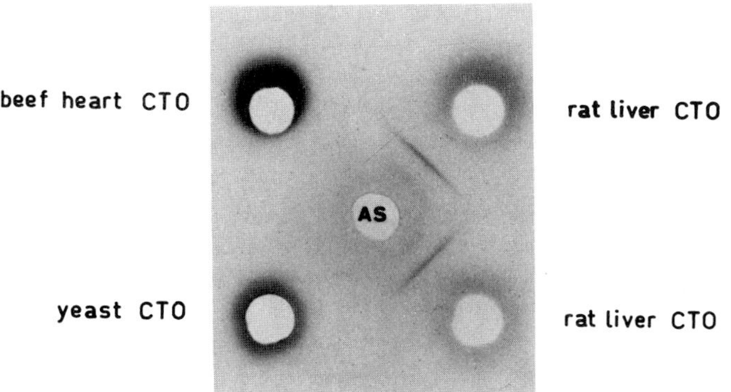

Fig. 4. Specificity of antiserum against cytochrome c oxidase as tested by immunodiffusion. The center well contained 10 μl of antiserum against rat liver cytochrome c oxidase. The peripheral wells, reading clockwise from the right, contained 12 μg of deoxycholate soluble cytochrome c oxidase from rat liver (8.5 nmol heme a per mg protein), 6.9 μg of purified rat liver cytochrome c oxidase (15.4 nmol heme a per mg protein), 10 μg of yeast cytochrome c oxidase (8.3 nmol heme a per mg protein), and 10 μg of beef heart cytochrome c oxidase (7.4 nmol heme a per mg protein)

from beef heart and from yeast mitochondria was found. Figure 5 shows the titration curve of rat liver cytochrome c oxidase with the antiserum. The increasing amount of precipitation at higher antiserum to enzyme ratios was paralleled by a larger inactivation of the enzyme by the antiserum. In a similar experiment yeast cytochrome c oxidase ac-

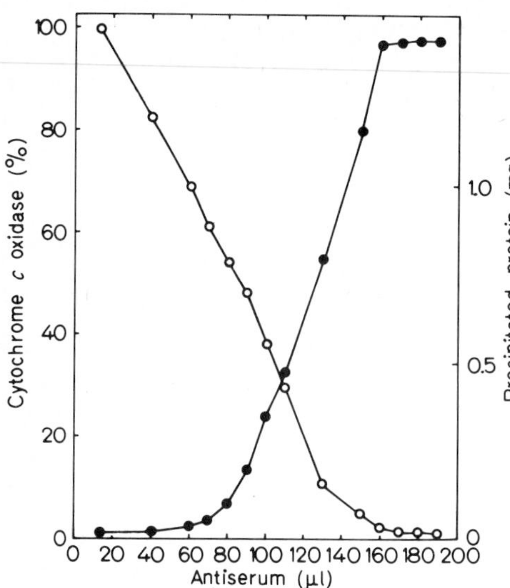

Fig. 5. Titration of deoxycholate soluble cytochrome c oxidase with specific antiserum. Purified, soluble rat liver cytochrome c oxidase (71.8 µg of protein, 10.2 nmol heme a per mg of protein) was incubated in a total volume of 200 µl with the indicated amounts of the antiserum at 37°C for 1 h. After 12 h at 0°C and centrifugation at 12000 g for 2 min, cytochrome c oxidase activity was determined with 10-µl aliquots of the supernatant. The value obtained in the absence of antiserum was taken as 100%. Similar amounts of control serum did not inhibit enzyme activity. The precipitate was washed four times with 0.4% sodium cholate in 0.14 M NaCl and used for protein determination. Cytochrome c oxidase (○—○); precipitated protein (●—●)

tivity was neither inhibited nor could it be precipitated with the anti-rat liver cytochrome c oxidase.

When the antigenicity of the seven polypeptides of rat liver cytochrome c oxidase was examined by the overlay technique described by Towbin et al. (1979), it was found that among the subunits IV-VII synthesized on cytoplasmic ribosomes only antibodies against subunit IV were present in the antiserum (not shown).

2.3 In Vitro Synthesis of a Larger Precursor for Subunit IV of Rat Liver Cytochrome c Oxidase

Either pulse-labeling of intact rat liver cells or in vitro translation of mRNA from polysomes may be used to study the biosynthesis of the cytochrome c oxidase polypeptides. Due to the fact that cytochrome c oxidase has such a low turnover, pulse-labeling of the enzyme in vivo was not considered.

Poly(A)RNA from phenol-extracted rat liver polysomes was translated in an in vitro cell-free wheat germ system into proteins. The RNA stimulated the incorporation of [^{35}S]methionine into proteins 20 to 30-fold. The labeled translation products were subjected to immunoprecipitation with the cytochrome c oxidase antiserum. After binding of the antigen-immunoglobulin complex to and elution from protein A Sepharose, sodium dodecylsulfate-polyacrylamide step gel electrophoresis and subsequent fluorography was carried out. Only one band migrating slower than all four cytoplasmically synthesized cytochrome c oxidase subunits IV to VII were visualized on the autoradiogram (Fig. 6, track 3). As mentioned above, cytochrome c oxidase has a long half-life and it is difficult to isolate radioactively labeled enzyme after in vivo labeling of adult rats. Therefore, the electrophoretic mobilities of the radioactively labeled proteins had to be compared to those of the stained cytochrome c oxidase subunit IV (Fig. 6, tracks 2 and 10). When the unlabeled individual subunits IV, V, VI, or VII, isolated from preparative sodi-

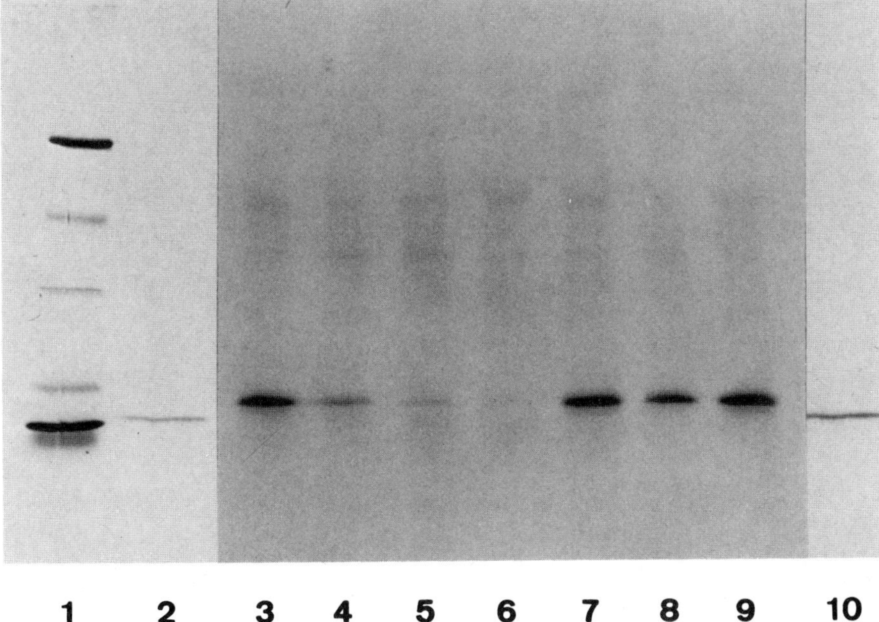

Fig. 6. Subunit IV of rat liver cytochrome c oxidase is synthesized as a larger form. Poly(A)RNA from rat liver polysomes was translated in an in vitro cell-free system, immunoprecipitated, separated by SDS-polyacrylamide slab gel electrophoresis and analyzed by autoradiography. *Track 1*, molecular weight markers: bovine serum albumin (68 000), ovalbumin (43 000), elastase (25 800), trypsin inhibitor (21 500), and myoglobin (17 800); *tracks 2* and *10*, 4.3 μg of subunit IV of rat liver cytochrome c oxidase stained with Coomassie blue R 250; *track 3*, in vitro synthesized precursor for cytochrome c oxidase subunit IV; *tracks 4-9*, same as *track 3* except that 2 μg (*track 4*), 5 μg (*track 5*), and 20 μg (*track 6*) of subunit IV, 20 μg of subunit V (*track 7*), 20 μg of subunit VI (*track 8*), and 20 μg of subunit VII (*track 9*) have been added to the in vitro translation mixture before immunoprecipitation and SDS-polyacrylamide gel electrophoresis

um dodecylsulfate-polyacrylamide gels, were added to the in vitro translation mixture, it was found that only subunit IV could compete with the binding of the precursor to the antibody (Fig. 6, tracks 3 to 9). The comparison of the electrophoretic mobilities of subunit IV and its precursor shows a difference in molecular weight of about 3000. Further support for the existence of a larger molecular weight precursor of subunit IV was provided by translation experiments with N-formyl-[^{35}S]methionine-tRNA$_f$, and by the comparison of two-dimensional tryptic fingerprints of subunit IV and the precursor to subunit IV (Schmelzer and Heinrich 1980).

Figure 7 shows that the subunit IV precursors, which are synthesized in the wheat germ system in the presence of either [^{35}S]methionine (track 5) or N-formyl-[^{35}S]methionine-tRNA$_f$ (track 6), exhibit identical electrophoretic mobilities. For comparison preproalbumin in vitro synthesized with [^{35}S]methionine (track 3), or N-formyl-[^{35}S]methionine-tRNA$_f$ (track 4) has been included in Fig. 7.

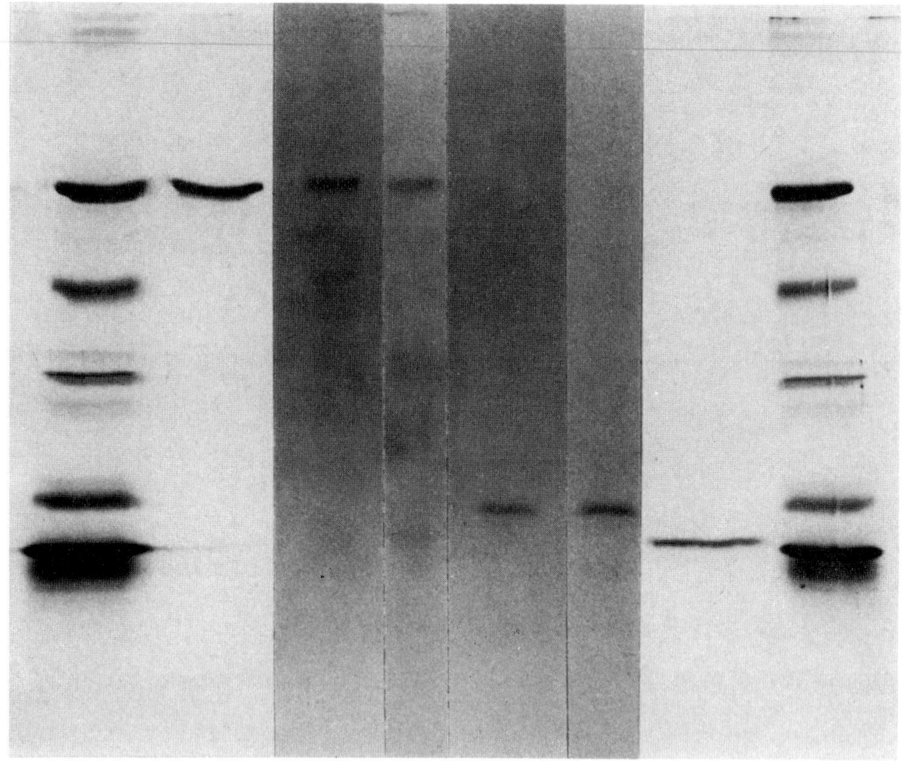

Fig. 7. Synthesis of a larger precursor for the subunit IV of rat liver cytochrome c oxidase in a cell-free wheat germ system. Rat liver poly(A)RNA was translated in the presence of [^{35}S]methionine (*3, 5*) and N-formyl-[^{35}S]methionine-tRNA$_f$ (*4, 6*). After immunoprecipitation with either albumin-antiserum (*3, 4*) or cytochrome c oxidase-antiserum (*5, 6*) and SDS-polyacrylamide gel electrophoresis, autoradiography was carried out. For comparison bovine serum albumin (*2*), mature cytochrome c oxidase subunit IV (*7*), and protein standards (*1, 8*) have been added. The protein molecular weight markers were the same as those used in the experiments of Fig. 6

Figure 8 shows two-dimensional fingerprints of the subunit IV precursor (B) and the mature subunit IV (A). A high degree of similarity between the mature subunit IV and its precursor can be visualized from Fig. 8.

The results presented are in contrast to the findings of Poyton and McKemmie (1976, 1979) and of Ries et al. (1978), where a high molecular weight polyprotein precursor to all four cytoplasmically translated cytochrome c oxidase subunits, from yeast and rat liver respectively, has been described.

In analogy to the cytoplasmically synthesized F1-ATPase subunits (Maccecchini et al. 1979), the cytochrome c oxidase subunits IV-VII are obviously synthesized in form of precursors for the individual subunits and not as a polyprotein. Similar data have been obtained very recently for the subunits V and VI (Lewin et al. 1980), and IV to VI (Mihara and Blobel 1980) of yeast cytochrome c oxidase.

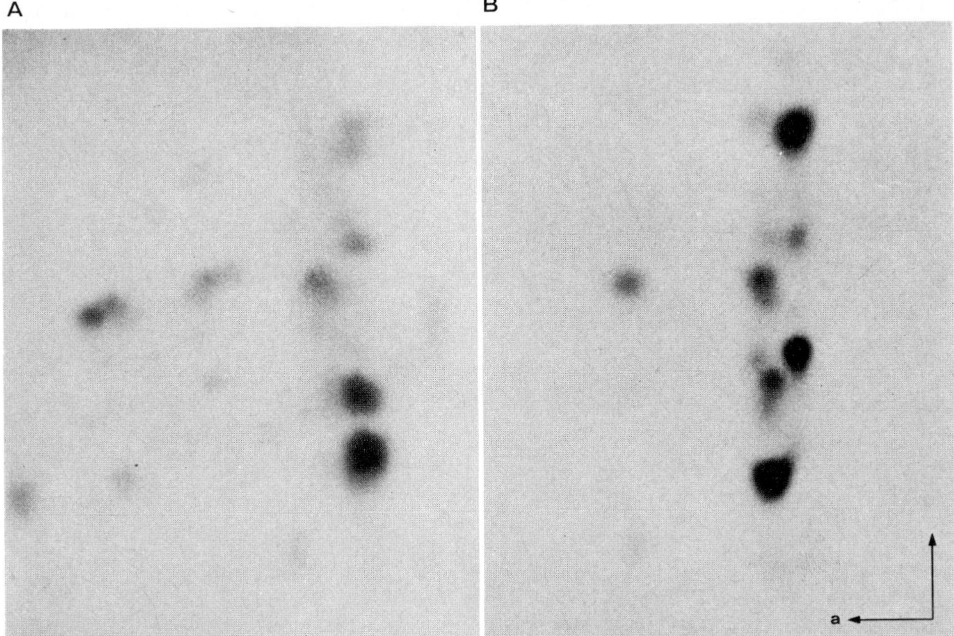

Fig. 8A, B. Two-dimensional tryptic fingerprints of the mature subunit IV (**A**) and the precursor (**B**) to subunit IV of cytochrome c oxidase. The two-dimensional tryptic fingerprinting was carried out as described by Lewin et al. (1980)

2.4 Studies on the Conversion of the Cytochrome c Oxidase Subunit Precursors

In the course of our experiments on the isolation of the cytochrome c oxidase "polyprotein" it turned out that the proteinase, originally localized in the inner mitochondrial membrane (Haas and Heinrich 1978), is an enzyme from mast cells, which are present in the connective tissue of the liver (Haas et al. 1979). It was found that the proteinase is associated with the mast cell granules, which cosediment with the mitochondria (Haas et al. 1979).

Therefore, the original concept of a highly specific inner mitochondrial membrane proteinase and its physiological substrate, the polyprotein precursor for the cytochrome c oxidase subunits IV to VII, has drastically changed to the following situation: not the polyprotein, but individual precursors for the cytochrome c oxidase subunits have to be processed by a highly specific proteinase. We have first tried to convert the in vitro synthesized radioactively labeled cytochrome c oxidase subunit IV precursor with unlabeled rat liver mitochondria. Unlike in the case of the precursor for the α, β, and γ-subunits of F_1-ATPase (Maccecchini et al. 1979), it was not possible to process the larger precursor to its mature form. It may well be that during the isolation of the mitochondria an important factor necessary for the conversion is lost. Further experiments along this line are presently in progress in our laboratory.

Table 2. Proteinase activity released from the rat liver mitochondrial fraction by hypotonic buffer solutions. Proteinase activity was measured with Bz-Ile-Glu-Gly-Arg-pNA as substrate in a final concentration of 1 mM at pH 7.5 and 30°C. 1 Unit corresponds to the cleavage of 1 μmol of chromogenic substrate per min measured at 405 nm ($\epsilon = 10.4 \times 10^3$ l/cm × mol)

Rat liver mitochondria	Specific proteinase activity (U/mg) $\times 10^{-3}$
Untreated	75
After 3 washes with 0.25 M sucrose in TKM buffer, pH 7.5	< 0.3
After digitonin treatment (1.75 mg per ml, 5 min, 0°C)	< 0.3
After osmotic shock (10 mM Tris/HCl buffer, pH 7.5, 60 min, 0°C)	130
After digitonin and osmotic shock	< 0.3

Table 3. Effects of proteinase inhibitors

Inhibitor	Final concentration (mM)	Inhibition of proteinase activity no EDTA (%)	0.1 mM EDTA (%)
Iodo-acetic acid	1	89	90
p-chloro-mercuri-benzoate	1	86	90
p-hydroxi-mercuri-benzoate	1	85	87
Co^{2+}	1	100	n.d.
Mn^{2+}	1	46	n.d.
Hg^{2+}	1	88	100
Ca^{2+}	1	0	0
Mg^{2+}	1	0	0
Tos-PheCH_2Cl	1	79	87
Tos-LysCH_2Cl	1	87	92
Benzamidine	1	0	20
ATP	5	0	36
ATP	10	0	50
Phenyl-methyl-sulfonyl-fluoride	1	0	0
	(mg/ml)		
Glycerol	200	81	n.d.
Rabbit serum	3	26	75
Soybean trypsin inhibitor	0.2	0	0
Kallikrein trypsin inhibitor	0.3	19	11
Pepstatin	0.03	0	0
Leupeptin	0.01	n.d.	90

We also tried to isolate proteolytic activity from rat liver mitochondria after osmotic shock. Although we were able to demonstrate the presence of a soluble proteinase in the supernatant after subjecting mitochondria to hypotonic buffer conditions, it was not possible to convert the cytochrome c oxidase subunit IV precursor into the mature subunit polypeptide by the action of this enzyme. The proteinase activity measured with the chromogenic substrate Bz-Ile-Glu-Gly-Arg-pNA was not detected in washed mitochondria (Table 2). The proteinase turned out to be a very labile enzyme. A loss of about 40% of its activity within 24 h at 0°C was observed. The enzyme is sensitive to SH-group reagents (Table 3). The pH optimum with Bz-Ile-Glu-Gly-Arg-p-NA as substrate is around 6. After gel filtration over Sephadex G 200 a molecular weight of about 145,000 was determined.

Recently a processing proteinase from yeast as well as from rat liver and rat heart mitochondria was discovered in the laboratory of G. Schatz (pers. comm.). The enzyme is sensitive to metal chelators, like o-phenanthroline and EDTA, and insensitive to phenylmethylsulfonylfluoride, tosyl-L-lysyl-chloromethylketone, tosyl-L-phenylalanyl-chloromethylketone, and pepstatin.

Acknowledgements. The authors would like to thank Prof. Dr. G. Schatz, Biocenter, Basel, for many helpful discussions. We would also like to thank Dr. A.S. Lewin, Biocenter, Basel, for his help with the preparation of N-formyl-[^{35}S]methionine-tRNA$_f$, and Dr. I. Gregor, Biocenter, Basel, for carrying out the two-dimensional tryptic fingerprints. This work was supported by a grant from the Deutsche Forschungsgemeinschaft (He 906) and a grant from the Stiftung Volkswagenwerk.

References

Buetow DE, Wood WM (1978) The mitochondrial translation system. Subcell Biochem 5:1-85

Haas R, Heinrich PC (1978) The localization of an intracellular membrane-bound proteinase from rat liver. Eur J Biochem 91:171-178

Haas R, Heinrich PC, Sasse D (1979) Proteolytic enzymes of rat liver mitochondria. Evidence for a mast cell origin. FEBS Lett 103:168-171

Heinrich PC, Raydt G, Puschendorf B, Jusic M (1976) Subcellular distribution of histone degrading activities from rat liver. Eur J Biochem 62:37-43

Jusic M, Seifert A, Weiss E, Haas R, Heinrich PC (1976) Isolation and characterization of a membrane-bound proteinase from rat liver. Arch Biochem Biophys 177:355-363

Koch G (1976) Synthesis of the mitochondrial inner membrane in cultured *Xenopus laevis* oocytes. J Biol Chem 251:6097-6107

Lewin AS, Gregor I, Mason TL, Nelson N, Schatz G (1980) Cytoplasmically made subunits of yeast mitochondrial F_1-ATPase and cytochrome c oxidase are synthesized as individual precursors, not as polyproteins. Proc Natl Acad Sci USA 77:3998-4002

Maccecchini M, Rudin Y, Blobel G, Schatz G (1979) Import of proteins into mitochondria: Precursor forms of the extramitochondrially made F_1-ATPase subunits in yeast. Proc Natl Acad Sci USA 76:343-347

Mason TL, Schatz G (1973) Cytochrome c Oxidase from Baker's Yeast. J Biol Chem 248:1355-1360

Mihara K, Blobel G (1980) The four cytoplasmically made subunits of yeast mitochondrial cytochrome c oxidase are synthesized individually and not as a polyprotein. Proc Natl Acad Sci USA 77:4160-4164

Nagasawa T, Nagasawa-Fujimori H, Heinrich PC (1979) Hydrophobic interactions of cytochrome c oxidase. Application to the purification of the enzyme from rat liver mitochondria. Eur J Biochem 94:31-39

Poyton RO, McKemmie E (1976) In: Bücher Th, Werner S, Neupert W (eds) Genetics and biogenesis of mitochondria and chloroplasts. Elsevier North Holland, Amsterdam New York, pp 107-114

Poyton RO, McKemmie E (1979) A polyprotein precursor to all four cytoplasmically translated subunits of cytochrome c oxidase from *S. cerevisiae*. J Biol Chem 254:6763-6771

Ries G, Hundt E, Kadenbach B (1978) Immunoprecipitation of a cytoplasmic precursor of rat liver cytochrome oxidase. Eur J Biochem 91:179-191

Rubin MS, Tzagoloff A (1973) Assembly of the mitochondrial membrane system. J Biol Chem 248: 4275-4279

Schmelzer E, Heinrich PC (1980) Synthesis of a larger precursor for the subunit IV of rat liver cytochrome c oxidase in a cell-free wheat germ system. J Biol Chem 255:7503-7506

Sebald W, Weiss H, Jackl G (1972) Inhibition of the assembly of cytochrome oxidase in *Neurospora crassa* by chloramphenicol. Eur J Biochem 30:413-417

Towbin H, Staehelin T, Gordon J (1979) Electrophoretic transfer of proteins from polyacrylamide gels to nitrocellulose sheets: Procedure and some applications. Proc Natl Acad Sci USA 76: 4350-4354

Discussion[2]

Blobel commented that the precursor proteins for the individual cytochrome c oxidase subunits cannot be derived by proteolytic cleavage from a polyprotein, since all of the 4 subunits in yeast cytochrome c oxidase are synthesized with the N-formyl-methionine at the beginning.

Heinrich mentioned that it was possible to demonstrate the presence of the "polyprotein" as an artefact when the immunoprecipitation was carried out at rather high antiserum concentrations, and without sodium dodecylsulfate.

Ballard asked whether it could be excluded that a synthesized polyprotein itself has proteolytic activity.

Heinrich answered that this possibility can be excluded in the case of yeast, because it has been shown that all 4 subunits are synthesized as N-formyl-methionine-peptides, but not yet in the case of rat liver.

Neurath raised a question concerning the inhibition pattern of the proteinase shown in the last table in comparison to the one found for the microsomal signal peptidase.

Heinrich pointed out that the properties of the signal peptidase, and in particular the inhibition pattern, have not yet been studied.

Blobel added that the proteinase may act only when the protein is translocated.

Heinrich mentioned the recent work by G. Schatz, where a soluble proteinase that cleaves precursors of yeast F1-ATPase has been discovered. This proteinase is inhibited by o-phenanthroline.

Cohen asked for the evidence that cytochrome c oxidase is composed of seven subunits and whether these subunits have been taken apart and put together again.

Heinrich stated that the number of subunits reported for purified cytochrome c oxidase from different tissues varies from 2 to 12, and that the same problem exists with all multi-subunit membrane proteins. A reconstitution of cytochrome c oxidase from the individual subunits has not yet been described.

---

2 Rapporteur: Peter C. Heinrich

Characterization and Possible Pathophysiological Significance of Human Erythrocyte Proteinases

A. De Flora, E. Melloni, F. Salamino, B. Sparatore, M. Michetti, U. Benatti, A. Morelli, and S. Pontremoli[1]

1 General Considerations

The mechanisms of intracellular proteolysis in human erythrocytes represent a topic of outstanding biochemical interest in view of the several physiological processes (e.g., aging of the circulating red cells) they are thought to mediate. An additional reason of interest concerns the expected relevance of proteolytic events to the cellular expression of some genetically determined disorders affecting the erythrocytes, notably thalassemias and enzyme defects. Among the last abnormalities, deficiency of glucose-6-phosphate dehydrogenase (G6PD) activity stands out for its high prevalence, but also the thalassemias are worldwide diseases and represent a serious problem of public health.

Table 1 shows in a schematic way the molecular pathology of the Mediterranean variety of G6PD deficiency as well as of the two commonest types of thalassemic syndromes, α and β. Although this matter is still not proven and mostly speculative, there is circumstantial evidence that an enhanced breakdown of the structural G6PD variant (Morelli et al. 1978) as well as a selective degradation of either excess globin chain, α or β (Bunn et al. 1977), are involved in the phenotypic expression of these genetic disorders.

Table 1. Biochemical mechanisms and role of proteolysis in the expression of G6PD deficiency and thalassemias

Disorder	Type	Molecular pathology	Correlation with intracellular proteolysis
G6PD deficiency	Mediterranean	Enhanced intracellular decay of mutant G6PD	Accelerated breakdown of mutant G6PD
Thalassemias	α Thalassemia	Decreased or absent synthesis of α globin chains	Degradation of excess β chains?
	β Thalassemia	Decreased or absent synthesis of β globin chains	Degradation of excess α chains?

1 Institute of Biochemistry, University of Genova, Viale Benedetto XV/1, 16132 Genova, Italy

For the foregoing reasons, defining the molecular and cellular abnormalities of G6PD deficiency and thalassemias rests, among other things, also on the elucidation of the proteolytic systems of the human red cell.

2 Identifying the Proteolytic Systems of Mature Human Erythrocytes

This task has been frustrated for several years by substantial contamination of the erythrocyte suspensions under study by leukocytes and platelets which are a rich source of several proteolytic enzymes. Accordingly, a technical prerequisite for the present investigations was the complete removal of leukocytes and platelets from the erythrocytes, according to a procedure developed by Beutler et al. (1976). The homogeneous erythrocyte suspensions so obtained were hemolyzed and investigated for their proteinase and peptidase activities, both in the membrane and in the soluble fraction.

Table 2 shows the nature, activity levels, and approximate molecular sizes of the proteolytic enzymes we were able to identify in the mature human erythrocyte (Pontremoli et al. 1979, 1980a, 1980b). The membrane proteinases were purified to homogeneity and partially characterized in their molecular and functional properties (Pontremoli et al. 1979, 1980a). They are three cathepsin D-like endopeptidases, with optimal activity in the acidic pH region; although displaying consistently different structural properties in terms of molecular size, amino acid composition, and primary structure, the three membrane endopeptidases do hydrolyze identical peptide bonds on common protein substrates and have a comparable susceptibility to inhibitors such as hemin, 2-mercapto-

Table 2A, B. Proteinases and peptidases of mature human erythrocytes [a]

Enzyme type	Activity (units/10 ml) packed RBC)	Approximate molecular weight
A. Membrane		
1. Acidic proteinase I	4 [b]	80,000
2. Acidic proteinase II	2.4 [b]	40,000
3. Acidic proteinase III	3.6 [b]	30,000
B. Cytosol		
1. Acidic Proteinase I	7	80,000
2. Acidic proteinase II	5	40,000
3. Acidic proteinase III	8	30,000
4. Neutral proteinase	23	110,000
5. Aminopeptidase (aspecific)	75	80,000
6. Aminopeptidase B	67	80,000
7. Dipeptidyl aminopeptidase II	240	80,000
8. Dipeptidyl aminopeptidase III	173	80,000

[a] Data taken from Pontremoli et al. (1979, 1980a, 1980b)
[b] Assayed after solubilization with butanol (Pontremoli et al. 1979)

ethanol, and others (Pontremoli et al. 1980a). The proteinases identified in the cytosol of mature erythrocytes were partially purified according to conventional procedures and characterized as well. As indicated in Table 2, three cathepsin D-like acidic endopeptidases are present in the soluble fraction, the properties of which are very similar to those of the corresponding membrane enzymes. In addition, a Ca^{2+}-stimulated neutral endopeptidase having an approximate molecular weight of 110,000 is also present. While no evidence for carboxypeptidase activities was found in the cytosol of mature human erythrocytes, the following exopeptidases were identified and characterized (Table 2): two aminopeptidases, one with broad substrate specificity and the other being an aminopeptidase B, respectively; and two dipeptidyl aminopeptidases, classified as dipeptidyl aminopeptidases II and III on the basis of their substrate specificities and pH optima (Pontremoli et al. 1980b).

3 Decay of Proteinase and Peptidase Activities During Erythrocyte Aging

The erythrocyte represents a suitable model for studies on cellular aging. Several enzyme activities are known to decline throughout the life-span of circulating erythrocytes (Piomelli and Corash 1976) and this general fact could be the result of intracellular proteolysis. Since nothing is known so far about such tentative correlation between proteolytic systems and the decay of enzyme proteins underlying the maturation of erythrocytes, the levels of the proteinase and peptidase activities listed in Table 2 were measured in erythrocyte fractions of increasing age.

Figure 1 shows the results obtained with the membrane endopeptidase activity which is distributed among three different enzyme proteins (Table 2). During erythrocyte aging, both the "total" and the "available" activities undergo a clearcut decline which is more extensive for the latter than for the former activity. Accordingly, the ratio of available to total proteolytic activity, which can be taken as an indication of the extent to which the membrane enzymes are freely accessible to their substrates, shows a progressive decrease. This fact compares well with previous findings obtained with similar proteolytic systems associated with the membrane of rabbit reticulocytes and erythrocytes (Pontremoli et al. 1979), and demonstrates that the three acidic endopeptidases become progressively more cryptic within the membrane as soon as the erythrocyte ages. Whether or not this modification in the molecular organization of the membrane is actually related to changes and (or) regulation of the intracellular proteolytic processes is still open to question. The behavior of the cytosolic proteinases and peptidases was also investigated in erythrocyte fractions of increasing age and Table 3 shows the relevant half-lives of enzyme activities. Again, the $t_{1/2}$ value of acidic endopeptidase activity is herein referred to the sum of the three distinct enzymes responsible for such activity rather than to each of the individual endopeptidases, because of difficulties inherent to the procedures of erythrocyte fractionation. However, gel chromatographic analyses indicated that Proteinase I affords the greatest contribution to the observed decline in the overall acidic endopeptidase activity during aging (Melloni et al. 1981).

As far as the other cytosolic enzymes are concerned, with the exception of dipeptidyl aminopeptidase III which keeps constant in the transition from the young to the old cells, also these proteinases and peptidases can be safely defined as "age-dependent enzymes" (Table 3).

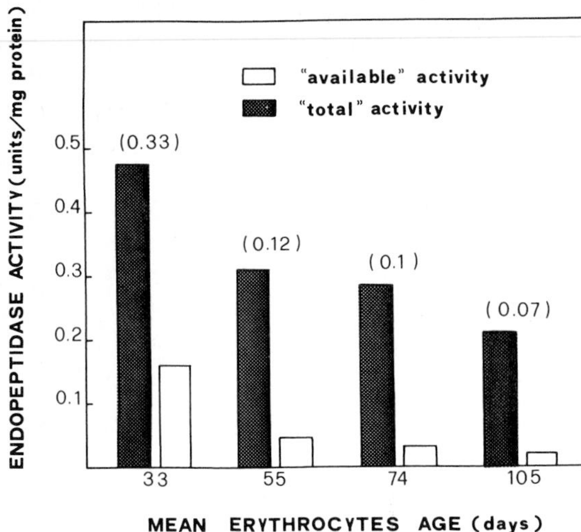

Fig. 1. Patterns of decay of the acidic endopeptidase activity associated with the membrane during aging of erythrocytes. Fractionation of human erythrocytes, preliminarily freed of leukocytes and platelets as described previously (Pontremoli et al. 1979), was obtained by centrifugation on a discontinuous density gradient of Stractan II (Corash et al. 1974). Four cell layers of increasing specific gravity were obtained which were designated as F_1 (*top*), F_2, F_3 (*at the interfaces*), and F_4 (*bottom*), respectively. The correlation between specific gravity of each erythrocyte layer and mean cellular age was estimated according to Piomelli et al. (1967, 1968). The erythrocyte layers were washed three times with 0.15 M NaCl and lysed by freezing-thawing as described (Melloni et al. 1981). The membrane pellets were then washed several times with 5 mM Na phosphate buffer, pH 7.5, until being completely hemoglobin-free. Assays of acidic endopeptidase activity associated with the membrane pellets were performed at pH 2.5, as described previously (Pontremoli et al. 1979), in the presence ("total" activity) and in the absence ("available" activity) of 4 mg/ml of Na deoxycholate. The *numbers in parentheses* refer to the ratios of "available" to "total" acidic endopeptidase activity

Table 3. Half-lives of cytosolic proteinase and peptidase activities of circulating human erythrocytes

Enzyme	In vivo $t_{1/2}$ [a]
Acidic endopeptidases [b]	87
Neutral endopeptidase	60
Aminopeptidase (aspecific)	133
Aminopeptidase B	49
Dipeptidyl aminopeptidase II	84
Dipeptidyl aminopeptidase III	

[a] Calculated according to Piomelli and Corash (1976)

[b] Assayed as described in the legend to Fig. 1 for the membrane-associated acidic endopeptidase activity. This activity is contributed by the three cytosolic acidic proteinases, I, II, and III, listed in Table 2

4 Correlations Between Erythrocyte Proteinases and Breakdown of α and β Globin Chains

The relevance of specific proteolytic systems of the erythrocyte to the cellular expression of G6PD deficiency is so far completely undetermined. The presence of pyroglutamic acid as the amino terminal residue of human G6PD (Yoshida 1972) restricts however the primary proteolytic attack on this enzyme protein to an endopeptidase action.

In an attempt to elucidate the mechanisms of intracellular proteolysis in thalassemias (see Table 1), preliminary experiments were started aiming at elucidating the susceptibility of the individual globin chains, α and β, to digestion by the above mentioned erythrocyte proteinases. Figure 2 shows the effect afforded by heme binding on the de-

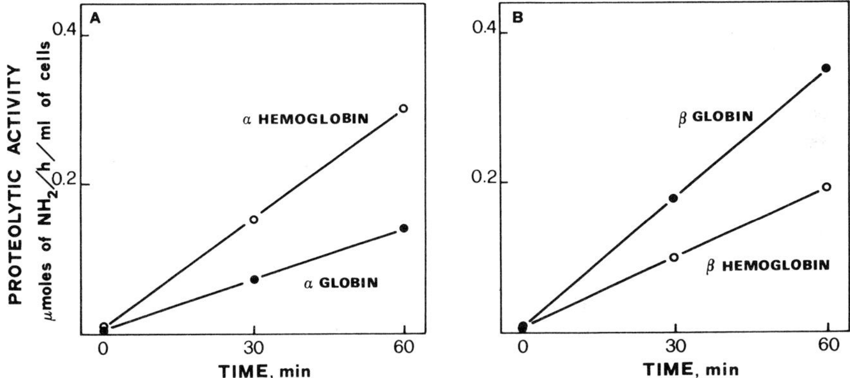

Fig. 2A, B. Distinctive susceptibility of α and β globin chains to in vitro digestion by purified membrane endopeptidases. Membrane proteinases I, II, and III were purified from human erythrocytes as described previously (Pontremoli et al. 1980a). Hemoglobin was prepared from human erythrocyte lysates and all cytosolic proteolytic enzymes were completely removed by absorption on DEAE Sephadex (5 g dry weight/g Hb), equilibrated in sodium phosphate buffer, pH 6.4. Resolution of α and β hemoglobin chains (containing bound heme) was obtained according to the procedure of Bucci and Fronticelli (1965). Heme-deprived α and β globin chains were prepared from the corresponding heme-containing globin chains by means of treatment with acetone-HCL (see Pontremoli et al. 1979). The mixtures were incubated at 37°C and contained, in a final volume of 1.0 ml, 100 mM Na acetate, pH 6.0, 2 mg of either heme-free or heme-containing α and β globin, 2.5 units (defined as described by Pontremoli et al. 1979) of purified membrane Proteinase I, 2.5 units of purified Proteinase II, and 2.5 units of purified Proteinase III. The extent of proteolysis was evaluated as described previously (Pontremoli et al. 1979) at the time intervals indicated

gradation of α globin chains (**A**) and of β globin chains (**B**) by a mixture of the three purified membrane endopeptidases. In these conditions, binding of heme to the isolated α chains stimulated nearly twofold their in vitro digestion. The opposite situation holds for degradation of β globin chains which is consistently lowered by presence of bound heme. It is interesting to note that these experiments were carried out at pH 6.0, a fact indicating that the membrane endopeptidases are still appreciably active at pH values far from their acidic optima and close to physiological values.

Similar findings were obtained as concerns digestion of the two isolated globin chains by the bulk of proteolytic enzymes present in the erythrocyte cytosol (Fig. 3). Thus, heme protected the β chains (**B**) while enhancing conversely the rate of degradation of α chains (**A**).

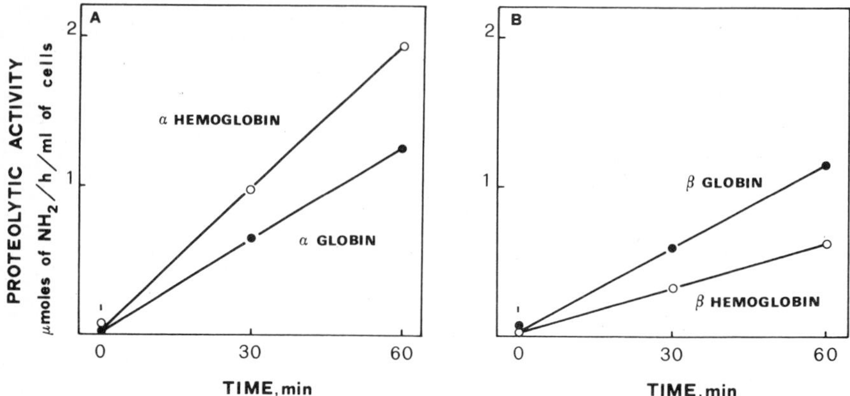

Fig. 3A, B. Distinctive susceptibility of α and β globin chains to in vitro digestion by cytosolic proteolytic enzymes. Heme-free and heme-containing α and β globin chains were prepared as described in the legend to Fig. 2. "Hemoglobin-free cytosolic fraction", containing the bulk of proteinase and peptidase activities of human erythrocytes, was prepared as described previously (Pontremoli et al. 1980b). The incubation mixtures (1.0 ml, final volume), kept at 37°C, contained 100 mM Na acetate, pH 6.0, α or β globin chains as reported in the legend to Fig. 2, and 0.2 ml of the "Hemoglobin-free cytosolic fraction", corresponding to 0.1 ml of packed erythrocytes

The data shown in Figs. 2 and 3 could be tentatively related to some known hematological features of thalassemias. With reference to this, it is relevant to remind that the circulating erythrocytes from β thalassemic subjects fail to accumulate the excess of synthesized α chains (Fessas and Loukopoulos 1964; Bank and O'Donnell 1969; Hanash and Rucknagel 1978). The present results suggest the possibility of the α chains being rapidly removed by proteolysis within the erythrocyte soon after their combination with heme, although alternative hypotheses such as "ineffective erythropoiesis", i.e., intracellular precipitation of sparingly soluble α chains leading to intramedullary destruction of erythroid cells, have been also advanced. On the other hand, in the erythrocytes from α thalassemic subjects the excess of heme-containing β globin chains is found intracellularly as the abnormal β_4 tetramer, also designated hemoglobin H, which appears to escape substantial degradation.

The above characteristics seem to indicate that the same proteolytic systems of the erythrocyte and specifically the cathepsin D-like endopeptidases endowed in the membrane are actively involved in the breakdown of heme-containing α globin chains occurring in β thalassemias, while leaving the heme-containing β globin chains undegraded in the α thalassemic erythrocytes. The distinctive digestibility of heme-containing α and β globin chains by the erythrocyte proteinases (Figs. 2 and 3) is consistent with such interpretation, which needs however further support through experiment to be pursued by means of intact red cell systems.

Conformation of the isolated α and β globin chains appears therefore to be a major determinant of their susceptibility to in vitro degradation by erythrocyte proteinases, although the enhanced breakdown of α chains following binding of heme may be opposite to expectations. The experiment illustrated in Fig. 4 lends further support to the view that

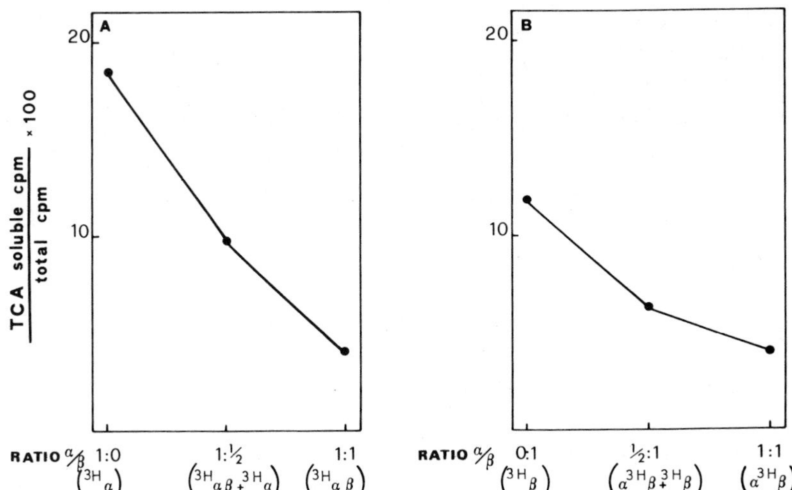

Fig. 4A, B. Comparative digestion of isolated vs in vitro reconstituted heme-containing globin chains by purified membrane endopeptidases. Membrane proteinases I, II, and III were purified from human erythrocytes as described previously (Pontremoli et al. 1980a). For synthesis of labeled hemoglobin, human reticulocytes (corresponding to F_1 fraction, prepared as described in the legend to Fig. 1) were suspended in 3 vol of Krebs-Ringer phosphate buffer, containing glucose (2 mg/ml), $FeSO_4$ (30 μg/ml) and 0.1 mM concentrations of each amino acid, excepting leucine. After 5 min at 37°C, 10 μCi/ml of [^{3}H]leucine (6 C/nmol) was added and the incubation was continued for 1 h. Cells were then washed three times with the Krebs-Ringer buffer containing 10 mM nonradioactive leucine. Resolution of [^{3}H]α and [^{3}H]β globin chains was obtained according to the procedure of Bucci and Fronticelli (1965). Non-radioactive, heme-containing α and β globin chains were similarly prepared from normal erythrocyte lysates. The incubation mixtures, kept at 37°C for 3 h, contained, in a final volume of 1.0 ml, 100 mM Na acetate buffer, pH 6.0, 2.5 units of each purified membrane endopeptidase (as reported in the legend to Fig. 2), and different combinations of radioactive and nonradioactive heme-containing α and β globin chains (2 mg/ml, final concentration), as indicated in the abscissa. The extent of digestion shown on the ordinates was equated to the amount of TCA-soluble radioactive material liberated after 3 h incubation

digestion of the individual globin chains is strongly affected by their actual conformations. In this case, the changes in the conformation of heme-binding α and β globin chains were induced by their reciprocal combination to achieve the oligomeric hemoglobin species shown in the abscissa. Thus, as shown in Fig. 4A, degradation of the [^{3}H]-labeled α chain by a mixture of purified membrane endopeptidases is progressively decreased by increasing amounts of unlabeled β chain. On the other hand, digestion of the labeled β chain, which is more limited in extent than that of the α chain alone as predictable by the results shown in Figs. 2 and 3, is similarly reduced by addition of the unlabeled α globin chain. Taken together, these results indicate that the normal $\alpha_2\beta_2$ tetramer (hemoglobin A) is most resistant to intracellular proteolysis, while the heme-containing α chain and the heme-free β chain show maximal susceptibility to digestion by the erythrocyte proteinases.

Preliminary experiments seem to rule out any significant difference in the complement, intracellular distribution, and activity levels of the various proteinases and peptidases between erythrocytes from healthy controls and thalassemic subjects, respectively.

Accordingly, the hematological features of α and β thalassemic erythrocytes and specifically their content of various hemoglobin species and globin pools are those expected on the basis of the differential susceptibility of these forms to in vitro digestion by the erythrocyte proteinases and especially by the membrane endopeptidases.

Acknowledgements. This work was supported by a grant from the Progetto Finalizzato di Medicina Preventiva del C.N.R., Subprogetto M.E.E.

References

Bank A, O'Donnell JV (1969) Intracellular loss of free α Chains in β Thalassemia. Nature (London) 222:295-296

Beutler E, West C, Blume KG (1976) The removal of leukocytes and platelets from whole blood. J Lab Clin Med 88:328-333

Bucci E, Fronticelli C (1965) A new method for the preparation of α and β subunits of human hemoglobin. J Biol Chem 240:PC 551-552

Bunn HF, Forget BG, Ranney HM (1977) Human Hemoglobins. W.B. Saunders Co., Philadelphia London Toronto, pp 140-192

Corash LM, Piomelli S, Chen HC, Seaman C, Gross E (1974) Separation of erythrocytes according to age on a simplified density gradient. J Lab Clin Med 84:147-151

Fessas P, Loukopoulos D (1964) Alpha-chain of human hemoglobin: Occurrence in vivo. Science 143:590-591

Hanash SM, Rucknagel DL (1972) Proteolytic activity in erythrocyte precursors. Proc Natl Acad Sci USA 75:3427-3431

Melloni E, Salamino F, Sparatore B, Michetti M, Morelli A, Benatti U, De Flora A, Pontremoli S (1981) Decay of proteinase and peptidase activities of human and rabbit erythrocytes during cellular aging. Biochim Biophys Acta (in press)

Morelli A, Benatti U, Gaetani GF, De Flora A (1978) Biochemical mechanisms of Glucose 6-phosphate dehydrogenase deficiency. Proc Natl Acad Sci USA 75:1979-1983

Piomelli S, Corash L (1976) Hereditary hemolytic anemia due to enzyme defects of glycolysis. In: Harris H, Hirschhorn K (eds) Advances in human genetics. Plenum Press, New York London, pp 165-240

Piomelli S, Lurinsky G, Wasserman LR (1967) The mechanism of red cell aging. I) Relationship between cell age and specific gravity evaluated by ultracentrifugation in a discontinuous density gradient. J Lab Clin Med 69:659-674

Piomelli S, Corash LM, Davenport DD, Miraglia J, Amorosi EL (1968) In vivo lability of Glucose 6-phosphate dehydrogenase in Gd^{A-} and $Gd^{Mediterranean}$ deficiency. J Clin Invest 47:940-948

Pontremoli S, Salamino F, Sparatore B, Melloni E, Morelli A, Benatti U, De Flora A (1979) Isolation and partial characterization of three acidic proteinases in erythrocyte membranes. Biochem J 181:559-568

Pontremoli S, Sparatore B, Melloni E, Salamino F, Michetti M, Morelli A, Benatti U, De Flora A (1980a) Differences and similarities among three acidic endopeptidases associated with human erythrocyte membranes. Molecular and functional studies. Biochim Biophys Acta 630:313-322

Pontremoli S, Melloni E, Salamino F, Sparatore B, Michetti M, Benatti U, Morelli A, De Flora A (1980b) Identification of proteolytic activities in the cytosolic compartment of mature human erythrocytes. Eur J Biochem 110;421-430

Yoshida A (1972) Micro method for determination of blocked NH_2-terminal amino acids of protein: Application to identification of acetylserine of phosphoglycerate kinase and pyroglutamic acid of Glucose 6-phosphate dehydrogenase. Anal Biochem 49:320-325

Discussion[2]

Ballard asked about the level of proteinases in erythrocytes and the ratio of proteolytic activities in reticulocytes and erythrocytes since some reticulocyte contamination could account for the proteinase activities.

De Flora agreed, but mentioned that there are no more than the usual 1% or 2% of reticulocytes in the erythrocyte preparations. The procedure of fractionation of erythrocytes of increasing age was reliable for measuring a number of biochemical parameters throughout the life-span of erythrocytes. Thus, the changes in proteinase and peptidase activities are really age-dependent. The age dependence of proteolytic enzymes could be better evaluated in the rabbit in which it was possible to induce a high reticulocytosis (up to 80%-90% reticulocytes) by administration of phenylhydrazine. The results obtained are in good agreement with the pattern of age-dependent decline observed in human erythrocytes, although such a decline was obviously more pronounced due to the high reticulocyte levels.

Holzer mentioned the reports on ATP-dependent proteolytic systems in several cell types including reticulocytes.

De Flora says that he was unable to find any ATP-dependence of the proteinases and peptidases in the mature circulating human erythrocytes.

2 Rapporteur: Peter C. Heinrich

Physiological Function and Catalytic Mechanism of Angiotensin Converting Enzyme

P. Bünning[1] and J.F. Riordan[2]

1 Introduction

Angiotensin converting enzyme (ACE) is part of the well-known renin-angiotensin system (Peach 1977; Reid et al. 1978). The physiological function of the enzyme is to convert the decapeptide angiotensin I to the hypertensive octapeptide angiotensin II and, therefore, the enzyme plays an important role in the control of blood pressure (Skeggs et al. 1954, 1956a, b; Soffer 1976). Recently, the clinical importance of ACE has been established by the discovery that synthetic inhibitors of this enzyme are potent drugs for the control of hypertension in man (Cushman et al. 1977; Gavras et al. 1978). The design of these inhibitors has been guided by a hypothetical model of the active site of ACE.

In addition to its physiological functions we were attracted to ACE because it contains a catalytically essential zinc atom (Das and Soffer 1975; Bünning and Riordan 1980) and, therefore, might possess mechanistic similarities to the much better characterized zinc proteases, bovine carboxypeptidase A and thermolysin. Since it has been suggested that these two metalloenzymes have similar active sites and, therefore, probably similar modes of action (Kester and Matthews 1977), ACE may also share their common features. In addition, it was of interest to investigate the mechanism of the enzyme activation by monovalent anions (Skeggs et al. 1954), a phenomenon which is not generally observed for proteolytic enzymes. We here describe the investigation of the kinetic characteristics of ACE including its activation by monovalent anions and the identification of the components of its active site by chemical modification reactions. These studies have enabled us to compare the enzyme with the much better characterized zinc proteases and, thereby, deduce a scheme for its mode of action.

2 The Physiological Function of ACE

The renin-angiotensin system plays an important role in maintaining the homeostasis of peripheral vascular resistance as well as volume and electrolyte composition of extracel-

---

Abbreviations: ACE, angiotensin converting enzyme; FA-, 2-furanacryloyl-; dansyl-, 5-dimethylamino-naphthalene-1-sulfonyl-; CMC, 1-cyclohexyl-3-(2-morpholinoethyl)-carbodiimide; FDNB, 1-fluoro-2,4-dinitrobenzene; Hepes, hydroxyethylpiperazine ethanesulfonate. All amino acids are of the L-configuration

1 Biochemisches Institut der Universität Freiburg, Hermann-Herder-Str. 7, 7800 Freiburg, FRG

2 Biophysics Research Laboratory, Department of Biological Chemistry, Harvard Medical School and the Brigham and Women's Hospital, Boston, MA 02115, USA

lular body fluids (Peach 1977; Reid et al. 1978). The biologically active component of the system, angiotensin II, is generated by a sequence of proteolytic cleavage steps from angiotensinogen, a plasma protein of the α_2-globulin fraction (Fig. 1). The system is initiated by the renin-catalyzed cleavage of angiotensinogen. Renin, an acid protease, is synthesized and stored in the juxtaglomerular apparatus of the kidney from where it is secreted into plasma upon appropriate stimulus. The renin substrate, angiotensinogen, is probably produced by the liver, secreted into the circulation and hydrolyzed by renin which results in the formation of angiotensin I. Subsequently, the C-terminal dipeptide of angiotensin I, histidyl-leucine, is removed by ACE, a dipeptidyl carboxypeptidase, to form the active octapeptide hormone, angiotensin II. ACE is associated with endothelial cell membranes of the vascular lining and it faces the lumen of the blood vessel (Ryan et al. 1975; Caldwell et al. 1976). The pulmonary vasculature is the major locus of the enzyme, and angiotensin I is almost completely converted to angiotensin II within a single passage through the lungs. In addition to the generation of angiotensin II ACE inactivates the hypotensive kinins (Yang et al. 1970) and, thereby, enhances the action of angiotensin II through the destruction of these opposite-acting agents. The inactivation of enkephalins has also been proposed to be mediated by ACE (Erdös et al. 1978) and, in principle, the enzyme has a broad specificity and could be expected to act on a wide range of oligopeptide substrates.

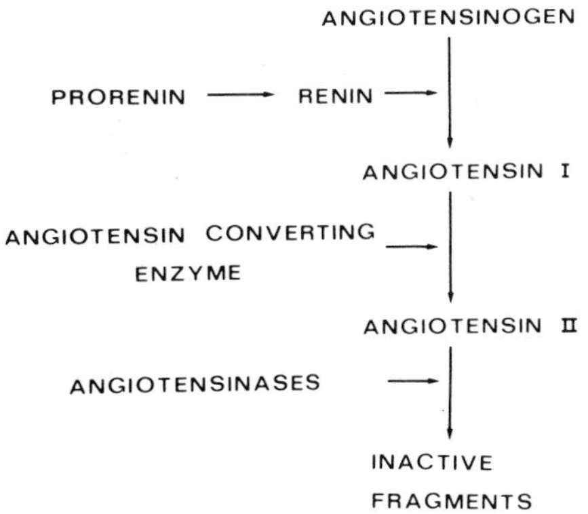

Fig. 1. The Renin-Angiotensin System

3 Kinetic Characteristics of ACE

To facilitate mechanistic investigations of ACE, we have developed a continuous spectrophotometric assay utilizing FA-tripeptides as substrates (Holmquist et al. 1979). The assay is based on a blue shift of the absorption spectrum that occurs upon hydrolysis of the substrate to produce a FA-blocked amino acid and a dipeptide. The examination of a variety of FA-tripeptides indicates that to a large extent the subsite specificity of ACE is determined by the antepenultimate amino acid residue. The enzyme preferentially cleaves peptides with hydrophobic residues in this position (Holmquist et al. 1979; Bünning et al. 1980). Among the substrates studied FA-Phe-Gly-Gly exhibits the most suitable properties for routine assays. A comparison of its kinetic parameters, $K_M = 3 \times 10^{-4}$ M and $k_{cat} = 19000\ min^{-1}$, to those for the hydrolysis of the physiological substrate angiotensin I, $K_M = 7 \times 10^{-5}$ M and $k_{cat} = 660\ min^{-1}$, shows that FA-Phe-Gly-Gly has a higher K_M value and that it is more rapidly turned over (Bünning et al. 1980).

Under first-order kinetics, i.e., $[S] < K_M$, the rate of FA-Phe-Gly-Gly hydrolysis varies markedly as a function of pH (Fig. 2). In the presence of 300 mM sodium chloride

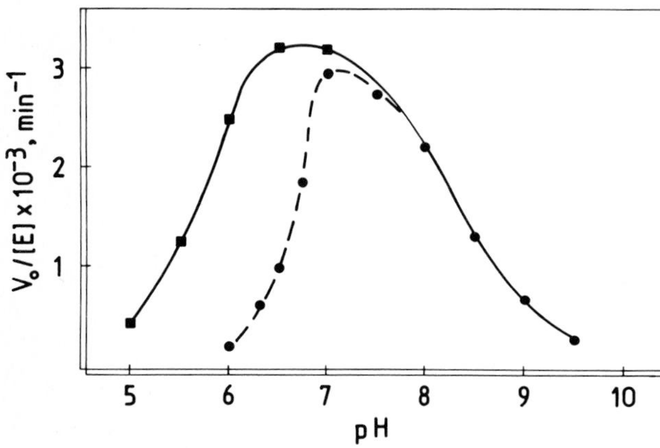

Fig. 2. The pH profile for the ACE-catalyzed hydrolysis of 5×10^{-5} M FA-Phe-Gly-Gly determined in 300 mM sodium chloride in the absence (●) or in the presence (■) of added zinc. Data from Bünning et al. (1980)

maximal substrate hydrolysis is observed at pH 7.0. Below pH 7 activity drops off markedly due to spontaneous loss of zinc from the enzyme. Addition of zinc to the assay mixture increases activity in the acidic pH region. When optimal zinc concentrations are employed the pH-rate profile for the hydrolysis of FA-Phe-Gly-Gly has apparent pK values for the free enzyme of 5.6 and 8.4 (Bünning et al. 1980). These pK values are close to those for other zinc proteases, 6.2 and 9.0 for carboxypeptidase A (Auld and Vallee 1970), and 5.6 and 7.5 for thermolysin (Feder and Schuck 1970; Holmquist and Vallee 1976).

One characteristic property of ACE is its almost absolute requirement of monovalent anions for catalytic activity (Fig. 3). Chloride is the most potent activator. In the

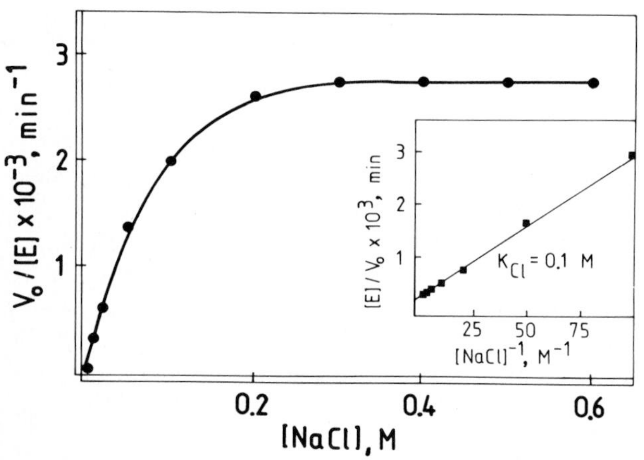

Fig. 3. Effect of sodium chloride on the activity of ACE measured with 5×10^{-5} M FA-Phe-Gly-Gly as substrate in 50 mM Hepes buffer, pH 7.5. Data from Bünning et al. (1980). *Inset:* The apparent chloride binding constant is obtained from a double reciprocal plot

absence of chloride and under first-order reaction conditions the enzyme is virtually inactive toward FA-Phe-Gly-Gly. With increasing chloride concentrations enzyme activity increases and reaches a maximum at about 300 mM chloride. The apparent chloride binding constant at pH 7.5 is 100 mM (Fig. 3, *inset*). A similar activation profile is observed for the hydrolysis of angiotensin I (Bünning and Riordan 1980).

The Lineweaver-Burk plots for the hydrolysis of FA-Phe-Gly-Gly at chloride concentrations from 20 to 300 mM show that with increasing chloride concentration K_M decreases but k_{cat} does not change (Fig. 4). This kinetic behavior fits an ordered bireactant mechanism

$$E + A \underset{}{\overset{K_A}{\rightleftharpoons}} EA + S \underset{}{\overset{K_S}{\rightleftharpoons}} EAS \xrightarrow{k_{cat}} E + P$$

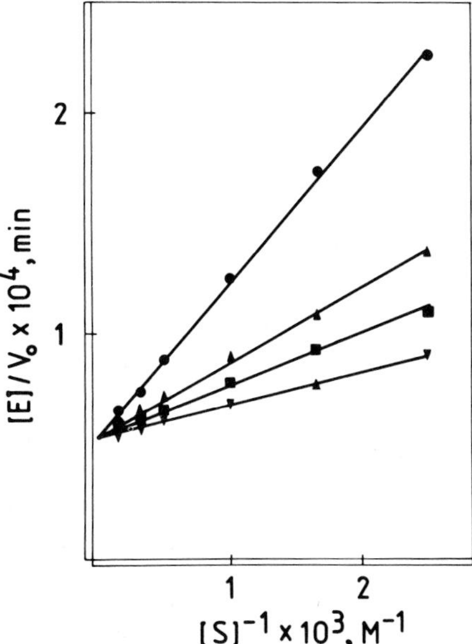

Fig. 4. Lineweaver-Burk plot for the ACE-catalyzed hydrolysis of FA-Phe-Gly-Gly in 50 mM Hepes buffer, pH 7.5, containing 20 (●), 50 (▲), 100 (■) or 300 (▲) mM sodium chloride. Data from Bünning and Riordan (1980)

in which the anion binds to the enzyme first and then substrate binds to the enzyme-anion complex followed by substrate hydrolysis (Segel 1975). Thus, the function of chloride is to mediate substrate binding. Direct evidence for such a mechanism was obtained by stopped-flow studies of ACE with a fluorescently labeled substrate. In the presence of chloride both binding and hydrolysis are observed, whereas in the absence of chloride the peptide does not bind to the enzyme (Bünning and Riordan 1980).

4 Components of the Active Site of ACE

Since ACE was known to contain zinc, essential for catalytic activity (Das and Soffer 1975), we determined its metal content and investigated the effects of chelating agents on enzymatic activity. The enzyme was purified under conditions that prevent contamination by extraneous metals (Thiers 1957) and its metal content was subsequently determined by atomic absorption spectroscopy. The enzyme contains one gram atom of zinc per mole (Bünning and Riordan 1980; Bünning et al. 1980). The functional necessity of the zinc atom was demonstrated by inhibition with the chelating agent 1,10-phenanthroline, whereas the nonchelating analog 4,7-phenanthroline is not an inhibitor (Fig. 5A).

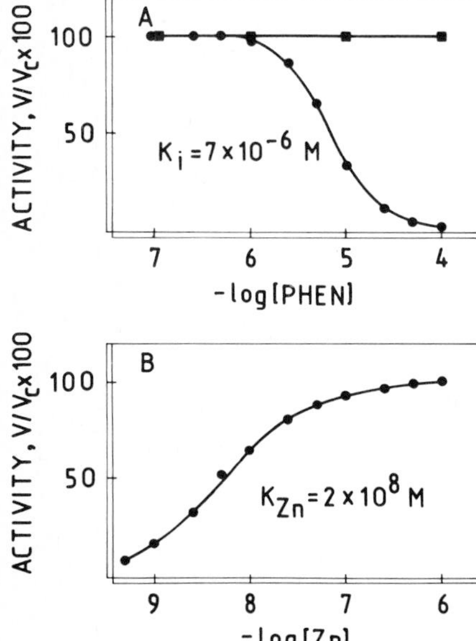

Fig. 5. A Effects of 1,10-phenanthroline (•) and 4,7-phenanthroline (▪) on ACE-catalyzed hydrolysis of FA-Phe-Gly-Gly, 5×10^{-5} M, in 50 mM Hepes buffer, pH 7.5. **B** Restoration of ACE activity from apoenzyme by addition of zinc in 50 mM Hepes buffer, pH 7.5, 300 mM sodium chloride. The apparent metal-enzyme stability constant is almost identical to the true stability constant since corrections for the binding of chloride and Hepes to zinc are very small

Dialysis of the 1,10-phenanthroline-inhibited enzyme to remove the chelating agent yields the metal-free apoenzyme which is virtually inactive. Readdition of zinc to the apoenzyme fully restores activity to that of the native enzyme, which demonstrates that ACE is a metalloenzyme that requires zinc for catalytic activity (Fig. 5B). The apparent K_i for 1,10-phenanthroline inhibition is approximately 7×10^{-6} M, much lower than generally observed for other zinc metalloenzymes (Riordan and Vallee 1976). The indication that metal binding to ACE is rather weak is evident from the low stability constant, 2×10^8 M. The relative ease of zinc dissociation also explains the spontaneous loss of the metal ion from the enzyme in the acidic pH region (Fig. 2). To elucidate the function of zinc further, we have carried out stopped-flow experiments with the apoenzyme (Bünning and Riordan 1980). In the presence of chloride a dansyl-peptide binds just as well to the apo- as to the zinc enzyme. However, the apoenzyme-peptide complex is stable and does not break down to products. Thus, the active site zinc atom is essential for catalysis but not for binding of the substrate.

In order to identify amino acid residues essential for catalytic activity, we have carried out a series of site-specific chemical modifications with ACE (Table 1; Bünning et al. 1978, 1979). The enzyme is inactivated by the carboxyl group reagent CMC. The rate of inactivation is not affected by the presence of the nucleophile glycine methylester, and

Table 1. Chemical modifications of functional residues in ACE. Effect of chloride on the rate of ACE inactivation

Reagent	$k \times 10^2$, min^{-1} ($-Cl^-$)	$k \times 10^2$, min^{-1} ($+ 0.3$ M Cl^-)
CMC [a]	3.9	4.1
Acetylimidazole	9.2	9.0
Tetranitromethane	2.0	2.1
Butanedione/Borate	1.7	1.6
FDNB [b]	5.8	2.7
Pyridoxal phosphate	3.5	< 0.1

[a] CMC, 1-cyclohexyl-3-(2-morpholinoethyl)-carbodiimide
[b] FDNB, 1-fluoro-2,4-dinitrobenzene

activity is not restored by hydroxylamine. These results are virtually identical to those observed for the CMC modification of carboxypeptidase A, where it has been demonstrated that the carbodiimide inactivates the enzyme by modification of an essential active-site glutamyl residue (Riordan and Hayashida 1970; Nau and Riordan 1975). Therefore, it is very likely that an essential carboxyl group is also present in ACE. Treatment with acetylimidazole rapidly inactivates the enzyme, and activity can be totally restored by addition of hydroxylamine. These results, together with the finding that treatment with the tyrosyl reagent tetranitromethane also decreases activity, indicate that the enzyme contains essential tyrosyl residues. ACE is also inactivated by arginyl reagents. Inactivation by butanedione and, in particular, the enhancing effect of borate are diagnostic of an essential arginyl residue. This is confirmed with another arginyl reagent, phenylglyoxal. The presence of essential arginyl residues in ACE is not unexpected since virtually every enzyme studied that is specific for anionic substrates has been found to be inactivated by arginyl reagents (Riordan et al. 1977). By reaction with pyridoxal phosphate the activity of ACE is decreased to about 50%, which points to essential lysyl residues. The pyridoxal phosphate modified enzyme is fully reactivated by dialysis, but there is no reactivation if the Schiff base is reduced by sodium borohydride prior to dialysis. Further evidence for the existence of an essential lysine was obtained by reacting the enzyme with FDNB.

When these chemical modifications are carried out in the presence of a competitive inhibitor of ACE, inactivations are prevented to a large extent (Bünning et al. 1978, 1979). This protection against loss of enzymatic activity during a chemical modification reaction by the presence of an inhibitor indicates that the essential amino acid residue is located in the active site of the enzyme. Thus, the essential carboxyl, tyrosyl, arginyl, and lysyl residues of ACE are most likely components of the active site of the enzyme.

The chemical modification reactions were also carried out in the presence of chloride (Table 1). The anion has almost no effect on the rate of the carboxyl, tyrosyl, and arginyl modifications. However, the rates for the FDNB and pyridoxal phosphate modifications are markedly reduced by chloride. This suggests that an essential lysyl residue is the site of chloride binding.

5 Discussion

The results of the chemical modifications, together with the demonstrated requirement of zinc for catalytic activity, have led us to compare ACE with the more extensively studied zinc proteases, carboxypeptidase A and thermolysin (Table 2). All three enzymes

Table 2. Comparison of zinc proteases [a]

	ACE	CPD	TLN
Zn, g-at/mol	1	1	1
Zinc ligands	Unknown	Glu, His, His	Glu, His, His
Catalytic residues			
Nucleophile	Glu or Asp	Glu-270	Glu-143
Proton donor	Tyr	Tyr-248	His-231
Carboxyl binding residue	Arg	Arg-145	–
Chloride binding residue	Lys	–	–

[a] ACE, angiotensin converting enzyme; CPD, bovine carboxypeptidase A; TLN, thermolysin

contain one gram atom of zinc per mole which is essential for catalytic activity. Whereas the zinc binding ligands of ACE are still unknown, those for carboxypeptidase A and thermolysin are identical, that is one glutamyl and two histidyl residues. However, the metal-enzyme stability constant for ACE, 2×10^8 M, is much lower than those observed for carboxypeptidase A, 3×10^{10} M (Coleman and Vallee 1960), and thermolysin, 2×10^{11} M (Voordouw et al. 1976), indicating that the zinc ligands in ACE may be different from those in carboxypeptidase A and thermolysin. The catalytic residues of these enzymes include a nucleophile which is the carboxyl group of a glutamate or aspartate in ACE, glutamate-270 in carboxypeptidase A, and glutamate-143 in thermolysin. A tyrosyl residue in ACE, analogous to tyrosine-248 in carboxypeptidase A and histidine-231 in thermolysin, probably functions as a proton donor. The binding site for the carboxyl group of the substrate is probably an arginine in ACE, analogous to arginine-145 in carboxypeptidase A. Thermolysin which is an endoprotease does not require such a binding site. Thus, all the catalytic and binding components of carboxypeptidase A and thermolysin are also present in ACE. In addition, ACE contains an essential lysyl residue which very likely is the chloride binding site.

The components of the active sites and the catalytic mechanisms of carboxypeptidase A and thermolysin have been investigated in depth by studies with chelating agents and chemical modification reagents as well as by X-ray crystallography. The similarities of these two zinc proteases and the active site of ACE enable us to offer an initial working hypothesis for the catalytic mechanism of ACE (Fig. 6). The substrate interacts through its carbonyl group with the active-site zinc atom to facilitate a general base or nucleophilic attack of the carboxyl group that leads to hydrolysis. Tyrosine donates a proton to the scissile peptide NH group, while specificity is determined by interaction of the terminal carboxyl group with an arginine and the hydrophobic R_1 group with a hydrophobic pocket. The one difference that has been found in comparison to the two

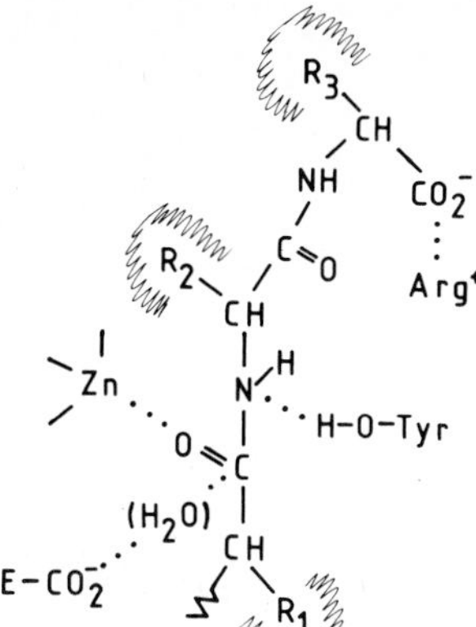

Fig. 6. Hypothetical scheme of the active site of ACE

other zinc proteases, the participation of an essential lysyl residue in the function of ACE, is related to the requirement of monovalent anions, notably chloride, for substrate binding. Nevertheless, it seems that the three zinc proteases have similar active sites. The close relationship between the active sites of carboxypeptidase A and thermolysin has recently been verified by X-ray crystallography and they likely represent the products of convergent evolution (Kester and Matthews 1977). It seems that a similar mechanistic process has also evolved for ACE. In fact, it may be that all of the zinc exo- and endoproteases share a common catalytic mechanism, much the same as the serine proteases share the well-known carboxylate-histidyl-serine triad.

Acknowledgements. The authors thank Dr. Bert L. Vallee for his continued advice and support and Dr. Barton Holmquist for many helpful discussions. This work was aided by NIH Grant HL 22387 from the Department of Health, Education and Welfare. Peter Bünning was the recipient of a fellowship from the Deutsche Forschungsgemeinschaft.

References

Auld DS, Vallee BL (1970) Biochemistry 9:4352-4359
Bünning P, Riordan JF (1980) Isr J Chem (in press)
Bünning P, Holmquist B, Riordan JF (1978) Biochem Biophys Res Commun 83:1442-1449
Bünning P, Holmquist B, Riordan JF (1979) In: Holzer H, Tschesche H (eds) Biological functions of proteinases. Springer, Berlin Heidelberg New York, pp 269-275
Bünning P, Holmquist B, Riordan JF (1980) Biochemistry (submitted)
Caldwell PRB, Seegal BC, Hsu KC, Das M, Soffer RL (1976) Science 191:1050-1051
Coleman JE, Vallee BL (1960) J Biol Chem 236:2244-2249
Cushman DW, Cheung HS, Sabo EF, Ondetti MA (1977) Biochemistry 16:5484-5491
Das M, Soffer RL (1975) J Biol Chem 250:6762-6768
Erdös EG, Johnson AR, Boyden NT (1978) Biochem Pharmacol 27:843-848
Feder J, Schuck JM (1970) Biochemistry 9:2784-2791
Gavras H, Brunner HR, Turini GA, Kershaw GR, Tifft CP, Cuttelod S, Gavras I, Vukovich RA, McKinstry DN (1978) N Engl J Med 298:991-995
Holmquist B, Vallee BL (1976) Biochemistry 15:101-107
Holmquist B, Bünning P, Riordan JF (1979) Anal Biochem 95:540-548

Kester WR, Matthews BW (1977) J Biol Chem 252:7704-7710
Nau H, Riordan JF (1975) Biochemistry 14:5285-5294
Peach MJ (1977) Physiol Rev 57:313-370
Reid IA, Morris BJ, Ganong WF (1978) Annu Rev Physiol 40:377-410
Riordan JF, Hayashida H (1970) Biochem Biophys Res Commun 41:122-127
Riordan JF, Vallee BL (1976) In: Prasad A (ed) Trace elements in human health and disease, vol I. Academic Press, London New York, pp 227-256
Riordan JF, McElvany KD, Borders Jr CL (1977) Science 195:884-886
Ryan JW, Ryan US, Schultz DR, Whitaker C, Chung A, Dorer FE (1975) Biochem J 146:497-499
Segel IH (1975) Enzyme kinetics. Wiley-Interscience, New York, pp 321-329
Skeggs LT, Marsh WH, Kahn JR, Shumway NP (1954) J Exp Med 99:275-282
Skeggs LT, Kahn JR, Shumway NP (1956a) J Exp Med 103:295-300
Skeggs LT, Kahn JR, Shumway NP (1956b) J Exp Med 103:301-307
Soffer RL (1976) Annu Rev Biochem 45:73-94
Thiers RE (1957) In: Glick D (ed) Methods of biochemical analysis, vol V. Interscience, New York, pp 273-335
Voordouw G, Milo C, Roche RS (1976) Anal Biochem 70:313-326
Yang HYT, Erdös EG, Levin Y (1970) Biochim Biophys Acta 214:374-376

Discussion[3]

Shaltiel suggested the possibility that chloride is needed to chase out "something".

Bünning answered that upon extensive dialysis of the enzyme, the chloride activation profile is identical to that observed before dialysis, which demonstrates that chloride does not remove a small molecular weight factor which might inhibit the enzyme.

Fischer commented that amylases are also activated by chloride, bromide, or nitrate.

Bünning mentioned that anion activation of the angiotensin-converting enzyme is similar to the activation of amylase and that in amylase the site of anion binding is also a lysyl residue.

3 Rapporteur: Peter C. Heinrich

Novel Aspects of Regulation

Regulation of Glutamine Synthetase Degradation

C.N. Oliver[1,2], R.L. Levine[2], and E.R. Stadtman[2]

1 Introduction

There is considerable evidence that the synthesis and degradation of proteins is a dynamic process which governs the steady-state level of enzymes in the cell. Although a great deal is known about the factors that regulate the rates of enzyme synthesis, relatively little is known about the mechanism involved in the regulation of enzyme degradation. It is known that various enzymes are degraded at different rates which are determined in part by undefined nutritional factors and that degradation is dependent upon a supply of metabolic energy (ATP) (Goldberg and St. John 1976; Holzer and Heinrich 1980). To gain more insight in to this problem, studies in our laboratory (Fulks 1977; Maurizi 1980; Oliver and Stadtman 1980) have been directed toward the development of cell-free enzyme preparations that catalyze the selective proteolysis of individual enzymes.

2 Effect of Carbon and Nitrogen Starvation on Enzyme Levels

To identify some of the enzymes whose levels are regulated by nutritional factors, changes in activities of 22 enzymes in *Escherichia coli* were monitored during the transition from exponential growth conditions to a steady-state growth condition provoked by either carbon or nitrogen limitation (Maurizi 1980; Pehlke and Stadtman unpublished data). It was found that the activities of about half of the enzymes examined decreased during the stationary growth phase, at rates ranging from 10% to 35% per h. That nutritional factors govern the loss of enzyme activities was indicated by the fact that variations in enzyme levels provoked by nitrogen starvation were not the same as those associated with carbon starvation. Nevertheless, efforts to observe losses of enzyme activities in cell free extracts of either nitrogen or carbon starved *E. coli* cells were unsuccessful.

3 Inactivation of Glutamine Synthetase in Cell-free Extracts

In similar studies with a strain of *Klebsiella aerogenes*, MK53, it was found that nitrogen starvation is associated with general protein degradation in vivo, and that among specific

1 Collaborative Program, Johns Hopkins University and Foundation for Advanced Education in the Sciences

2 Laboratory of Biochemistry, National Heart, Lung, and Blood Institute, National Institutes of Health, Bethesda, MA 20205, USA

enzymes examined, glutamine synthetase and aspartate kinase were rapidly degraded. This in vivo degradation was inhibited by uncouplers of oxidative phosphorylation (dinitrophenol), and was greatly accelerated by addition of chloramphenicol (Fulks 1977). Cell-free extracts of *K. aerogenes* MK53 or MK900 also catalyzed rapid inactivation of endogenous glutamine synthetase as well as a highly purified preparation of glutamine synthetase from *E. coli* which was added exogenously. However, in contrast to results with intact cells, the inactivation reaction in these extracts was not associated with proteolytic degradation, as judged by failure to detect a disappearance of cross-reactive material to antibodies prepared against *E. coli* glutamine synthetase. On the assumption that the inactivation reaction might represent an initial step in degradation of the enzyme in vivo, the properties of the reaction were investigated.

As shown in Table 1 (column 1), it was found that the inactivation reaction is dependent upon the presence of O_2, NADH or NADPH, Fe(III), and is inhibited by catalase.

Table 1. Glutamine synthetase (GS) inactivation by catalase-free enzyme preparations

	Source of inactivation system		
Conditions	*K. aerogenes* GS lost/ 10 min	*P. putida* GS lost/ 10 min	*E. coli* GS lost/ 15 min
	%	%	%
Complete system	53	28	61
– ATP	–	19	42
– $FeCl_3$	3	11	–
– NADPH	3	1	1
– O_2	0	0	–
+ Catalase	2	1	3
– Enzyme preparation	–	3	1

The *Klebsiella* extract from strain MK9000 was obtained by French pressure cell disruption followed by two centrifugation steps. The resulting supernatant fraction contained 15-18 mg/ml protein as determined by the Coomassie Blue method (Bradford 1976), and contained less than 1 unit/ml of catalase (Beers et al. 1952). Similarly prepared *E. coli* extracts possess 350 units/ml of catalase. To measure inactivation of endogenous glutamine synthetase in the *Klebsiella* extract, reaction mixtures (0.250 ml) contained 20 mM 2-methyl imidazole buffer (pH 8.0), 10 mM $MgCl_2$, 0.1 mM $FeCl_3$, 1.0 mM NADPH, up to 0.2 ml extract, and where indicated, 0.57 μM catalase. Reactions were started by the addition of nucleotide, incubated at 37°C in a shaking water bath, and at various times aliquots were removed, diluted into cold 10 mM imidazole buffer (pH 7.0), 1 mM $MnCl_2$, 100 mM KCl and assayed for glutamine synthetase activity by the pH 7.57 γ-glutamyltransferase assay (Stadtman et al. 1979)

E. coli and *P. putida* extracts were prepared by French pressure cell disruption or sonication. The *E. coli* strain used in these studies was a heme-deficient mutant obtained by neomycin selection (Sasarman et al. 1968) which possessed tenfold less catalase activity than wild type. The *P. putida* cells used in this study were not grown in the presence of camphor. Partial purification of the extracts to remove residual catalase activity involved streptomycin sulfate treatment, mercaptoethanol treatment, ammonium sulfate fractionation, DEAE chromatography, followed by concentration to 1-6 mg/ml protein by Coomassie Blue assay (Bradford 1976). Details will be published elsewhere. These *E. coli* and *P. putida* preparations were assayed for the capacity to inactivate exogenous purified *E. coli* glutamine synthetase (GS). Reaction mixtures contained 200 μg glutamine synthetase, 1 mM ATP, 1 mM NADPH, 50 mM Tris buffer (pH 7.4), 50 μM $FeCl_3$, 10 mM $MgCl_2$, 80 μg *E. coli* extract, or 53 μg of *P. putida* extract, and where indicated, 0.10 μM bovine catalase. Reaction conditions and sampling were the same as described above, except the sampling buffer was 50 mM Tris (pH 7.4) 10 mM $MnCl_2$

In view of the marked inhibition by catalase, it is fortuitous that we were able to observe inactivation of glutamine synthetase in these extracts. For unknown reasons the particular strains of *K. aerogenes* (i.e., M53 or MK9000) used in these studies produce little or no catalase when grown under our conditions. Nevertheless, in view of these results it seemed possible that the *E. coli* extracts failed to catalyze inactivation of glutamine synthetase because they contain high levels of catalase. This possibility was supported by the finding that glutamine synthetase inactivation in extracts of *K. aerogenes* could be prevented by the addition of *E. coli* extract. Based on its catalase content, the amount of *E. coli* extract required to prevent the inactivation reaction was only slightly less than the amount of bovine catalase needed to activate a comparable inhibition. In further studies (Oliver and Stadtman 1980) it was shown that after partial purification to remove catalase, protein fractions of either *E. coli* or *Pseudomonas putida* will catalyze the O_2 and NADPH-dependent inactivation of glutamine synthetase. In addition, it was found that these inactivation reactions were stimulated by ATP (Table 1).

4 Inactivation of Glutamine Synthetase by Mixed Function Oxidase Systems

Since NADPH, O_2, and Fe(III) are all involved in a number of mixed function oxidase catalyzed oxidation reactions, we examined the ability of two well-defined mixed function oxidation systems to catalyze inactivation of pure *E. coli* glutamine synthetase preparations. One of the systems was comprised of cytochrome C reductase and the phenobarbital-induced P-450 (LM2) from rabbit liver microsomes; the other was a three-component system consisting of putidaredoxin reductase, putidaredoxin (a nonheme electron carrier protein), and P-450_c, all derived from camphor grown cells of *P. putida.*

As shown by the data in Tables 2, 3, and 4, both mixed function oxidase systems catalyze inactivation of glutamine synthetase in the presence of NADPH (or NADH) and O_2. Inactivation by the microsomal system is dependent also upon the presence of both the cytochrome C reductase and P-450 (Table 2). The data in Table 3 show that

Table 2. Inactivation of glutamine synthetase by rabbit liver microsomal P-450 system [a]

Conditions	% Inactivation in 30 min
Complete system	43
– Dilauroylphosphatidylcholine	40
– NADPH	3
– Cytochrome C reductase	1
– P-450	3
Glutamine synthetase only	0

The complete inactivation mixture (0.2 ml) contained 200 μg of glutamine synthetase, 0.3 μM cytochrome C reductase, 0.15 μM P-450 (LM2), 20 μg dilauroylphosphatidylcholine, 1 mM NADPH, 50 mM Tris buffer (pH 7.4), 50 μM $FeCl_3$, 10 mM $MgCl_2$. Reaction mixtures were incubated and assayed for glutamine synthetase activity as described for Table 1

[a] The highly purified preparations of rabbit liver microsomal cytochrome C reductase and the phenobarbital induced P-450 (LM2) were generously provided by M.J. Coon

Table 3. Inactivation of glutamine synthetase by *Pseudomonas* reductase-redoxin system [a]

Conditions	% Inactivation in 10 min
Complete system	43
– NADH	3
– Redoxin	0
– Reductase	2

Inactivation mixtures of *Pseudomonas* reductase-redoxin contained 130 μg glutamine synthetase, 0.5 μM reductase, 10 μM redoxin, 1 mM NADPH, 50 mM Hepes buffer (pH 7.4), 5 mM $MgCl_2$. Sampling and assay were described in Table 1
[a] The highly purified preparation of putidaredoxin reductase and putidaredoxin were generously provided by I.C. Gunsalus

Table 4. Effect of redoxin concentration on requirement for P-450 in inactivation reaction [a]

Redoxin concentration	% Inactivation	
	$-P450_c$	$+P450_c$
μM		
5	50	87
10	72	74

The complete inactivation mixture was the same as described for Table 3, except that 5 μM redoxin and 0.5 μM P-450 were used. Sample and assay were described in Table 1
[a] The highly purified preparation of putidaredoxin reductase, putidaredoxin, and camphor induced $P\text{-}450_c$ were generously provided by I.C. Gunsalus

redoxin reductase together with high concentrations of redoxin (20 μM) will catalyze the NADH-dependent inactivation of glutamine synthetase. Under these conditions, the further addition of $P\text{-}450_c$ is without effect (data not shown). However, as shown in Table 4, when lower concentrations of redoxin (5 μM) are used, the inactivation reaction is greatly stimulated by $P\text{-}450_c$, whereas little or no stimulation of inactivation by $P\text{-}450_c$ occurs when 10 μM redoxin is used.

4.1 Properties of the Various Inactivation Systems

Some other properties of the inactivation system are summarized in Table 5. In addition to those described above, a nonenzymic system consisting of ascorbic acid, Fe(III), and O_2 is also included because it was shown by Udenfriend et al. (1954) that this system will catalyze many reactions that are catalyzed by mixed function oxidase system. It can be seen from the data in Table 5 that the five inactivation systems have many properties in common. In all cases, inactivation of glutamine synthetase is dependent upon O_2, is stimulated by Fe(III), and is inhibited by catalase, horseradish peroxidase, and by Mn(II). Stimulation of the inactivation reaction by azide in the *K. aerogenes* system is likely attributable to inhibition of small amounts (barely detectable) of catalase or peroxidases

Table 5. Properties of the glutamine synthetase inactivation systems

	Inactivation systems				
	Klebsiella [a]	*Ascorbate* [a]	*Microsomal* [a]	*Pseudomonas*	
Additions				$-P\text{-}450_c$	$+P\text{-}450_c$
None [b]	100	100	100	100	100
Inert gas	0	2	3	20	8
Catalase	19	0	0	4	3
Peroxidase	11	0	14	–	–
NaN_3	168	–	100	106	95
$MnCl_2$	2	0	0	0	0
$FeCl_3$	126	240	135	122	114
EDTA	0	0	0	2	0
EDTA + $FeCl_3$	0	15	0	160	22
o-Ph [c]	9	0	7	0	5
o-Ph + $FeCl_3$	75	154	178	88	98

[a] These data were taken from Levine et al. (1981)
[b] For each system, the extent of inactivation without additions is assumed to be 100
[c] *o*-Ph, *o*-phenanthroline
Preparation of the *Klebsiella* extract was the same as that described in Table 1. The *Klebsiella* inactivation system was the same as in Table 1, except that 1 mM NADPH was replaced by 6.5 mM NADH and $FeCl_3$ was not added. Other additions, where indicated, were: 0.57 μM catalase, 0.078 μM peroxidase (horseradish), 2 μM superoxide dismutase, 1 mM $MnCl_2$, 100 μM $FeCl_3$, 1 mM EDTA, 1 mM EDTA plus 1.5 mM $FeCl_3$, 180 μM *o*-phenanthroline, 180 μM *o*-phenanthroline plus 270 μM $FeCl_3$, 500 μM NaN_3. The ascorbate inactivation mixtures contained 0.48 μM glutamine synthetase, 15 mM ascorbate, 100 mM 2-methylimidazole (pH 7.2), 10 mM $MgCl_2$. Other additions, where indicated, were: 0.61 μM catalase, 0.56 μM peroxidase, 4 mM $MnCl_2$, 100 μM *o*-phenanthroline, 100 μM *o*-phenanthroline plus 500 μM $FeCl_3$. The concentrations of EDTA, $FeCl_3$, and NaN_3 were the same as above. The microsomal inactivation mixture was the same as described for Table 2, except that $FeCl_3$ was not added, and the concentrations of other substances, where indicated, were the same as described for the ascorbate system. Inactivation mixtures of the *Pseudomonas* reductase-redoxin were same as described for Table 3. The *Pseudomonas* reductase-redoxin-P-450 inactivation mixture was the same as described for Table 4. Where indicated, the concentrations of other substances were the same as described for the microsomal system, except that 200 μM $FeCl_3$, 700 μM EDTA plus 900 μM $FeCl_3$ were used. Sampling and assay were described in Table 1

The highly purified preparation of putidaredoxin reductase, putidaredoxin and camphor induced $P\text{-}450_c$ were generously provided by I.C. Gunsalus

The highly purified preparations of rabbit liver microsomal cytochrome C reductase and the phenobarbital induced P-450 (LM2) were generously provided by M.J. Coon

present in these extracts. In every case, the inactivation reaction is inhibited by the metal ion chelators *o*-phenanthroline and EDTA. In every instance, the inhibition by *o*-phenanthroline is reversed by a 0.5-fold excess of Fe(III), whereas with one exception (the P-450 independent redoxin-dependent activity) an excess of Fe(III) fails to reverse the inhibition by EDTA. It appears from the data in Table 6 that different mechanisms are involved in the putidaredoxin-catalyzed inactivation reaction in the presence and absence of $P\text{-}450_c$, since superoxide dismutase, dimethylsulfoxide, and histidine all inhibit the reaction in the absence but little or not at all in the presence of $P\text{-}450_c$. It should be noted that an exceptionally high concentration of superoxide dismutase (140 μM) was

Table 6. Effect of activated oxygen scavengers on the *Pseudomonas* inactivation systems [a]

Addition	Concentration	*Pseudomonas* systems Reductase + redoxin	Reductase + redoxin + P-450$_c$
	mM	% of control	% of control
None	–	100	100
Superoxide dismutase	0.14	36	90
Dimethyl sulfoxide	94	50	95
Mannitol	100	83	85
Histidine	10	50	85

Conditions for *Pseudomonas* reductase-redoxin and reductase-redoxin-P-450$_c$ inactivation reactions were the same as described for Tables 3 and 4. Reaction mixtures were incubated and assayed for glutamine synthetase as described for Table 1

[a] The highly purified preparation of putidaredoxin reductase, putidaredoxin, and camphor induced P-450$_c$ were generously provided by I.C. Gunsalus

used in these studies, therefore it is uncertain whether the effect of the enzyme preparation is due solely to destruction of superoxide anion.

4.2 Regulation of the Inactivation Reaction

If our working hypothesis is correct that the inactivation of glutamine synthetase is an initial step in the degradation of glutamine synthetase, then this might be the step at which degradation is regulated. This possibility is supported by the data in Fig. 1, showing that the inactivation of glutamine synthetase is affected by both the state of adenylylation of the enzyme and by the presence of two of the enzyme's substrates, ATP and glutamate. The data in Fig. 1 show that in the absence of substrates the susceptibility of the enzyme to inactivation by the microsomal mixed function oxidase system decreases

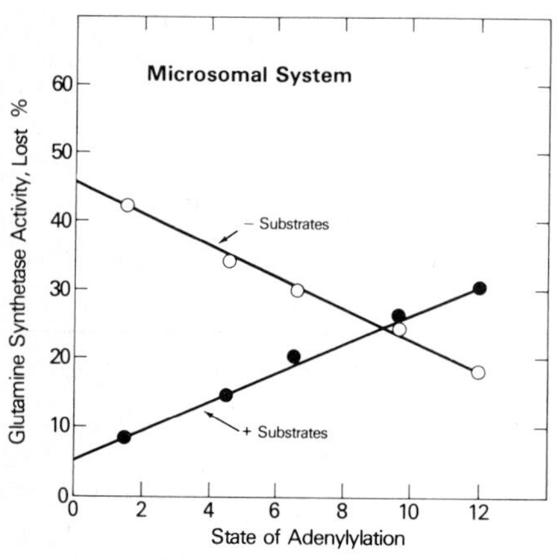

Fig. 1. Effect of substrate and state of adenylylation on inactivation of glutamine synthetase by rabbit liver microsomal P-450 system. Reaction mixtures contained 200 μg of glutamine synthetase of various states of adenylylation and, where indicated, the cosubstrates 1 mM ATP and 30 mM glutamate. The P-450 components were the same as described in Table 2. Reaction mixtures were incubated and assayed for glutamine synthetase activity as described in Table 1

somewhat as the state of adenylylation increases. However, the reverse is true when ATP and glutamate are both present. Thus, substrates protect unadenylylated enzyme from inactivation, but stimulate inactivation of the adenylylated enzyme. Similar results (not shown) were obtained when either the putidaredoxin or the ascorbate-catalyzed inactivation systems were used.

5 Effect of Inactivation on Proteolytic Degradation

That inactivation of glutamine synthetase may be a controlling step in its degradation is supported by the data in Table 7, which shows that after its inactivation by either the ascorbate system (experiment 1) or the putidaredoxin system (experiment 2), glutamine synthetase is more readily degraded by partially purified protease preparations from *E. coli*. The protease preparation used in experiment 1 apparently contained appreciable inactivation activity since its ability to catalyze degradation of native enzyme was greatly stimulated by NADPH and ATP, and this stimulated activity was inhibited by catalase.

Table 7. Proteolysis of native and inactivated glutamine synthetase (GS) by partially purified *E. coli* extracts

		Proteolysis	
Experiments	Additions	Native GS	Inactivated GS
	None	0	0
	Extract	9.0	61.3
	Extract + ATP	60.0	73.7
1 [a]	Extract + NADPH	33.0	63.9
	Extract + ATP + NADPH	72.2	95.8
	Extract (treated 60°C, 5') + ATP + NADPH	4.4	25.0
	Extract (treated 100°C, 5') + ATP + NADPH	2.8	1.7
	Extract + ATP + NADPH + catalase	13.9	66.2
	None	0	0
2	Extract	0	24.2
	Extract + ATP	8.9	36.7

Reaction mixtures in experiment 1 contained 200 μg of native or ascorbate-inactivated glutamine synthetase (GS), 23 μg of catalase-free *E. coli* extract similar to preparation described in Table 1, 50 mM Tris buffer, (pH 7.4), 50 μM $FeCl_3$, 10 mM $MgCl_2$, and where indicated, 1 mM ATP, 1 mM NADPH, and 0.1 μM bovine catalase. Reaction mixtures in experiment 2 contained 200 μg of native or *Pseudomonas* reductase-redoxin inactivated-reisolated GS, 83 μg of a different preparation of catalase-free *E. coli* extract, 50 mM Tris (pH 7.4), 10 μM $FeCl_3$, 5 mM $MgCl_2$, and where indicated, 1 mM ATP. Experiment 1 was incubated for 30 min and experiment 2 for 60 min at 37°C. The incubation mixtures were sampled and treated with SDS buffer, and then analyzed by high performance liquid chromatography, using a calibrated Toyo Soda SW3000-size exclusion column. The column effluent was monitored at 215 nm in experiment 1 and at 290 nm in experiment 2. Proteolysis was quantitated by determining the loss of glutamine synthetase subunit with time by peak area integration. Appropriate corrections were made for the contribution of the extract protein at each time point under identical incubation conditions

[a] These data were taken from Levine et al. (1981)

For the data in Table 7, degradation of glutamine synthetase was followed by measuring the loss of subunit protein by means of high performance liquid chromatography of reaction mixtures in the presence of sodium dodecylsulfate (SDS). In other experiments (Oliver et al. 1981), proteolysis was measured by observing the disappearance of glutamine synthetase subunits by means of SDS slab gel electrophoresis. These experiments showed that glutamine synthetase which had been inactivated by the microsomal P-450 system was readily degraded by *E. coli* protease, whereas native enzyme was not.

6 Discussion

The data presented here are compatible with the hypothesis that the degradation of glutamine synthetase in *E. coli* and *K. aerogenes* involves first its inactivation by a mixed function oxidase catalyzed reaction, the nature of which remains to be determined, followed by proteolytic attack of the inactivated enzyme. Whether or not this mechanism is fundamental to the regulation of glutamine synthetase and possibly other enzyme levels in the cell, remains to be ascertained. If it represents a normal mechanism for the regulation of glutamine synthetase turnover, then it is obvious that inhibition of the inactivation reaction by catalase must be explained. Either catalase must be separated from components of the inactivation reaction, by cellular compartmentation, or the capacity of catalase to inhibit the inactivation reaction must also be regulated. In this regard it may be significant that the levels of catalase in *E. coli* are modulated by growth conditions (Hassan and Fridovich 1978), and in plants proteins capable of inhibiting catalase activity have been isolated (Tsaftaris et al. 1980).

One should consider also the possibility that one of the main physiological functions of catalase in *E. coli* is to prevent the inactivation of glutamine synthetase and perhaps similar inactivation of other enzymes as well. In this case, the high sensitivity of inactivated glutamine synthetase to proteolytic degradation might be regarded as a part of the normal sanitation system which is designed to prevent the accumulation of nonfunctional protein, viz the accumulation of inactive glutamine synthetase, that had inadvertently escaped the protective influence of catalase. It should be pointed out, however, that if the mixed function oxidase-catalyzed inactivation of glutamine synthetase is merely a rare event that occurs only because of imperfect protection by catalase, there would appear to be no obvious role for the effects of ATP, glutamate, and the state of adenylylation of glutamine synthetase in the regulation of the inactivation reaction. In any event, the results of this study suggest the possibility that the mixed function oxidase-catalyzed inactivation of glutamine synthetase may be an obligatory step in the metabolite-directed degradation of the enzyme, which together with the modulation of enzyme synthesis is concerned with regulation of its intracellular concentration. It is tempting to speculate further that the glutamine synthetase system may be the prototype of a more general mechanism utilized in the regulation of normal enzyme degradation. If so, mixed function oxidation of the enzyme (which might or might not lead to inactivation) would be a prerequisite for proteolytic degradation, and modulation of the oxidation reaction by metabolites or covalent modification could be an important mechanism of control. Presumably, oxidation of functional groups on the enzyme by the mixed function oxidase would lead to conformational changes that expose recognition sites on the enzyme (perhaps hydrophobic regions) that are susceptible to attack by proteases. According to this model, a relatively few nonspecific mixed function oxidases and proteases working together could catalyze the degradation of many different enzymes. The observed variations in the relative rates of degradation of different enzymes might simply reflect highly specific interaction of each enzyme with its own allosteric effectors and substrates which modulate its susceptibility to oxidation and proteolysis.

References

Beers RF, Jr, Sizer IW (1952) A spectrophotometric method for measuring the breakdown of hydrogen peroxide by catalase. J Biol Chem 195:133-140

Bradford MM (1976) A rapid and sensitive method for the quantitation of microgram quantitites of protein utilizing the principle of protein-dye binding. Anal Biochem 72:248-254

Fulks RM (1977) Regulation of glutamine synthetase degradation in *Klebsiella aerogenes*. Fed Proc (Abstr) 36:919

Goldberg AL, St John AC (1976) Intracellular protein degradation in mammalian and bacterial cells. Annu Rev Biochem 45:747-803

Hassan HM, Fridovich I (1978) Regulation of the synthesis of catalase and peroxidase in *Escherichia coli*. J Biol Chem 253:6445-6450

Holzer H, Heinrich PC (1980) Control of proteolysis. Annu Rev Biochem 49:63-91

Levine RL, Oliver CN, Fulks RM, Stadtman ER (to be published 1981) Turnover of bacterial glutamine synthetase: Proteolysis is preceded by oxidative inactivation. Proc Natl Acad Sci USA

Maurizi MR (1980) Degradation of specific enzymes in *Escherichia coli*. Fed Proc (Abstr) 39:403

Oliver CN, Stadtman ER (1980) Inactivation and proteolysis of glutamine synthetase by *Escherichia coli* extracts. Fed Proc (Abstr) 39:402

Oliver CN, Levine RL, Stadtman ER (to be published 1981) Regulation of glutamine synthetase degradation. In: Ornston LN (ed) Experiences in biochemical perception. Academic Press, London New York

Sasarman A, Surcleanu M, Sźegli G, Horodniceanu T, Greceanu V, Dumitrescu A (1968) Hemin-deficient mutants of *Escherichia coli* K-12. J Bacteriol 96:570-572

Stadtman ER, Davis JN, Smyrniotis PZ, Wittenberger ME (1979) Enzymatic procedures for determining the average state of adenylylation of *Escherichia coli* glutamine synthetase. Anal Biochem 95:275-285

Tsaftaris AS, Sorenson JC, Scandalios JG (1980) Glycosylation of catalase inhibitor necessary for activity. Biochem Biophys Res Commun 92:889-895

Udenfriend S, Clark CT, Axelrod J, Brodie BB (1954) Ascorbic acid in aromatic hydroxylation. I. A model system for aromatic hydroxylation. J Biol Chem 208:731-739

Discussion[3]

P-450 hydroxylates some drugs in a manner not inhibited by catalase but with other substrates the system is inhibited by catalase, a fact which indicates that P-450 can generate hydrogen peroxide. However, other hydrogen peroxide-generating systems do not work in the inactivation of glutamine synthetase. The action of P-450 could be related with the reduction of iron: reduced iron autooxydizes to generate hydrogen peroxide that could react with additional reduced iron to produce hydroxyl radicals or activated oxygen that react with a component of the glutamine synthetase to produce inactivation.

3 Rapporteur: Carlos Gancedo

Since degradation of glutamine synthetase depends on the degree of adenylylation of the enzyme, certain subunits in the oligomeric enzyme might be degraded and others not; there is, however, no dissociation-reassociation of the subunits of glutamine synthetase and therefore dismutation of partially modified molecules to totally unmodified enzyme and totally degraded material is not likely to occur.

Mono ADP-ribosylation and Poly ADP-ribosylation of Proteins

H. Hilz, P. Adamietz, R. Bredehorst, and K. Wielckens[1]

1 Introduction

Postsynthetic modification of proteins by transfer of ADP-ribosyl groups from NAD has been shown to occur in numerous systems. Besides ADPR transferase reactions associated with the action of bacterial toxins and viruses, ADP ribosylation reactions were also observed in eukaryotic cells (cf. Hilz and Stone 1976; Hayaishi and Ueda 1977; Purnell et al. 1980).

A nuclear system from chick liver able to form poly(ADPR) was first described by Mandel and co-workers (Chambon et al. 1966), and independently also by Hayaishi's (Reeder et al. 1967) and Sugimura's (Fujimura et al. 1967) groups. In this system transfer of ADPR residues from NAD can occur to two acceptors (Fig. 1):

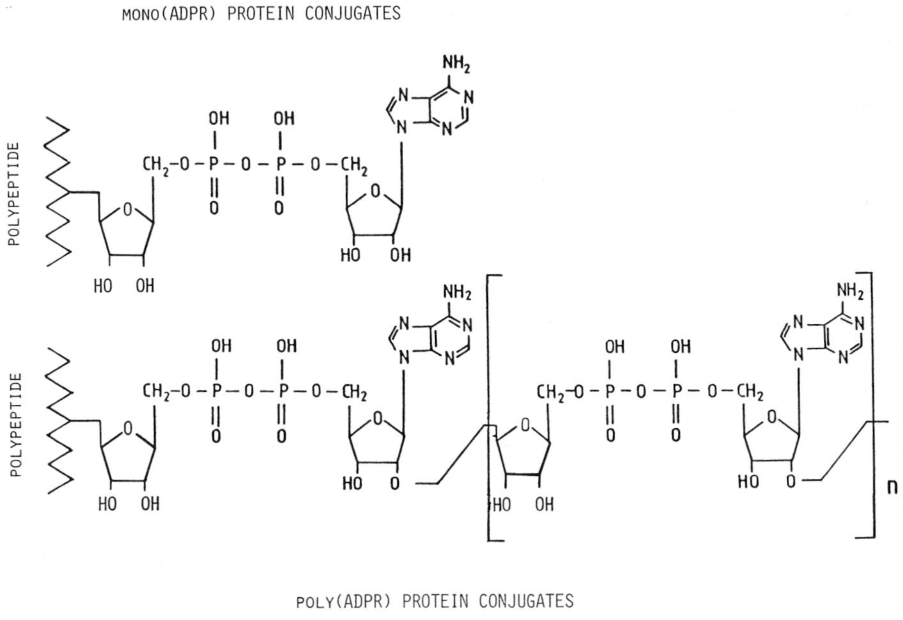

MONO(ADPR) PROTEIN CONJUGATES

POLYPEPTIDE

NH_2 OH OH $CH_2-O-P-O-P-O-CH_2$ O O HO OH HO OH

POLYPEPTIDE

NH_2 OH OH $CH_2-O-P-O-P-O-CH_2$ O O HO O HO OH $[\ldots]_n$

POLY(ADPR) PROTEIN CONJUGATES

Fig. 1. Mono(ADPR) and poly(ADPR) protein conjugates

1 Institut für Physiologische Chemie, Universität Hamburg, Martinistraße 52, 2000 Hamburg 20, FRG

1. transfer to proteins forming mono(ADPR) protein conjugates
2. transfer to a preceding ADPR moiety forming a homopolymer – (poly(ADPR)) – which, too, is linked covalently to proteins (Nishizuka et al. 1969; Adamietz and Hilz 1976).

This paper summarizes experiments of our group dealing with the isolation and characterization of ADP-ribosylated proteins and the quantitation of three $(ADPR)_n$ protein conjugate subfractions from intact tissues. On this background possible functions of ADP-ribosylation reactions in eukaryotic cells will be discussed.

2 Results and Discussion

2.1 Acceptor Proteins of Nuclear ADP-ribosylation

Due to the heterogeneity in acceptor proteins, in widely varying chain lengths of the modifying poly(ADPR) groups, and because of rather low levels, isolation of the ADP-ribosylated conjugates proved difficult. Recently, we succeeded in obtaining the ADP-ribosylated proteins formed in EAT cell nuclei in pure form and with a yield of > 85% (Adamietz et al. (1979). Isolation of ADPR protein conjugates free of unmodified proteins was accomplished by chromatography of the perchloric acid insoluble fraction on boronate matrices. Due to the presence in poly(ADPR) of ribosyl residues containing an unsubstituted cis diol configuration, the ADPR conjugates at pH 8.2 were retained by the matrix and could be separated from DNA, RNA, and unrelated proteins. Release of the ADPR conjugates was then effected by lowering the pH. SDS-gel electrophoretic analysis of these conjugates revealed the following data (Fig. 2):

There are multiple bands of modified proteins. At least in the high resolving region between 20 and 40 K, a regular band pattern was observed. Autoradiography of the $[^3H]$-labeled ADPR residues showed the same pattern as protein stain, though with rather different intensities. This is in accordance with the notion of a preferential ADP-ribosylation of certain acceptor proteins. Identification of the proteins as ADP-ribosylated conjugates was performed by specific removal of ADPR residues either by treatment with phosphodiesterase I, or by alkaline treatment. After both treatments, all proteins had lost the affinity to the boronate column, indicating that the ADPR conjugate fraction did not contain contaminating glycoproteins. Furthermore, release of the $(ADPR)_n$ residues led to a shift of the protein bands to positions of lower molecular weights, most clearly seen in the high-resolving region of the gel between 20 and 40 K (Fig. 2, *lane D*, vs *lane E*).

When the isolated conjugates were subjected to ion exchange separation as applied for the isolation of histones from nonhistone proteins (Levy et al. 1972), only a relatively small proportion of the conjugates was found in the "histone fraction" (Fig. 2, *lane B*). The protein bands in this fraction moved slightly behind authentic histones, indicating that these conjugates carried only 1-3 ADPR residues. Release of the modifying groups shifted the proteins to the positions of the histones (Fig. 2, *lane C*). The introduction of more negatively charged $(ADPR)_n$ residues into the basic histones, however,

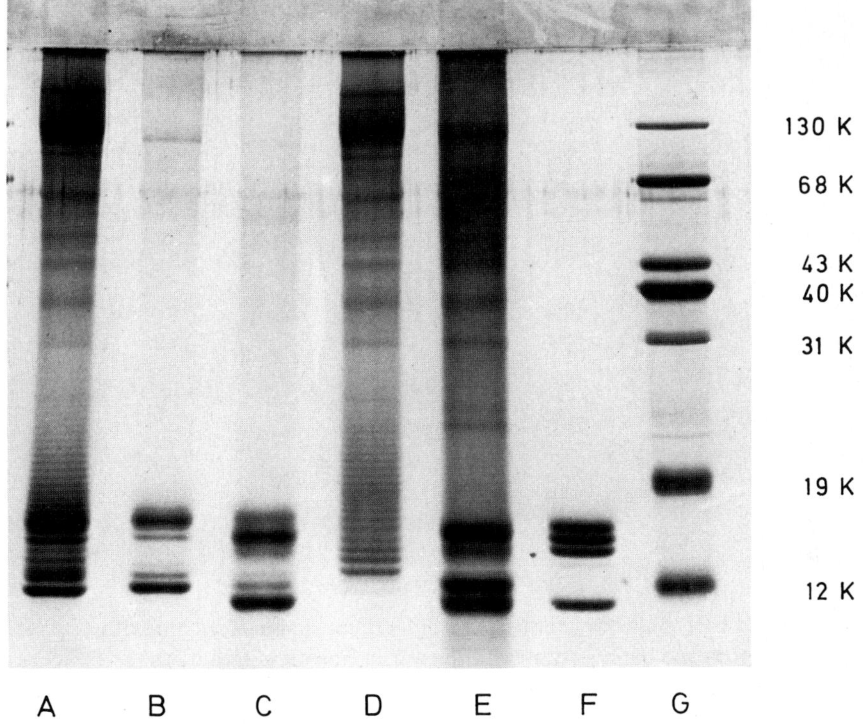

Fig. 2. SDS-gel electrophoresis of (ADPR)n protein conjugates from EAT cell nuclei (Adamietz et al. 1979). *A*, Total $(\mathrm{ADPR})_n$ protein conjugates; *B*, Histone fraction derived from A; *C*, Histone fraction after alkaline treatment; *D*, Nonhistone fraction derived from A; *E*, Nonhistone fraction after alkaline treatment; *F*, Core histones from EAT cells; *G*, Molecular weight markers: β-galactosidase (130 K), bovine serum albumin (68 K), ovalbumin (43 K), aldolase A subunit (40 K), DN'ase I (31 K), β-lactglobuline (18.7 K), cytochrome C (11.8 K)

changed the properties of these acceptors to such an extent that they behaved like nonhistone proteins. The data in Fig. 2 show that treatment with phosphodiesterase released from the nonhistone fractions a much larger amount of histones than treatment of the true histone fraction. These data show that ADP-ribosylation can change fundamentally the properties of acceptor proteins.

A quantitative analysis of the released proteins revealed that both, histones and nonhistone proteins, serve as acceptors. Under the conditions applied, and in these nuclei, about 60% of the ADP-ribosylated proteins were nonhistone acceptors.

We have also isolated the ADPR conjugates of a single acceptor protein (Braeuer et al. 1981). It is histone H1 which was not included in the above study because of the solubility of H1 (Johns and Butler 1962) and its ADPR conjugates (Adamietz et al. 1978) in 5% perchloric acid.

$(\text{ADP-ribosyl})_n$ histone H1 was purified from EAT cell nuclei by RioRex 70 and boronate chromatography of perchloric acid extracts, which yielded the conjugates in pure form (Fig. 3). Specific release of the ADPR residues resulted in a shift of the mul-

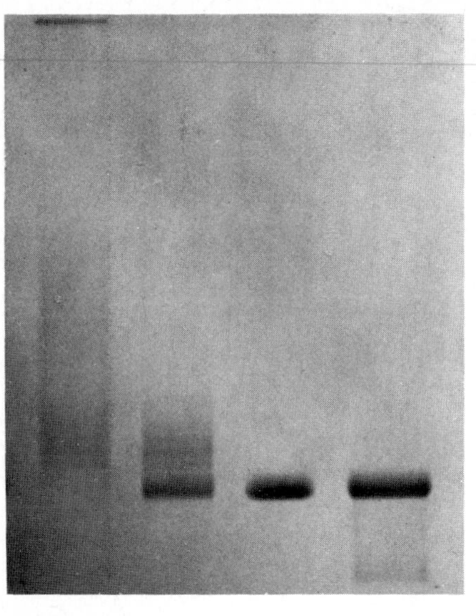

Fig. 3. ADP-ribosylated histone H1 conjugates isolated from EAT cell nuclei (Braeuer et al. 1980). *A*, native conjugates; *B*, treated with 0.1 M NaOH for 20 min at 56°C; *C*, histone H1 control; *D*, conjugates after treatment with phosphodiesterase I (10 μg/ml) for 10 h at 37°C

tiple protein bands to a single protein that was indistinguishable from authentic EAT cell histone H1. From the band pattern and the average chain length of the $(ADPR)_n$ residues = 2.5 it can be deduced that each of the protein band differs from the next one by a single ADPR residue.

In order to study the distribution of ADPR residues over the histone H1 molecules, splitting of H1 conjugates by bromosuccinimide at a single tyrosine residue (about residue 72) was performed. Two interesting details were obtained:

1. Most ADPR residues are associated with the C-terminal fragment.
2. The splitting yielded a high percentage of N-terminal fragments that were completely devoid of ADPR residues. Likewise, there were C-terminal fragments that appeared to be devoid of ADPR residues. The data suggest that as much as half of the molecular population may carry ADPR residues exclusively either at the C-terminal *or* the N-terminal fragment. The biological significance of this "uneven" modification is not known.

Recently, it has been reported (Burzio et al. 1979; Ogata et al. 1980) that linkage of ADPR to rat liver histone H1 occurs through NH_2OH-sensitive ester glycoside bonds involving two (Burzio et al. 1979) or three (Ogata et al. 1980) glutamate residues, and the terminal carboxyl group (Ogata et al. 1980).

With the histone H1 conjugates from EAT cells, about two-thirds of the bonds resisted treatment with neutral NH_2OH. At present, no solid information concerning the nature of the NH_2OH-resistant linkage is available.

2.2 Mono and Poly(ADP-ribosyl) Protein Conjugate Levels in Intact Tissue

So far, most data in the field were obtained with in vitro systems, with isolated nuclei, chromatin or permeabilized cells. However, there is accumulating evidence that isolation of nuclei leads to an artifactual, 10 to 30-fold activation of a latent form of a poly(ADPR) polymerase that may "override" the in vivo activities.

When a comparative study of histone H1 modified in intact HeLa cells and in isolated nuclei obtained from the same cultures was performed (Adamietz et al. 1978), fundamental differences in histone H1 conjugates formed in vivo versus in vitro became apparent (Table 1): The differences pertain to types of linkage, to a preponderance of poly-

Table 1. Comparison of histone H1 of HeLa cells ADP-ribosylated in vivo and in vitro (Adamietz et al. 1978)

Parameter	Measurement	
	in vivo	in vitro
	with precursor	
	[^{3}H]adenosine	[^{3}H]NAD
ADP-ribose residues:		
Total ADP-ribose/histone H1 (pmol/mg)	218	1620
ADP-ribose/histone H1 (mol/mol)	0.005	0.04
Monomeric ADP-ribose/histone H1 (pmol/mg)	174	243
Polymeric ADP-ribose/monomeric ADP-ribose	0.25	5.67
	%	%
Types of binding to histone H1:		
NH_2OH-sensitive	non (< 3)	70
pH-10.5-sensitive	70	100
pH-10.5-resistant	30	none
HCl-sensitive (30°C)	none	none
High-M_r complex	–	+
Histone H1-bound $(\text{ADP-ribose})_n$ as fraction of total protein-bound $(\text{ADP-ribose})_n$	2.5	0.6

meric ADPR chains in vitro, and to the absolute amounts of ADPR residues bound to H1. It should also be mentioned that less than 1% of the H1 population carried ADPR residues, and that less than 2.5% (in vivo) or 1% (in vitro) of the total protein-bound ADPR residues were associated with histone H1.

The high divergence of findings obtained with isolated nuclei led us to develop procedures that would allow to quantify the amounts of ADPR protein conjugates from intact tissues. These procedures are based on the specific conversion of ADPR to 5'-AMP and of poly(ADPR) to PR-AMP, followed by the quantitation of these products by highly specific radioimmuno-assays (Bredehorst et al. 1978a; Bredehorst et al. 1978b; Bredehorst et al. 1981a; Wielckens et al. 1981). They allow to differentiate the following types of conjugates (Fig. 4): Conjugates carrying mono or poly (ADPR) residues that are linked by NH_2OH sensitive bonds, and conjugates with NH_2OH resistant linkage. Recent data from our laboratory (Wielckens et al. 1981) indicate that, at least in rat liver, most of the poly(ADPR)-carrying conjugates were sensitive to NH_2OH. In the fraction of the mono(ADPR) protein conjugates, about half of the modifying groups were linked by NH_2OH-sensitive, the rest by NH_2OH-resistant bonds.

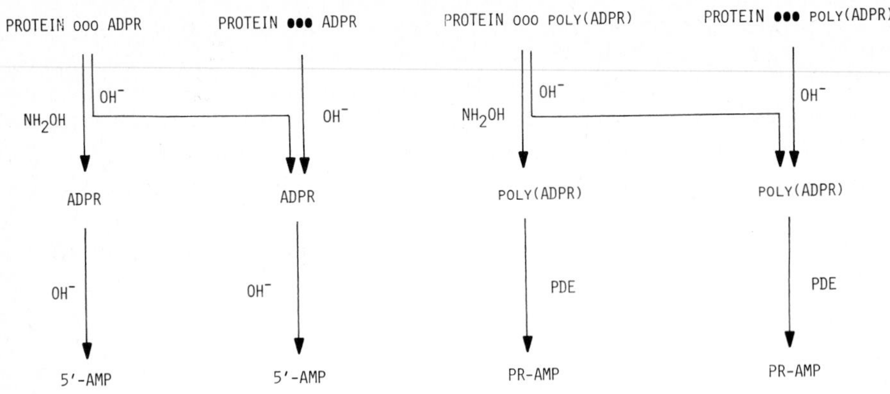

Fig. 4. Release of mono(ADPR) and poly(ADPR) residues from different types of conjugates

Several tissues were analyzed for their endogenous levels of ADPR conjugates (Wielckens et al. 1981). The data show that by far most proteins modified by ADP-ribosylation carry single ADPR groups, only a very small fraction (< 1% of the total acceptor sites) containing oligomeric or polymeric $(ADPR)_n$ residues. The different ADPR protein subfractions appear to serve different functions. This is indicated by independent changes, as seen for instance during development of liver (Table 2): when neonatal rat livers at

Table 2. Alterations of protein-bound monomeric and polymeric ADPR residues in rat liver during development (Wielckens et al. 1980b)

Age of rats	ADPR residues (pmol/mg DNA ± s.d.) Monomeric	Polymeric	Poly / Mono
1 day	2 300 ± 240	32.5 ± 9.3	0.014
17 days	3 180 ± 100	9.7 ± 1.8	0.003
6 months	12 570 ± 3 230	39.3 ± 5.3	0.093

day 1 and 17 were compared with the adult status, mono(ADPR) conjugates showed a steady increase, reaching six times the neonatal values in the adult. Poly(ADPR) conjugates changed differently. They decreased at day 17 (the time of the most rapid postnatal liver growth), and rose again in the adult state, although they hardly surpassed the neonatal levels. Independent changes were also seen in mouse kidney after withdrawal of testosterone (Gartemann et al. 1981). An analysis of the two mono(ADPR) conjugate *subfractions*, the NH_2OH-sensitive and -resistant mono(ADPR) protein conjugates, was performed during ontogenesis of rat liver (Hilz et al. 1980b; Bredehorst et al. 1981b): In the fetal stage, the NH_2OH-resistant fraction was practically absent from fetal liver. Significant amounts were seen around birth. High levels were only found when the organ had reached its final metabolic capacity. Quite different the NH_2OH-sensitive fraction. Significant amounts were present already in the fetal stage. At birth, they had increased by a factor of 6. However, at days 15-17 an inverse change was seen. Since this is the period of the most rapid liver growth, the findings indicated an inverse relationship between growth rate and the extent of ADP-ribosylation involving the NH_2OH-sensitive mono(ADPR) conjugates. By contrast, the NH_2OH-resistant mono(ADPR) conjugates could be related to the degree of differentiation, reaching their highest values when the liver had reached its full metabolic competence (Bredehorst et al. 1980).

Additional support of the concept that higher levels of NH_2OH-resistant mono (ADPR) protein conjugates are indicative of a higher degree of differentiation came from two other systems. Analysis of the slime mold *Dicytostelium discoidium* (Bredehorst et al. 1980) showed (Fig. 5) that the transition of the unicellular growing cells to unicellular

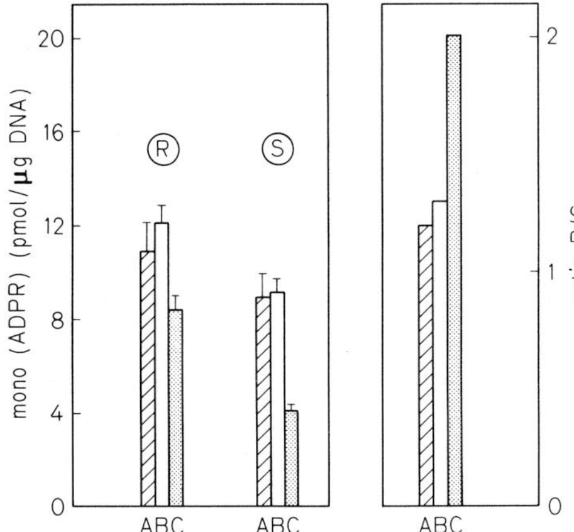

Fig. 5. Hydroxylamine-sensitive and -resistant mono(ADPR) protein conjugates in the growth phase (*A*), the aggregation competent stage (*B*) and the grex stage (*C*) of *D. discoideum*. Data represent means and standard errors as calculated in each case from six determinations. *R*, NH_2OH-resistant residues; *S*, NH_2OH-sensitive residues (Bredehorst et al. 1980a)

aggregation competent cells was not associated with significant changes of either subfraction. However, when the multicellular sorocarp state was formed, it was again the NH_2OH-resistant conjugate subfraction that whowed a relative increase compared to the NH_2OH-sensitive subfraction.

Correlation of the NH_2OH-resistant mono(ADPR) protein conjugate subfraction with the degree of differentiation was also observed in the system of human lymphocytes and lymphocytes from patients with chronic lymphocytic leukemia (CLL) (Wielckens et al. 1980b). Both cell types, normal and tumor cells, do not proliferate. The CLL cells, however, exhibit defects in differentiation, e.g., an extended life span (Dameshek 1967), a retarded or missing response to mitogens (Havemann and Rubin 1968), and a changed pattern of protein kinases (Weber et al. 1981). When analyzing the ADPR conjugates (Table 3), it became apparent that there was a marked decrease in mono(ADPR) conjugates in the leukemic lymphocytes, and this decrease concerned mainly the NH_2OH-resistant subfraction. Thus a higher degree of differentiation appears to be associated with a higher degree of mono ADP-ribosylation of proteins of the NH_2OH-resistant type.

The quantitative data obtained by detailed analysis of ADP-ribosylated protein levels under various conditions indicate that the different subfractions of the $(ADPR)_n$ protein conjugates serve independent functions. This notion receives support from recent data on the intracellular localization of the subfractions (Adamietz et al. 1981). These analyses suggest that a substantial part, if not the majority, of the mono(ADPR)

Table 3. Mono(ADPR) protein conjugate subfractions in peripheral lymphocytes from normal donors and from patients with chronic lymphocytic leukemia (CLL). (Wielckens et al. 1980)

Parameter	Lymphocytes		
	Normal	Leukemic	Leukemic/Normal
Mono(ADPR) residues: (pmol/mg DNA)			
total	537.3 ± 146.8	145.9 ± 59.1	0.26
NH_2OH-sensitive	101.7 ± 53.7	41.7 ± 20.2	0.41
NH_2OH-resistant	435.6 ± 138.6	104.2 ± 53.5	0.23

conjugates is *not* localized in the nucleus but associated with other subcellular compartments.

Poly ADP-ribosylation seems to be associated with processes involving DNA fragmentation. In permeabilized cells, but not in isolated nuclei, poly(ADPR) transferase is present in a latent form that can be activated by the addition of nucleases, or pretreatment of cells with alkylating agents or X-rays (Berger et al. 1978; Halldorsson et al. 1978; Miller 1975; Bredehorst et al. 1979). Treatment with repair-inducing agents is also associated with a fall in the substrate NAD – well known for many years (cf. Roitt 1956; Holzer et al. 1958; Hilz et al. 1963). A transient increase in poly(ADPR) upon treatment of SV 40-transformed 3T3 cells with N'-methyl-N'-nitro-nitroso guanidine has also been reported (Juarez-Salinas et al. 1979). A direct involvement of poly-ADP-ribosylation in repair processes, however, has not been shown so far. In addition, two recent papers report that nicotinamide, an effective inhibitor of the poly(ADPR)-forming transferase, drastically *in*creased repair replication (Berger and Sikorski 1980; Althaus et al. 1980). Further, benzamide, another inhibitor of the polymerizing transferase, even at high concentrations had no synergistic effect on growth inhibition by DNA-fragmenting cytostatics (unpublished experiments). Poly ADP-ribosylation has also been implicated in the regulation of cell proliferation (cf. Hilz and Stone 1976; Hayaishi and Ueda 1977; Purnell et al. 1980) and in differentiation processes (cf. Caplan and Rosenberg 1975; Berger et al. 1980). However, the available evidence for an essential role of poly ADP-ribosylation reaction in growth and in differentiation is circumstantial, and more data are required to substantiate the proposed involvements.

3 Summary

Multiple proteins – histone and nonhistone proteins – can serve as acceptors for the transfer of mono and poly(ADPR) residues in isolated nuclei as well as in permeabilized cells. Modification by $(ADPR)_n$ groups profoundly influences the properties of the acceptor proteins. Thus, histones can be shifted into the nonhistone fraction.

Histone H1 proved to be a minor acceptor in EAT and in HeLa cells in vitro and in vivo. An analysis of the isolated ADPR conjugates from EAT nuclei showed preferential ADP-ribosylation of the C-terminal fragment, and the modification of a large fraction of H1 molecules by $(ADPR)_n$ residues exclusively at the C-terminal *or* the N-terminal fragment.

Quantitation of ADPR conjugates in intact tissues revealed that in all tissues so far analyzed more acceptor sites were occupied by mono(ADPR) residues than by poly(ADPR) chains. The $(ADPR)_n$ protein conjugates could be further divided in two subclasses on the basis of their sensitivity toward neutral hydroxylamin. In three systems the relative levels of the NH_2OH-resistant mono(ADPR) conjugate subfraction related to the degree of differentiation. Mono(ADPR) protein conjugates, the majority of which appears to be localized extranuclearly, varied independently from the conjugates carrying poly-(ADPR) residues under different conditions. Poly ADP-ribosylation has been implicated in DNA repair, regulation of cell proliferation, and in differentiation. More experiments, however, are required before ADP-ribosylation can be considered an integral part of these processes.

Acknowledgements. This work was supported by grants from the Deutsche Forschungsgemeinschaft and Stiftung Volkswagen-Werk.

References

Adamietz P, Hilz H (1976) Poly(ADPR) is covalently linked to nuclear proteins by two types of bonds. Hoppe-Seyler's Z Physiol Chem 357:527-534

Adamietz P, Bredehorst R, Hilz H (1978) ADP-ribosylated histone H1 from HeLa cultures. Eur J Biochem 91:317-326

Adamietz P, Klapproth K, Hilz H (1979) Isolation and partial characterization of the ADP-ribosylated nuclear proteins from Ehrlich ascites tumor cells. Biochem Biophys Res Commun 91;1232-1238

Adamietz P, Wielckens K, Bredehorst R, Hilz H (1981) Subcellular distribution of different $(ADPR)_n$ protein conjugate fractions in rat liver. Biochem Biophys Res Commun (in press)

Althaus FR, Lawrence SD, Sattler GL, Pitot HC (1980) The effect of nicotinamide on unscheduled DNA synthesis in cultured hepatocytes. Biochem Biophys Res Commun 95:1063-1070

Berger NK, Weber G, Kaichi AS (1978) Characterization and comparison of poly(adenosine diphosphoribose) synthesis and DNA synthesis in nucleotide permeable cells. Biochim Biophys Acta 519:87-104

Berger NA, Sikorski GW (1980) Nicotinamide stimulates repair of DNA damage in human lymphocytes. Biochem Biophys Res Commun 95:67-72

Berger NA, Sikorski GW, Petzold SJ (1980) Association of poly (ADP-ribose) synthesis with DNA strand breaks in replication and repair. In: Smulson ME, Sugimura T (eds) Novel ADP-ribosylations of regulatory enzymes and proteins. Elsevier/North Holland, Amsterdam New York, p 185

Braeuer H, Adamietz P, Nellessen U, Hilz H (1981) ADP-ribosylated histone H1. – Isolation from Ehrlich ascites tumor cell nuclei and partial characterization. Eur J Biochem 114:63-68

Bredehorst R, Ferro AM, Hilz H (1978a) Determination of ADP-ribose and poly(ADP-ribose) by a new radioimmunoassay. Eur J Biochem 82:115-121

Bredehorst R, Wielckens K, Gartemann A, Lengyel H, Klapproth K, Hilz H (1978b) Two different types of bonds linking single ADP-ribose residues covalently to proteins. Eur J Biochem 92: 129-135

Bredehorst R, Goebel M, Renzi F, Kittler M, Klapproth K, Hilz H (1979) Intrinsic ADP-ribose transferase activity versus levels of mono(ADP-ribose) protein conjugates in proliferating Ehrlich ascites tumor cells. Hoppe Seyler's Z Physiol Chem 360:1737-1743

Bredehorst R, Klapproth K, Hilz H, Scheidegger C, Gerisch G (1980) Protein-bound mono(ADP-ribose) residues in differentiating cells of Dicytostelium discoideum. Cell Differ 9:95-103

Bredehorst R, Schlüter M, Hilz H (1981a) Determination of 5'-AMP in the presence of excess 3'(2')-AMP with the aid of antibodies raised against N^6-carboxymethyl-5'-AMP conjugates. Biochim Biophys Acta 652:16-28

Bredehorst R, Wielckens K, Steinhagen-Thiessen E, Adamietz P, Lengyel H. Hilz H (1981b) Divergent alterations of $(ADPR)_n$ protein conjugate subfractions during liver development. Eur J Biochem (in press)

Burzio LO, Riquelme PT, Koide SS (1979) ADP-ribosylation of rat liver nucleosomal core histones. J Biol Chem 25:3029-3037

Caplan A, Rosenberg N (1975) Interregulationship between poly (ADP-ribose) synthesis, intracelllar NAD levels and muscle or cartilage differentiation from mesodermal cells of embryonic chick limb. Proc Natl Acad Sci USA 72:1852-1857

Chambon P, Weill JD, Doly J, Strosser MT, Mandel P (1966) On the formation of a novel adenylic compound by enzymatic extracts of liver nuclei. Biochem Biophys Res Commun 25:638-643

Dameshek W (1967) Chronic lymphocytic leukemia – an accumulative disease of immunologically incompetent lymphocytes. Blood 29:566-584

Fujimura S, Hasegawa S, Sugimura T (1967) Nicotinamide mono-nucleotide-dependent incorporation of ATP into acid insoluble material in rat liver nuclei preparation. Biochim Biophys Acta 134:496-499

Gartemann A, Bredehorst R, Strätling WH, Wielckens K, Hilz H (1981) ADP-ribosylation of proteins in mouse kidney following castration and testosterone treatment. Biochem J 198:37-44

Halldorsson H, Gray DA, Shall S (1978) Poly(ADP-ribose)polymerase activity in nucleotide permeable cells. FEBS Lett 85:349-352

Havemann K, Rubin A (1968) The delayed response of chronic lymphocytic leukemia lymphocytes to phytemagglutinin in vitro. Proc Soc Exp Biol Med 127:668-671

Hayaishi O, Ueda K (1977) Poly(ADP-ribose) and ADP-ribosylation of proteins. Annu Rev Biochem 46:95-116

Hilz H, Hlavica P, Bertram B (1963) Die entscheidende Bedeutung der DPNase-Aktivierung für den DPN-Abfall in bestrahlten Ascites-Tumor-Zellen. Biochem Z 338:283-299

Hilz H, Stone P (1976) Poly(ADP-ribose) and ADP-ribosylation of proteins. Rev Physiol Biochem 76:1-58

Hilz H, Bredehorst R, Wielckens K, Adamietz P (1980a) Quantitation of mono(ADPR) protein conjugates from intact tissues at various stages of gene expression. In: Smulson ME, Sugimura T (eds) Novel ADP-ribosylations of regulatory enzymes and proteins. Elsevier/North Holland, Amsterdam New York, p 143

Holzer H, Glogner P, Sedlmayr G (1958) Zum Mechanismus der Glykolysehemmung durch carcinostatisch wirkende Äthyleniminverbindungen. Biochem Z 330:59-72

Johns EW, Butler JAV (1962) Further fractionations of histones from calf thymus. Biochem J 82: 15-18

Juarez-Salinas H, Sims JL, Jacobson MK (1979) Poly(ADP-ribose) levels in carcinogen-treated cells. Nature (London) 282:740-741

Levy S, Simpson RT, Sober HA (1972) Fractionation of chromatin components. Biochemistry 11: 1547-1554

Miller EG (1975) Stimulation of nuclear poly(adenosine diphosphate-ribose) polymerase activity from HeLa cells by endonucleases. Biochim Biophys Acta 395:191-200

Nishizuka Y, Ueda K, Yoshihara K, Yamamura H, Takeda M, Hayaishi O (1969) Enzymic adenosine diphosphoribosylation of nuclear proteins. Cold Spring Harbor Symp Quant Biol 34:781-786

Ogata N, Ueda K, Kagamiyama H, Hayaishi O (1980) ADP-ribosylation of histone H1. J Biol Chem 255:7616-7620

Purnell MR, Stone PR, Whish WJD (1980) ADP-ribosylation of nuclear proteins. Biochem Soc Trans 8:251-227

Reeder RH, Ueda K, Honjo T, Nishizuka Y, Hayaishi O (1967) Studies on the polymer of adenosine diphosphate ribose. Characterization of the polymer. J Biol Chem 242:3172-3179

Roitt JM (1956) The inhibition of carbohydrate metabolism in Ascites tumor cells by ethyleneimines. Biochem J 63:300-307

Weber W, Schwoch G, Gartemann A, Wielckens K, Hilz H (1981) cAMP receptor proteins and protein kinase in human lymphocytes: Fundamental alterations in chronic lymphotic leukemia cells. Eur J Biochem (in press)

Wielckens K, Garbrecht M, Kittler M, Hilz H (1980) ADP-ribosylation of proteins in normal lymphocytes and in low-grade malignant Non-Hodgkin lymphoma cells. Eur J Biochem 104:279-287

Wielckens K, Bredehorst R, Adamietz P, Hilz H (1981) Protein-bound polymeric and monomeric ADPR residues in hepatic tissues: Comparative analysis using a new procedure for the quantitation of poly(ADPR). Eur J Biochem 117:69-74

Discussion[2]

In ADP-ribosylated proteins the linkage is sensitive to hydroxylamine when ADP ribose is linked to aspartate or glutamate, the bond is hydroxylamine-resistant when arginine is implicated. The transferase activity observed is not similar to the mono ADP ribosyltransferase from pigeon erythrocytes which is specific for arginine, since in a crude system both hydroxylamine-resistant and -sensitive bonds are found.

The single nuclear ADP-ribosylating system is not specific for a protein, but can act on different histones and possibly on a high mobility protein, on non-histone proteins, and even on the polymerase itself.

If the results were associated with an aminolysis of the ester bond by lysyl residues, amines in vitro should reduce the transfer of carboxyl groups. However, no effect of amines was observed in these experiments.

2 Rapporteur: Carlos Gancedo

ADP-ribosylation of Nonhistone Chromatin Proteins in Vivo and of Actin in Vitro and Effects of Normal and Abnormal Growth Conditions and Organ-specific Hormonal Influences

E. Kun, A.D. Romaschin, R.J. Blaisdell, and G. Jackowski[1]

1 Introduction

In contrast to a large variety of covalent modifications of both histone and nonhistone proteins by phosphorylation, acetylation, methylations etc., reactions that appear to exhibit a relatively broad selectivity, poly ADP-ribosylation by $(ADP\text{-}R)_n$ in small rodents in vivo is confined to an extent of at least 99% to nonhistone chromosomal proteins, as determined by an $(ADP\text{-}R)_n > 4$ specific IgG antiserum assay (Minaga et al. 1979). Similar results were recorded in another laboratory in varying biological material by different experimental methods (Bradehorst et al. 1978; Bradehorst et al. 1979). It should be noted that ADP-ribosylation of histone H_1 in vitro is readily demonstrable when larger quantities of the ADP-R acceptor histone H_1 are added to the purified poly ADP-R synthetase (Kawaichi et al. 1980), but this is a typical in vitro enzymological result. Quantitatively misleading results can be obtained in vivo or, in cell cultures, when only radiochemical assays of ADP-R are the basis of analyses without consideration of the absolute quantities of $(ADP\text{-}R)_n$ (Giri et al. 1978; Jump et al. 1979). It is obvious that measurement of radioactivity alone does not distinguish between small quantities of highly radioactive $(ADP\text{-}R)_n$ or larger amounts of $(ADP\text{-}R)_n$ with lower specific activity, both parameters being related to turnover (Kun et al. 1979). There is no doubt that only about 0.1% to 0.6% of the total $(ADP\text{-}R)_n > 4$ is associated in vivo with histones and shorter as well as longer homopolymers exist as covalent complexes with histones (Minaga et al. 1979). However, the overwhelming targets of ADP-ribosylation are *nonhistone proteins* that have been identified by the pioneering experiments of Paul and Gilmour (1968) to play an as yet not well understood role in the regulation of differentiated cellular functions in eukaryotes. The previously proposed correlation between transcriptionally active chromatin and the degree of ADP-ribosylation (Mullins et al. 1977) is presently doubtful, since more recently no connection was found between the localization or activity of poly ADPR polymerase and transcriptional activity of chromatin (Yukioka et al. 1978).

Since positive correlation between a variety of differentiated cell functions as well as differentiation itself and the regulatory role of nonhistone proteins is probable (Paul and Gilmour 1968), we pursued the problem of the regulatory function of poly ADP-ribosylation in experimental models that express the operation of development and differentiation: early phase of carcinogenesis, age-dependent and hormone-influenced

1 Departments of Pharmacology, Biochemistry, and Biophysics and the Cardiovascular Research Institute, the University of California, San Francisco, Surge Building, Room 101, San Francisco, CA 94143, USA

organ-specific nuclear responses. The present paper is concerned with ongoing experimental work in this area. Results are consistent with the working hypothesis that predicts a major cellular regulatory function of poly ADP-ribosylation by way of modification of nonhistone proteins. This modification process can be influenced either at the transcriptional level, resulting in variations of the synthesis of enzymatic components of the poly ADP-R synthetase systems, or by changes in the concentration of target nonhistone proteins, or both. A unique metabolic effect on chromatin function through this system is suggested by the variation of ADP-ribosylations due to metabolically induced changes in NAD^+/NADH ratios (Kun and Chang 1976) possibly contributing to the initiation signal of stress-induced organ hypertrophy (Zak and Rabinowitz 1979).

2 Evaluation of Methodology for in Vivo Studies

The advantage of in vivo experimentation is its obvious relevance to physiologic function. Physiologic interpretation of in vitro enzymology in a relatively undeveloped field requires constant reevaluation of in vitro systems, since it is uncertain whether or not the chosen in vitro system contains all regulatory components present in the intact cell. This plurality of experimental approach is reflected in experimental methods, developed in our laboratory. Notable advances in related areas were made also in other laboratories, but the present paper will be confined to the analysis of techniques used in our research group.

2.1 Immunochemistry of Poly(ADP-R)$_n$ > 4

The quantitative analysis of the homopolymer poly(ADP-R) by radiochemistry is hampered by the absence of a specific precursor. Labeled ribose, or [^{32}P] are suitable and [^{14}C]-ribose is preferable because of its convenience of handling. Adenine is an unsuitable precursor in vivo, probably because preferential pathways other than those leading to poly(ADP-R) obscure the labeling of the homopolymer. Development of a highly specific anti-poly(ADP-R) IgG rabbit globulin that reacts with (ADP-R) > 4 (Ferro et al. 1978) significantly aided our experiments. An improved extraction procedure was also developed that was used to determine the distribution of the polymer among various types of nuclear proteins (Minaga et al. 1979). Combination of immunochemical analyses and in vivo labeling (Minaga et al. 1979; Kun et al. 1979) gave a $t_{\frac{1}{2}}$ for NAD^+ and of poly(ADP-R) of 2.7 and 3.1 h respectively, in terms of ribose incorporation, the slight discrepancy representing experimental variation only.

2.2 Microchemical Methods

The disadvantage of the immunochemical technique is twofold. First, the preparation of the antibody may be variable and second, the method does not detect oligomers smaller than tetramers. For these reasons entirely chemical micro procedures were developed. Poly and mono ADP-ribosylated nuclear proteins were quantitatively isolated on a highly substituted affinity column: m-aminophenylboronic acid glutaryl-hydrazide poly acrylamide (Romaschin et al. 1981), that exhibits better than 90% recovery of ADP-R-proteins, in contrast to 50%-70% of polysaccharide-based columns containing no spacers. The isolated ADP-ribosylated nuclear proteins, containing in vivo administered labeled precursors (either [^{14}C]-ribose for ADPR, or labeled amino acids as protein markers) can be directly analyzed by HPLC-molecular sieve columns (Romaschin et al. 1981) or hydrolyzed to free $(ADPR)_n$ and proteins. In case the protein was not labeled, [^{3}H]-dansylation is performed and the macromolecular protein profile of the ADPR-free target proteins of ADP–ribosylation are subjected to HPLC-molecular filtration, combined with isotope analyses (Romaschin et al. 1981). For more detailed peptide separations the proteins are labeled with [^{125}I] and subjected to one-dimensional electrofocusing and SDS electrophoresis in the second dimension. Protein-free ADPR oligomers are degraded by purified poly(ADPR) glycohydrolase, and ADPR analyzed as a fluorescent derivative on an anion exchange column by HPLC-fluorometry, using a phosphate gradient and combined with isotope dilution with [^{14}C]-ADPR. Less than picomol quantities are readily measurable. Alternative methods were developed for the determination of the macromolecular profile of ADPR oligomers. Either HPLC-molecular filtration (when the homopolymer contains [^{14}C]-ribose) was performed to determine the newly formed oligomers, detected by ribose labeling (representing mostly chain elongations and internal randomization of [^{14}C]-ribose with the unlabeled ribose pool), or by [^{3}H]-dimethylsulfate labeling of *all* $(ADPR)_n$ species, followed by gel electrophoresis, radioautography, and densitometry. A combination of the above techniques is employed in our laboratory for the quantitative assay and macromolecular association of both monomeric and oligomeric $(ADPR)_n$-protein adducts. Good agreement between immunochemical and chemical analyses was obtained (Minaga et al. 1979; Kun et al. 1979; Romaschin et al. 1981). It was also confirmed that in normal animal tissues there are no free ADPR oligomers unless artificially produced by the action of intranuclear proteases during in vitro manipulations.

3 Quantitative Studies in Vivo During the Early Precancerous State Induced by Dimethyl-nitrosamine Treatment of Syrian Hamsters

The first model related to pathophysiological alterations of differentiation in animals was the early stage of hepatocarcinogenesis in Syrian hamsters treated with small doses of dimethylnitrosamine (weekly IP injection of 0.5 mg per 100 g body weight for 1 or 2 months; Herrold 1967). As described in the paper by Romaschin et al. (1981), treatment of 1 month-old hamsters for 1 month (four injections of 0.5 mg dimethylnitrosamine/100 g) resulted in a fivefold increase in chemically determined poly(ADPR) and a significant increase in the incorporation of in vivo administered [^{14}C]-ribose (50 μCi/100 g) into both poly(ADPR) and NAD^+, indicating an increased turnover of the pathway $NAD^+ \rightarrow$ poly(ADPR). There is simultaneously a decrease in the steady-state concentration of NAD^+ (~25%), consistent with its increased utilization for the protein-ADP-ribosylating pathway. Continued treatment of hamsters for 2 months diminished the difference between steady-state concentrations of poly(ADPR) between normal and hepatocarcinogen-treated hamsters, but further increased the rate of flux between NAD^+ and poly(ADPR). Notably, the turnover of ribose in 2 months-old animals diminished, indicating an age-dependent variation of NAD^+ and poly(ADPR) metabolism, however the carcinogen induced increase is independent from this age-related change and is only dependent on the dose and time of administration of the carcinogen. These results are summarized in Table 1. Separation of labeled poly ADP-ribosylated proteins, prepared from 1 month precancerous hamster liver by CsCl density ultracentrifugation is shown in Fig. 1. It is apparent that the three macromolecular components, DNA, RNA, and

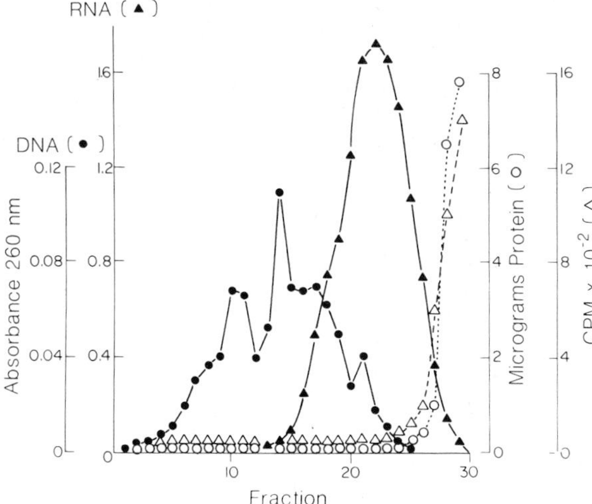

Fig. 1. Separation of DNA (60 μg), RNA (400 μg), and ADPR-protein adducts by ultracentrifugation in CsCl gradient. Time of centrifugation = 40 h at 125,000 x g (av.) at 4°. (●—●), 60 μg DNA; (▲—▲), 400 μg RNA; (○—○), [^{14}C]-ribose $(ADPR)_n$-protein counts; (△—△), protein content. *DNA*, thymus DNA; *RNA*, obtained from yeast

poly(ADP-R)-protein adducts are readily separable by this physical procedure. Separation and determination of the macromolecular profile in vivo [^{14}C]-ribose labeled protein-poly(ADPR) adducts in 1.5 N guanidine formate solution by HPLC is illustrated in Fig. 2. Both normal and precancerous (1 month treated) hamster livers were analyzed and the molecular profile (on HPLC) was monitored in a TSK-3000 molecular sieve. The molecular masses (in kilodaltons: kd) are plotted against fraction numbers. In these experiments no definite determination of the absolute quantities of poly(ADPR) is possible

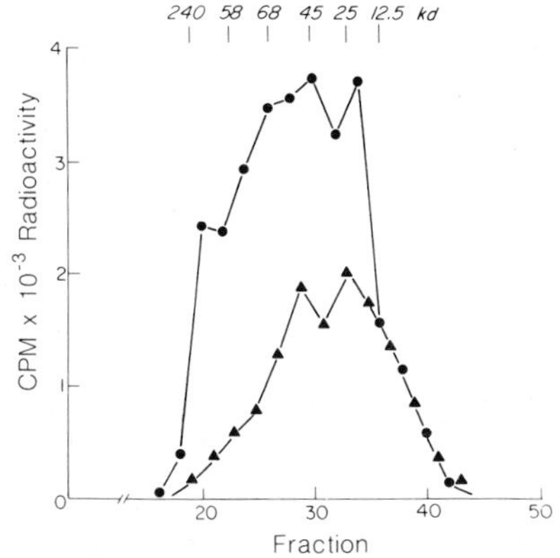

Fig. 2. Molecular mass distribution of protein-$(ADPR)_n$ adducts isolated from hamster nuclei (control and treated with DNMA* for 1 month) by the affinity chromatography method on a TSK-3000 SW HPLC molecular sieve. Developing solvent: 1.5 M guadinine-formate, pH 5.0, flow rate 0.4 ml/min. ▲—▲, control; ●—●, dimethylnitrosamine-treated

Table 1. Poly ADPR content, NAD^+ content and [^{14}C]-ribose pulse labeling in normal and precancerous hamster liver

No.	Treatment of hamsters	$(ADPR)_n$ nmol ADPR/mg DNA	$(ADPR)_n$ spec. act. μCi/mmol ADPR	NAD^+ content nmol/g liver	[^{14}C]-ribose spec. act. of NAD^+, μCi/mmol
1	Control (1 month)	22 ± 5	34 ± 8	646 ± 62	44 ± 8
2	DMNA treatment for 1 month	112 ± 38	58 ± 19	400 ± 40	83 ± 15
3	Control (2 months)	23 ± 5	2.9 ± 0.7	649 ± 60	5.3 ± 0.7
4	DMNA treatment for 2 months	38 ± 8	11.0 ± 2.6	550 ± 44	17.4 ± 3

Eight male Syrian hamsters were selected on the basis of equal body weight within the chosen age groups. The hamsters (two animals in each group) were deprived of food for 16 h, then pulse-labeled with 50 μCi [^{14}C]-labeled ribose (58 mCi/mmol, ICN Pharmaceuticals) by intraperitoneal injection. Three hours after labeleing liver homogenates were prepared in citric acid and nuclei were isolated. The total radioactive material per g liver homogenate, prepared from the control and the DMNA-treated hamsters, was identical. Quantitative isolation of $(ADPR)_n$-protein adducts was carried out by boronate affinits chromatography. Analyses of $(ADPR)_n$ were performed after either chemical or enzymatic hydrolysis of (ADPR)-protein adducts, followed by HPLC-fluorometry of characteristic nucleotides. NAD^+ was determined in the postnuclear supernatant after deproteinization with 5% $HClO_4$ and isolation by HPLC. The body weight of groups 1 and 2 was 80-100 g, and of 3 and 4, 140-160 g. Analyses were done in triplicates and results show arithmetic mean values and standard deviation

because pulse labeling in vivo randomly labels elongating homopolymers and is also recycling with endogenous poly(ADPR).

Quantitative assessment is always based on direct chemical analyses (see Table 1). The protein-poly(ADP-R) adducts were hydrolyzed, the polymer-free proteins labeled with [^{3}H]-dansyl-chloride and both proteins and homopolymers subjected to HPLC-molecular mass analyses. The homopolymers were separated on a TSK-2000 molecular sieve. Results are shown in Fig. 3. It is evident that in precancerous liver nuclei poly ADP-ribo-

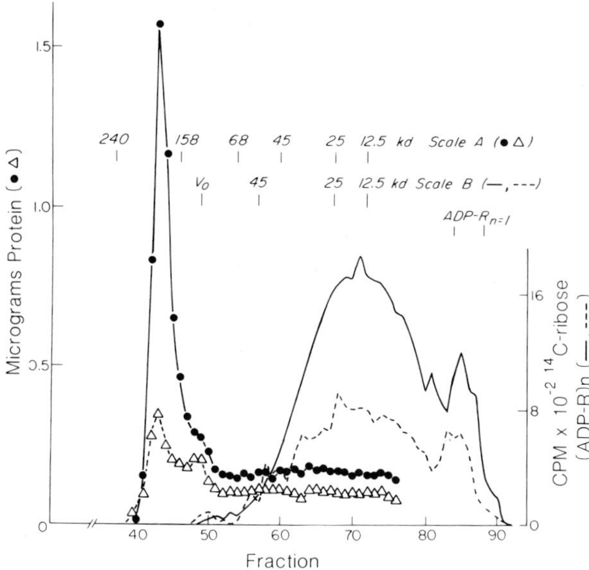

Fig. 3. Distribution of molecular mass of proteins (△- - -△, control; ●—●, treated for 1 month with DMNA) and of free ADPR-oligomers (— — —, controls; ——, treated for 1 month with DNMA). Proteins were separated as described in the legend of Fig. 2, and protein-free ADPR-oligomers on TSK-2000 HPLC-molecular sieve with 100 mM ammonium phosphate (pH 6.0) as developing solvent (0.4 ml/min). Authentic samples of $(ADPR)_n$ of known chain length served as standards for the estimation of molecular masses of ADPR-oligomers. *Scale A*, standards for protein masses; *Scale B*, standards for $(ADPR)_n$-oligomers calculated in molecular masses

sylation of proteins of a mass about 160 kd is greatly increased. The analysis of the molecular profile of free homopolymers indicated that only about 10% to 15% of in vivo mono ADP-ribosylation takes place and at least 80% to 90% of the homopolymer is larger than monomeric. These results extend and support previous analyses obtained by the $(ADPR)_n > 4$ specific antiserum (Minaga et al. 1979) but, in contrast to the immunochemical method that does not assay smaller than tetramers, the chemical procedures reveal a detectable quantity of mono ADP-ribosylation. However, the monomers are still only 10% to 15% of the oligomers.

In vitro pulse labeling of both normal and precancerous hamsters with [^{3}H]-thymidine and analysis of incorporation into DNA revealed that in the precancerous animals the rate of thymidine incorporation increased about fourfold. Alkaline sucrose density gradient analysis of DNA species indicated no increase of DNA species of short chain length, no typical fragmentation, as observed in tissue cultures treated with high concentrations of alkylating agent (2 mM dimethylsulfate, cf. Durkacz et al. 1980), therefore at this early precancerous stage, where no morphologic evidence of malignancy is detectable, there is no evidence suggesting a correlation between DNA fragmentation and an increase of poly(ADPR) metabolism. It is probable that 2 mM alkylating drug in vitro in tissue cultures is a highly toxic dose, and DNA fragmentation reflects nonspecific toxicity. It is known that the subtle transformation process to malignancy does not necessarily involve gross chromosomal alterations (Dipaolo 1977) but coincides with much less obvious modifications of DNA. Our model, that focuses on an early stage of malignant transformation, is consistent with this view (Dipaolo 1977), therefore the large increase in poly(ADPR) turnover and nonhistone protein modification *precedes* any gross alteration (fragmemtation) of DNA, and is probably a very early signal of the precancerous state. This mechanism and its relationship to DNA repair is an open question.

4 In Vitro ADP-ribosylation of Actin-like Proteins of Isolated Nuclei and of Added Actin

It is known that nuclear actin is a powerful and specific physiologic inhibitor of nuclear DNA-ase I (Lazarides and Lindberg 1974). It has also been reported that a variant form of actin appears to be synthesized during chemically induced malignant transformation (Leavitt and Kakunaga 1980). Since actin is a well defined nonhistone chromosomal protein possessing a critical regulatory function in DNA breakdown, we have studied its ADP-ribosylation so far in vitro.

When nuclear sonicates prepared from normal and precancerous hamster livers are incubated with [^{14}C]-NAD^{+} and ADP-ribosylated proteins separated by SDS-gel electrophoresis, following autoradiography and densitometry, the protein association of newly formed labeled ADPR oligomers with proteins are readily determined, as shown in Fig. 4. It is seen from Fig. 4.I that a small but distinct ADP-ribose containing protein peak is present in normal nuclei, which by its apparent molecular weight, and by the addition to the chromatin incubate of authentic actin is identifiable with actin. Addition of actin also increases ADP-ribosylation of proteins between 65.000 and 150.000 molecular weight (Fig. 4.II). The apparent ADP-ribosylation of actin-like protein is greatly increased in chromatin incubates prepared from precancerous livers (Fig. 4.III), and addition of pure actin to these systems only slightly increases ADP-ribosylation of actin, suggesting that the reaction is already maximal. If this were the case, then it would be expected that there should be a larger amount of ADP-ribosylated actin present in nuclei prepared from precancerous livers.

This possibility was tested by isolating ADPR protein adducts with the affinity chromatography technique (Romaschin et al. 1981) and subjecting the mixture of ADP-ribosylated proteins to molecular filtration by HPLC. Results are summarized in Fig. 5. Both radioactivity and absorbance of 280 nm were monitored simultaneously. It is expected that the A_{280} assay would detect only relatively large quantities of protein ADPR adducts, whereas [^{14}C]-analyses, indicating newly formed ADPR oligomers, would signal the presence of small amounts of proteins. In agreement with in vivo labeling (Fig. 3), large-molecular weight proteins were most heavily ADP-ribosylated and absorbance at 280 nm (*dotted line*) shows a significant amount of protein in this molecular mass region (around 160 kd). In chromatin from precancerous hamsters a second protein peak is also detectable by A_{280} at the range of molecular mass around 48 kd. Due to anomalous hydrodynamic properties of actin, this protein assumes a somewhat higher molecular mass in the molecular filter column, therefore the larger amount of protein at 48 kd is

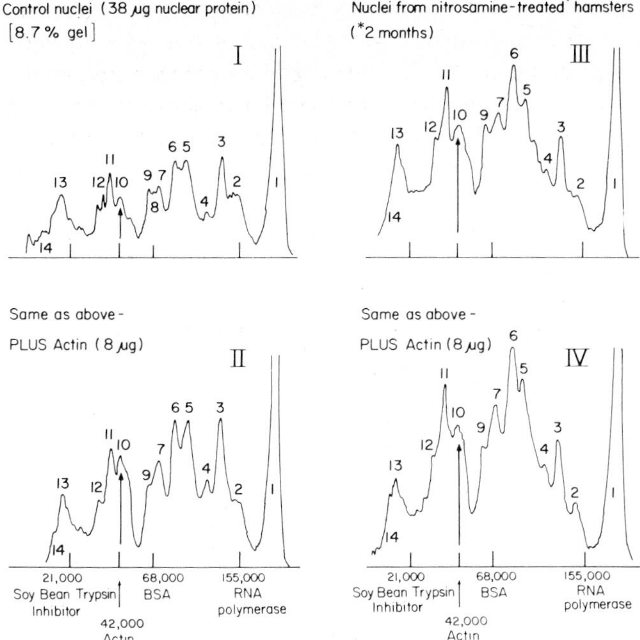

Fig. 4. In vitro ADP-ribosylated chromatin proteins were prepared by incubation of sonicated chromatin with [^{14}C]-NAD$^+$ as described in the legend of Fig. 7. Proteins were separated by SDS poly acrylamide gel, radioactivity localized by autoradiography and subsequent densitometry

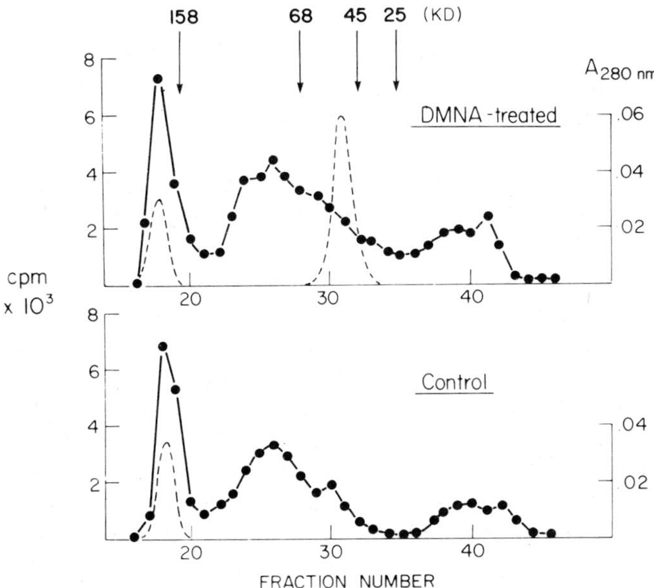

Fig. 5. Separation of ADP-ribosylated proteins from in vitro incubated sonicated chromatin (see legend of Fig. 7) by HPLC molecular filtration on TSK-3000 molecular filter. ADP-ribosylated proteins were first isolated on the boronate affinity column (Romaschin et al. 1981) then developed in 1.5 M guadinine-formate. Both radioactivity and A_{280} were simultaneously monitored. Fraction volume = 0.2 ml

most probably an actin-like protein. More definitive molecular studies to ascertain the identity of ADP-ribosylated nonhistone proteins is presently carried out. As a further confirmation of the ADP-ribosylation of authentic added actin by normal chromatin, we reisolated both ADP-ribosylated actin and ADP-ribosylated histones from in vitro incubation systems. Results are shown in Table 2. On a molar basis added actin is about ten times more ADP-ribosylated than endogenous histones.

Table 2. Poly(ADP-ribosylation) of endogenous histones and of added actin by chromatin of hamster liver

		m mol ADPR per mol		Histone content
No.	Preparation	Actin	Histone	μg/test
1	Control + 6.4 μg actin	60.1	5.6	17.9
2	Control (no additions)	–	7.5	17.0
3	DMNA-treated + 7.1 μg actin	85.9	9.1	18.1
4	DMNA-treated (no additions)	–	11.1	17.7

Labeled actin and histones were reisolated from SDS poly acrylamide gel after electrophoretic separation and newly formed ADPR-oligomers determined radiochemically

5 Rates of Poly(ADP-R) Synthesis in Cardiocyte Nuclei as a Function of Age and Organ-specific Hormonal Influences in Vivo on Rates of Poly(ADPR) Synthesis in Isolated Cardiocyte Nuclei, in Vitro Effect of Aldosterone

We have previously shown that poly(ADPR) synthesis in cardiac nuclei is two to three times higher than in livers (Ferro and Kun 1976) of the same animals, suggesting a possible physiologically prominent function of ADP-ribosylation in cardiocytes. With the aid of a specific separation technique cardiocyte nuclei were isolated from a mixed nuclear population (Jackowski and Liew 1980) and the effect of aging – an expression of development – on rates of in vitro poly(ADP-R) synthesis was determined. Striking results were obtained, as shown in Fig. 6. Rates of poly(ADPR) synthesis in vitro were much higher in myocardial nuclei isolated from 5 day-old than from 65 day-old rats (Long Evans strain). These results are in apparent variance with an earlier report (Claycomb 1976) that claimed higher poly(ADPR) synthesis in nuclei of older rats. Unfortunately, Claycomb (1976) did not separate myocardial nuclei from nuclei of nonmyocytes – as done in the present work (Jackowski and Liew 1980) – and it is likely that the fortuitous relative enrichment of myocyte nuclei containing higher enzymatic activity than cardiocyte nuclei signaled a higher apparent total enzymatic activity. The possibility of this type of artifact has been actually considered by Claycomb (1976). In general, when rates of poly(ADPR) synthesis are increased, the steady-state concentra-

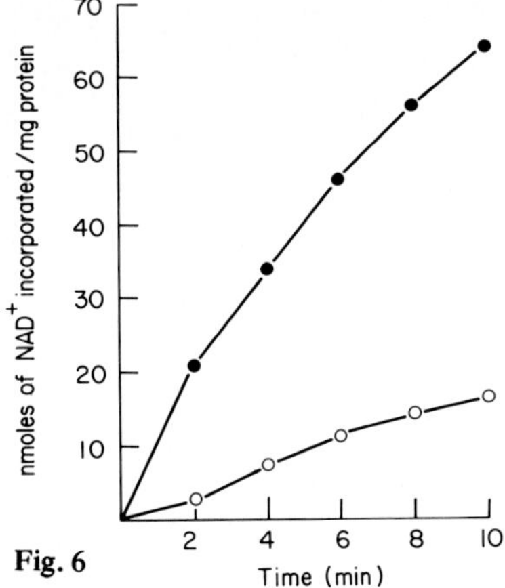

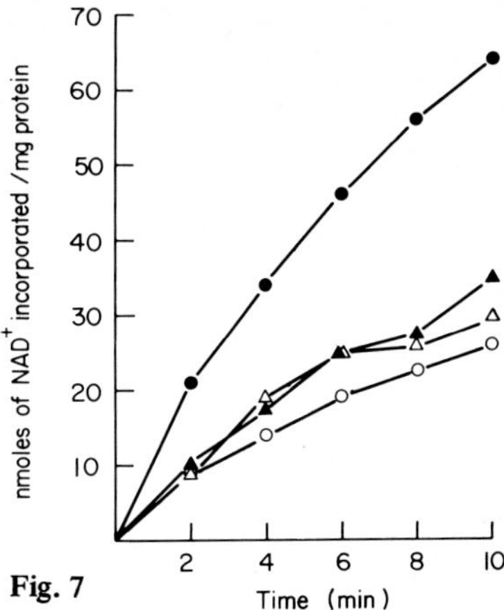

Fig. 6 **Fig. 7**

Fig. 6. Changes in Myocardial Nuclear Poly(ADP-R) Synthetase Activity with Age. Myocardial nuclei isolated from 5 day-old and 65 day-old Long Evans rats were incubated with [^{14}C]-NAD$^+$ and the incorporation of [^{14}C]-NAD$^+$ into an acid insoluble product was monitored with time, as described in legend of Fig. 7. ●, Myocardial nuclei (5 day-old); ○, Myocardial nuclei (65 day-old)
Fig. 7. Poly(ADPR) synthetase activity of myocardial and liver nuclei from cortisol-treated and control rats. Hydrocortisone sodium phosphate (Hydrocortone, MSD, 5 mg/rat) was injected intraperitoneally into 30 day-old Long Evans rats (100 g body weight) daily for 4 days, prior to the isolation of nuclei; a control group received isotonic saline at the same time. Poly(ADPR) synthetase activity in the isolated nuclei was monitored by measuring the rate of incorporation of [^{14}C]-NAD$^+$ into an acid insoluble product. The reaction mixture, at 25°C in a volume of 235 μl contained 100 mM Tris HCl pH 8.2, 2 mM DTT, 20 mM $MgCl_2$, 0.1 mM PMSF, 0.5 mM EDTA, 0.5 mM NAD$^+$, 2.8×10^7 dpm of [^{14}C]-NAD$^+$, 80 μg of nuclear material. ●, Myocardial nuclei (control); ▲, liver nuclei (control); △, liver nuclei (cortisol-treated); ○, myocardial nuclei (cortisol-treated)

tion of nuclear NAD$^+$ decreases. Claycomb (1976) observed a decrease in NAD$^+$ content of cardiac tissue of young animals, an observation consistent with our results. The kinetic data shown in Fig. 6 support the view that *aging* markedly influences the poly ADP-ribosylation pathway.

In further studies we tested the apparent effects of endocrines on nuclear poly(ADPR) synthetase activity. The most striking phenomenon observed was *organ specificity*. Only myocardial nuclei exhibited a hormone sensitivity, whereas liver nuclei were unresponsive. As illustrated in Fig. 7, pharmacologic dose of cortisol (5 mg/rat) severely depressed the poly(ADPR) synthetase activity of myocardial nuclei, but not of liver nuclei. The relatively large effective dose of cartisol made it suspect that possibly a side effect of this hormone (mineralocorticoid-like action) was observed. The enzymatic activity in myocardial nuclei of hypophysectomized rats was markedly diminished – as would be anticipated – and cortisol at 5 mg/100 g dose and aldosterone at *5 μg*/100 g dose further depressed activity, indicating that indeed aldosterone was the more powerful inhibitory agent in vivo, and the large dose of cortisol mimicked the effect of the typical mineralocorticoid. These in vivo effects are shown in Fig. 8. Adrenalectomy significantly diminished poly(ADPR) synthetase activity of myocardial nuclei, and cortisol (5 mg/100 g rat) further depressed activity (Fig. 9). The effects of pituitary and adrenal hormones are clearly complex. Presence of both pituitary and adrenal glands are required for opti-

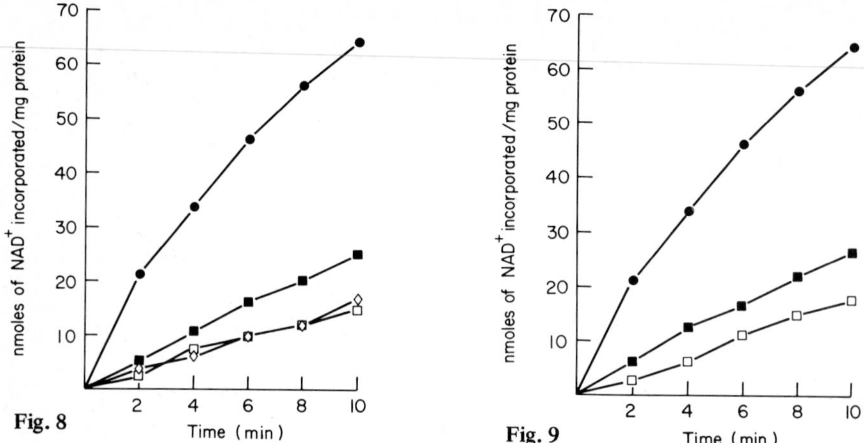

Fig. 8. The effect of hypophysectomy, cortisol and aldosterone treatment of hypophysectomized rats on myocardial nuclear poly(ADPR) synthetase activity. Hydrocortisone sodium phosphate (Hydrocortone, MSD; 5 mg/100 g body weight) was injected intraperitoneally into hypophysectomized Long Evans rats daily for 4 days, prior to the isolation of nuclei. Aldosterone (Sigma; 5 μg/100 g body weight) was injected intraperitoneally into hypophysectomized rats following the same injection protocol employed for hydrocortisone; a control group consisting of normal rats and hypophysectomized rats received sterile isotonic saline. Poly(ADPR) synthetase activity was monitored as described in Fig. 7. •, myocardial nuclei (control); ▪, myocardial nuclei (hypophysectomized); ◇, myocardial nuclei (hypophysectomized plus aldosterone); □, myocardial nuclei (hypophysectomized plus cortisol)

Fig. 9. The effect of adrenalectomy and cortisol-treatment of adrenalectomized rats on myocardial nuclear poly(ADPR) synthetase activity. Hydrocortisone sodium phosphate (Hydrocortone, MSD; 5 mg/100 g body weight) was injected intraperitoneally into adrenalectomized Long Evans rats daily for 4 days, prior to the isolation of nuclei; a control group consisting of normal rats and adrenalectomized rats received sterile isotonic saline. Poly(ADPR) synthetase activity was monitored as described in Fig. 7. •, Myocardial nuclei (control); ▪, myocardial nuclei (adrenalectomized); □, myocardial nuclei (adrenalectomized plus cortisol)

mal enzymatic activity in myocardial nuclei, but cortisone in large doses or aldosterone in near physiological doses is inhibitory. The identification of molecular mechanisms for these phenomena depends on detailed analytical and enzymological work, that is presently pursued. However, the organ-specific age and endocrine effects strongly support the view that poly(ADPR) metabolism has direct relevance to differentiated organ functions in higher animals.

The organ-specific in vivo inhibitory effect of cortisol and especially of aldosterone that is demonstrable in normal hypophysectomized and adrenalectomized rats suggested a direct regulatory influence of steroids on nuclear poly(ADPR) metabolism. If this is a direct effect, the inhibitory influence should be reproducible in vitro, which was indeed the case. Preincubation of isolated myocardial nuclei with 5.96 nM aldosterone for 30 min at 0° inhibits poly(ADPR) synthesis as treatment of rats in vivo with a daily dose of 5 μg aldosterone, indicating that a true organ-specific inhibitory effect of aldosterone is observed both in vivo and in vitro. These results are shown in Table 3. It is unlikely that aldosterone actually acts directly as an inhibitor on the poly(ADPR) polymerase

Table 3. Inhibitory effect of aldosterone in vivo and in vitro on the poly(ADPR) synthetase activity of myocardial nuclei

	Poly(ADPR) Synthetase Activity (nmol/mg protein/min)	% Inhibition
In vivo		
Control	8.8	
Aldosterone (5 μg/100 g rat)	5.6	36%
In vitro		
Control	5.6	
Aldosterone (5.9 nM)	3.1	45%

In vivo effect. Aldosterone was injected (5 μg/100 g body weight) intraperitoneally into 30 day-old Long Evans rats for 4 days, prior to the isolation of nuclei. Control received isotonic saline at the same time. Poly(ADPR) synthetase activity was determined as described in Fig. 7. In all the poly-(ADPR) synthetase assay the content of nuclear DNA per assay was the same.
In vitro effect of aldosterone on myocardial poly(ADPR) synthetase activity. Myocardial nuclei from 30 day-old Long Evans male rats were preincubated at 0° for 30 min in the presence or absence of 5.9 nM aldosterone. The system contained 100 mM Tris-HCl pH 8.2, 2 mM DTT, 20 mM $MgCl_2$, 0.1 mM PMSF, 0.5 mM EDTA. At the end of preincubation poly(ADPR) synthesis was initiated by the addition of labeled NAD^+ (0.5 mM) and the synthesis carried out at 25°C as described in legend of Fig. 7

enzyme. Our present working hypothesis predicts a specific interaction between aldosterone and a regulatory nonhistone protein that participates in the control of the poly-(ADPR) polymerase system. It is interesting that aldosterone given in vivo increases cardiac RNA-polymerase (Liew et al. 1972). We presume that the poly(ADPR) polymerase-nonhistone protein modification system plays an intermediary role in the control of organ-specific transcription.

6 Summary: Outline of Regulatory Sites and Their Probable Function

A deterministic role of specific nonhistone proteins in the regulation of selective transcription of DNA segments by RNA-polymerases is a plausible hypothesis that may explain the function of the large array of nonhistone chromatin proteins in eukaryotes. The study of poly ADP-ribosylation affords a selective probe (Minaga et al. 1979) in this field. The modifying reagent NAD^+ is the only nucleotide which possesses rapid chemical responsiveness toward microenvironmental changes (i.e., oxidation/reduction). Stress-induced reduction of NAD^+ will necessarily diminish protein ADP-ribosylation, as we have observed it in the hypertrophic part of the cardiac muscle that exhibits significant diminution of poly(ADPR) content as compared with the normal part of the same cardiac muscle (Minaga and Kun unpublished experiments). This regulatory effect of decreased poly ADP-ribosylation, which is presumably equivalent to a form of physiologic de-repression, coincides with an increased synthesis of all cellular macromolecules, (except DNA) resulting in hypertrophy. Much greater complexity exists when the effects of a carcinogen or of hormones are being followed. Because of the large number of nonhistone chromatin proteins, each potentially regulating a discrete transcription of DNA segments, no uniform relationship between the regulatory effects of individual nonhistone proteins and their degree of ADP-ribosylation is likely. It is possible that activating effects of certain nonhistone proteins (on transcription) may depend on ADP-ribosylation, whereas with other proteins ADP-ribosylation is inhibitory. It follows that it is necessary to determine ADP-ribosylations of all nonhistone proteins and establish an organ-specific pattern related to function before meaningful experimental questions can be formulated.

Our present information regarding the role of ADP-ribosylations in sterol hormone action suggests a possible connecting link between the sterol-specific nuclear receptors (Jensen 1979) and the selective activation of RNA-polymerases. Our results with aldosterone in myocardial nuclei projects a feasible experimental approach.

Acknowledgements. This work was supported by research grants of the American Cancer Society (BC-304; RD-89) and the United States Air Force Office of Scientific Research (AFOSR 3698). Ernest Kun is the recipient of the Research Career Award of the United States Public Health Service, Alexander D. Romaschin of a Fellowship of the National Science Research Council (Canada), and George Jackowski of a Postdoctoral Scholarship of the Canadian Heart Association.

References

Bradehorst R, Weickens K, Gartemann A, Lengyel H, Klapproth K, Hilz H (1978) Two different types of bonds linking single ADP-ribose residues covalently to proteins: Quantification in eukaryotic cells. Eur J Biochem 92:129-135

Bradehorst R, Goebel M, Renzi F, Kittler M, Klapproth K, Hilz H (1979) Intrinsic ADP-ribose transferase activity versus levels of mono(ADP-ribose) protein conjugates in proliferating Ehrlich ascites tumor cells. Hoppe Seyler's Z Physiol Chem 360:1737-1743

Claycomb WC (1976) Poly(adenosine diphosphate-ribose) polymerase activity and nicotinamide adenine dinucleotide in differentiating cardiac muscle. Biochem J 154:387-393

Dipaolo JA (1977) Chromosomal alterations in carcinogen transformed mammalian cells. In: Ts'o POP (ed) The molecular biology of the mammalian genetic apparatus, vol 2, ch 18. North Holland Publ Co, Amsterdam New York Oxford, pp 205-227

Durkacz BM, Omidiji O, Gray DA, Shall S (1980) $(\text{ADP-ribose})_n$ participates in DNA excision repair. Nature (London) 283:593-596

Ferro AM, Kun E (1976) Macromolecular derivatives of NAD^+ in heart nuclei: Poly adenosinediphospho-ribose and adenosine diphosphoribose-proteins. Biochem Biophys Res Commun 71:150-154

Ferro AM, Minaga T, Piper WN, Kun E (1978) Analysis of larger than tetrameric poly(adenosine diphosphoribose) by a radioimmunoassay in nuclei separated in organic solvents. Biochim Biophys Acta 519:291-305

Giri CP, West MHP, Smulson M (1978) Nuclear protein modification and chromatin substructure I. Differential poly(adenosine diphosphate) ribosylation of chromosomal proteins in nuclei versus isolated nucleosomes. Biochemistry 71:3495-3500

Herrold KM (1967) Histogenesis of malignant liver tumors induced by dimethylnitrosamine. An experimental study in Syrian hamsters. J Natl Cancer Inst 39:1099-1111

Jackowski G, Liew CC (1980) Fractionation of rat ventricular nuclei. Biochem J 188:363-373

Jensen VE (1979) Interaction of steroid hormones with the nucleus. Pharmacol Rev 30:477-491

Jump DB, Butt TR, Smulson M (1979) Nuclear protein modification and chromatin substructure 3. Relationship between poly(adenosine diphosphate) ribosylation and different functional forms of chromatin. Biochemistry 18:983-990

Kawaichi M, Ueda K, Hayaishi O (1980) Initiation of poly(ADP-ribosyl) histone synthesis by poly-(ADP-ribose) synthetase. J Biol Chem 255:816-819
Kun E, Chang ACY (1976) Protein-ADP-ribosylating system of mitochondria. In: Shaltiel S (ed) Metabolic interconversion of enzymes 1975. Springer, Berlin Heidelberg New York, pp 156-160
Kun E, Romaschin AD, Blaisdell RJ (in press) Subnuclear localization poly ADP-ribosylated proteins. In: Sugimura T, Smulson M (eds) Conference on novel ADP-ribosylations of regulatory enzymes and proteins. John Fogarthy International Center for Advanced Study in Health Sciences, National Institute of Health
Lazarides E, Lindberg V (1974) Actin is the naturally occurring inhibitor of deoxyribonuclease I. Proc Natl Acad Sci USA 71:4742-4746
Leavitt J, Kakunaga T (1980) Expression of a variant form of actin and additional polypeptide changes following chemical-induces in vitro neoplastic transformation of human fibroblasts. J Biol Chem 225:1651-1661
Liew CC, Liu DK, Gornall AG (1972) Effects of aldosterone on RNA polymerase in rat heart and kidney nuclei. Endocrinology 90:488-495
Minaga T, Romaschin AD, Kirsten E, Kun E (1979) The in vivo distribution of immunoreactive larger than tetrameric polyadenosine diphosphoribose in histone and non-histone protein fractions of rat liver. J Biol Chem 254:9663-9668
Mullins DW, Giri CP, Smulson M (1977) Poly(ADP-ribose) polymerase: The distribution of a chromosome-associated enzyme within the chromatin substructure. Biochemistry 16:506-513
Paul J, Gilmour S (1968) Organ-specific restriction of transcription in mammalian chromatin. J Mol Biol 34:305-316
Romaschin AD, Kirsten E, Jackowski G, Kun E (1981) Increased adenosine-diphosphoribosylation in vivo of nuclear proteins in precancerous liver. J Biol Chem (in press)
Yukioka M, Okai Y, Hamada T, Inoue H (1978) Non-preferential localization of poly(ADP-ribose) polymerase activity in transcriptionally active chromatin. FEBS Lett 86:85-88
Zak R, Rabinowitz M (1979) Molecular aspect of cardiac hypertrophy. Annu Rev Physiol 41:539-552

Discussion[2]

The possible regulatory role of ADP ribosylation is not clear since different results are obtained depending on whether nuclei from embryonic or adult cells are used.

Tubulin does not appear to be ADP ribosylated.

The isolation of nuclei does not seem to introduce artifacts in the experiments with hormones, since the increase in activity observed in isolated nuclei correlates with an increase of ADP ribosylation in vivo. The experiments have been done with aldosterone on cardiocytes, the myocardial nuclei being separated by the technique of Jackowski and Liew (1980).

The extraction of nuclei with citric acid at pH 2.5 appears to be sufficient to stabilize the state of ADP ribosylation of proteins, so that no addition of inhibitors of ribosylation-deribosylation reactions is necessary.

2 Rapporteur: Carlos Gancedo

ADP-ribosylation of Ribonucleases

E. Leone, B. Farina, M.R. Faraone Mennella, and A. Mauro[1]

1 Introduction

Covalent modification of enzymes can be effected by ADP-ribosylation reactions. Examples in this regard are represented, in prokaryotes, by elongation factor 2 (EF-2) of protein synthesis, a peptidyl transferase that is inhibited when acted upon in the presence of NAD^+ by diphteria toxin, which is endowed with pApase activity (Honjo et al. 1971); as a result, protein synthesis is blocked. A similar mechanism operates in the infection by *Pseudomonas aeruginosa*, the toxin of which also promotes ADP-ribosylation of EF-2 (Iglewski and Kabat 1975). Also T-4 phage-infected *E. coli* RNA-polymerase is modified, and thereby inhibited, by ADP-ribosylation at arginine residues of the α subunits (Goff 1974). A modification of the adenylate cyclase system has been recently found in a number of instances; it would appear that ADP-ribosylation of the GTP-binding protein component of that system leads to reduction of GTP hydrolysis, and thereby to enhanced adenylate cyclase activity (Gill et al. 1979). In eukaryotes, a Ca^{++}, Mg^{++}-dependent endonuclease of rat liver nuclei has been reported by Yoshihara et al. (1975) to be ADP-ribosylated by an enzyme preparation from the same tissue.

We have directed our attention to the ADP-ribosylation of ribonucleases. We have previously found a high pApase activity to be a characteristic feature of mammalian testes, bulls showing the highest level (Farina et al. 1979). We have used this enzyme to investigate whether seminal RNAase, an alkaline RNAase isolated from bull seminal plasma (Leone and D'Alessio 1977), as well as pancreatic RNAase A, could be modified by ADP-ribosylation. The present report deals with the results we have obtained so far.

2 Materials and Methods

Pancreatic RNAase A (Sigma type XII A), venom phosphodiesterase (Boehringer), and alkaline phosphatase from calf intestine (Sigma type XX), were commercial products. Bull semen RNAase was prepared according to D'Alessio et al. (1972); samples of pancreatic RNAases of mouse, rat, and hamster were kindly provided by Dr J. Beintema, Biochemistry Laboratory, University of Groningen, The Netherlands. 5'-UTP-Sepharose was from Sigma, SP-Sephadex C-50 from Pharmacia, cellulose acetate strips (Cellogel) from Pratiga, Milano. Dowex 1-X2 resin was from Bio-Rad.

Abbreviations: RNA, ribonucleic acid. RNAase, ribonuclease. ADPR, adenosine diphosphate ribose. ADPR-RNAase, ADP-ribosylated RNAase. pApase, poly(ADPR)polymerase. DTT, dithiothreitol.

1 Istituto di Chimica Organica e Biologica, Facoltà di Scienze, Università di Napoli, Italy

Yeast RNA (Sigma) was purified prior to RNAase assays according to Shortman (1951); cyclic 2',3'-CMP and NAD^+ were from Sigma; U-[^{14}C]NAD^+, specific activity 266 mC/mmol, was a product of the Radiochemical Centre, Amersham, England.

Protein concentration was determined according to Lowry et al. (1951). Differential spectra were determined with a Cary mod. 118 C spectrophotometer. Cellulose acetate electrophoresis was performed at pH 2.2 at 220 V and a supply of 5 mA per strip, for 40 min at room temperature.

RNAase activity was determined by the spectrophotometric procedure of Kunitz (1946) with yeast RNA as a substrate; in some instances also the spectrophotometric method of Crook et al. (1960), using cyclic 2',3'-CMP as a substrate, was employed. pApase activity was determined by the procedure of Farina et al. (1979), which is based on the measure of [^{14}C] incorporation into TCA-insoluble fractions of incubation mixtures with [^{14}C]-NAD; radioactivity was measured with a Beckman LS-133 liquid scintillation counter.

pApase was purified from bull testes according to Farina et al. (1979) up to the 1 M NaCl extract, with specific activities ranging from 3.5 to 5.0. Rat liver pApase was prepared essentially by the same procedure.

ADPR-RNAase isolation from incubation mixtures varied depending on whether it started from RNAases added to purified pApase or from isolated nuclei of bull seminal vesicles. In the former instance, RNAase, from 35 to 45 nmol, was added to an incubation mixture composed, in a total volume of 10 ml, of [^{14}C]-NAD^+, 3.7 nmol (2.372×10^6 cpm), Tris-HCl pH 8.0 900 nmol, NaF 50 nmol, DTT 20 nmol, plus pApase, 40 mUnits (about 10 mg protein). After 30 min at 25°C the incubate was dialyzed against H_2O for 12 h and lyophilized. The dry preparation, dissolved in 200 μl of 0.025 M piperazine-HCl buffer pH 5.3, was taken for affinity chromatography to a column, equilibrated with the same buffer, of UTP-Sepharose (Smith et al. 1978).

In the latter instance, seminal vesicles nuclei were incubated in the same conditions reported above. Nuclei from 50 g bull seminal vesicles, isolated according to Utakoji et al. (1968), were suspended in 4 ml 20 mM Tris-HCl buffer pH 8.0 containing 1 mM $MgCl_2$, $ZnCl_2$ and $CaCl_2$. After a number of washings (five to six), until no RNAase activity was present in the supernatant, they were incubated, in a final volume of 50 ml, with 4.5 μmol Tris-HCl pH 8.0, 100 nmol DTT, 250 nmol NaF, 18 μmol NAD^+ and 1.95 nmol of [^{14}C]-NAD^+ (1.25×10^6 cpm). After 30 min at 25°C the mixture was homogenized with a Potter glass homogenizer with Teflon pester and centrifuged at 4400 g for 10 min. The supernatant, dialyzed against H_2O overnight, was 60% saturated with ammonium sulfate; after standing 12 h at 4°C the supernatant from centrifugation at 24,000 g was dialyzed and lyophilized. The lyophilized preparation, dissolved in 0.005 M Tris-HCl buffer pH 8.0 containing 0.1 M NaCl, was taken to a column of SP-Sephadex C-50 (cm 1.6×17) equilibrated with the same solution. After initial elution by the 0.1 M NaCl, NaCl concentration was raised to 0.2 M and then to 0.5 M. The fraction eluted by 0.5 M NaCl, containing ADPR-RNAase, after dialysis and lyophilization, was taken to the UTP-Sepharose column, from which elution, as formerly described, was first effected by the piperazine-HCl buffer, which in this case gave at the breakthrough a fair amount of labeled protein, which was devoid of RNAase activity; on changing to 0.25 M Na phosphate buffer pH 5.45, ADPR-RNAase was eluted.

3 Results

3.1 In Vitro ADP-ribosylation of RNAases

Incubation of pApase with alkaline RNAases leads to inhibition of the nucleases, while pApase appears to be activated on the basis of increased incorporation of radioactivity from [^{14}C]-NAD$^+$ into the acid-insoluble fraction of incubation mixtures. Figure 1 shows the results obtained on incubation of seminal RNAase (a dimeric enzyme with the monomer primary structure closely resembling RNAase A) with a partially purified prepara-

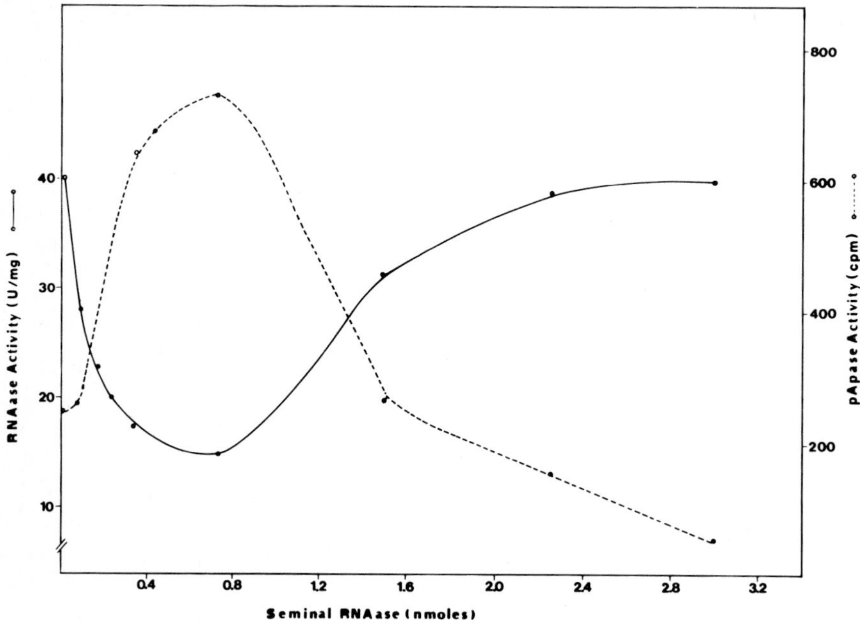

Fig. 1. RNAase and pApase activities on incubation of both enzymes in the presence of NAD$^+$ with increasing amounts of RNAases. RNAase activity (——) is expressed as specific activity (Kunitz Units/mg protein), pApase activity (– – –) as counts per whole incubation mixture. Each mixture contains in a final volume of 125 μl, bull testis pApase, 0.06 mUnits (12 μg), [^{14}C]-NAD$^+$, 39 pmol (25,000 cpm), 90 mM Tris-HCl buffer pH 8.0, 5 mM NaF, 2 mM DTT, and seminal RNAase as indicated. Incubation at 25°C for 5 min

tion of pApase from bull testis in the presence of NAD$^+$; it can be seen that as the amount of RNAase is increased, with pApase held at a constant level, the nuclease activity is inhibited, maximum inhibition being effected by 0.73 nmol of RNAase; on further increase of RNAase there is a gradual return to the initial level. Conversely, incorporation of [^{14}C]-NAD$^+$ into the acid-insoluble fractions of incubate shows an increase, which reaches a peak at the same RNAase concentration where maximum RNAase inhibition occurs, and is then followed by a steady decline.

Essentially the same results are obtained with pancreatic RNAase A, as well as with pancreatic RNAases from other species like mouse, rat, hamster.

Table 1 gives the results obtained on incubation of bull testis pApase with seminal RNAase and with RNAase A. It is seen that seminal RNAase is almost as effectively inhibited as RNAase A, and that heat-denatured seminal RNAase is still capable of acting as ADPR acceptor. Experiments with rat liver pApase, not shown in Table 1, have demonstrated that also this enzyme is able to carry out ADP-ribosylation of RNAases.

Table 1. pApase activation and RNAase inhibition following pApase incubation with [^{14}C]-NAD$^+$ in the presence of RNAases [a]

		pApase		RNAase
		activity	activation %	inhibition %
pApase		245		
"	plus sem. RNAase 0.35 nmol	644	162	56
pApase		245		
"	plus sem. RNAase 0.73 nmol	728	197	67
pApase		155		
"	plus denatured sem. RNAase 0.44 nmol	269	73	
pApase		288		
"	plus RNAase A 0.34 nmol	415	44	48

[a] pApase utilized in these experiments was a partially purified preparation from bull testis, spec. activ. 5 mU/mg. pApase activity is expressed as cpm/125 μl of incubation mixtures. Incubation mixtures composition, final volume, and experimental conditions as in the legend to Fig. 1

Determination of pApase and RNAase activities at different intervals of incubation gives the time course curves reported in Fig. 2. While seminal RNAase shows a steady decline to minimal values, which are reached after 30 min, pApase activity increases, with a definite larger increase for the incubation mixture containing RNAase; maximum values are attained after 20-30 min. The slight decrease which follows is probably due to poly(ADPR) degradation, either spontaneous or by hydrolyzing enzymes still present as contaminants in the pApase preparation.

In order to isolate and characterize ADPR-RNAase, a purification procedure was set up (described in Sect. 2) in which, starting from an incubation mixture made up of bull testis pApase, 40 mUnits (10 mg), [^{14}C]-NAD$^+$, 3.7 nmol (2.372 $\times$ 10^6 cpm) and 38.5 nmol of seminal RNAase, seminal ADPR-RNAase was isolated by affinity chromatography on UTP-Sepharose (Fig. 3). The protein so obtained proved to be homogenous on cellulose acetate electrophoresis, where it showed a decreased mobility, relative to the nonmodified enzyme (Fig. 4). The differential UV spectrum showed a peak of maximum absorption at 260 nm, characteristic of ADPR (Fig. 5).

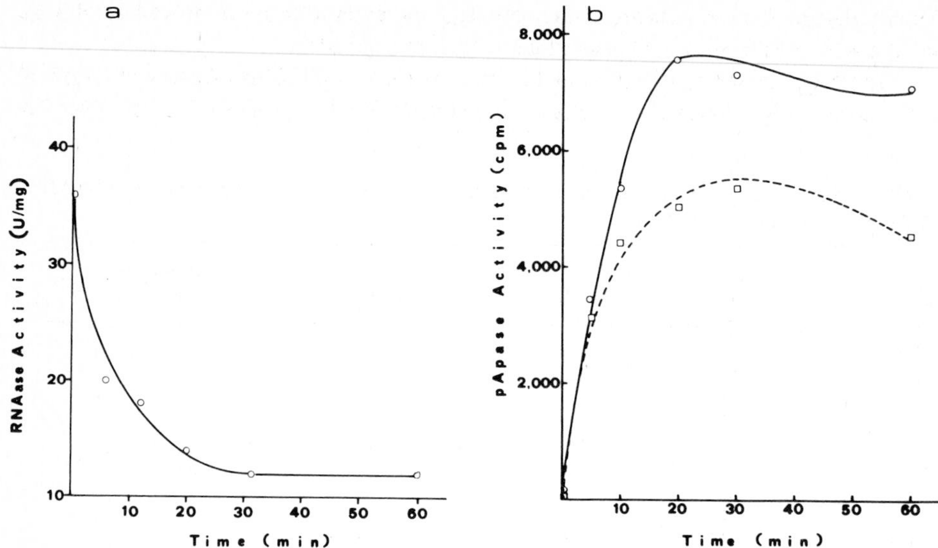

Fig. 2a, b. Time course of RNAase and pApase activities on incubation with NAD^+. In **a** RNAase is reported as specific activity; in **b** pApase activity without (– – –) and with (——) added seminal RNAase, 0.73 nmol. Incubation mixtures as in the legend to Fig. 1. Incubation at 25°C for 5 min

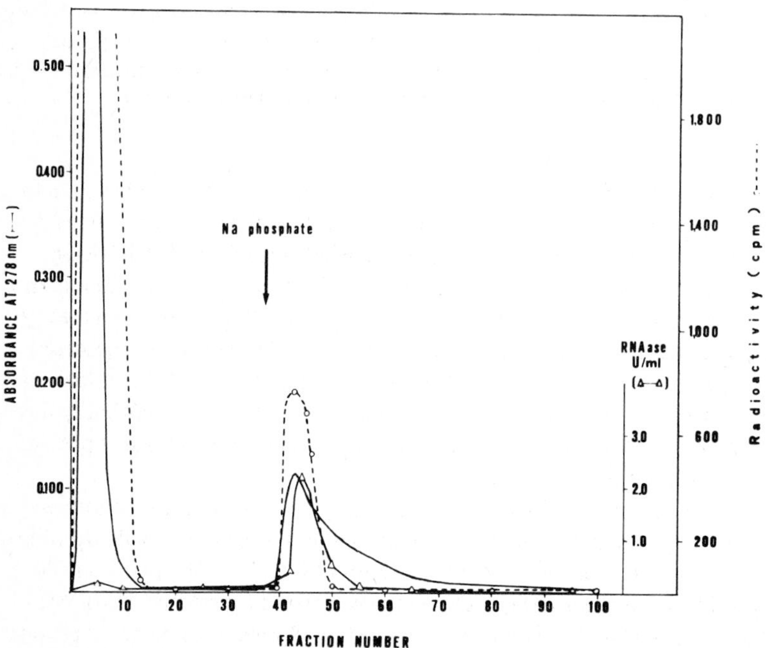

Fig. 3. Affinity chromatography of seminal ADPR-RNAase. Incubation mixture as described in Sect. 2, with bull testis pApase and 38.5 nmol of seminal RNAase. Elution from the UTP-Sepharose column (cm 0.5 × 1.5) by 0.025 M piperazine-HCl buffer pH 5.3, substituted at *arrow* (fraction 38) by 0.25 M Na phosphate buffer pH 5.45. Fraction volume 0.6 ml. RNAase activity (U/ml) and radioactivity refer to levels per fraction

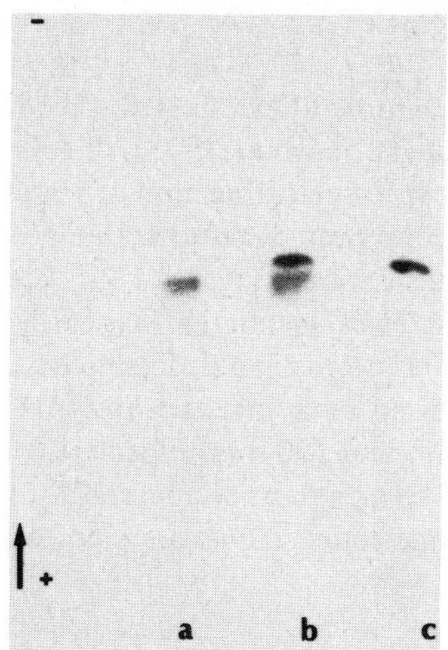

Fig. 4a-c. Cellulose acetate electrophoresis of ADPR-RNAase. **a**, seminal ADPR-RNAase; **b**, mixture of seminal RNAases, before and after ADP-ribosylation; **c**, seminal RNAase. Proteins were deposited at the starting line (*at the bottom*, in coincidence with the *+ sign*) in amounts of 0.5-2.5 μg each. Electrophoresis (pH 2.2, 200 V, 5 mA/strip) run for 40 min

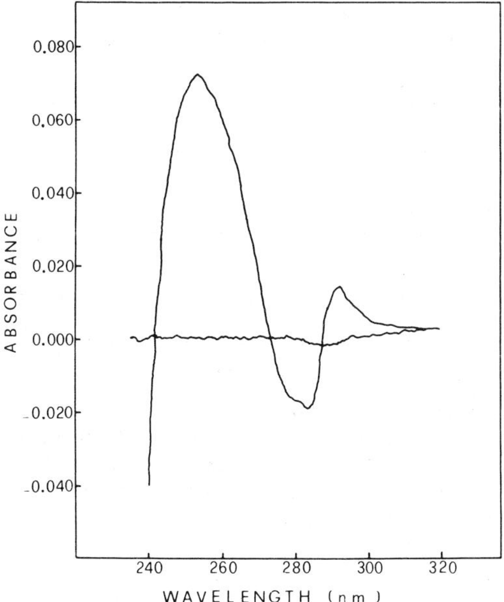

Fig. 5. UV differential spectrum of ADPR-RNAase

When ADPR-RNAase was made alkaline to pH 10.3 and left at 0°C for 17 h, an increase of RNAase activity was observed, which corresponded to 50% reactivation of the inhibited enzyme. It is known that at alkaline pH the ADPR residues are split off the acceptor protein molecules (Nishizuka et al. 1969).

Identification of breakdown products after snake venom phosphodiesterase digestion of ADPR-RNAase lead to the identification of 2'(5"-phosphoribosyl)-5'-AMP (*iso*ADPR) and 5'-AMP, further confirmed by identification of products from alkaline phosphatase digestion (Sugimura 1973), as expected for ADPR structure; furthermore a polymer length of two to four ADPR residues could be calculated.

3.2 ADP-ribosylation of Endogenous RNAase in Isolated Nuclei

From bull seminal vesicles nuclei were obtained by the procedure of Utakoji et al. (1968). Since these glands are the major source of seminal RNAase (Farina et al. 1973), the cytoplasm fraction, represented by the supernatant from low-speed centrifugation of whole tissue homogenate, had a high content of the enzyme. After repeated centrifugations until no more RNAase was present in the washings, the nuclei were assayed for RNAase and pApase activities. Since both were found at significant levels, incubation was carried out of such nuclei in order to ascertain whether, at the levels normally existing in vivo of the two enzymes, ADP-ribosylation of RNAase would occur in the presence of NAD^+. In fact, this was shown by the following results: after incubation (30 min) of isolated nuclei with NAD^+, the endogenous RNAase was inhibited to an extent of about 22% of the original level. This appears from Fig. 6 in which the time course curves are reported

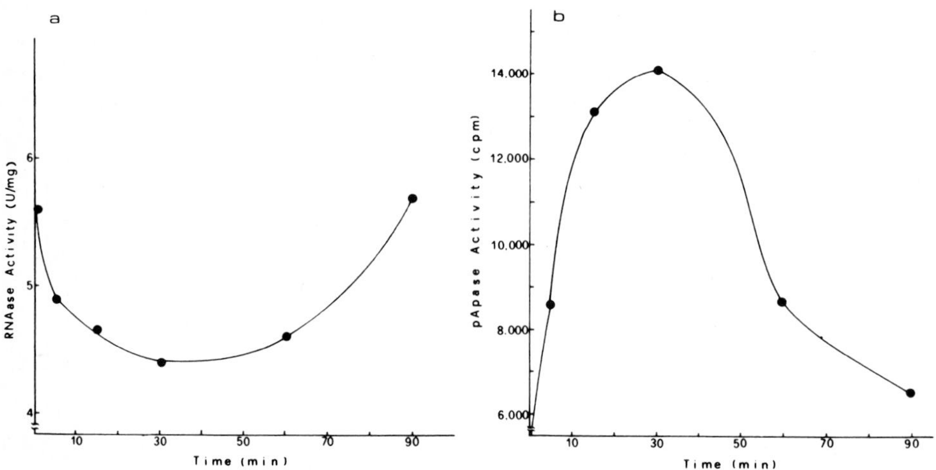

Fig. 6a, b. Time course of RNAase and pApase activities on incubation of isolated nuclei from bull seminal vesicles with [^{14}C]-NAD^+. In **a**, RNAase (spec. activity); in **b**, pApase activity. Isolated nuclei (0.31 mg) were incubated in a final volume of 2.0 ml with [^{14}C]-NAD^+, Tris-HCl buffer, NaF and DTT being present in the same proportions as in the legend to Fig. 1. At the given times, 5 μl aliquots were taken for RNAase, and 125 μl for pApase determination

for both pApase and RNAase activities; it can be seen how, unlike the curves of Fig. 2, RNAase, after reaching a maximum inhibition at 30 min, rises back to the starting level after 80 min, and also pApase after maximum activation at 30 min falls almost to initial levels.

Isolation of the endogenous RNAase after incubation with NAD^+ yielded a homogenous protein, which could be identified as ADPR-RNAase by the same line of evidence found for the experiments previously reported, i.e., electrophoretic mobility lower than native, non-ADP-ribosylated RNAase, recovery of RNAase activity after alkali treatment of ADPR-RNAase, UV differential spectrum with maximum absorption at 260 nm, proof of ADPR structure by analysis of phosphodiesterase and phosphatase digests.

4 Discussion

Alkaline RNAases, like pancreatic RNAase A and seminal RNAase, appear to act as efficient ADPR-acceptor proteins by the experiments just reported; more important, ADP-ribosylation leads to partial inactivation of the two enzymes. Their similarity to histones for such properties as isoelectric point and molecular size makes these effects more understandable, since histones are considered to be among the most suitable substrates, in the field of nuclear proteins, for ADPR-ribosylation (Hayaishi and Ueda 1977).

Inhibition of enzymic activity in ADP-ribosylated RNAases is of special interest, since it may have a specific physiological meaning. Nuclear RNAases have been implicated in the sizing of rRNA (Sierakowska and Shugar 1977), and in this context the possibility of modulation of such, or closely related, kind of activities by ADP-ribosylation must be considered. The finding that ADP-ribosylation can occur also in isolated nuclei of bull seminal vesicles, where endogenous pApase and RNAase interact in the presence of NAD^+ to give ADPR-RNAase, gives further support to such a possibility.

Acknowledgment. This work has been done with the financial help of Consiglio Nazionale delle Ricerche, C.N.R., grant 79.01159.85 of Progetto Finalizzato della Biologia della Riproduzione.

References

Crook EM, Mathias AP, Rabin BR (1960) Spectrophotometric assay of bovine pancreatic ribonuclease by the use of cyclic 2',3'CMP. Biochem J 74:234-238

D'Alessio G, Floridi A, De Prisco R, Pignero A, Leone E (1972) Bull semen ribonucleases. 1. Purification andphysico-chemical properties of the major component. Eur J Biochem 26:153-161

Farina B, Quesada P, Carsana A, Parente A, De Prisco R, Leone E (1973) Ribonucleasi di vescichette seminali di toro. Boll Soc Ital Biol Sper 49:1445-1449

Farina B, Faraone Mennella MR, Leone E (1979) Nucleic acids, histones and spermiogenesis: the poly-(adenosine diphosphate ribose)polymerase system. In: Salvatore F, Marino G, Volpe P (eds) Macromolecules in the functioning cell. Plenum Press, New York, p 283

Gill DM, Enomoto K, Meren R (1979) ADP-ribosylation of membrane proteins catalysed by cholera toxin and *E. coli* enterotoxin. In: Fogarty Int Center Conf on novel ADP-ribosylations of regulatory enzymes and proteins. Washington DC:GPO

Goff CG (1974) Chemical structure of a modification of the *Escherichia coli* ribonucleic acid polymerase α polypeptides induced by bacteriophage T_4 infection. J Biol Chem 249:6181-6190

Hayaishi O, Ueda K (1977) Poly(ADP-ribose) and ADP-ribosylation of proteins. Annu Rev Biochem 46:95-116

Honjo T, Nishizuka Y, Kato I, Hayaishi O (1971) Adenosine diphosphate ribosylation of aminoacyl transferase II and inhibition of protein synthesis by diphteria toxin. J Biol Chem 246:4251-4260

Iglewski BH, Kabat D (1975) NAD-dependent inhibition of protein synthesis by *Pseudomonas aeruginosa* toxin. Proc Natl Acad Sci USA 72:2284-2288

Kunitz M (1946) A spectrophotometric method for the measurement of ribonuclease activity. J Biol Chem 164:563-568

Leone E, D'Alessio G (1977) Structure and properties of seminal ribonuclease. Biochem Soc Trans 5:466-470

Lowry OH, Rosebrough NJ, Farr AL, Randall RJ (1951) Protein measurement with the Folin phenol reagent. J Biol Chem 193:265-275

Nishizuka Y, Ueda K, Yoshihara K, Yamamura H, Takeda M, Hayaishi O (1969) Enzymic adenosine diphosphoribosylation of nuclear proteins. Cold Spring Harbor Symp Quant Biol 34:781-786

Shortman K (1951) Studies on cellular inhibitors of ribonuclease. 1. The assay of the ribonuclease-inhibitor system and the purification of the inhibitor from rat liver. Biochim Biophys Acta 51: 37-49

Sierakowska H, Shugar D (1977) Mammalian nucleolytic enzymes. Prog Nucleic Acid Res Mol Biol 20:60-130

Smith GK, Schray KJ, Schaffer W (1978) Use of 5'-UTP-Agarose for ribonuclease affinity chromatography. Anal Biochem 84:406-414

Sugimura T (1973) Poly(adenosine diphosphate ribose). Prog Nucleic Acid Res Mol Biol 13:127-151

Utakoji T, Muramatsu M, Sugano H (1968) Isolation of pachytene nuclei from the Syrian hamster testes. Exp Cell Res 53:447-458

Yoshihara K, Tanigawa Y, Burzio L, Koide SS (1975) Evidence for adenosine diphosphate ribosylation of Ca^{2+}, Mg^{2+}-dependent endonuclease. Proc Natl Acad Sci USA 72:289-293

Discussion[2]

The RNAse found in the nuclei is unlikely to be a contaminant due to the high activity found and the extensive washings used in the preparation. This RNAse is not found in the form of RNAse inhibitor complexes.

2 Rapporteur: Carlos Gancedo

Supramolecular Organization of Regulatory Proteins into *Calcisomes:* A Model of the Concerted Regulation by Calcium Ions and Cyclic Adenosine 3':5'-Monophosphate in Eukaryotic Cells

J. Haiech[1] and J.G. Demaille[2]

1 Introduction

Living organisms are capable of maintaining homeostasis or steady state in spite of extreme variations of their environment. To this purpose cells must receive, code, and analyze informations carried by molecules called messengers. The number of intercellular messengers or first messengers, such as hormones, neuromediators or neuropeptides, increases with the complexity of the organism. In contrast, very few intracellular or second messengers are found in all eukaryotic cells. Among them calcium ions and cyclic nucleotides are the most extensively studied (Robison et al. 1971; Rasmussen 1970). Each of them exhibits a common scheme of action in all eukaryotic species. Individual cells and tissues respond however differently to an increase in the concentration of a given second messenger as a consequence of various enzyme equipments resulting from differentiation.

Both second messengers do not act independently but rather in a concerted manner, by which Ca^{2+} ions and cAMP modulate the level of each other and exert their effects on the same metabolic pathways (Rasmussen and Goodman 1977; Fischer et al. 1976). The aim of this paper is the description of a general scheme of action of the second messengers, with special emphasis on the calcium pathway and on the supramolecular organization of regulatory proteins into *calcisomes.*

2 The Generator-Sensor-Suppressor Cycle

Upon external stimulation (e.g., hormone or nerve impulse) of a cell, a transient increase in the concentration of a given compound occurs. When a mechanism is present within the cell that transforms this transient increase into a physiological event, such a compound is called a second messenger. The following elements must therefore be present:

1. a *generator* of second messenger coupled to an external signal;
2. a *sensor* that will bind the second messenger reversibly upon its transient increase and thereby trigger a physiological response, such as glycogenolysis or contraction;
3. a *suppressor* which allows return to steady state (or resting state).

This general scheme is illustrated in Fig. 1 and may be abbreviated as GSS cycle.

---

1 Present address: Laboratory of Biochemistry, NCI, NIH, Bethesda, MA 20014, USA

2 Faculté de médecine, FRA INSERM "Biochimie des régulations" et Centre de Recherche de Biochimie Macromoléculaire du CNRS, BP 5051, 34033-Montpellier, France

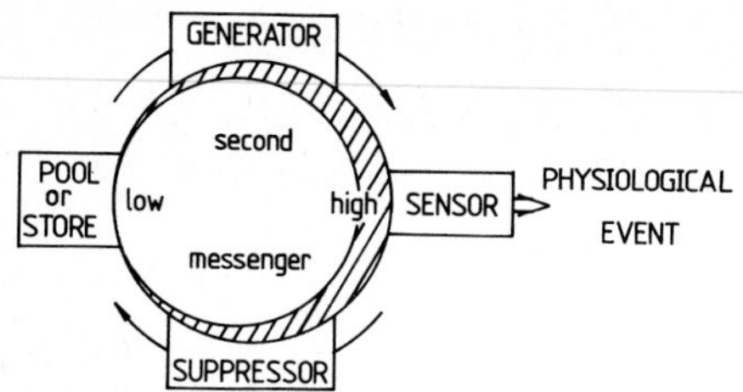

Fig. 1. The generator-sensor-suppressor cycle of second messengers.

Examples of	Are:
Generator:	Adenylate cyclase, Ca^{2+} channels;
Sensor:	Regulatory subunit of cAMP-dependent protein-kinase, calmodulin, troponin C;
Physiological event:	Glycogenolysis, contraction, secretion;
Suppressor:	cAMP-phosphodiesterase, Ca^{2+} pumps, parvalbumins;
Pool or store:	Nucleotide pool, Ca^{2+} stores (extracellular, sarcoplasmic reticulum)

3 The GSS Cycle of Calcium Ions

Calcium ions are stored either in the extracellular compartment (pCa ~ 3) or within intracellular sequestrating devices such as sarcoplasmic reticulum vesicles (up to pCa 2). Upon stimulation of, e.g., heart cells, free Ca^{2+} concentration increases from 0.26 μM (quiescent ventricle) to as high as 10 μM during contractures (Marban et al. 1980).

The generator is a calcium channel through membranes, that is coupled to external stimuli, e.g., membrane depolarization. The sensor may be different in various cell types, for instance troponin C in sarcomeric muscles vs calmodulin in smooth muscle and non-muscle cells (Perry 1979). In all cases, the sensor belongs to the family of intracellular calcium-binding proteins which all derive from a common ancestor, as shown by the evolutionary tree depicted in Fig. 2. Among them, calmodulin, the ubiquitous and multifunctional Ca^{2+}-dependent regulator (see Cheung 1980; Means and Dedman 1980 for reviews) is the closest to the four-domain ancestor (Goodman et al. 1979; Demaille et al. 1980). It meets the criteria of a sensor: presence of at least one Ca^{2+}-specific binding site (Goodman et al. 1979) with K_d in the range of the Ca^{2+} transient, Ca^{2+}-induced transconformation, and propagation of this information to another interacting macromolecule.

The ultimate suppressor is made of Ca^{2+}-activated Mg^{2+}-ATPases that extrude Ca^{2+} ions from the cytosol to internal or external stores at the expense of ATP hydrolysis. When faster relaxation than the one provided by the pumps alone is required, fast tissues (fast twitch muscle or nervous tissue) contain soluble suppressors such as parvalbumins which exhibit high affinity Ca^{2+}-Mg^{2+}-binding sites or relaxing sites (Haiech et al. 1979). These sites will bind Ca^{2+} ions with high affinity after the kinetic delay introduced by the dissociation of Mg^{2+} ions (Birdsall et al. 1979). This brief description of the Ca^{2+} cycle raises two major questions. First, the coupling mechanism of the external stimulus to the Ca^{2+} channel opening is as yet unknown. Second, the mechanism of Ca^{2+} sensor/

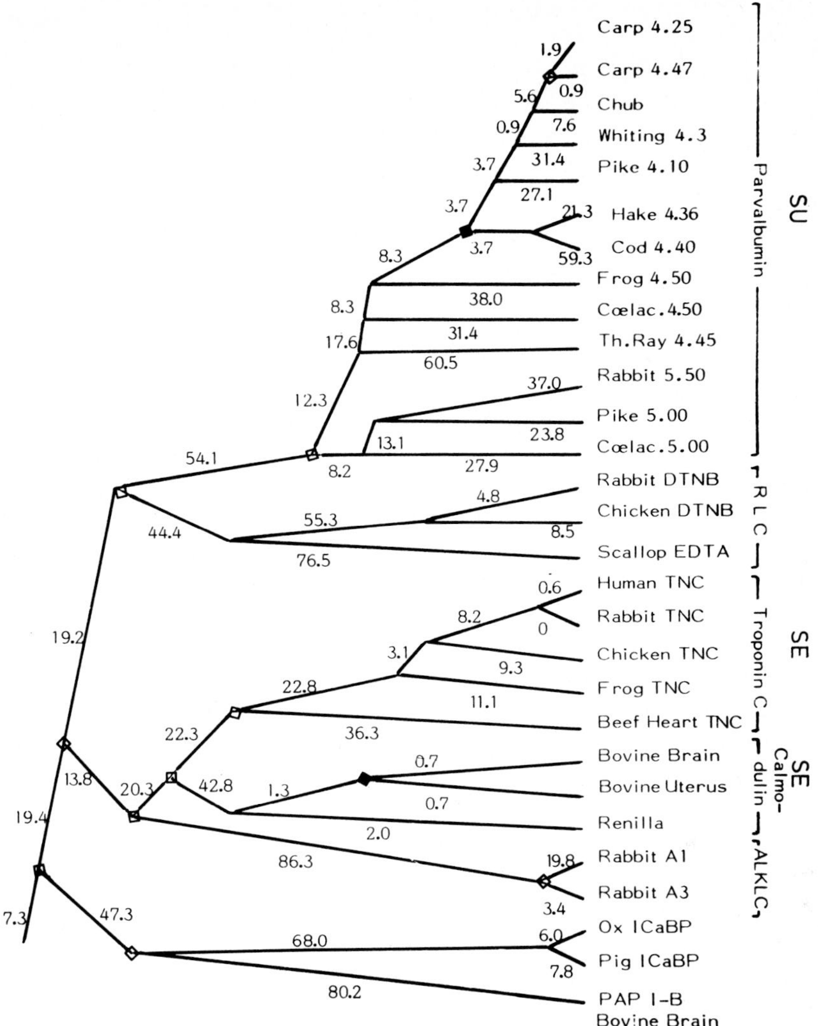

Fig. 2. The evolutionary tree of intracellular calcium-binding proteins. *Numbers on the links* are nucleotide replacements per 100 codons. Symbol for an obvious gene duplication is ◇. ◆ is used whenever there is only circumstantial evidence for gene duplication. *SE,* sensor proteins that obey the following criteria: one or more Ca^{2+}-specific sites with pK_d between 5 and 7, and protein-protein interaction. *SU,* suppressor proteins that obey the following criteria: one or more high affinity Ca^{2+}-Mg^{2+} sites, resulting in a kinetic delay in Ca^{2+} binding. Unlabeled proteins are not yet classified as either sensor or suppressor proteins. Abbreviations: RLC, myosin regulatory light chain, released by either DTNB or 5,5'-dithio bis-(2-nitrobenzoic acid), or EDTA, ethylene diamine tetraacetic acid; TNC, troponin C; ALKLC, myosin alkali light chains; ICaBP, intestinal Ca^{2+}-binding protein, PAP I-B, one of the components of brain S-100 protein. For references of the primary structures determinations, see Goodman et al. 1979 and Demaille et al. 1980

physiological event coupling is also poorly understood and will be dealt with further in the following section.

4 Calmodulin Activation of Calmodulin-dependent Enzymes

Calmodulin is able to interact with and activate a number of enzymes in a Ca^{2+}-dependent manner (Cheung 1980; Means and Dedman 1980). Four calcium binding sites are predictable from the primary structure (Watterson et al. 1980). Studies by Teo and Wang (1973), Wolff et al. (1977) and others have indeed shown that calmodulin binds four mol Ca^{2+} per mol with high affinity. However, some discrepancies are found in these reports with respect to the values of the affinity constants and to the effects of other cations on the Ca^{2+}-binding properties. This is why these properties were recently reexamined (Haiech et al. submitted for publication). Calmodulin binds four Ca^{2+} ions in a sequential and ordered manner. Calcium-binding sites also bind K^+ and Mg^{2+} ions, but K^+ and Mg^{2+} binding is not sequential. If A, B, C, and D stand for the four Ca^{2+}-binding sites recognized in the primary structure from the NH_2- to the COOH-terminus, the Ca^{2+}-binding sequence is either B → A → C → D or B → C → A → D. Terbium ions, when used as a probe for mammalian calmodulin calcium-binding sites, were found to bind first to sites that lack tyrosyl residues, i.e., sites B and A (Kilhoffer et al. 1980a). When the same investigation was carried out on octopus calmodulin, which contains a single tyrosine in domain D, it became obvious that the third Tb^{3+} ion binds to domain C, and the fourth to the lower affinity domain D (Kilhoffer et al. 1980b). Assuming that the sequence of Tb^{3+} binding is similar to the sequence of Ca^{2+} binding, the latter can be written B → A → C → D. The minimal kinetic scheme (Huang et al. 1980) for the binding of Ca^{2+} to calmodulin and the activation of calmodulin-dependent enzymes is depicted in Fig. 3. Enzymes fall into five subsets according to the kinetic pathway (0 to IV respectively) followed by the activation of each enzyme. Assuming that steps involving calcium-calmodulin-enzyme ternary complexes (steps 1' to 8') are more rapid than steps 1 to 8,

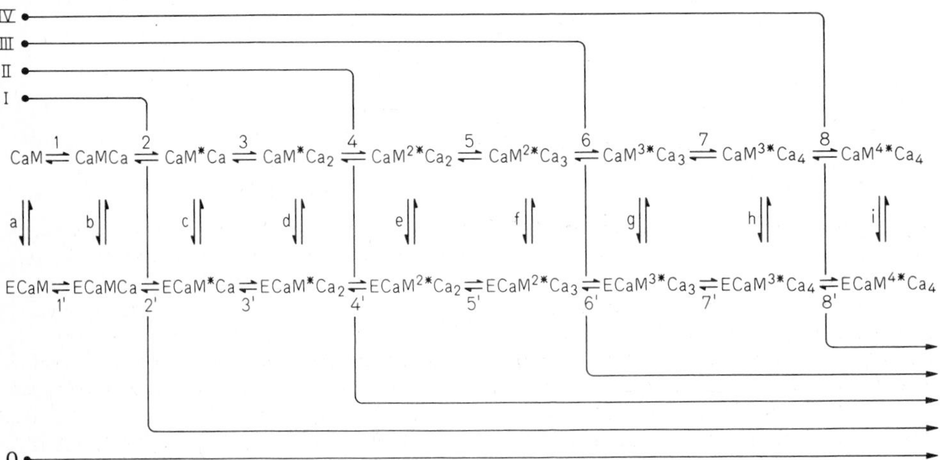

Fig. 3. A general scheme for the sequential activation of calmodulin-dependent enzymes. *E*, the activatable enzyme; *CaM*, calmodulin; *CaM^{n*}*, the transconformed calmodulin after binding of the nth Ca^{2+} ion. *1, 2, . . . 8* represent the sequential steps of Ca^{2+} binding to free calmodulin, and *1', 2', . . . 8'* the sequential steps of Ca^{2+} binding to the calmodulin-enzyme complex. *a, b, . . . i* stand for equilibrium steps between calcium-calmodulin complexes and ternary complexes. *0 to IV* are preferential kinetic pathways that define enzyme subsets E_0 to E_{IV}, respectively

which describe Ca^{2+} binding to free calmodulin, calmodulin is now able to transform a quantitative entry of Ca^{2+} into qualitatively different responses. The first enzyme subset that is activated upon a limited entry of Ca^{2+} is E_o, exemplified by glycogen phosphorylase *b* kinase. In E_o, calmodulin (here the δ-subunit) is attached to the enzyme even in the absence of Ca^{2+} (Cohen et al. 1978). Then the enzyme subsets E_I to E_{IV} will be activated in this order.

This sequential activation of enzyme subsets upon calcium release into the cytosol is shown in Fig. 4. Since the on and off rates of Ca^{2+} binding to calmodulin and calmodulin binding to the enzymes are as yet unknown, assumptions had to be made as to their hypothetical values. The qualitative representation of Fig. 4 stands, however, true for a vaste range of constants, provided the lower kinetic pathway is more rapid than the upper one, which describes Ca^{2+} binding to free calmodulin.

The Ca^{2+} occupancy of the *active* Ca^{2+}-calmodulin-enzyme ternary complex is still a matter of speculation. There is evidence, however, that activation of phosphodiesterase (Crouch and Klee 1980) and of myosin light chain kinase (Blumenthal and Stull 1980) requires at least three bound Ca^{2+} ions.

In conclusion, the sequential and ordered binding of Ca^{2+} to calmodulin generates the possibility of a sequential activation of different calmodulin-dependent enzyme subsets.

5 Deactivation of Ca^{2+}-Calmodulin-Enzyme Complexes

Proteins have been described that are able to bind calmodulin in the presence of Ca^{2+} ions and do not exhibit any known enzymatic activity. Among these calmodulin-binding proteins or inhibitory proteins, leiotonin A (Ebashi 1980) is found in smooth muscle, and calcineurin is abundant in nervous tissues (Wang and Desai 1977; Klee and Krinks 1978; Wallace et al. 1979). Calcineurin low-molecular weight B subunit binds Ca^{2+} with high affinity (Klee et al. 1979a). The large-molecular weight A subunit binds calmodulin (Klee and Haiech 1980), explaining the inhibition of calmodulin-dependent enzymes by calcineurin. Such inhibitory proteins may play a crucial role in the deactivation of calmodulin-dependent enzymes. When Ca^{2+} triggers both the synthesis and the degradation of a compound such as cAMP by adenylate cyclase and phosphodiesterase respectively, a futile cycle can only be prevented if deactivation of the cyclase occurs when phosphodiesterase is activated, i.e., if sequential deactivation occurs after sequential activation.

Let us assume that inhibitory proteins exhibit the following properties:

1. They interact with calmodulin in a Ca^{2+}-dependent manner,
2. this interaction blocks the activation of some or all calmodulin-dependent enzymes; these first two properties are the basis for the isolation and characterization of inhibitory proteins;
3. they interact with calmodulin even when the latter is bound to a calmodulin-dependent enzyme. Therefore calmodulin interacts with both the enzyme and the inhibitory protein at different sites. Such possibility was recently suggested on evolutionary (Perry et al. 1979) and experimental grounds (Cohen et al. 1978; Cohen et al. 1979) for calmodulin and troponin C. The formation of the quaternary complex inhibitory protein-Ca^{2+}/calmodulin-enzyme allows fast deactivation of calmodulin-activated enzymes,

A

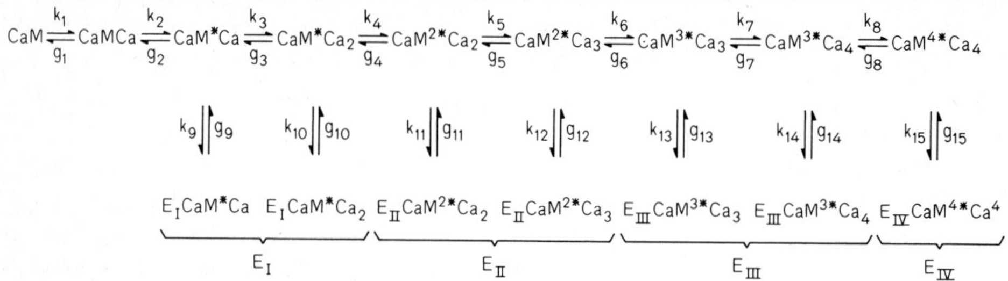

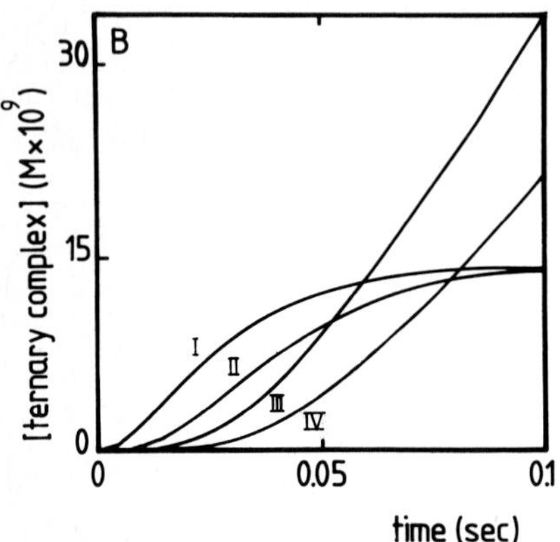

Fig. 4A, B. Computer simulation of the kinetics of formation of selected Ca^{2+}-calmodulin-enzyme ternary complexes. **A** Kinetic scheme used in the simulation. Forward rate constants are $k_1 \ldots k_{15}$, backward rate constants are $g_1 \ldots g_{15}$. E_I stands for the sum of the concentrations of E_I CaM* + E_I CaM*Ca_2, since kinetic recognition is based upon calmodulin conformation and not upon occupancy by Ca^{2+} ions. It was hypothesized that the steps between ternary complexes are not rate limiting and therefore that the concentration of E_I reflects the final activation of the E_I enzyme subset. **B** Time course of changes in the concentration of ternary complexes $E_{I \ldots IV}$ upon abrupt increase in Ca^{2+} concentration to pCa 4. $E_{I \ldots IV}$ are defined as above. Total enzyme concentrations were $[E_I] = [E_{II}] = [E_{III}] = [E_{IV}] = 1\ \mu M$.

Values for the rate constants were assumed as follows:

$k_1 = k_3 = k_5 = k_7 = 10^8\ M^{-1}.\ s^{-1}$

$k_9 = k_{10} = k_{...} = k_{15} = 10^5\ M^{-1}.\ s^{-1}$

These values were chosen as representative of diffusion-limited kinetics.

$k_2 = k_4 = k_6 = k_8 = 70\ s^{-1}$

This value was arbitrarily taken as representative of a calcium-binding protein transconformation.

$g_1 = g_3 = g_5 = g_7 = 70\ s^{-1}$

$g_2 = 7\ s^{-1}$, $g_4 = 14\ s^{-1}$, $g_6 = 60\ s^{-1}$ and $g_8 = 90\ s^{-1}$

The product $g_1 \times g_2$ was computed from macroscopic equilibrium constants (Haiech et al. submitted for publication). Within this product, g_1 was arbitrarily taken as $70\ s^{-1}$. Similar assignments were performed on the other products $g_n \times g_{n+1}$.

$g_9 = g_{10} = g_{...} = g_{15} = 10^{-4}\ s^{-1}$.

This value was obtained assuming that the affinity constant for the activation of enzymes by calmodulin is $10^9\ M^{-1}$

4. deactivation follows activation after a kinetic delay. This last assumption is not as yet substantiated by any experimental result. Two hypotheses can be proposed to account for such a delay.

The first one, recently proposed by Klee and Haiech (1980), takes mostly into account the relative concentrations of calmodulin and calcineurin within the cell and their relative affinities for Ca^{2+} ions. Assuming similar on rates for Ca^{2+} binding to both calmodulin and calcineurin B, Ca^{2+} ions will first bind to calmodulin, the concentration of which is at least one order of magnitude higher than the one of calcineurin, and thereby activate calmodulin-dependent enzymes. Ca^{2+} ions are then rapidly mobilized from calmodulin to calcineurin B, which exhibits a much higher affinity for Ca^{2+} ions (Klee et al. 1979a), with concomitant deactivation of the calmodulin-dependent enzyme.

Interaction of calmodulin with a target enzyme increases its affinity for Ca^{2+} (Crouch and Klee 1980; Huang et al. 1980). Under these conditions, calcineurin B affinity for Ca^{2+} may have to be ca. three orders of magnitude higher than the one displayed by calmodulin, either permanently or as a result of calmodulin binding to the inhibitory protein, to induce efficient calcium removal from enzyme-bound calmodulin. At any rate, with such pKd ca. 8, return to resting state would only occur if calcineurin interacts with Ca^{2+} pumps (pK_d ca. 7).

The second hypothesis (Fig. 5) postulates the existence of quaternary complexes calcineurin – Ca^{2+} – calmodulin – enzyme. Calcineurin binding to calmodulin is Ca^{2+}-

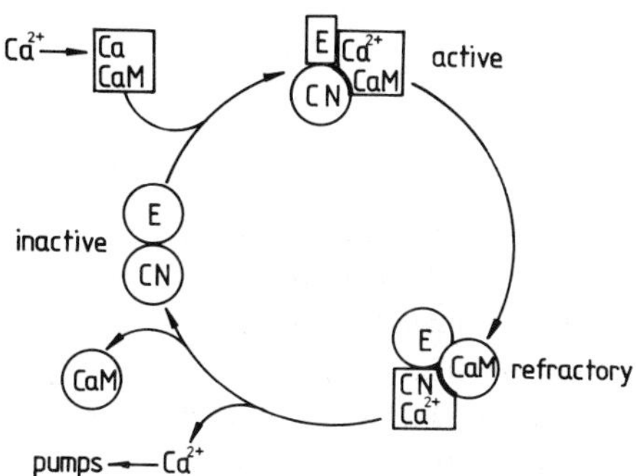

Fig. 5. Hypothetical scheme of the activation-deactivation of an elementary calcisome including the calmodulin-dependent enzyme (*E*), calcineurin (*CN*) and calmodulin (*CaM*). *Square boxes* designate active species and *circles* inactive species

dependent, as a result of calcium binding to calmodulin *and* to calcineurin B. At first, Ca^{2+} binds mainly to calmodulin and the affinity of calcineurin for the enzyme-bound calmodulin is moderate. Enzyme inhibition only occurs when Ca^{2+} binds to calcineurin B thereby increasing the affinity of calcineurin A for calmodulin.

This scheme only holds if calcineurin B binds Ca^{2+} after a kinetic delay, introduced for instance by Mg^{2+} – Ca^{2+} exchange at high affinity Ca^{2+}-Mg^{2+} sites, as was shown for parvalbumins (Haiech et al. 1979; Birdsall et al. 1979). Calcineurin B was indeed shown to be similar to parvalbumins in terms of affinity for Ca^{2+} and of amino acid composition (Klee and Haiech 1980).

Such a scheme would result in a fast activation-deactivation of calmodulin-dependent enzymes after a Ca^{2+} wave. However, sequential deactivation of the different enzyme subsets (see Fig. 3) would only be possible through a mechanism similar to the one proposed in Sect. 4 for the activation. Calcineurin B binds 3-4 mol of Ca^{2+} per mol (Klee and Haiech 1980). On theoretical grounds (Reid and Hodges 1980), binding is likely to be sequential. Sequential deactivation is therefore conceivable if the energy needed for the calcineurin A – induced enzyme inhibition is provided by the Ca^{2+} binding to calcineurin B.

6 The Calcisome Concept

While the above discussion deals with isolated proteins studied in vitro, the in vivo behavior of regulatory proteins is highly dependent on their assembly and localization within the cell. It is well established that Ca^{2+} diffusion in the cytosol is restricted, leading to localized increases in Ca^{2+} level as was detected in maps of spatial and temporal Ca^{2+} distribution (Taylor et al. 1980). Upon influx of calcium, e.g., into nerve cells, the increase in free intracellular Ca^{2+} concentration is confined to the cell periphery where the machinery responsible for short-term buffering of Ca^{2+} ions is localized (Tillotson and Gorman 1980). Only in this area will the Ca^{2+} signal be transformed into physiological events. This is performed in and by *calcisomes*, defined as a specific assembly of calcium sensors, target enzymes and inhibitory proteins, able to respond specifically to a transient and localized rise in Ca^{2+} concentration.

The properties of calcisomes can be predicted as follows: It is a "particle" either associated with or in close vicinity to the membrane, which contains the Ca^{2+} pumps and channels. The time scale of the dissociation of the calcisome components is very large compared to the time scale of the Ca^{2+} transient. Myofibrils on the one hand, and glycogen particles on the other hand (Heilmeyer et al. 1970) are known examples of calcisomes.

Strong protein-protein interactions within calcisomes explain the difficulties encountered in purifying calmodulin-dependent enzymes from, for instance, the inhibitory proteins (Klee et al. 1979b) as well as copurification of several calmodulin-dependent activities in the first purification step (Haiech unpublished). Also, their extreme susceptibility to proteolysis may result from disruption of protein-protein interactions leading to the exposure to solvent of previously shielded hinges between domains.

7 Covalent Regulation Within Calcisomes

A number of calmodulin-dependent enzymes are phosphorylatable by cAMP-dependent protein kinase. This covalent modification results in either an increase of their catalytic activity (e.g., phosphorylase kinase; Cohen 1973), or a decrease in activity (e.g., smooth muscle myosin light chain kinase; Adelstein et al. 1978), or no significant change in activity (e.g., cyclic nucleotide phosphodiesterase; Sharma et al. 1980). Only in the case of myosin light chain kinase has the activity change been explained by a decrease in the affinity of the enzyme for calmodulin (Conti and Adelstein 1980). It is reasonable to postulate that protein kinase and phosphoprotein phosphatase are present in calcisomes, close to the membrane-bound adenylate cyclase. The same organelle may therefore mediate the concerted effects of both second messengers.

8 Conclusion

This manuscript introduces the hypothetical concept of a supramolecular coordinated organelle, the calcisome, that forms the quantum of cAMP and Ca^{2+}-dependent transient regulations. The sequential activation-deactivation of different enzymes present in calcisomes provides a better understanding of the mechanism by which the only ionic signal within the cell, a transient and localized increase in Ca^{2+} level, triggers qualitatively different responses, and by which futile cycles are prevented.

This concept applies only to reversible departures from steady state in homeostatic short-term regulations. Long-term regulation of differentiation, growth, and multiplication or trophic effects are not likely to be mediated by calcisomes.

The scheme presented herein provides a frame to unify and classify the abundant experimental findings reported in the last years, and should suggest further approaches to the understanding of metabolic regulation. Among them, studies of the mechanism of calcium sensor-target protein interaction, of the molecular effects of target protein phosphorylation, of the localization and composition of calcisomes within the cell, and finally of the regulation of calcisome synthesis and assembly at the genomic level, should be fruitful avenues of future research.

Acknowledgements. This work was supported in part by grants from CNRS (ATP Modulation de l'action des hormones au niveau cellulaire), INSERM (CRL 78.4.086.1, 79.4.151.3, 80.5.032 and ATP 63.78.95, Fibre musculaire cardiaque), DGRST (ACC Biologie et Fonction du Myocarde et Parois artérielles), NATO (grant no 1688) and Fondation pour la Recherche Médicale Française. J.H. acknowledges receipt of fellowships from DGRST and the Fogharty Center. The authors are indebted to Ms D. Waeckerlé for expert editorial assistance, and to Dr. Claude B. Klee, NIH, Bethesda, for stimulating discussions.

References

Adelstein RS, Conti MA, Hathaway DR, Klee CB (1978) Phosphorylation of smooth muscle myosin light chain kinase by the catalytic subunit of adenosine 3':5'-monophosphate-dependent proteinkinase. J Biol Chem 253:8347-8350

Birdsall WJ, Levine BA, Williams RJP, Demaille JG, Haiech J, Pechère JF (1979) Calcium and magnesium binding by parvalbumin. A proton magnetic resonance study. Biochimie 61:741-750

Blumenthal DK, Stull JT (1980) Activation of purified myosin light chain kinase from skeletal muscle. Fed Proc Fed Am Soc Biol 39:2707a

Cheung WY (1980) Calmodulin plays a pivotal rile in cellular regulation. Science 207:19-27

Cohen P (1973) The subunit structure of rabbit skeletal muscle phosphorylase kinase, and the molecular basis of its activation reactions. Eur J Biochem 34:1-14

Cohen P, Burchell A, Foulkes JG, Cohen PTW, Vanaman TC, Nairn AC (1978) Identification of the Ca^{2+}-dependent modulator protein as the fourth subunit of rabbit skeletal muscle phosphorylase kinase. FEBS Lett 92:287-293

Cohen P, Picton C, Klee CB (1979) Activation of phosphorylase kinase from rabbit skeletal muscle by calmodulin and troponin. FEBS Lett 104:25-30

Conti MA, Adelstein RS (1980) Phosphorylation by cyclic adenosine 3':5'-monophosphate-dependent protein kinase regulates myosin light chain kinase. Fed Proc Fed Am Soc Exp Biol 39: 1569-1573

Crouch TH, Klee CB (1980) Positive cooperative binding of calcium to bovine brain calmodulin. Biochemistry 19:3692-3698

Demaille JG, Haiech J, Goodman M (1980) Calcium-binding proteins: From diversification of structure to diversification of function. In: Protides of the biological fluids, vol 28. 95-98

Ebashi S (1980) Regulation of muscle contraction. Proc R Soc London Ser B 207:259-286

Fischer EH, Becker JU, Blum HE, Byers B, Heizmann C, Kerrick GW, Lehky P, Malencik DA, Pocinwong S (1976) Concerted regulation of glycogen metabolism and muscle contraction. In: Heilmeyer L et al. (eds) Molecular basis of mobility. Springer, Berlin Heidelberg New York, pp 137-158

Goodman M, Pechère JF, Haiech J, Demaille JG (1979) Evolutionary diversification of structure and function in the family of intracellular calcium-binding proteins. J Mol Evol 13:331-352

Haiech J, Derancourt J, Pechère JF, Demaille JG (1979) Magnesium and calcium binding to parvalbumins. Evidence for differences between parvalbumins and an explanation of their relaxing function. Biochemistry 18:2752-2758

Heilmeyer Jr LMG, Meyer F, Haschke RH, Fischer EH (1970) Control of phosphorylase activity in a muscle glycogen particle II. Activation by calcium. J Biol Chem 245:6649-6656

Huang CY, Chau V, Chock PB, Sharma RK, Wang JH (1980) On the mechanism of Ca^{2+} activation of calmodulin-regulated cyclic nucleotide phosphodiesterase. Fed Proc Fed Am Soc Exp Biol 39: 288a

Kilhoffer MC, Demaille JG, Gerard D (1980a) Terbium as luminescent probe of calmodulin calcium binding sites. Domains I and II contain the high affinity sites. FEBS Lett 116:269-272

Kilhoffer MC, Gerard D, Demaille JG (1980b) Terbium binding to Octopus calmodulin provides the complete sequence of ion binding. FEBS Lett 120:99-103

Klee CB, Haiech J (1980) Concerted role of calmodulin and calcineurin in calcium regulation. Ann NY Acad Sci (in press)

Klee CB, Krinks MH (1978) Purification of cyclic 3',5'-Nucleotide phosphodiesterase inhibitory protein by affinity chromatography on activator protein coupled to sepharose. Biochemistry 17: 120-126

Klee CB, Crouch TH, Krinks MH (1979a) Calcineurin: A calcium- and calmodulin-binding protein of the nervous system. Proc Natl Acad Sci USA 76:6270-6273

Klee CB, Crouch TH, Krinks MH (1979b) Subunit structure and catalytic properties of bovine brain Ca^{2+}-dependent cyclic nucleotide phosphodiesterase. Biochemistry 18:722-729

Marban E, Rink TJ, Tsien RW, Tsien RY (1980) Free calcium in heart muscle at rest and during contraction measured with Ca^{2+}-sensitive microelectrodes. Nature (London) 286:845-850

Means AR, Dedman JR (1980) Calmodulin – an intracellular calcium receptor. Nature (London) 285:73-77

Perry SV (1979) The regulation of contractile activity in muscle. Biochem Soc Trans 7:593-617

Perry SV, Grand RJA, Nairn AC, Vanaman TC, Wall CM (1979) Calcium-binding proteins and the regulation of contractile activity. Biochem Soc Trans 7:619-622

Rasmussen H (1970) Cell communication, calcium ion, and cyclic adenosine monophosphate. Science 170:405-412

Rasmussen H, Goodman DBP (1977) Relationships between calcium and cyclic nucleotides in cell activation. Physiol Rev 57:421-509

Reid RE, Hodges RS (1980) Co-operatively and calcium/magnesium binding to troponin C and muscle calcium binding parvalbumin: An hypothesis. J Theor Biol 84:401-444

Robinson GA, Butcher RW, Sutherland EW (1971) Cyclic AMP. Academic Press, London New York

Sharma RK, Wang TH, Wirch E, Wang JH (1980) Purification and properties of bovine brain calmodulin-dependent cyclic nucleotide phosphodiesterase. J Biol Chem 255:5916-5923

Taylor DL, Blinks JR, Reynolds G (1980) Contractile basis of ameboid movement VIII Aequorin luminescence during ameboid movement, endocytosis and capping. J Cell Biol 86:599-607

Teo TS, Wang JH (1973) Mechanism of activation of a cyclic adenosine 3':5'-monophosphate phosphodiesterase from bovine heart by calcium ions. Identification of the protein activator as a Ca^{2+} binding protein. J Biol Chem 248:5950-5955

Tillotson D, Gorman ALF (1980) Non-uniform Ca^{2+}-buffer distribution in a nerve cell body. Nature (London) 286:816-817

Wallace RW, Lynch TJ, Tallant EA, Cheung WY (1979) Purification and characterization of an inhibitor protein of brain adenylate cyclase and cyclic nucleotide phosphodiesterase. J Biol Chem 254:377-382

Wang JH, Desai R (1977) Modulator binding protein. Bovine brain protein exhibiting the Ca^{2+}-dependent association with the protein modulator of cyclic nucleotide phosphodiesterase. J Biol Chem 252:4175-4184

Watterson DM, Sharief F, Vanaman TC (1980) The complete amino acid sequence of the Ca^{2+}-dependent modulator protein (calmodulin) of bovine brain. J Biol Chem 255:962-975

Wolff DJ, Brostrom MA, Brostrom CO (1977) Divalent cation binding sites of CDR and their role in the regulation of brain cyclic nucleotide metabolism. In: Wasserman RH et al. (eds) Calcium binding proteins and calcium function. North Holland, New York, pp 97-106

Discussion[3]

Cohen supported the conclusion that the trimethyllysine residue in calmodulin is not essential for activation of phosphorylase kinase. In collaboration with his group, Dr. Drabikowski in Warsaw has cleaved calmodulin into two fragments (residues 1-77 and 78-148). It was found that both fragments activate the enzyme, each about 20-fold less efficiently than calmodulin itself; in other words, the N-terminal fragment lacking trimethyllysine is as good an activator as the C-terminal fragment containing the trimethyllysine.

Hofmann asked if calcisomes might exist only in brain and skeletal muscle, or in other tissues as well. Demaille answered that the maximum Ca^{2+} buffering capacity is localized in the vicinity of the cell membrane. From the affinity constant, one would assume that removal of Ca^{2+} from a Ca^{2+}-dependent enzyme would require several minutes. Since there must be a Ca^{2+} control mechanism for rapid activation and deactivation of such enzymes, calcisomes must be present in the vicinity of the cell membrane of most eukaryotic cells.

Severin inquired about a possible role of calmodulin in Ca^{2+} channels of nervous tissue. Demaille did not know about such a role in nervous tissue. He said that in cardiac tissue, however, calmodulin plays a definite role in the increase of the Ca^{2+} uptake rate of cardiac sarcoplasmic reticulum vesicles. The pump activator, phospholamban, is phosphorylated not only by cAMP-dependent protein kinase, but also by a Ca^{2+}-calmodulin-dependent membrane-bound protein kinase, resulting in an increase in the rate of Ca^{2+} uptake.

On request by Larner, Demaille commented on the mode of action of chlorpromazine and related drugs: from Dan Storm's work, binding of trifluoroperazine occurs on a hydrophobic patch which is formed on the surface of calmodulin upon Ca^{2+} binding. Troponin I, which binds to calmodulin in the presence of Ca^{2+}, prevents the binding of trifluoroperazine or ANS or anthroylcholine to calmodulin.

3 Rapporteur: Wolfgang Burgermeister

Inhibition of Catalytic Subunit by Fragments of cAMP-dependent Protein Kinase Regulatory Subunit I and II

F. Hofmann, H. Wolf, R. Mewes, and M. Rymond[1]

1 Introduction

Regulatory subunits of cAMP-dependent protein kinase I and II possess two important properties which are closely interrelated: firstly, to bind the catalytic subunit and, thereby, to inhibit its phosphotransferase activity and, secondly, to bind cAMP which process leads to activation of the enzyme by inducing the dissociation of the catalytic subunit from the regulatory subunits [for a review see ref. 1, 2]. The structural requirements and the localization of these closely connected properties on the regulatory subunits are still largely unknown, although various proteolytic fragments of regulatory subunit I and II ranging from 39,000 to 10,000 dalton have been studied [3-13]. Limited proteolysis of regulatory subunit I from porcine skeletal muscle generated a lage fragment (M_r~35,000), which bound cAMP but lost its ability to dimerize [3, 4], and a smaller fragment (M_r~11,000) which did not bind cAMP but retained the ability to dimerize [3, 4]. An intermediate size fragment (M_r~16,000) obtained from rabbit skeletal muscle regulatory subunit I has been reported to bind cAMP [5]. Limited proteolysis of regulatory subunit II obtained either in vitro or in vivo from porcine skeletal muscle or bovine cardiac muscle generated a large fragment (M_r ~39,000-37,000) that bound cAMP but was unable to dimerize [3,6-10], and a small fragment (M_r~ 17,000) [5, 6, 8, 9]. Conflicting data on the presence of a cAMP-binding site in this small fragment have been reported [5, 8]. In addition, a partial amino terminal sequence of the large fragment from porcine skeletal muscle regulatory subunit I and II [4, 8], the amino sequence around one phosphorylation site of regulatory subunit II [1, 5], the amino sequence of the "hinge"region between the two binding domains for cAMP and catalytic subunit of regulatory subunit II [11], and the amino sequence around one cAMP binding site [12] have been reported. With one exception [13], all the fragments tested had lost their ability to inhibit the activity of the catalytic subunit. This latter conclusion was usually derived from qualitative data. So far, quantitative data on the possible changes in affinity of each binding domain of the regulatory subunits by proteolytic fragmentation have not been reported. On the other hand, those data appeared necessary to understand the nature of these binding sites and their close functional relationship. In this communication evidence is presented that the binding of cAMP and the inhibition of the catalytic activity can be retained in the same proteolytic fragment of regulatory subunit I and II, although fragmentation reduces mainly the affinity for the catalytic subunit.

1 Pharmakologisches Institut der Universität Heidelberg, Im Neuenheimer Feld 366, D-6900 Heidelberg, FRG

2 Methods

Assays. Binding of cAMP was determined at room temperature as described [14, 15]. Inhibition of catalytic activity by the fragments was determined as described [15] using catalytic subunit, sodium chloride, and magnesium acetate at concentrations of 50 pM, 150 mM, and 1 mM, respectively. The phosphorylation reaction was carried out under the same conditions. Sucrose density gradients and SDS-gel electrophoresis were carried out as described [14]. Protein was determined by the Lowry procedures [15].

Purification of Proteins. Catalytic subunits, regulatory subunit I and II were purified from rabbit skeletal and bovine heart muscle, respectively, as described [14, 15]. Samples of regulatory subunits I and II, in which after sortage for several months at 4°C proteolysis had occureed, were pooled and dialyzed against 5 mM Mes-buffer, pH 6.5, containing 10 mM NaCl, 0.1 mM EDTA, and 15 mM 2-mercaptoethanol. The dialyzed samples were next charged on two identical DEAE-cellulose columns (0.6 x 30 cm). The columns were eluted by a linear solium chloride gradient (200 ml) from 10-200 mM. Fragments of the regulatory subunit I and II were identified by SDS-gel electrophoresis and their ability to bind cAMP. Fractions containing pure, large-molecular weight fragments were pooled and concentrated on a small DEAE-cellulose column. The low-molecular weight fragment of regulatory subunit I was not retained by the DEAE-cellulose column [8]. It was concentrated from the nonabsorbed fraction by lyophilization. All fragments used were stored in 5 mM Mes buffer, pH 6.5, containing 0.2 mM EDTA, 0.2 mM benzamidine, 150 mM NaCl, and 0.5 mM DTT.

3 Results and Discussion

As reported previously by several laboratories [3, 6, 7], preparations of pure regulatory subunit I and II are slowly degraded to smaller fragments by a contaminating protease when stored for several months at 4°C. Aged preparations of these subunits yielded on SDS-disc gels, in addition to the native subunit, two major stained bands (approximate M_r's 37,000-32,000 and 16,000-11,000), and several minor stained bands. Samples in which degradation of the regulatory subunits had occurred were used for preparation of fragments having an approximate M_r of 37,000-32,000 and of 10,000. These fragments were purified on DEAE-cellulose columns. The large fragments were eluted from the column well ahead of the native regulatory subunits and yielded, on SDS-disc gels, one stained band corresponding to a M_r of 37,000 and 32,000 for the fragment of regulatory subunit I and II, respectively (Fig. 1). Variation of the amount of protein applied to the gels from 1 to 10 µg did not reveal additional stained bands. On sucrose density grandients both fragments sedimented as symmetrical cAMP-binding peaks with an identical $S_{20,w}$ value of 2.7, suggesting that they were present as monomers. In control tubes native regulatory subunit II sedimented as a dimer ($S_{20,w}$ 4.4) (see Table 1). Both fragments were also tested for their ability to serve as substrate protein for the catalytic subunit. Although stoichiometric concentrations of fragments and catalytic subunit (1.4 µM each) were incubated for up to 120 min in the absence and

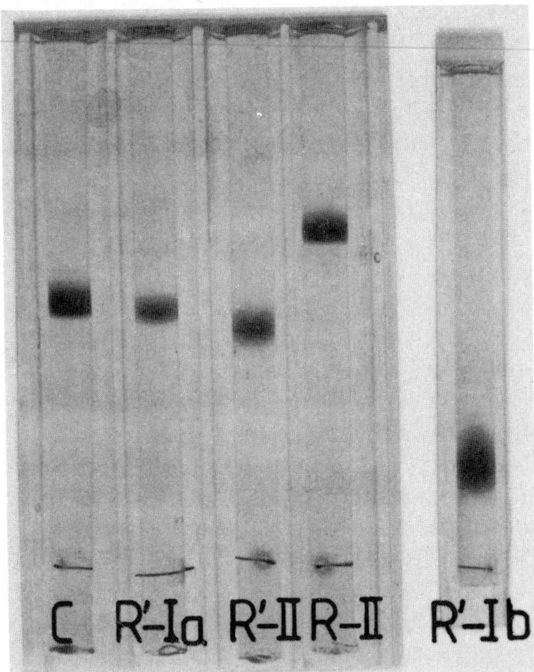

Fig. 1. Purity of fragments used. The purity of fragments used was determined by polyacrylamide (7.5%) gel electrophoresis in the presence of 0.1% sodium dodecyl sulfate and 2-mercaptoethanol. Electrophoresis was from top to bottom and following proteins were applied (from left to right): catalytic subunit (*C*; 5.7 μg); fragment of regulatory subunit I (*R'-Ia*; 5.2 μg); fragment of regulatory subunit II (*R-II*; 5.1 μg), fragment of regulatory subunit I (*R'-Ib*; 7.5 μg). The last gel (*R'-Ib*) was electrophoresed separately from the other gels

Table 1. Properties of proteolytic fragments of regulatory subunit I and II

	R'-I	R'-II	R-II
M_r (SDS-PAGE)	37,000	32,000	56,000
$S_{20,W}$	2.7	2.7	4.4
Mol phosphate/Mol "R"	0.06	0.16	0.92
Mol cAMP bound/Mol "R"	0.92	0.85	1.75
Approx. K_D for cAMP (nM)	17.0	12.5	8.5
Approx. K for "C" (nM)	1000.0	20.0	0.26

In all experiments, each of the three proteins was used at the same time under identical conditions. In other experiments [15] identical values to those shown for native regulatory subunit II (R-II) have been obtained for native regulatory subunit I.

Only the M_r determined in the presence of SDS was 48,000. R'-I, R'-II, and C stand for fragment of regulatory subunit I and II, and catalytic subunit, respectively. The approx. K for C equals that concentration of R'-I, R'-II, or R-II that inhibits 50% of the activity of the catalytic subunit added at a concentration of 50 pM

presence of 0.1 mM cAMP, only minimal amounts of phosphate were incorporated into the fragments (Fig. 2). In the same experiments, 0.92 mol phosphate was incorporated per mol of native regulatory subunit II. Identical values as shown in Table 1 were obtained when aliquots of incubated samples were subjected to SDS-disc gel electrophoresis. After gel electrophoresis only such slices of individual gels contained

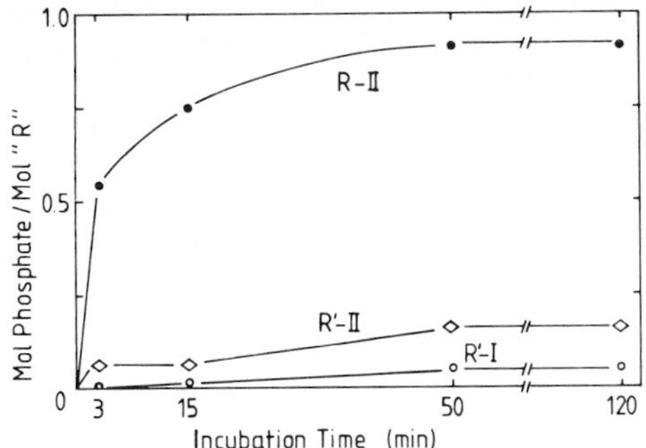

Fig. 2. Phosphorylation of native regulatory subunit II and the large fragments by catalytic subunit. Phosphorylation was carried out as given in Sect. 2 in the presence of 0.1 mM cAMP. Stoichiometric amounts (72 pmol) of catalytic subunit and native regulatory subunit II (*R-II*, •), the large fragments of regulatory subunit I (*R'-I*, ○) and II (*R'-II*, ◇) were used. Almost identical results were obtained in the absence of cAMP

radioactivity that corresponded to the stained bands of the fragments or the native regulatory subunit. Similar results of the inability of catalytic subunit to phosphorylate fragments of regulatory subunit II have been obtained by some [6, 7] but not by all laboratories [8]. The fragments of regulatory subunit I and II, and native regulatory subunit II, bound 0.9, 0,85, and 1.75 mol of cAMP per mol of protein, respectively. The app. K_D's for cAMP varied from 8.5 to 17.0 nM. These values represent the extremes found in several experiments. In three out of five experiments the approx. K_D's differed only by a factor of 1.25, suggesting that the affinity of at least one cAMP binding site was essentially unaffected by the proteolytic fragmentation (Table 1).

In contrast to these results, proteolytic fragmentation of the regulatory subunits had a marked effect on their ability to inhibit the catalytic subunit. As is evident from Fig. 3, the activity of the catalytic subunit was inhibited half-maximally by 0.26 nM of native regulatory subunit II, by 20.0 nM of the fragment of regulatory subunit II, and by 1,000 nM of the fragment of regulatory subunit I. The inhibition was specific for the regulatory subunit or their fragments, since no inhibition was observed in the presence of cAMP. The possibility that the differences in the inhibitory activity between native regulatory subunit II and its fragment was caused by a 1% contamination of the fragment preparation by the native subunit was ruled out by the following experiment. The fragment of regulatory subunit II (144 pmol) was incubated for 50 min in the presence of 150 pmol of catalytic subunit, cAMP (0.1 mM), magnesium (1 mM), and [γ-^{32}P] ATP (0.1 mM; specific activity 620 cpm/pmol). The incubation was terminated by boiling the sample in the presence of SDS and 2-mercaptoethanol. Thereafter, 140 pmol of native SDS-denaturated regulatory subunit II were added, and the proteins were separated on SDS-disc gels. After staining and destaining of the gels, stained bands corresponding to native regulatory subunit II and its fragment were cut out, dried, and counted. No radioactivity was associated with native regulatory subunit II, whereas 0.16 mol of phosphate per mol of fragment was associated with the fragment. Assuming a detection limit of 100 cpm per band and stoichiometric phosphorylation of the assumed contaminating native regulatory subunit, less than 0.1% of contamination would have been detected. Therefore, it appears unlikely that the 75-fold lower potency of the fragment to inhibit the catalytic subunit as compared to that of the native subunit II was caused by a contamination of the fragment preparation by the native molecule.

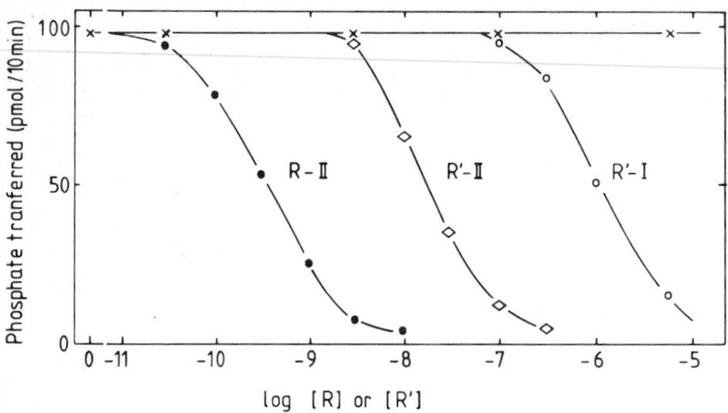

Fig. 3. Inhibition of catalytic subunit by fragments of regulatory subunit. Inhibition of the activity of pure catalytic subunit (50 pM) by native regulatory subunit II (*R-II*), and the fragments of regulatory subunit II (*R'-II*), and I (*R'-I*) was determined in the absence of cAMP (●, ◇, ○) and in the presence of 1 mM magnesium acetate and 150 mM sodium chloride in the same assay. Only a few of the values determined in the presence of cAMP (*x*) are shown. The concentration given for *R* or *R'* refers always to the total concentration of regulatory subunit or fragment present during the incubation

In previous studies no quantitative relationship between the degree of fragmentation and the change in the two important properties of regulatory subunit were reported, presumably due to the fact that the previously used preparations of regulatory subunits contained cAMP. The advantage of the fragments used here was that they were derived from regulatory subunits which were free of cAMP [14, 15]. As is evident from the data, degradation of regulatory subunit I and II yielded fragments in which at least one site for cAMP binding is unaltered as suggested by the unchanged K_D's. In contrast to the cAMP binding data, the affinity for the catalytic subunit was markedly reduced but not abolished. This indicates that the fragments have lost some but not all structural requirements for the interaction with catalytic subunit. The data also suggest that the site, which is easily phosphorylated in regulatory subunit II, is not essential for the binding of catalytic subunit to the reulatory subunit, since no correlation between phosphorylation and fragmentation has been observed.

Interestingly, the fragment of regulatory subunit I had the lowest affinity for catalytic subunit although it contained a greater part of the native molecule (M_r 48,000) than the fragment of regulatory subunit II. It was tempting to sepculate that the lost fragment of approx. 11,000 dalton contains a greater part of the determinants for high-affinity binding to the catalytic subunit as suggested by Srivastava and Stellwagen [13]. Therefore, a low-molecular weight fragment of regulatory subunit I was tested for its ability to inhibit the activity of the catalytic subunit. In the presence of SDS and 2-mercaptoethanol, this fragment had an approximate M_r of 11,000 (Fig. 1). In agreement with previous results [3-5], this fragment did not bind cAMP as determined by equilibrium dialysis or by the filtration method, and did not inhibit the activity of the catalytic subunit present at a concentration of 50 pM although a 40,000-fold excess of this small fragment over that of the catalytic subunit was used. This would indicate that high-affinity binding of catalytic subunit to regulatory subunit I and concomitant inhibition of the catalytic activity is essentially dependent either on the complete structure of regulatory subunit I or on that region of the molecule that was cleft by the protease. However, this interpretation does not preclude the possibility that low-affinity binding and inhibition of catalytic subunit can be retained by the lager fragment. In general, the shown results suggest that the use of cAMP-free regulatory subunit is advantageous for studies designated to elucidate the structural requirements for the functional properties of regulatory subunits of cAMP-dependent protein kinase isoenzyme, since their use allows the determination of quantitative data. The importance of quantitative data for investigations of site specificity is well recognized. A similar approach has been used previously [17] to establish the primary sequence of a protein in order to be a optimal substrate for the catalytic subunit of cAMP-dependent protein kinase.

Acknowledgements. The expert technical assistent of Mrs. U. Kitz and P. Janko is gratefully acknowledged. This work was supported by a grant from DFG.

References

1. Krebs EG, Beavo JA (1979) Annu Rev Biochem 48: 923-959
2. Rosen OM, Rangel-Aldao R, Erlichman J (1977) Curr Top Cell Regul 12: 39-74
3. Potter RL, Stafford PH, Taylor SS (1978) Arch Biochem Biophys 190: 174-180
4. Potter RC, Taylor SS (1979) J Biol Chem 254: 2413-2418
5. Rannels SR, Corbin JD (1979) J Biol Chem 254: 8605-8610
6. Corbin JD, Sugden PH, West C, Flockhart DA, Lincoln TM, McCarthy D (1978) J Biol Chem 253: 3997-4003
7. Weber W, Hilz H (1978) Eur J Biochem 83: 218-225
8. Potter RL, Taylor SS (1979) J Biol Chem 254: 9000-9005
9. Flockhart DA, Watterson DM, Corbin JD (1980) J Biol Chem 255:4430-4440
10. Weber W, Hilz H (1979) Biochem Biophys Res Commun 90: 1073-1081
11. Takio K, Walsh KA, Neurath A, Smith SB, Krebs EG, Titani K (1980) FEBS Lett 114: 83-88
12. Kerlavage AR, Taylor SS (1980) Fed Proc 39: 2117
13. Srivastava AK, Stellwagen RH (1978) J Biol Chem 253: 1752-1755
14. Schwechheimer K, Hofmann F (1977) J Biol Chem 252: 7690-7696
15. Hofmann F (1980) J Biol Chem 255: 1559-1564
16. Lowry OH, Rosebrough NJ, Farr AL, Randall RJ (1951) J Biol Chem 236: 1372-1379
17. Kemp BE, Graves DJ, Benjamin E, Krebs EG (1977) J Biol Chem 252: 4888-4894

Discussion[2]

Demaille appreciated that Dr. Hofmann for the first time has obtained the enzyme from heart with a high specific activity. In his own lab, an enzyme with a very low specific activity was prepared which, however, had the nice feature to be regulated by cAMP-dependent phosphorylation decreasing the affinity for calmodulin. The ideal would be to get an enzyme which would exhibit both high affinity and cAMP-dependent modulation.

Stadtman remarked that the interpretation from the Hill plot may be incorrect that calmodulin did not bind to the enzyme in the absence of Ca^{2+}. As in the case

2 Rapporteur: Wolfgang Burgermeister

of adenylate cyclase, calmodulin may bind in absence of Ca^{2+}, but the expression of enzymatic activity may be dependent on occupation of the Ca^{2+} binding sites.

Hofmann answered that the results obtained with the calmodulin Sepharose affinity column confirmed his interpretation. In order to bind the enzyme to this column one needs about 3 to 4 mol Ca^{2+} per mol calmodulin on the column.

Control of Mg^{2+} and Ca^{2+} Binding to Troponin C and Calmodulin by Phosphorylation of the Respective Holocomplexes Troponin and Phosphorylase Kinase

L.M.G. Heilmeyer, jr., B. Koppitz, K. Feldmann, J.E. Sperling, U. Jahnke, K.P. Kohse, and M.W. Kilimann[1]

1 Introduction

The degree of phosphorylation of a protein should correlate with a meaningful change of a functional property if this process is of physiological importance (Krebs and Beavo 1979). Thus, it would require that the introduced phosphate group perturbs interaction of peptide domaines within a polypeptide or between subunits. The phosphate group itself may be used as a reporter group to monitor such perturbations since it yields a [^{31}P] NMR signal which is sensitive to its environment. Therefore the [^{31}P] NMR spectra of phosphorylated skeletal muscle troponin and myosin will be compared with that of phosphorylase *a* (see also Sperling et al. 1979; Koppitz et al. 1980). It will be shown that these phosphoproteins can be classified into two types according to the appearance of their [^{31}P] NMR signals.

Phosphorylation may trigger similar mechanisms in the two Ca^{2+} binding proteins, phosphorylase kinase and troponin. Both contain Ca^{2+} receptor proteins, calmodulin (Cohen et al. 1978) and troponin C (for review see Ebashi, 1980) respectively, which show a high degree of sequence homology (Vanaman et al. 1977) and similar Ca^{2+}-and Mg^{2+}-binding properties (Heilmeyer et al. 1980a). These respective Ca^{2+} receptor proteins interact in both native protein complexes with three heterologous subunits: α, β, and γ in phosphorylase kinase (Cohen et al. 1978) and two molecules troponin I and one molecule troponin T in troponin (Sperling et al. 1979). Finally, both holoproteins can be phosphorylated on two kinds of these heterologous subunits, namely on the α and β, or on the T and I subunits, respectively (for review see Heilmeyer et al. 1980b; Sperling et al. 1979).

Phosphorylation-dephosphorylation of troponin yields changes of the metal-binding properties; the effect of phosphorylation on the metal-binding properties of phosphorylase kinase can be deduced from the Ca^{2+}- and Mg^{2+}-dependence of three partial phosphorylase kinase activities.

The comparison of the behavior of these two proteins suggests that as a mechanism common to both proteins phosphorylation primarily influences the Mg^{2+}-binding properties which in turn modify Ca^{2+}-binding and consecutively Ca^{2+}-dependent functions.

---

1 Institut für Physiologische Chemie, Lehrstuhl I, Ruhr-Universität, Postfach 102148, 4630 Bochum 1, FRG

2 Results and Discussion

2.1 Comparison of the [^{31}P] NMR Spectra of Troponin, Myosin and Phosphorylase *a*

Isolated rabbit skeletal muscle troponin is composed of one mol subunit T, two mols I, and one mol C (Sperling et al. 1979) and contains one mol phosphate, bound to the N-terminal serine of the T subunit (abbreviated: [P^1] TI_2C) (Pearlstone et al. 1976; Sperling et al., 1979). The latter group can be detected in the [^{31}P] NMR spectrum showing a narrow line width of ~3 Hz (Fig. 1). Usually, a minor peak (less than 20%) is also observed at higher magnetic field. By their chemical shift both signals are identified as phosphoserine. Furthermore, their chemical shifts change upon variation of the pH identically to those of free phosphoserine. Saturation of troponin C with Ca^{2+} or Mg^{2+} does not influence this phosphoserine signal.

Freshly isolated myosin from the same source also contains a phosphate group at serine-14 oder-15 of the light chain 2 (so called "P light chain") (Perrie et al. 1973), which yields a [^{31}P] NMR signal with a chemical shift identical to that of free phosphoserine (Fig. 2). Again, this signal changes with pH ad does that of free phosphoserine; it shows however, a ca. tenfold increased line width in comparison nith the signal of troponin.

The influence of pH on the chemical shift of both phosphoserine, that of troponin T and of the P light chain, indicated that these phosphate groups are exposed to the

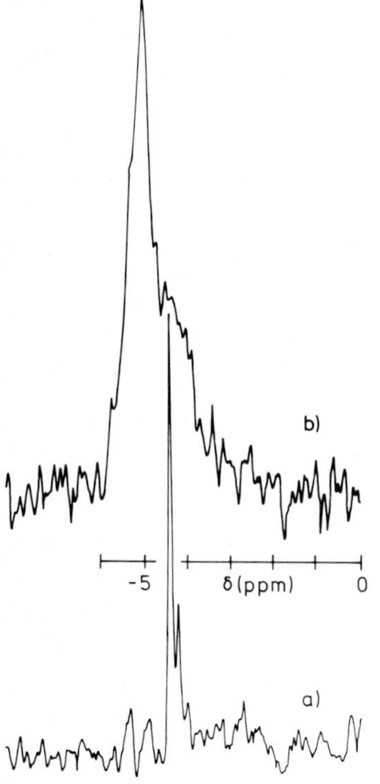

Fig. 1 a, b. [^{31}P] NMR spectra of troponin [P^1] TI_2C (**a**) and phosphorylase a (**b**). Troponin (17 mg/mol) was dialyzed against 0.5 M KCl, 80 mM EGTA, 2 mM Tris/HCl, pH 7.0. The spectrum represents 63,342 scans with an acquisition time of 0.679 s (pulse delay 0.7 s) and a pulse width of 20 μs. Phosphorylase *a* (16 mg/mol) reconstituted with pyridoxal 5′-deoxymethylenephosphonate was dissolved in 100 mM MOPS, 2 mM EDTA, 10 mM 2-mercaptoethanol, 0.2 M NaCl, pH 7.03. The spectrum represents 40,800 scans

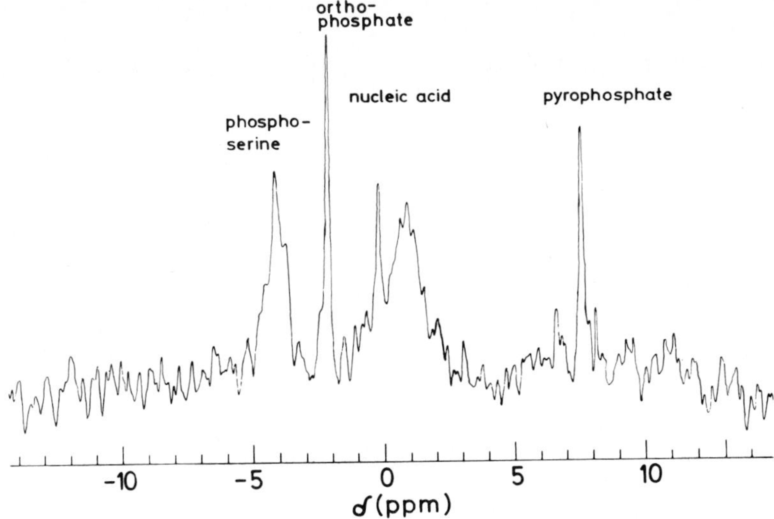

Fig. 2. [^{31}P] NMR spectrum of freshly isolated rabbit skeletal muscle myosin. Myosin (35 mg/mol) was dissolved in 0.5 M KCl, 50 mM Tris/HCl, 15 mM 2-mercaptoethanol, 2 mM EGTA, pH 7.5, 0°C. The spectrum represents 26,000 scans. Orthophosphate, nucleic acids and pyrophosphate are contaminants of the preparation

solvent. The line width of the troponin T signal is identical to that of free phosphoserine, which shows that this phosphate group can freely rotate. The tenfold broader line width of the P light chain signal indicates a somewhat reduced mobility. However, the Stokes-Einstein relationship predicts that τ_c and, hence, the line width is proportional to the molecular weight when the phosphate group is tightly bound to a globular protein. For this case a [^{31}P] line width of ~140 Hz using a M_r of ca. 500,000 and the data of Schnackerz et al. (1979) would be expected. However, myosin is an elongated molecule which would reduce the averaged τ_c value. Therefore, it cannot be decided if the observed line width of ~40 Hz is due to the deviation of myosin from the globular structure or to a significant internal mobility.

The chemical shift of phosphoserine -14 in phosphorylase *a* is expected to be similar to that of the cofactor, pyridoxalphosphate. Therefore, to be able to observe the phosphoserine signal in phosphorylase *a* the apoenzyme was reconstituted with pyridoxal 5′-methylenephosphonate, the chemical shift of which does not interfere with that of phosphoserine-14 (for details see Hörl et al. 1979).

The main part of this phosphoserine signal is observed at a lower magnetic field if measured under the same conditions as troponin (Fig. 1). Furthermore, it is insensitive to pH changes. Part of the phosphoserine of phosphorylase *a* shows a signal of approximately the same chemical shift as that of free phosphoserine at the corresponding pH. However, the signal shows a broader line width.

The difference in line width and chemical shift of the main signal of phosphorylase *a* from that of free phosphoserine can be explained by the formation of an intramolecular salt bridge probably to arginine-96. The broad line width of this signal is mainly due to the fixation of the residue to the protein. Only a minor contribution (< 5 Hz) results from the slow exchange with a portion of unliganded phosphoserine. This can be deduced from spectra taken at pH 6.5 where only the "salt-bridged" form is found with a line width of ca. 35 Hz (not shown).

Thus phosphate groups introduced into proteins may function in two alternative ways, as schematically shown in Fig. 3. Like in phosphorylase *a*, the phosphate group may anchor the N-terminal domain to the protein by an intramolecular salt bridge. In contrast, like in myosin and troponin, a phosphate group may prevent an interaction of the N-terminal domain with the proteins; splitting of the phosphoserine may allow then such an interaction of the N-terminal domain with the protein.

2.2 Comparison of Ca^{2+} and Mg^{2+} Binding to Isolated Calmodulin from Phosphorylase Kinase and to Holophosphorylase Kinase $(\alpha\,\beta\,\gamma\,\delta)_4$

Homogeneous calmodulin, isolated from phosphorylase kinase, probably possesses six divalent cation binding sites (Fig. 4) which can be observed in buffer of low ionic strength (Table 1). Increase of the salt concentration allows to differentiate four binding sites into two Ca^{2+}/Mg^{2+} high-affinity and two Ca^{2+}-specific low-affinity sites (Fig. 4). The two Ca^{2+}/Mg^{2+} sites show positive cooperative interaction, the two Ca^{2+}-specific sites are observable in presence of 10 mM Mg^{2+}.

Two Mg^{2+}-specific low-affinity sites have been demonstrated to exist on troponin C (Potter and Gergely 1975). By analogy, similar sites may be present on isolated calmodulin, as can be concluded from the total binding capacity of six divalent cations (Fig. 4). The binding constants are given in Table 1.

Calmodulin integrated into holophosphorylase kinase shows a ca. 200-fold increased affinity for Ca^{2+} at the Ca^{2+}/Mg^{2+} sites and no positive cooperative interaction as measured in buffer of high ionic strength (Table 1).

The affinity of the Ca^{2+}-specific sites seems to be reduced to such an extent that they cannot be observed in absence of Mg^{2+} in a Ca^{2+} concentration range up to 20 μM (Fig. 4). In presence of 10 mM Mg^{2+}, which possibly binds to the Mg^{2+}-specific sites of calmodulin, the Ca^{2+}-specific low-affinity sites are induced (Fig. 4). These

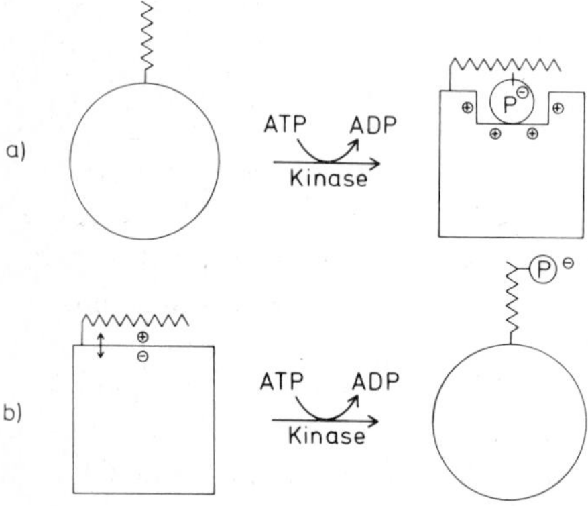

Fig. 3a, b. Schematic representation of interactions of phosphorylated and dephosphorylated N-terminal peptide domains within a polypeptide. The phosphate group may either assure **(a)** or prevent **(b)** an interaction of an N-terminal domain with the protein. For discussion see text

Table 1. Comparison of Ca^{2+} binding to calmodulin in the isolated form or integrated into phosphorylase kinase[a]

	Calmodulin				Holophosphorylase kinase (calculated for $\alpha, \beta, \gamma, \delta$)	
Conditions	Site	Number	K_D (Ca^{2+}) (M)	n_H	Number	K_D (Ca^{2+}) (M)
Low ionic strength		4	5.8×10^{-8}	–	3 1	1.6×10^{-8} 5.9×10^{-7}
1 mM sodium-glycerol 2-phosphate, pH 7.0		2	2.6×10^{-5}		2	2.6×10^{-5}
High ionic strength	Ca^{2+}/Mg^{2+}	2	4.6×10^{-6}	2.0	2	1.8×10^{-8}
1 mM sodium-glycerol 2-phosphate, 180 mM NH_4Cl, pH 7.0	$Ca_{specific}$	2	5.0×10^{-5}		–	–
1 mM sodium-glycerol 2-phosphate, 155 mM NH_4Cl	Ca^{2+}/Mg^{2+}	2	6.6×10^{-5} [b]	2.0	2	3.3×10^{-7}
10 mM $MgCl_2$, pH 7.0	$Ca_{specific}$	2	2.8×10^{-5}		2	2.9×10^{-6}
			K_D (Mg^{2+}) 7.5×10^{-4}			K_D (Mg^{2+}) 5.9×10^{-4}

[a] The experimental data were fitted to the model of two Ca^{2+}/Mg^{2+}, two Ca^{2+}-specific and two binding sites which were observed as Ca^{2+}-binding sites at low ionic strength but which physiologically are probably Mg^{2+}-binding sites. The program is described by Kalbitzer and Stehlik (1979)
[b] Calculated from the given K_D (Mg^{2+}) value

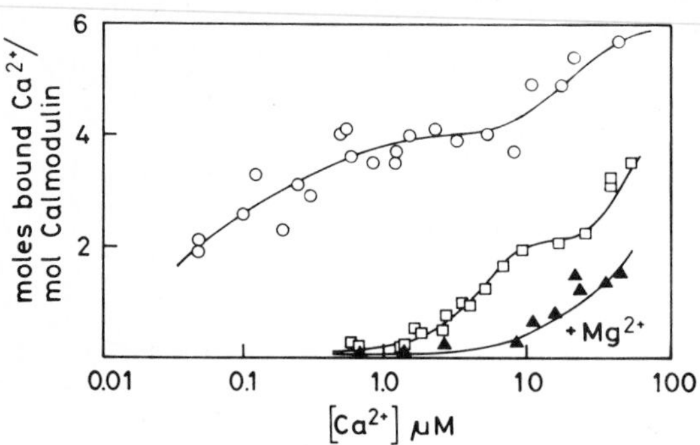

Fig. 4. Effect of Mg^{2+} on the Ca^{2+}-binding properties of isolated calmodulin and phosphorylase kinase. Ca^{2+} binding to calmodulin was carried out in analogy as described by Kilimann and Heilmeyer (1977). In buffer of low ionic strength, 1 mM sodium β-glycerophosphate, (–o–), 4 Ca^{2+} are bound strongly (K_D = 59 nM) and approximately 2 Ca^{2+} weakly (K_D = 26 μM). Addition of 180 mM NH_4Cl (–□–) allows to differentiate two high-affinity Ca^{2+}-binding sites (K_D = 4.6 μM) designated as Ca^{2+}/Mg^{2+} sites since they show Mg^{2+} competition (–▲), and additionally positive cooperativity (Hill coefficient, n_H = 2.0) from two low-affinity sites (K_D = 50 μM) designated as Ca^{2+}-specific sites since they are observed in presence of 10 mM Mg^{2+} and then show a K_D value of 28 μM.
Phosphorylase kinase binds in buffer of high ionic strength (1 mM sodium β-glycerophosphate, 180 mM NH_4Cl, pH 6.8) 2 mol $Ca^{2+}/\alpha\beta\gamma\delta$ (–□–) to the Ca^{2+}/Mg^{2+} high-affinity sites with a K_D value of 18 nM. In presence of additionally 10 mM Mg^{2+} (–▲–) the affinity of these sites is reduced (K'_D = 0.33 μM) and two Ca^{2+}-specific low-affinity sites ($K_{0.5}$ = 2.9 μM) are induced. (Data replotted according to Kilimann and Heilmeyer 1977)

latter sites then show a K'_D value of 3 to 5 μM (Kilimann and Heilmeyer 1977), which is ca. tenfold lower than the corresponding K'_D value of 28 μM of isolated calmodulin (compare legend to Fig. 4). The competition of Ca^{2+} and Mg^{2+} on the high-affinity sites allows to calculate a K_D (Mg^{2+}) value which is found to be remarkably similar in isolated calmodulin, and calmodulin integrated into holophosphorylase kinase (Table 1).

2.3 Comparison of Ca^{2+} and Mg^{2+} Binding to Isolated Troponin C and Holotroponin $[P^1]$ TI_2C

It was shown by Potter and Gergely (1975) that isolated troponin C possesses four Ca^{2+} binding sites which can be subdivided into two Ca^{2+}/Mg^{2+} high-affinity and two Ca^{2+}-specific low-affinity sites (Fig. 5). In addition, two Mg^{2+} specific sites are present (not shown). Holotroponin shows an increased affinity of both the Ca^{2+}/Mg^{2+} and Ca^{2+}-specific sites by factors of ca. 7 and 5, respectively (Jahnke and Heilmeyer 1980). Additionally, the Ca^{2+}/Mg^{2+} sites show apparent positive cooperativity (Hill coefficient $n_H = 2.0$), which is not observed on isolated troponin C (Potter and Gergely 1975). Mg^{2+} seems to influence the cooperative binding behavior of Ca^{2+} in troponin; after saturation of troponin with Mg^{2+}, Ca^{2+} binding to the Ca^{2+}/Mg^{2+} high-affinity sites occurs hyperbolically (Fig. 5).

2.4 Correlation of Ca^{2+} and Mg^{2+} Binding with Function in Phosphorylase Kinase

A basal activity level called partial activity A_o of phosphorylase kinase is expressed between 0.4 nM and 0.02 μM Ca^{2+}, i.e., at Ca^{2+} concentrations at which no binding of this metal to the enzyme occurs. This activity comprises only a small part (20-30 mU/mg) of the enzyme's total potential activity (Fig. 6). With augmentation of the

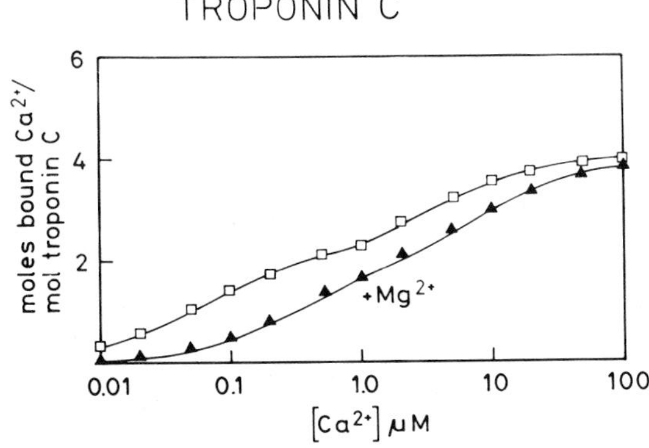

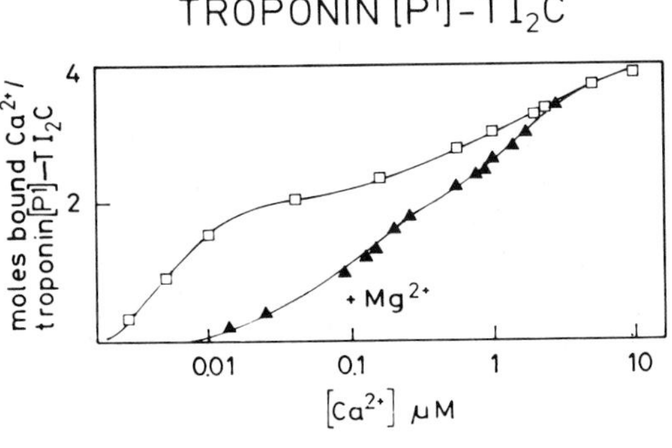

Fig. 5. Effect of Mg^{2+} on the Ca^{2+}-binding properties of troponin C and holotroponin $[P^1]$ TI_2C. The curves of troponin C were calculated with the K_D values given by Potter and Gergely (1975).
Troponin $[P^1]$ TI_2C (–□–) binds 2 mol Ca^{2+} to the Ca^{2+}/Mg^{2+} sites with high affinity ($K_{0.5}$ = 5.3 nM). They show positive cooperative interaction (Hill coefficient, $n_H = 2.0$). 2 mol Ca^{2+} are bound to the Ca^{2+}-specific sites with lower affinity (K_D = 1.1 μM). In presence of 3.5 mM Mg^{2+} (–▲–) the affinity of the Ca^{2+}/Mg^{2+} sites is reduced to 0.13 μM and the Hill coefficient to 1.3. Increase of Mg^{2+} to 10 mM reduces the Hill coefficient to 1.1 (not shown)

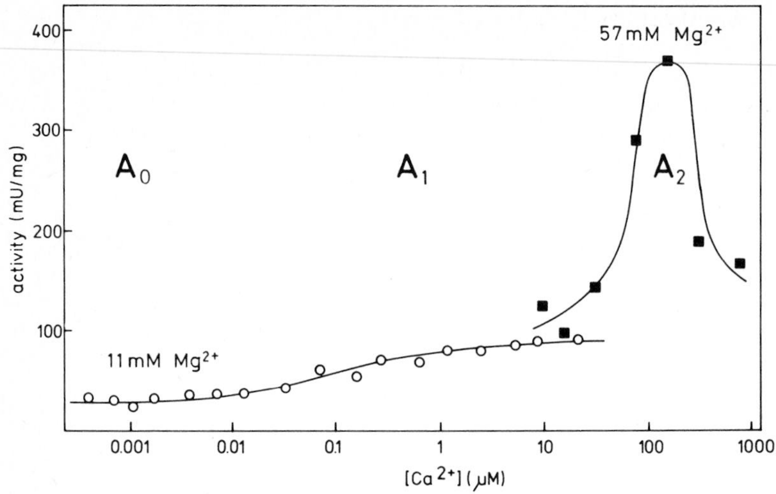

Fig. 6. The three partial activities of phosphorylase kinase as function of Ca^{2+} and Mg^{2+}. The activity of phosphorylase kinase was assayed as function of $[Ca^{2+}]$ and $[Mg^{2+}]$: (○) 11 mM, (■) 57 mM Mg^{2+}, both at pH 6.8 and a total ionic strength of I = 200 mM. Half-activation of the activity A_1 is obtained at ca. 1 μM, and that of A_2 at ca. 70 μM Ca^{2+}

free Ca^{2+} concentration, activity increases in up to two steps, designated A_1 and A_2, (Fig. 6). Most of the total activity (80-150 mU/mg) at pH 6.8 and 10 mM Mg^{2+} is contributed by the partial activity A_1. The $[Ca^{2+}]$ required for half-maximal stimulation is a function of $[Mg^{2+}]$ due to Mg^{2+}/Ca^{2+} competition. It allows the conclusion that Ca^{2+} saturation of the Ca^{2+}/Mg^{2+} high-affinity sites stimulates the partial activity A_1. Increase of Mg^{2+} above 20 mM (at pH 6.8) or pH above 7.5 (at 10 mM Mg^{2+}) induces an additional low-affinity Ca^{2+}-dependent activity A_2. Its K'_m for Ca^{2+} is 10-70 μM (dependent on Mg^{2+} and pH) and the activity is inhibited by $Ca^{2+} >$ 200 μM, thus it constitutes a narrow activity maximum. The induction of this partial activity A_2 by high $[Mg^{2+}]$ correlates with the induction of the Ca^{2+}-specific low-affinity binding sites. Therefore, it is concluded that their Mg^{2+} induction and consecutive Ca^{2+} saturation lead to the expression of the partial activity A_2.

2.5 Correlation of Ca^{2+} and Mg^{2+} Binding with Function in Troponin

Saturation of the Ca^{2+}-specific low-affinity sites coincides with the deinhibition of the actomyosin ATPase (Potter and Gergely 1975). No function was known until now for the Ca^{2+}/Mg^{2+} high-affinity sites. However, the cAMP-dependent protein kinase specifically phosphorylates the troponin I subunit (Sperling et al. 1979). The rate of this phosphorylation is inhibited with Ca^{2+} saturation of the Ca^{2+}/Mg^{2+} high-affinity sites. If Ca^{2+} is replaced by Mg^{2+} on these sites, this troponin I phosphorylation is activated (Fig. 7). A further stimulation of troponin I phosphorylation is observed by an additional increase of the Mg^{2+} concentration which may be correlated to Mg^{2+} binding to their specific sites (Fig. 7).

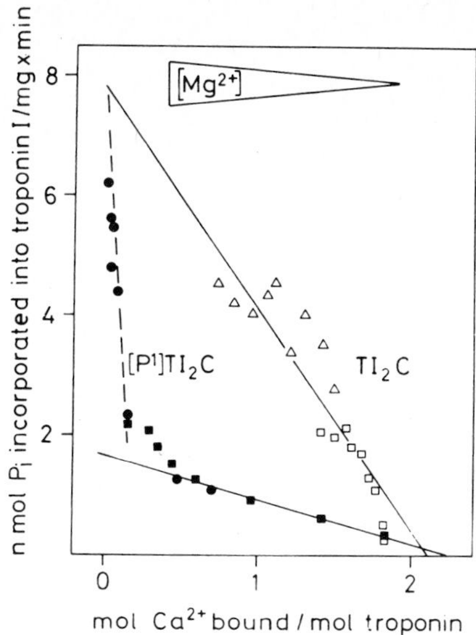

Fig. 7. Deinhibition of troponin I phosphorylation by replacement of Mg^{2+} for Ca^{2+} at the Ca^{2+}/Mg^{2+} high-affinity sites and stimulation by Mg^{2+} binding to the Mg^{2+}-specific sites. The initial phosphate incorporation rate into troponin I by Sperling et al. (1979). Troponin $[P^1]$ TI_2C (●, ■) or dephosphotroponin TI_2C (△, □) were incubated in 50 mM sodium β-glycerophosphate, 15 mM 2-mercaptoethanol, 0.35 mM γ $[^{32}P]$ ATP, pH 6.9, with 15 μg/mol catalytic subunit of the cyclic AMP-dependent protein kinase. Free $[Ca^{2+}]$ was established with 1 mM Ca^{2+}/EGTA buffer: (●) 3-7 x 10^{-9} M, (□, ■) 5-8 x 10^{-8} M, (△) 0.9-2.5 x 10^{-8} M. The free $[Mg^{2+}]$ was varied from 0.2 x 10^{-6} to 14.5 x 10^{-3} M. The amount of Ca^{2+} bound was calculated from the free $[Ca^{2+}]$ and $[Mg^{2+}]$ with the $K_{0.5}$ values for Ca^{2+} and Mg^{2+} given in the legend of Fig. 4

2.6 Phosphorylation of Phosphorylase Kinase and Troponin

Both proteins, phosphorylase kinase and troponin, can be phosphorylated on several hydroxy amino acids. Ca. 1 mol phosphate can be incorporated into each of the α and β subunits of phosphorylase kinase by the cAMP-dependent protein kinase. Into the same polypeptides a higher amount, ca. 20 mol phosphate, can be incorporated autocatalytically in a Ca^{2+}-dependent reaction which is catalyzed by the partial acitivity A_1. In this instance Ca^{2+} binding to the high-affinity sites not only activates the partial activity A_1 but also increases the final level of phosphate incorporation, i.e., it determines the availability of serine or threonine residues for phosphorylation (Heilmeyer et al. 1980a).

Troponin I is phosphorylated by the cAMP-dependent protein kinase and also by the partial activity A_O of phosphorylase kinase only if troponin C contains no bound Ca^{2+} (Kilimann and Heilmeyer 1979; Sperling et al. 1979). Ca^{2+} saturation of phosphorylase kinase on the low-affinity sites activates the partial activity A_2, which then can phosphorylate the troponin T subunit (Kilimann and Heilmeyer 1979). The presence of other protein substrates does not interfere with the rate of troponin T phosphorylation, thus A_O, A_1, and A_2 may represent different catalytical centers (Dickneite et al. 1978; Kilimann and Heilmeyer 1979).

Additionally, both freshly isolated proteins, troponin and phosphorylase kinase, contain phosphate endogenously. Phosphorylase kinase contains approx. 2 mol per α, β, γ, δ (Kilimann and Heilmeyer, unpublished; cf. Mayer and Krebs 1970), skeletal muscle troponin ca. 1 mol per troponin $[P^1]$ TI_2C, which is bound to the N-terminal serine of troponin T (Pearlstone et al. 1976; Sperling et al. 1979).

2.7 Function of Phosphorylation of Phosphorylase Kinase

Phosphorylase kinase in which predominantly the β subunit was phosphorylated by the cAMP dependent protein kinase shows a selective enhancement of the V_{max} value of the low-affinity Ca^{2+}-dependent activity A_2 (Fig. 8).

The two other partial activities A_0 and A_1 are essentially uninfluenced. Alternatively, the α and β subunits were phosphorylated autocatalytically to an extent of ca. 3 and 1 mol phosphate, respectively. It results in an increase of both V_{max} values of A_1 and A_2, again A_0 remaining essentially uninfluenced. Both kinds of phosphorylations increase the V_{max} values; the respective K'_m values for Ca^{2+} activation are essentially identical to those values determined for the nonmodified enzyme (Heilmeyer et al. 1980b).

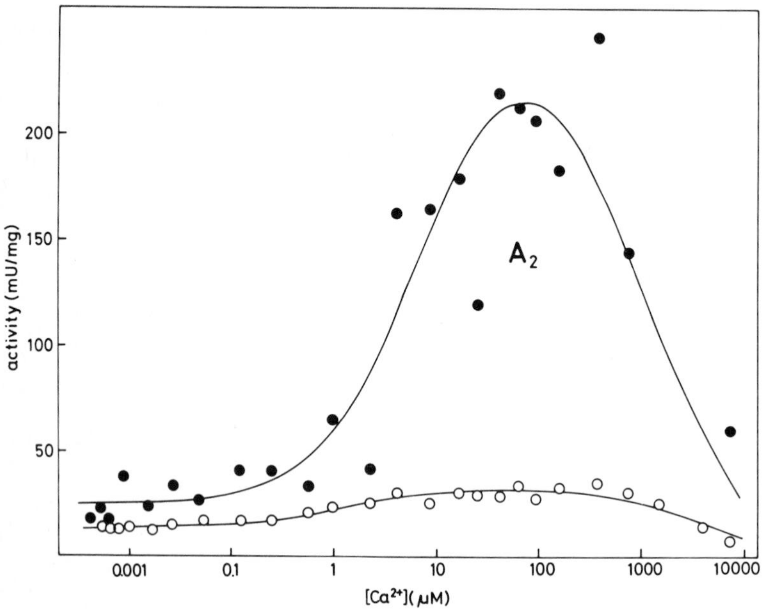

Fig. 8. Activation of the partial activity A_2 by β subunit phosphorylation. Phosphorylase kinase was phosphorylated by the catalytic subunit of the cAMP dependent protein kinase to an extent of 0.44 mol/β and 0.08 mols/α subunit. The activity of this enzyme (•) was compared to that of the nonmodified enzyme (○) as function of [Ca^{2+}] in presence of 10 mM free Mg^{2+}, 6 mM ATP-Mg at pH 6.5

An increase of the V_{max} value of the partial activity A_2 can also be brought about by an increase of the Mg^{2+} concentration (Fig. 6). An increase of pH results also in an increase of the V_{max} of A_2 (Heilmeyer et al. 1980b) due to an enhancement of the Mg^{2+} affinity. Therefore, one might conclude that phosphorylation of the enzyme leads to an increase of the Mg^{2+} affinity, which consecutively induces the low-affinity Ca^{2+}-specific sites. Their saturation with Ca^{2+} results in an expression of the partial activity A_2.

2.8 Function of the N-terminal Phosphoserine in Troponin T

Phosphoserine -1, which is present in freshly isolated troponin T, can be hydrolyzed by alkaline phosphatase (Sperling et al. 1979). The Ca^{2+}-binding properties of the $[P^1]TI_2C$ and TI_2C troponin complexes do not differ significantly when determined in absence of Mg^{2+}. However, Mg^{2+} competes effectively with Ca^{2+} on the high-affinity Ca^{2+}/Mg^{2+} sites of phosphotroponin only (Fig. 9). The dephosphorylated complex TI_2C shows a much weaker Mg^{2+} competition. In agreement with this observation, Mg^{2+} binding at the Ca^{2+}/Mg^{2+} sites of dephosphotroponin TI_2C is ca. 30-fold weaker than that of phosphotroponin $[P^1]TI_2C$. The affinity of the Mg^{2+}-specific sites is approximately identical in both troponin complexes (Fig. 9). The calculated dissociation constants are given in Tables 2 and 3.

The observed difference in Mg^{2+} affinity explains the differential influence of this ion on the troponin I phosphorylation of phospho- and dephosphotroponin catalyzed by the cAMP-dependent protein kinase. Upon increase of the Mg^{2+} concentration in phosphotroponin first Ca^{2+} is replaced by Mg^{2+} at the Ca^{2+}/Mg^{2+} high-affinity sites, which deinhibits troponin I phosphorylation, and then Mg^{2+} binds to th Mg^{2+}-specific sites, which stimulates troponin I phosphorylation. In contrast to that, in dephosphotroponin TI_2C first the Mg^{2+}-specific sites are saturated and then

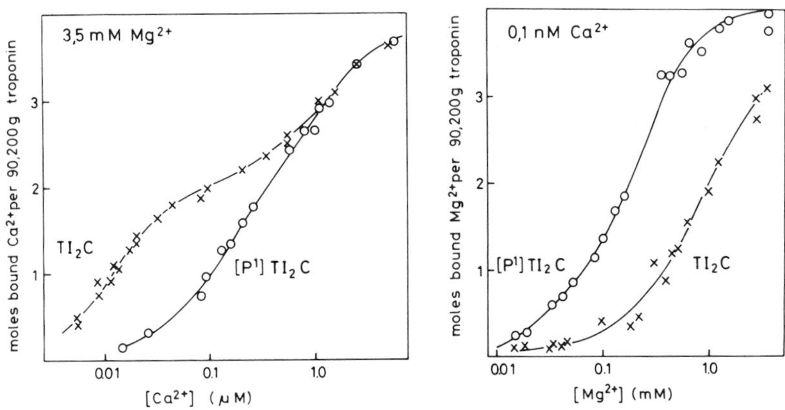

Fig. 9. Comparison of Ca^{2+} and Mg^{2+} binding to phospho- and dephosphotroponin $[P^1]TI_2C$ and TI_2C. Ca^{2+} and Mg^{2+} binding to troponin $[P^1]$ TI_2C (○) and dephosphotroponin TI_2C (x) as well as the preparation of the troponin complexes was carried out as described by Jahnke and Heilmeyer (1980)

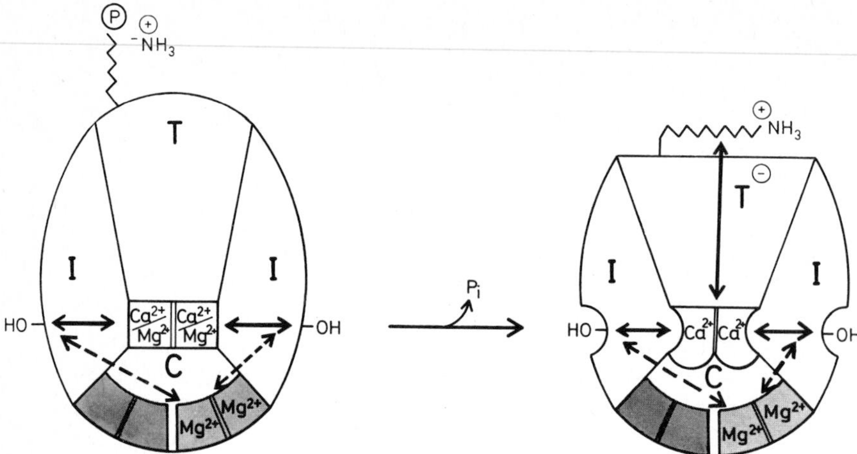

Fig. 10. Intramolecular interaction in phosphotroponin [P^1]-TI_2C and dephosphotroponin TI_2C. For explanation see text

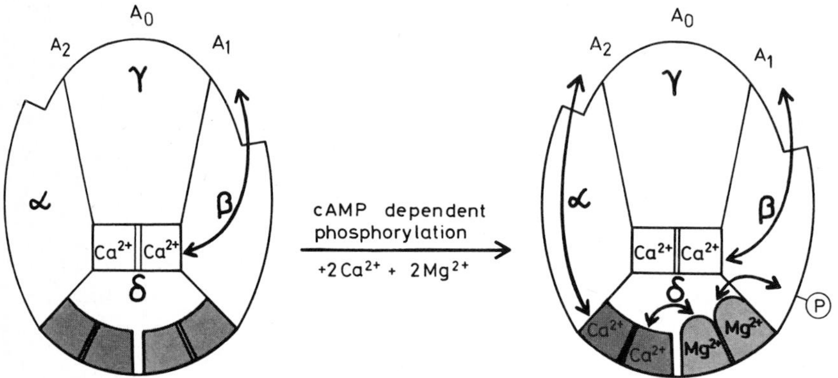

Fig. 11. Intramolecular cascade in phosphorylase kinase from β subunit phosphorylation to A_2 activation. For explanation see text

Ca^{2+} is replaced by Mg^{2+} on the Ca^{2+}/Mg^{2+} high-affinity sites. Therefore, in dephosphotroponin Mg^{2+} stimulation and deinhibition of troponin I phosphorylation ar observed concomitantly and in phosphotroponin separately (cf. Fig. 7).

2.9 Intramolecular Cascades

The phosphate group at serine -1 of troponin T determines the Mg^{2+} affinity of the Ca^{2+}/Mg^{2+} high-affinity sites, i.e., it regulates Ca^{2+} binding by the effectivity of Mg^{2+} for competition on these sites. Since Ca^{2+} binding at these sites limits the degree of phosphorylatability of troponin I, troponin T can regulate the troponin I phosphorylation (Fig. 10).

In case of phosphorylase kinase phosphorylation of the β subunit enhances the V_{max} value of the partial activity A_2. This seems to be mediated by an increase of Mg^{2+} affinity (Fig. 11). Thus, phosphorylation of the β subunit may induce a change of the Mg^{2+} affinity which controls Ca^{2+} binding at the low-affinity sites, and by this way expression of the partial activity, A_2.

Acknowledgements. This work was supported by the Minister für Wissenschaft und Forschung des Landes Nordrhein-Westfalen and the Fonds der Chemie and a grant from the Deutsche Forschungsgemeinschaft He 595/10 and 11.

References

Cohen P, Antoniw JF (1973) The control of phosphorylase kinase phosphatase by "second site phosphorylation"; a new form of enzyme regulation. FEBS Lett 34: 43

Cohen P, Burchell A, Folkes JG, Cohen PTW, Vanaman TC, Nairn AC (1978) Identification of the Ca^{2+} dependent modulator protein as the fourth subunit of rabbit skeletal muscle phosphorylase kinase. FEBS Lett 92: 287

Dickneite G, Jennissen HP, Heilmeyer LMG Jr. (1978) Differentiation of two catalytic sites on phosphorylase kinase for phosphorylase *b* and troponin T phosphorylation. FEBS Lett 87: 297

Ebashi S (1980) Regulation of muscle contraction (The Croonian Lecture), Proc R Soc London Ser B 207: 259

Heilmeyer LMG Jr, Jahnke U, Kilimann MW, Kohse KP, Sperling JE, Varsanyi M (1980a) Troponin C and calcium dependent regulator protein, two ancestral skeletal muscle calcium binding proteins, BioSystems 12: 317

Heilmeyer LMG Jr, Gröschel-Stewart U, Jahnke U, Kilimann MW, Kohse KP, Varsanyi M (1980b) Novel aspects of skeletal muscle protein kinase and protein phosphatase regulation by Ca^{2+}, Adv Enzyme Regul 18: 121

Hörl M, Feldmann K, Schnackerz KD, Helmreich EJM (1979) Ionization of pyridoxal 5′-phosphate and the interactions of AMP-S and Thiophosphoseryl residues in native and succinylated rabbit muscle glycogen phosphorylase *b* and *a* as inferred from ^{31}P NMR spectra, Biochemistry 18: 2457

Jahnke U. Heilmeyer LMG Jr (1980) Comparison of the Mg^{2+} and Ca^{2+} binding properties of troponin [P^1]-TI_2C and TI_2C, Eur J Biochem 111:325

Kalbitzer HR, Stehlik D (1979) On the analysis of competitive binding of various ligands to cooperative and independent binding sites of macromolecules. Z Naturfosch 34 c: 757

Kilimann MW, Heilmeyer LMG Jr (1977) The effect of Mg^{2+} on the Ca^{2+} binding properties of non-activated phosphorylase kinase. Eur J Biochem 73: 191

Kilimann MW, Heilmeyer LMG Jr (1978) Multiple enzymatic activities of phosphorylase kinase. Hoppe Seyler's Z Physiol Chem 359: 1104

Kilimann MW, Heilmeyer LMG Jr (1979) Different protein substrate specificities of the three enzymatic activities of phosphorylase kinase. Hoppe Seyler's Z Physiol Chem 360: 299

Koppitz B, Feldmann K, Heilmeyer LMG Jr (1980) The P-light chain of rabbit skeletal muscle myosin: A ^{31}P NMR study. FEBS Lett 117: 199

Krebs EG, Beavo JA (1979) Phosphorylation-dephosphorylation of enzymes, Annu Rev Biochem 48: 923

Mayer SE, Krebs EG (1970) Studies on the phosphorylation and activation of skeletal muscle phosphorylase and phosphorylase kinase in vivo. J Biol Chem 245: 3153

Pearlstone JR, Carpenter MR, Johnson P, Smilie LB (1976) Amino acid sequence of tropomyosin binding component of rabbit skeletal muscle troponin. Proc Natl. Acad Sci USA 73: 1902

Perrie WT, Smillie LB, Perry SV (1973) A phosphorylated light chain component of myosin from skeletal muscle. Biochem J 135: 151

Potter JD, Gergely J (1975) The calcium and magnesium binding sites on troponin and their role in the regulation of myofibrillar adenosine triphosphatase. J Biol Chem 250: 4628

Schnackerz KD, Feldmann K, Hull WE (1979) Phosphorus-31 nuclear magnetic resonance study of D-serine dehydratase: Pyridoxal phosphate binding site. Biochemistry 18: 1536

Sperling JE, Feldmann K, Meyer H, Jahnke U. Heilmeyer LMG Jr (1979) Isolation, characterization and phosphorylation pattern of the troponin complexes TI_2C and I_2C. Eur J Biochem 101:581

Vanaman TC, Sharief F, Watterson DM (1977) Structural homology between brain modulator protein and muscle TnCs, In: Wassermann RH et al. (eds) Calcium-binding proteins and calcium function. North Holland, New York, p 107

Discussion[2]

Betz remarked that concentrations of 100 μM Ca^{2+} or 50 mM Mg^{2+} as required for stmulation of the phosphorylase kinase activities A_1 and A_2 could hardly be reached under physiological conditions. Heilmeyer answered that 50 mM Mg^{2+} is only required to stimulate the A_2 activity in the nonphosphorylated enzyme whereas in the phosphorylated enzyme 3 mM Mg^{2+} is sufficient for stimulation. Ten to 30 μM Ca^{2+} is sufficient for half-saturating the partial activity A_2. This concentration may be reached, e.g., in tetanic stimulated muscle.

Cohen mentioned that his own results on the effect of phosphorylation on the Ca^{2+} sensitivity don't seem to agree with Dr. Heilmyer's. The apparent activation constant is 20 μM Ca^{2+} in the dephospho form of the enzyme. However, when the enzyme is fully phosphorylated, a 15-fold increase in V_{max} and a 15-fold decrease in the activation constant for Ca^{2+} take place. Dr. Cohen's interpretation is that the dephospho form is activated by $(Ca^{2+})_4$-calmodulin, and the phosphorylated enzyme is activated by $(Ca^{2+})_2$-calmodulin. Heilmeyer showed some additional data obtained with an enzyme phosphorylated to a higher degree. As one could see, not only the partial activity A_2 but also A_1 increased upon phosphorylation of the enzyme. He suggested that the apparent increase in affinity for Ca^{2+} seen by Dr. Cohen is due to the increase in the partial activity A_1.

Demaille mentioned that according to M. and K. Barany, phosphorylation of myosin light chain could play a role in actin-myosin interaction through interaction of the phosphate group with the actin-bound Ca^{2+}. With the NMR technique one should be able to test this hypothesis by using actomyosin containing a phosphorylated light chain. Heilmeyer answered that unfortunately rabbit actomyosin is insoluble in low salt concentrations; therefore it is difficult to obtain NMR signals.

2 Rapporteur: Wolfgang Burgermeister

Bacterial Prohistidine Decarboxylase: Kinetics of Conversion to the Active Enzyme

P.A. Recsei and E.E. Snell[1]

1 Introduction

The histidine decarboxylase from *Lactobacillus sp. 30a* was first isolated in pure form by Rosenthaler et al. (1965). Although it lacked pyridoxal-5′-phosphate as coenzyme, it was inactivated by carbonyl group reagents. The essential carbonyl group was shown to be present in an amide-bound pyruvoyl group (Riley and Snell 1968), which participates in the catalytic cycle of the enzyme (Recsei and Snell 1970).

This decarboxylase is the prototype of several amino acid decarboxylases that have since proved to be pyruvoyl enzymes: (1) a very similar histidine decarboxylase from *Micrococcus n.sp.* (Alekseeva and Prozorovskii 1976); (2) the membrane-bound phosphatidylserine decarboxylase from *Escherichia coli* B (Satre and Kennedy 1978); (3) adenosyl-methionine decarboxylase from *E. coli* W. (Wickner et al. (1970), yeast (Cohn et al. 1977), and mammalian liver (Demetriou et al. 1978); and (4) aspartate α-decarboxylase from *E. coli* (Williamson and Brown 1979). Substantial interest attaches to the mechanism by which the protein-bound pyruvoyl group functions in place of the pyridoxyl-5′-P present at the catalytic center of the more familiar amino acid decarboxylases, and how this pyruvoyl residue arises during biosynthesis of the enzyme.

Our limited knowledge in this area has been reviewed previously (Snell et al. 1976; Snell 1977). Briefly, the pyruvoyl residue of histidine decarboxylase arises without dilution from a serine residue (Riley and Snell 1970) of a single precursor peptide subunit (the π subunit, $M_R \cong 37{,}000$) of an inactive proenzyme (Recsei and Snell 1973) simply on incubation at 37°C at pH values >7.0. During this process, each π subunit gives rise to the pyruvoyl peptide (the α subunit, $M_R \cong 28{,}000$) and the β subunit ($M_R \cong 9{,}000$) present in the active enzyme. Since no other cleavage products except ammonia could be detected, the activation process can be formulated as shown in Fig. 1. This figure has been published before (Snell 1977), but unfortunately still summarizes our knowledge of the overall activation process. However, we have extended our knowledge of the kinetics and some other characteristics of the conversion process, and it is these results that I shall describe today. I should add that accumulation of prohistidine decarboxylase (as opposed to the active enzyme) has so far been observed only in a mutant strain of *Lactobacillus 30a,* and kinetic results to be described here were necessarily obtained with this mutant proenzyme. Since the nature of the mutation is not known, the findings may or may not pertain to the corresponding process in the parent culture.

1 Departments of Microbiology and Chemistry, The University of Texas, Austin, TX 78712, USA

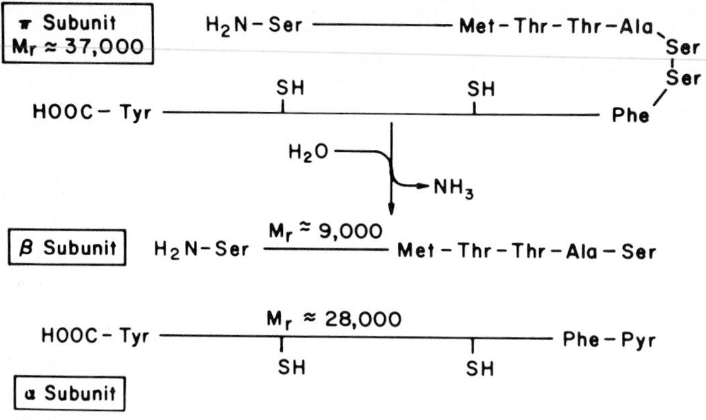

Fig. 1. Chemical changes during the overall conversion of a single (π) subunit of prohistidine decarboxylase to the light (β) and heavy (α) subunits of the active enzyme. Snell (1977)

2 Kinetics of Activation of Prohistidine Decarboxylase

The conversion of prohistidine decarboxylase to the active enzyme is activated by monovalent metal ions and is sensitive to pH. The initial rate of activation of the purified proenzyme at pH 7.6 and 37°C determined over a 300-fold range of protein concentration, using both the radioactive and manometric assays for decarboxylase activity, is directly proportional to the protein concentration (Fig. 2), and is therefore first order with respect to total protein. In 0.45 M potassium phosphate[2], pH 7.6, 37°C, the initial rate of activation follows the equation $V_{initial} = 0.08\ h^{-1}$. [protein].

The time course for activation of the proenzyme in varying concentrations of potassium phosphate at pH 7.6, 37° C, is shown in Fig. 3. In each case the proenzyme

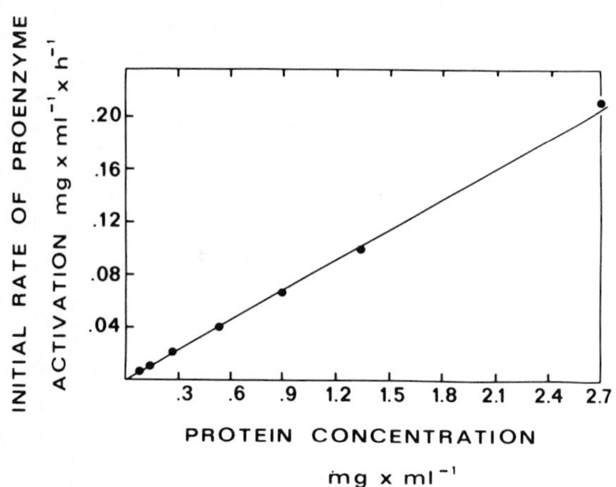

Fig. 2. First-order relationship between initial rate of proenzyme activation and the protein concentration. Activation was followed at 37°C in potassium phosphate buffer (0.45 M in K^+) at pH 7.6

2 Concentrations of all solutions are given in terms of the molarity of the monovalent cation

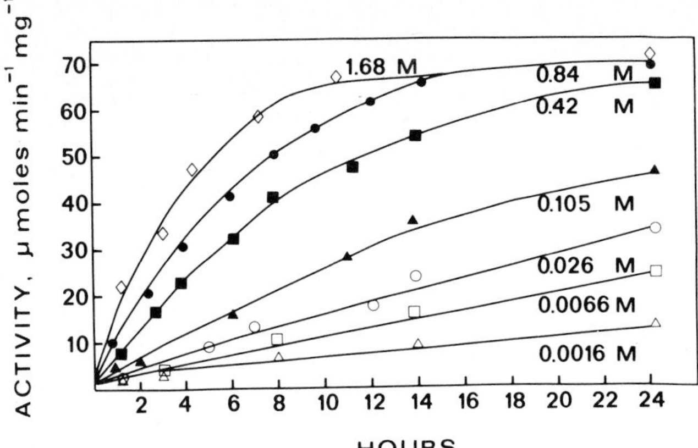

Fig. 3. Effect of concentration of potassium ion (the molarity is given on each curve) on the rate of conversion of proenzyme (1.0 mg ml^{-1}) to active histidine decarboxylase at pH 7.6 and 37°

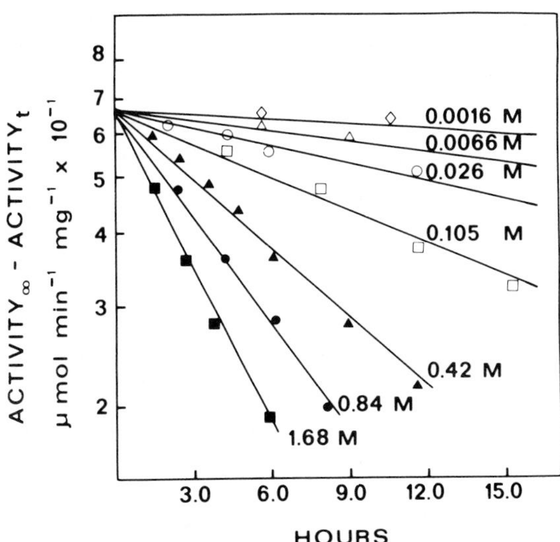

Fig. 4. Replot of data of Fig. 3 showing the first-order relationship between proenzyme activation and proenzyme concentration at all concentrations of K^+

is completely activated after a sufficient time of incubation. Plots of the logarithm of the difference between final activity and activity at time t vs time are linear to at least 98% activation at all concentrations of potassium phosphate tested (Fig. 4 and Table 1). Activation is therefore first order with respect to proenzyme. In 0.42 M potassium phosphate, pH 7.6, 37°C the rate of activation follows the equation V = 0.10 h^{-1} . [proenzyme]

The apparent first-order rate constants calculated from the data shown in Table 1 are proportional to the square root of the potassium ion concentration over the 1050-fold concentration range studied (Fig. 5). Activation is therefore half order with respect

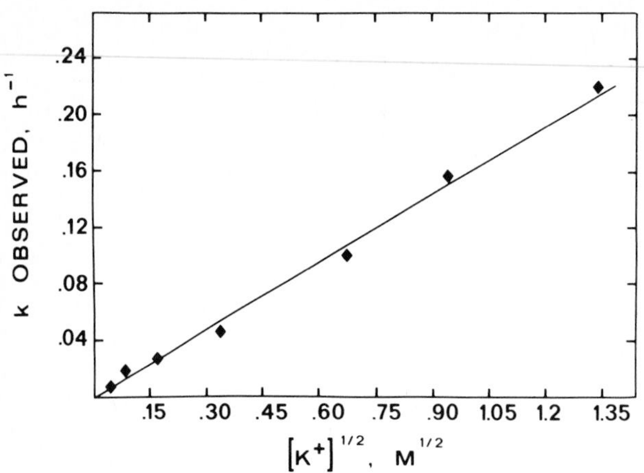

Fig. 5. Half-order relationship between the observed rate constants for proenzyme activation (calculated from Fig. 4) and the concentration of potassium ion

Table 1. Activation of prohistidine decarboxylase in potassium phosphate at pH 7.6 and 37°C. Least squares analysis of kinetic data fitted to the equation, $\text{Log}\ (A_\infty\text{-}A_t) = -\frac{\text{kobs}}{2.303}\ t$

Potassium ion concentration	No. of points	Negative slope[a]	r[b]
0.0016 M	7	.0032 ± .0002	.992
0.0066 M	6	.0065 ± .0002	.998
0.026 M	6	.0104 ± .0003	.998
0.105 M	7	.0186 ± .0008	.995
0.42	8	.040 ± .001	.998
0.84 M	7	.065 ± .002	.997
1.68 M	6	.094 ± .003	.998

[a] Slope ± standard deviation of the slope of least squares line
[b] Correlation coefficient of least squares line

to potassium ion and one and a half order overall. In potassium phosphate, pH 7.6, 37°, the rate follows the equation $V = 0.17\ h^{-1}$. [proenzyme] $[K^+]^{1/2}$. At the concentrations of K^+ so far tested, no saturation is observed, and there is therefore no direct evidence for complex formation between proenzyme and K^+ prior to activation.

Rates of proenzyme activation in the presence of potassium phosphate, potassium pyrophosphate, potassium chloride, ammonium phosphate, ammonium sulfate, and sodium phosphate under varying conditions of pH and temperature are given in Table 2.

In each case the rate of activation can be fitted to the equation V = k (proenzyme) $(\text{monovalent cation})^{1/2}$. The nature of the counter ion in solution (i.e., phosphate, sulfate, chloride, pyrophosphate) has little effect on the rate, a result that demonstrates that stimulation of proenzyme activation is not related simply to the ionic strength of the medium. This is also clear from the fact that plots of log k_{vs} the square root of the ionic strength are not linear, even at low ionic strengths, whereas linearity is required by the Debye-Hückel equation ($\log k = \log k_o + 1.04\ Z_A Z_B\ [\text{ionic strength}^{1/2}]$),

Table 2. Dependence of rate of activation of prohistidine decarboxylase upon monovalent cation concentration. Least Squares Analysis of Kinetic Data Fitted to $k_{obs} = k\ [Cation^+]^{1/2}$

Conditions of activation		No. of points	Slope[a] $hr.^{-1}\ M^{-1/2}$	r[b]
Potassium phosphate, 37°C, pH	6.1	5	.029 ± .003	.981
	6.6	7	.055 ± .002	.997
	7.1	7	.15 ± .01	.987
	7.6	7	.166 ± .004	.998
	8.0	7	.128 ± .007	.992
Potassium pyrophosphate, 37° pH	8.5	3	.05 ± .01	.96
Potassium phosphate, pH 7.6,	10°	3	.014 ± .01	.995
	20°	4	.044 ± .004	.992
	30°	4	.114 ± .005	.998
	37°	7	.166 ± .004	.998
	45°	3	.24 ± .05	.982
	50°	1	.29	
Ammonium phosphate, pH 7.6,	37°	7	.137 ± .005	.997
Ammonium sulfate, pH 7.6,	37°	5	.155 ± .006	.997
Potassium chloride, pH 7.6,	37°	6	.149 ± .002	.999
Potassium phosphate, pH 7.6,	37°	7	.166 ± .004	.998
Sodium phosphate, pH 7.6,	37°	6	.035 ± .005	.958

[a] Slope ± standard deviation of the slope of least squares line
[b] Correlation coefficient of least squares line

relating ionic strength to the rate of reaction between two species with charges Z_A and Z_B.

Activation of the proenzyme in the presence of various monovalent cations (0.9 M solutions of the chloride salt in 7 mM sodium phosphate, pH 7.6, 37°) shows the order of effectiveness $K^+ > NH_4^+ > Rb^+ > Na^+ \sim Cs^+ > Li^+$. This order correlates well with the crystal ionic volume (Fig. 6) of the cations but not with their hydrated volume. Either an increase or decrease in unhydrated ion size from that of K^+ leads to a progressive loss of effectiveness. No such relationship is observed between the size of the hydrated ion and activator effectiveness. For example, monovalent potassium, cesium, and rubidium have very similar hydrated radii but have dissimilar effects on the rate of activation. It is therefore most likely that the unhydrated ion interacts with the proenzyme in the rate-limiting step of activation. Thallous ion, with an unhydrated ionic volume between that of K^+ and Rb^+, is also effective in promoting activation, but partially inactivates the activated proenzyme, possibly due to its reactivity with sulfhydryl groups. Mg^{2+} and Ca^{2+} do not stimulate activation.

The effect of temperature on the rate of activation in potassium phosphate, pH 7.6, plotted according to the Arrhenius equation (Fig. 7) shows an energy of activation of 15.2 kcal x mol^{-1}; at 37°, the enthalpy of activation ($\Delta H^{\neq}$) equals 14.6 kcal x mol^{-1}, and the entropy of activation ($\Delta S^{\neq}$) equals −31 e.u. The high negative entropy change may reflect a rate-limiting conformational change of the proenzyme or the requirement for ordering of reactants involved in a rate-limiting bond-making or bond-breaking step.

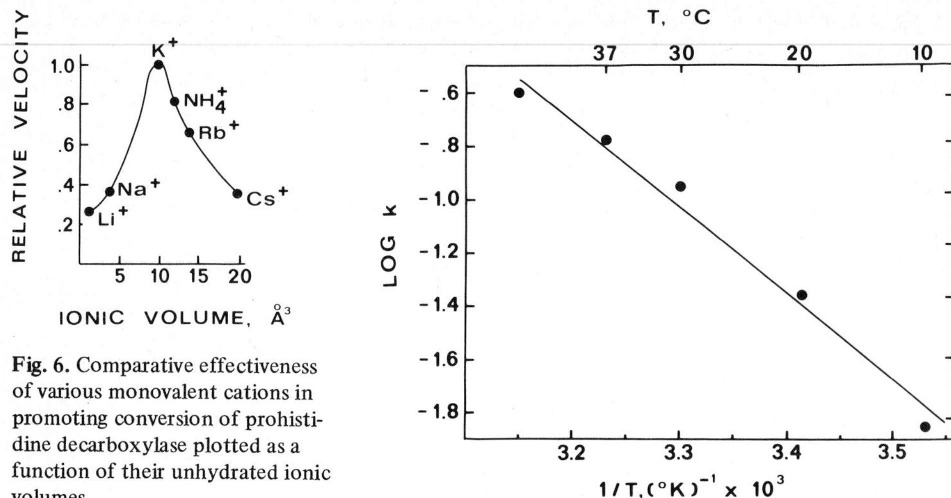

Fig. 6. Comparative effectiveness of various monovalent cations in promoting conversion of prohistidine decarboxylase plotted as a function of their unhydrated ionic volumes

Fig. 7. Arrhenius plot showing the rate of proenzyme activation as a function of termperature

The effect of pH on the rate of proenzyme activation at 37° in potassium phosphate (pH 6.1 to 8.0) and potassium pyrophosphate (pH 8.5) is shown in Fig. 8. Optimum activation occurs around pH 7.6. Slopes (n) of the linear ascending and descending portions of the curve indicate that the acidic titration of a group with pK near 6.8 and the basic titration of a group with pK near 8.2 lead to an appreciable decrease in rate.

Single crystals of proenzyme have been isolated, washed thoroughly with buffer, and activated under the usual conditions following dissolution. The result indicates that pure proenzyme free from contaminating protein retains the capacity to activate.

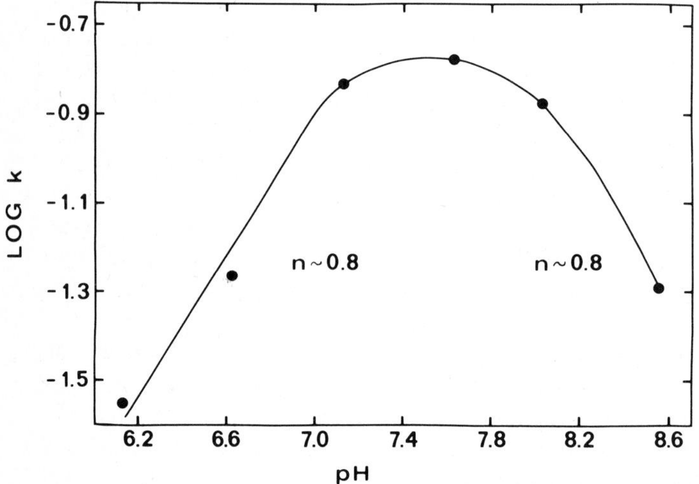

Fig. 8. Effect of pH on the rate of activation of prohistidine decarboxylase

Table 3. Summary of kinetic properties of immobilized and free proenzyme preparations[a]

Sample	Observed rate constant for activation	Kinetic properties of activated preparation Km	V_{max}
Proenzyme	0.17 h^{-1}	0.4 mM	65 μmol/min/mg
Activated proenzyme coupled to Sepharose[b]		0.4 mM	19 μmol/min/mg
Proenzyme coupled to Sepharose	0.20 h^{-1}	0.3 mM	11 μmol/min/mg
Proenzyme coupled to Sepharose, treated with 9M urea[b]	0.43 h^{-1}	0.3 mM	3.8 μmol/min/mg

[a] Activation was in 0.94M potassium phosphate, pH 7.6, 37°, for 24 h
Activity was assayed in 0.2M ammonium acetate, pH 4.8
[b] 5 mg CNBr and 0.39 mg of protein were added per ml of Sepharose
Coupling was in 0.05 M sodium citrate, pH 6.2

This result is confirmed by the observation in Table 3, which shows that proenzyme immobilized to cyanogen bromide-activated Sepharose also is converted to active enzyme by incubation with monovalent cations. Activation is first order with respect to proenzyme with a rate similar to that observed with free proenzyme. Up to 60% of the Sepharose-bound protein can be removed by treatment with 9 M urea in 0.2 M ammonium acetate, pH 4.8 for 50 h at 23°. Following gradual dilution of the urea solution with buffer, the remaining Sepharose-bound protein activates under the usual conditions. Activation is again first order with respect to proenzyme but with a two-fold increase in rate. These and related results suggest (1) that monomeric and/or intermediate oligomeric forms of the proenzyme that remain attached to the matrix after removal of associated subunits with urea activate faster than the proenzyme itself, possibly as a result of increased flexibility of the proenzyme subunit; (2) since activation does not occur at pH 7.6 in 9 M urea containing added K^+, that some tertiary structure of the proenzyme is required for activation; and (3) that each proenzyme molecule activates itself. The kinetics of activation (first order with respect to total protein and proenzyme) and the absence of in vitro complementation between histidine decarboxylaseless and wildtype cell extracts are consistent with this interpretation. It is possible that the histidine decarboxylase gene has evolved by fusion of a gene encoding a proenzyme and a gene(s) encoding an activating enzyme(s).

To summarize the kinetic results, activation of prohistidine decarboxylase is first order with respect to proenzyme and occurs optimally at pH 7.6 by an apparently intramolecular process that proceeds with a high negative entropy of activation and shows half-order dependency on K^+ or other monovalent cations. Although saturation with such cations has not been observed, the size preference indicates that an unhydrated metal ion must interact with the proenzyme in the rate-limiting step of activation. A mechanism in which two active sites are formed per rate-limiting step involving one unhydrated monovalent cation is consistent with the observed kinetic order of activation. While the nature of the interaction between proenzyme and cation is not yet known, the conversion of a serine residue to a pyruvate residue during activation suggests that an α, β-elimination reaction of some type occurs during the conversion. It is therefore of interest to note that pyridoxal-5′-phosphate-dependent enzymes which catalyze the conversion of serine to pyruvate (i.e., tryptophanase from *E. coli*, the B protein of tryptophan synthetase from *E. coli*, β-tyrosinase from *E. intermedia*, serine dehydratase from *E. coli*) also require potassium or ammonium ions for maximal activity, for reasons still unknown.

3 Revised Subunit Structure of Histidine Decarboxylase

Both native histidine decarboxylase (Chang and Snell 1968; Riley and Snell 1970) and the proenzyme (Recsei and Snell 1973) were found to have a molecular weight by sedimentation equilibrium of 190,000. These observations, together with the molecular weights of the separate subunits (see Fig. 1) indicated a subunit stoichiometry of $(\pi)_5$ for the proenzyme and $(\alpha\beta)_5$ for the active enzyme. Recent findings derived from crystallographic and ultracentrifugal studies, supplemented by electron micrography have forced a revision in this structure (Hackert et al. 1981).

We found that the wildtype enzyme, the proenzyme, and the activated proenzyme were all readily obtainable in a variety of crystal forms. Prohistidine decarboxylase and the activated proenzyme were obtained as trigonal or hexagonal plates (from polyethyleneglycol solutions by vapor equilibration at 4°C) that belonged to the trigonal space group P321 with a = b = 100 Å and c = 164 Å. The space group symmetries and unit cell contents of these crystals indicated 32 point group symmetry for the subunit structure of these enzymes, a structure that is not compatible with any arrangement of five subunits, but is compatible with the presence of six subunits. This conclusion was confirmed by ultracentrifugal work, which showed that histidine decarboxylase formed a rapidly dissociating-associating system with a nearly homogeneous particle of M_R = 208,000 at pH 4.8 at both high (I = 0.20) and low (I = 0.02) ionic strengths. At pH 7.0 this same particle predominated at high ionic strength, but at low ionic strength an additional species of approximately one half that molecular weight (M_R = 104,000) was obtained (Fig. 9). Finally, electron micrographs of enzyme negatively stained at low ionic strength near neutral pH show a predominant particle approx. 60 Å on each edge, while similar micrographs of enzyme cross-linked with glutaraldehyde before staining show a dumbbell-shaped particle approx. 60 Å in width and 120 Å in length. Assuming usual protein densities, the size of these particles correspond approx. to M_R values of 104,000 and 208,000, respectively, which agree in turn with an $(\alpha\beta)_3$ structure for the smaller particle, and an $(\alpha\beta)_6$ structure for the native enzyme at its pH optimum. This indicates in turn that the proenzyme has the subunit composition $(\pi)_6$, and that stable $(\alpha\beta)_3$ and $(\pi)_3$ particles exist under certain conditions. The latter findings relate our enzyme more closely to the histidine decarboxylase from *Micrococcus sp. n.* originally described by Semina and Mardashev (1965),

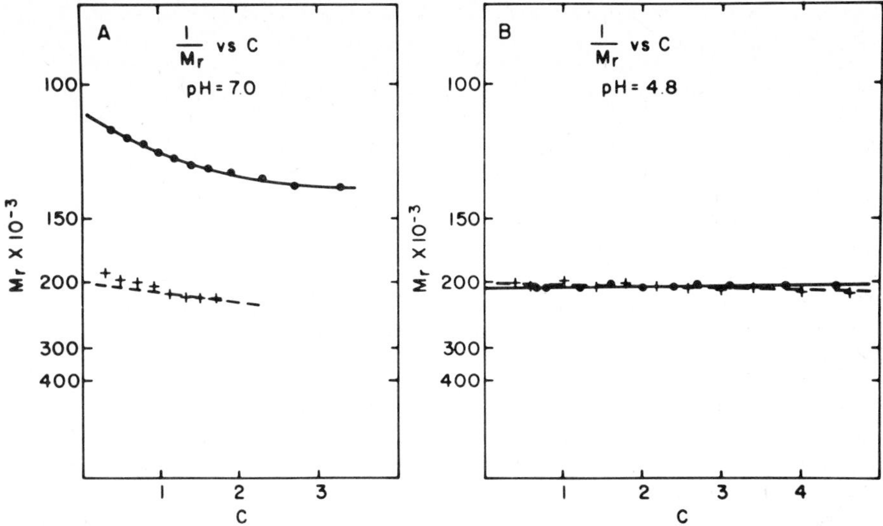

Fig. 9A, B. Effect of pH, ionic strength, and protein concentration (*C*) on the apparent molecular weight of native histidine decarboxylase. **A** 0.01 M sodium phosphate, pH 7.0 (I = 0.02) or **B** 0.02 M sodium acetate, pH 4.8 (I = 0.02). *Solid circles* show molecular weights in the presence of no additional salt; *crosses* represent results obtained in the presence of 0.2 M NaCl (I = 0.22). Hackert et al. (1981)

which more recently also has been found to contain two dissimilar subunits, one of which has an N-terminal pyruvoylphenyl-alanyl group (Alekseeva and Prozorovskii 1976), and to exist as a trimer of $\alpha\beta$ subunits (L'vov and Gonchar 1978). It seems likely that the two enzymes are closely related.

Experiments to determine the predominant subunit structure of the proenzyme at pH 7.6 in the presence of high concentrations of activating cations are, of course, planned but remain to be carried out. Whatever the result, however, it is unlikely to throw much light on the question of primary interest: what series of chemical events are involved in the intriguing and apparently unprecedented activation reaction, in which both conversion of a serine residue to a pyruvate residue and peptide chain cleavage occur under such mild conditions?

Acknowledgements. Unpublished work reported here was supported by research grants (AM 19898 and AI 13940) from the National Institutes of Health, U.S. Public Health Service.

References

Alekseeva AE, Prozorovskii VN (1976) Location of pyruvic acid in the histidine decarboxylase of *Micrococcus* sp. n. Biokhimiya 41: 1584-1587

Chang GW, Snell EE (1968) Histidine decarboxylase of *Lactobacillus* 30a. III. Composition and subunit structure. Biochemistry 7: 2012-2020

Cohn MS, Tabor CW, Tabor H (1977) Identification of a pyruvoyl residue in S-adenosylmethionine decarboxylase from *Saccharomyces cerevisiae.* J. Biol Chem 252: 8212-8216

Demetriou AA, Cohn MS, Tabo CW, Tabor H (1978) Identification of pyruvate on S-adenosylmethionine decarboxylase from rat liver. J Biol Chem 253: 1684-1686

Hackert ML, Meador WE, Oliver RM, Salmon JB, Recsei PA, Snell EE (1981) Crystallization and subunit structure of histidine decarboxylase from *Lactobacillus* 30a. J Biol Chem (in press)

L'vov YuM, Gonchar NA (1978) Determining the molecular weight and radius of gyration of histidine decarboxylase from *Lactobacillus* 30a and from *Micrococcus* sp. n. Biofizika 23: 381-382

Recsei PA, Snell EE (1970) Histidine decarboxylase of *Lactobacillus* 30a. VI. Mechanism of action and kinetic properties. Biochemistry 9: 1492-1497

Recsei PA, Snell EE (1973) Prohistidine decarboxylase from *Lactobacillus* 30a. A new type of zymogen. Biochemistry 12: 365-371

Riley WD, Snell EE (1968) Histidine decarboxylase from *Lactobacillus* 30a. IV. The presence of covalently bound pyruvate as the prosthetic group. Biochemistry 7: 3520-3528

Riley WD, Snell EE (1970) Histidine decarboxylase of *Lactobacillus* 30a. V. Origin of enzyme-bound pyruvate and separation of nonidentical subunits. Biochemistry 9: 1485-1491

Rosenthaler J, Guirard BM, Chang GW, Snell EE (1965) Purification and properties of histidine decarboxylase from *Lactobacillus* 30a. Proc Natl Acad Sci USA 54: 152-158

Satre M, Kennedy EP (1978) Identification of bound pyruvate essential for the activity of phosphatidylserine decarboxylase of *Escherichia coli.* J Biol Chem 253: 479-483

Semina LA, Mardashev SR (1965) Purification and crystallization of microbial histidine decarboxylase. Biokhimiya 30: 100-106

Snell EE (1977) Pyruvate-containing enzymes. Trends Biochem Res 2: 131-135

Snell EE, Recsei PA, Misono H (1976) Histidine decarboxylase from *Lactobacillus* 30a: Nature of conversion of proenzyme to active enzyme. In: Shattiel S (ed) Metabolic interconversion of enzymes 1975. Springer, Berlin Heidelberg New York pp 213-219

Wickner RB, Tabor CW, Tabor H (1970) Purification of adenosylmethionine decarboxylase from *Escherichia coli* W: Evidence for covalently bound pyruvate. J Biol Chem 245: 2132-2139

Williamson JM, Brown GM (1979) Purification and properties of L-aspartate-α-decarboxylase, an enzyme that catalyzes the formation of β-alanine in *Escherichia coli.* J Biol Chem 254: 8074-8082

Discussion[3]

In the discussion with Drs. Shaltiel, Cohen, Fischer, Stadtman and Holzer the following points were raised:

a) β-Elimination of the decarboxylase may result from the phosphorylation of the neighboring serine. The mutant then could host an enzyme which is a poor β-eliminator or a poor substrate for the respective phosphokinase; alternatively, a highly specific kinase could be defective. Several other possibilities have, however, also to be considered.

b) The sequences of this enzyme and of the histidine decarboxylase studied by the Russian group are not yet determined. The Pyruvoyl-phenylalanine moiety is present in both enzymes.

c) So far, there is no evidence for a negative cooperativity of the decarboxylase with respect to potassium binding.

d) For Tabor's enzyme and the *Micrococcus* histidine decarboxylase, proenzymes have not been reported. Nevertheless, they may exist.

3 Rapporteur: Heinrich Betz

Sulfation of a Cell Surface Component Correlates with the Developmental Program During Embryogenesis of *Volvox carteri*

M. Sumper and S. Wenzl[1]

1 Introduction

The colonial green flagellates of the genus Volvox present an interesting model for studying the control of cellular differentiation (Powers 1907, 1908, Barth 1964). With only two different types of cells, Volvox is one of the most primitive multicellular organisms. Asexual colonies of *Volvox carteri* are differentiated into several thousands of somatic cells with no potential for division and only 8 to 16 reproductive cells (gonidia). Each reproductive cell of the spheroid undergoes a series of cleavages to form a new individual. At the end of embryogenesis each mother spheroid contains up to 16 daughter spheroids. Finally, the mature daughter spheroids are released through a rupture in the matrix material of the mother spheroid. In the developing asexual embryo differentiation into somatic and reproductive cells is seen at the division from 32 to 64 cells (Kochert 1968; Starr 1969, 1971). At this stage 16 out of the 32 cells undergo unequal cleavage, forming a small somatic and a large reproductive initial. While cell division ceases in the reproductive initials, the remaining cells continue their divisions, and finally differentiate into somatic cells (Fig. 1). Obviously, some sort of a counting mechanism exists in the developing embryo telling a cell that the embryo is in the 2-, 4- . . . or 32-cell stage. This counting mechanism exhibits a certain degree of flexibility: Under less than optimal growth conditions, the differentiating cell cleavage may already occur at the division from 16 to 32 cells. In this case, 8 out of the 16 cells undergo unequal cleavage, thus producing only 8 reproductive initials. On the other hand, a delay of the differentiating cleavage is observed under the influence of a specific glycoprotein, the sexual inducer (Starr and Jaenicke 1974). For instance, in the strain HK 10 of *Volvox carteri* the differentiating cell cleavage by action of the inducer is delayed up to the division from 64 to 128 cells. At this stage, again only half of the embryonic cells (32) undergo unequal cleavage forming 32 reproductive initials. These reproductive cells differentiate to egg cells producing a sexual (female) spheroid with its typical appearance (Fig. 1).

Besides this cell-counting problem, the embryogenesis of Volvox presents the problem of pattern formation in a fascinating simplicity. The spatial arrangement of the reproductive initials within the developing embryo is exactly controlled. During the whole process of embryogenesis the reproductive initials remain in the same relative positions where they were formed during early embryogensis, i.e., the relative

1 Institut für Biochemie, Genetik und Mikrobiologie, Lehrstuhl Biochemie I, Universität Regensburg, Universitätsstraße 31, 8400 Regensburg, FRG

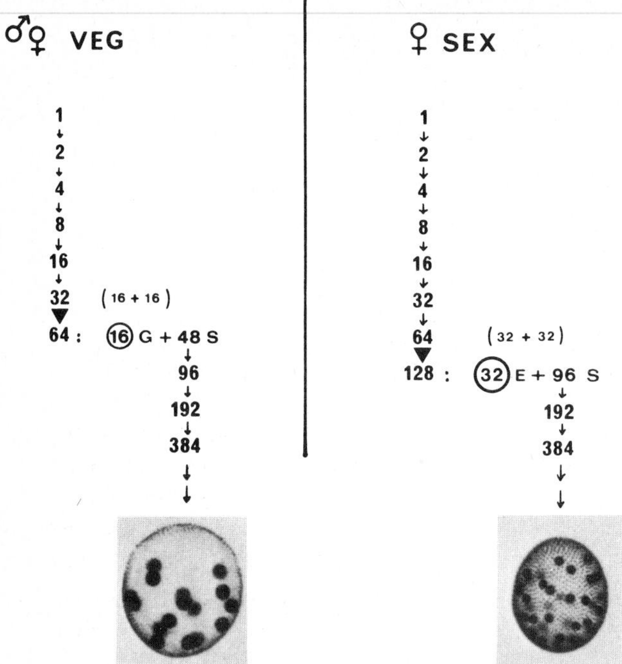

Fig. 1. Asexual and sexual development of a new individual from a reproductive cell in *Volvox carteri.* Somatic and reproductive cells are differentiated during the 32 → 64 and 64 → 128 division, respectively. *G* gonidia (reproductive cells); *S* somatic cells; *E* egg cells

positions remain exactly the same in the mature spheroid. The fidelity of this pattern formation in Volvox embryogenesis is documented in Fig. 2, comparing photographs of spheroids containing 8, 12, and 16 reproductive cells, respectively. Eight gonidia are always arranged in two parallel rectangles, turned by 45° against each other. Twelve gonidia are similarily arranged in three parallel rectangles. Sixteen gonidia are arranged in two parallel rectangles (8 gonidia) with the residual 8 gonidia positioned on a zigzag line in between the rectangles.

2 A Biochemical Model Explaining Pattern Formation During Embryogenesis

In the following we would like to propose a biochemical model which is able to predict correctly both the counting system and the pattern formation during embryogenesis of Volvox (Sumper 1979). Finally, we will describe some of the experiments we performed in order to test the model.

In its most straightforward version, the model requires only two assumptions, which both are experimentally verified in a number of differentiating cell systems (for review, see Frazier and Glaser 1979):

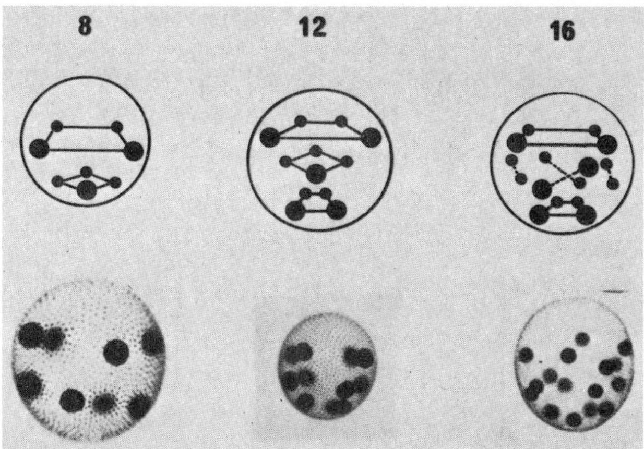

Fig. 2. Spatial arrangement of reproductive cells in Volvox spheroids containing *8, 12,* and *16* reproductive cells, respectively

1. The reproductive cell in limiting amounts produces a specific cell surface component which is able to mediate cell to cell contacts.
2. The absence of this component on the surface of a given embryonic cell signals the differentiating cell cleavage.

On the basis of these assumptions the schematic drawings of Fig. 3 demonstrate how the embryonic cells become sorted out into two subclasses during early embryogenesis: The cell surface of the uncleaved gonidium contains a pool of the contact-forming component (denoted as A-component in the following). As soon as septum formation of the first division is initiated, a certain fraction of the A-molecules becomes trapped within the contact region simply by random lateral diffusion and contact formation. In the second division two more cell-to-cell contacts are formed, again consuming a given amount of the free (uncomplexed) A-molecules. Since it is reasonable to assume that the cell-to-cell contact area of the first cleavage is not divided by the cleavage planes of the second division, the cells of the 4-cell embryo will be arranged as shown in Fig. 3. Redrawing of the four cells in a more symmetrical configuration results in an arrangement typical of that seen by light microscopic observation: two cells touch one another while the other two have no contact area in common. Light microscopic observation reveals that the 8-cell embryo is shaped like a cup: The four cells at the posterior end of the embryo remain in close contact, while the four anterior cells do not touch each other and form a pore. The model explains this characteristic configuration: the four additional contact sites established during the third division are shown in the schematic drawing of Fig. 3. Folding up this two-dimensional representation of the 8-cell embryo to a hollow sphere necessarily creates a pore, because no contacts have been established between the four anterior cells.

Since the number of cell contacts increases exponentially during embryogenesis, the pool of the free A-molecules becomes exhausted at a sharply defined stage of division. If, for instance, this titration point is reached at the stage of the 8-cell embryo, then the embryonic cells become sorted out into two subclasses by the subsequent

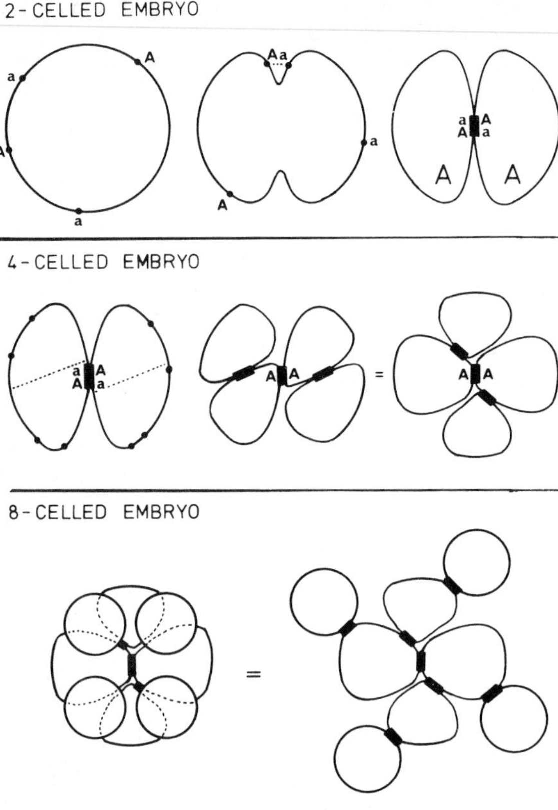

Fig. 3. Cell-to-cell contact formations in early embryogenesis. For details see text

cleavage: 8 cells of the 16-cell embryo are equipped with contacts containing the A-molecules, whereas the other 8 cells lack the A-molecules in their contacts. As assumed by the model, the absence of the A-molecule on a cell surface triggers the differentiating cell cleavage. Therefore only 8 cells of the 16-cell embryo can undergo the differentiating cell cleavage. In addition, the model is able to predict the relative positions of those embryonic cells which undergo the differentiating cell cleavage. The cells lacking the A-surface component (and thus producing gonidial initials) are marked with a *black dot* in the two-dimensional representation of the 16-cell embryo in Fig. 4. By transforming again this two-dimensional pattern into a hollow sphere it follows that the 8 gonidial initials will be arranged in two parallel rectangles, turned by 45° against each other. One rectangle will be located near the equator and while the other rectangle will be located in the anterior half of the sphere, the posterior half of the embryo will contain only somatic cells. This is exactly the pattern found in those Volvox spheroids containing only 8 gonidia.

Development of spheroids containing only 8 gonidia occurs under less than optimal growth conditions. Under favorable conditions nearly all Volvox spheroids contain 16 gonidia. This means in terms of the model that more A-surface component is synthesized under favorable conditions. Thus the pool of free A-contact component

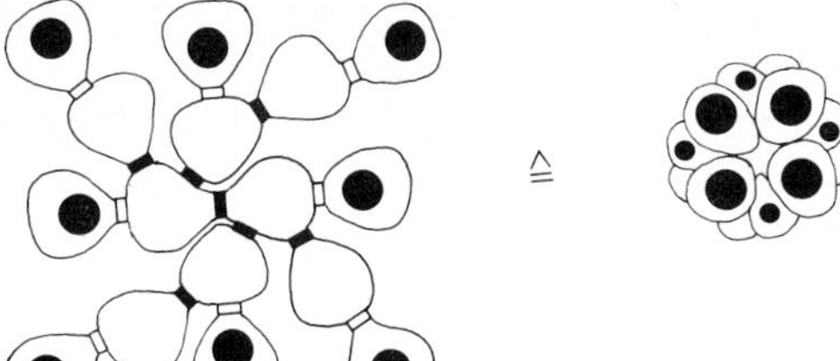

Fig. 4. Positioning of embryonic cells having no surface component A within the 16-cell embryo. These are labeled with a *black dot*. Filled cell-to-cell contacts contain surface component A, open contacts lack the A-component. For details see text

becomes exhausted only at a stage beyond the 8-cell stage, for instance, at the stage of the 16-cell embryo. Then, in the subsequent cleavage to the 32-cell embryo, the embryonic cells become subdivided into 16 cells equipped with A-component and another 16 cells lacking A-component. The relative positions of those cells lacking the A-component and thus producing gonidial initials can be estimated from the schematic drawing of the 32-cell embryo in Fig. 5. Transformation of this two-dimensional pattern into a hollow sphere results in a spatial arrangement of the 16 gonidia, as it is observed in Volvox spheroids.

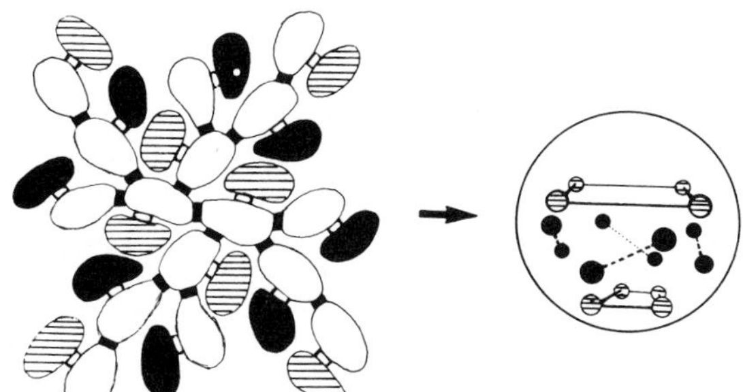

Fig. 5. Positioning of the 16 embryonic cells having no surface component A within the 32-cell embryo. These are labeled either in *black* or with *black lines* in order to facilitate pattern recognition. Filled cell-to-cell contacts contain surface component A, open contacts lack the A-component. For details see text

3 Experimental Test of the Model

3.1 The Developmental Program is Disturbed by Cell Surface Modifying Agents

In order to check experimentally the assumptions made by the model, isolated gonidia were treated with substances known to modify cell surface components. If isolated gonidia were treated with the protease subtilisin (40 μg/ml) for a short period (60 min) during any time of their maturation period, they developed completely normally to Volvox spheroids. In sharp contrast, the developmental program was strongly disturbed if the subtilisin treatment was performed after the initiation of cell division. Although viable Volvox colonies developed, the number as well as the spatial arrangement of gonidia was found to be completely irregular (Wenzl and Sumper 1979). This experimental result provides evidence for a surface protein involved in the control of differentiation in Volvox embryogenesis.

3.2 A Sulfated Surface Component Appears to be Involved in the Control of Embryogenesis

Since the gonidia became sensitive to protease treatment only during the cleavage period, a corresponding change in the membrane protein pattern should be detectable at this time. For this reason, a highly synchronously growing Volvox population was pulse-labeled in order to determine the patterns of membrane proteins synthesized at the different developmental stages. Figure 6 shows fluorograms of SDS-polyacrylamide gels obtained by using $[H[^{14}C]O_3^-$ as the radioactive label. It is obvious that the membrane protein patterns obtained at different developmental stages do not differ very much from each other. In sharp contrast, a quite distinct change is observed during development when the same type of pulse-labeling is with $[^{35}S]O_4^{2-}$ as the radioactive label. At the beginning of cell cleavage two highly labeled membrane components appear. Production of these components stops after the end of the cleavage period. The more prominent component of the two has an apparent molecular weight of 185000. Mild acid hydrolysis of these $[^{35}S]$-labeled components liberates $[^{35}S]O_4^{2-}$ in nearly quantitative yield, indicating that the labeled components are sulfated molecules. Partially purified 185 K component is found to be sensitive to protease treatment. Digestion with subtilisin or pronase reduces the apparant molecular weight to 150 K. However, the resulting sulfated core material is resistant even to prolonged protease treatment.

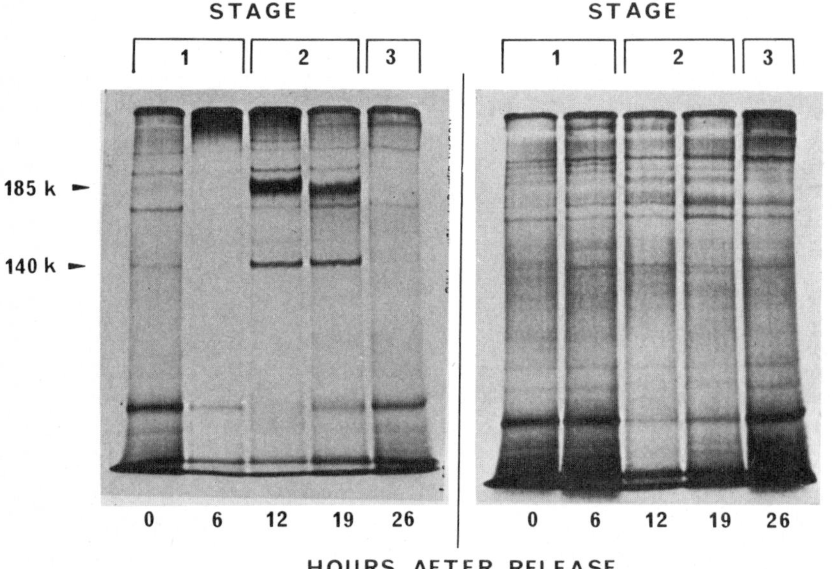

Fig. 6. SDS polyacrylamide gel electrophoresis of membrane components pulse-labeled in vivo at different developmental stages

Stage 1, period of gonidial enlargement (12 h). *Stage 2*, beginning of gonidial cleavage to end of inversion (period of embryogenesis). *Stage 3*, end of embryogenesis to release of daughter spheroids. (*Left*) Patterns obtained by pulse-labeling with $[^{35}S]O_4^{2-}$. (*Right*) Patterns obtained by pulse-labeling with $[H[^{14}C]O_3^-$.Reproduced from Sumper and Wenzl (1980)

3.3 The Sulfated 185 K Molecule is a Cell Surface Component

In order to examine the localization of the sulfated 185 K component, intact Volvox colonies were treated with subtilisin. Subtilisin most effectively dissociates colonies of Volvox to single cells. The viability of the reproductive cells is not affected by subtilisin treatment since cell cleavage continues after removal of the protease by washing. Thus, subtilisin treatment appears to be an adequate procedure to digest only surface-associated proteins. In the experiment of Fig. 7, Volvox colonies containing 4-cell embryos were pulse-labeled with $[^{35}S]O_4^{2-}$. The colonies were then dissociated by subtilisin treatment and the resulting suspension of somatic cells and of embryos centrifuged at low speed. The cell-free supernatant as well as the sedimented cells were analyzed by SDS polyacrylamide gel electrophoresis. As shown by the fluorography of Fig. 7 nearly all of the sulfated 185 K component is recovered in the supernatant fraction (as its 150 K derivative), nearly no 185 K (or 150 K derivative) remains associated with the cellular fraction. Therefore, the 185 K component is a cell surface-associated molecule.

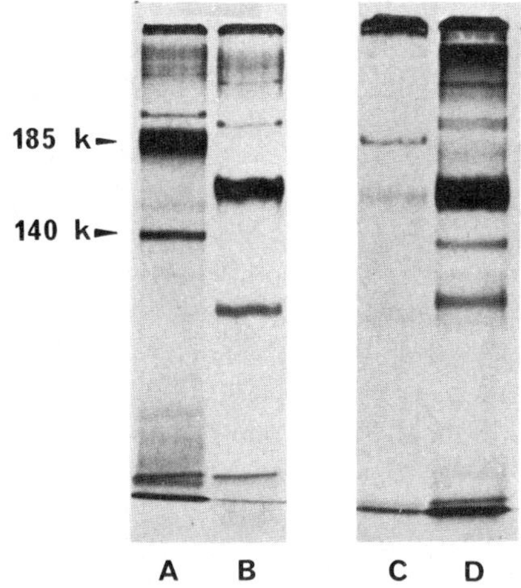

Fig. 7A-D. 185 K component is a cell surface component: Volvox spheroids containing 4-cell embryos were pulse-labeled with $[^{35}S]O_4^{2-}$. **A**, analysis of the isolated membrane fraction by SDS polyacrylamide gel electrophoresis and fluorography. **B**, as (**A**), after treatment of the membrane fraction with subtilisin. **C** and **D**, the pulse-labeled Volvox spheroids were dissociated into single cells by subtilisin treatment. After centrifugation, the cellular fraction (**C**) and the cell-free supernatant (**D**) were analyzed by SDS-gel electrophoresis and fluorography

3.4 The Sulfate Residues of the 185 K Component Show a High Turnover Rate

The $[^{35}S]$-sulfate residues of the 185 K component turn over with an unusually high rate, as revealed by pulse-chase experiments. The half-life is less than 20 min, as estimated from the experiment of Fig. 8. Unlabeled sulfate (10 mM) or p-nitrophenyl-sulfate (10 mM) added during the chase period effectively inhibits the degradation of the sulfated 185 K component. This observation may indicate the action of a sulfatase.

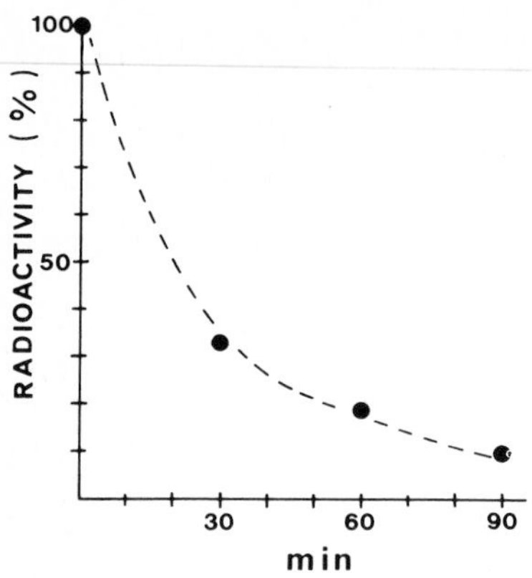

Fig. 8. The sulfated 185 K component in vivo is very short-lived. Volvox spheroids containing embryos were pulse-labeled with $[^{35}S]O_4^{2-}$. Zero time marks the beginning of a chase experiment. After a chase period of 30, 60, and 90 min, the radioactivity of the 185 K component was analyzed after isolation by SDS-gel electrophoresis

3.5 Is the Sulphated 185 K Component Involved in the Control of Differentiation?

Short $[^{35}S]O_4^{2-}$ pulse-labeling experiments were performed over the whole period of embryogenesis to investigate whether the production (or degradation) of the sulfated 185 K component correlates in any way with the developmental program. For experimental reasons, sexually induced embryos were used for this type of experiment. Cell division occurs every 50-60 min, therefore a pulse length of 30 min was selected. The results of these pulse-labeling experiments using sexually induced embryos of a female strain (HK 10) are shown in Fig. 9. At the beginning of cleavage, production of the sulfated 185 K component is initated. The net production reaches a maximum level during the 8-cell stage and sharply declines as division proceeds. A minimum level of net $[^{35}S]O_4^{-}$ incorporation is reached at the 32-cell stage, i.e., one stage of division before the differentiating cleavage will be initiated, which in induced female embryos occurs at the division from 64 to 128 cells. Beyond the 64-cell stage, the net incorporation increases again. This pattern of $[^{35}S]O_4^{2-}$ incorporation is reproducible if the Volvox population develops highly synchronized. The correlation of net 185 K production and developmental stage suggests a role of this sulfation-desulfation reaction in the control of differentiation.

4 Discussion

The experimental results obtained so far demonstrate that the sulfated 185 K component meets at least three of the properties postulated for the hypothetical membrane component A of our model:

1. The sulfated 185 K component is localized on the cell surface.
2. Its production is essentially limited to the period of embryogenesis.
3. Its net production reaches a minimum level exactly one stage of division before the differentiating cleavage.

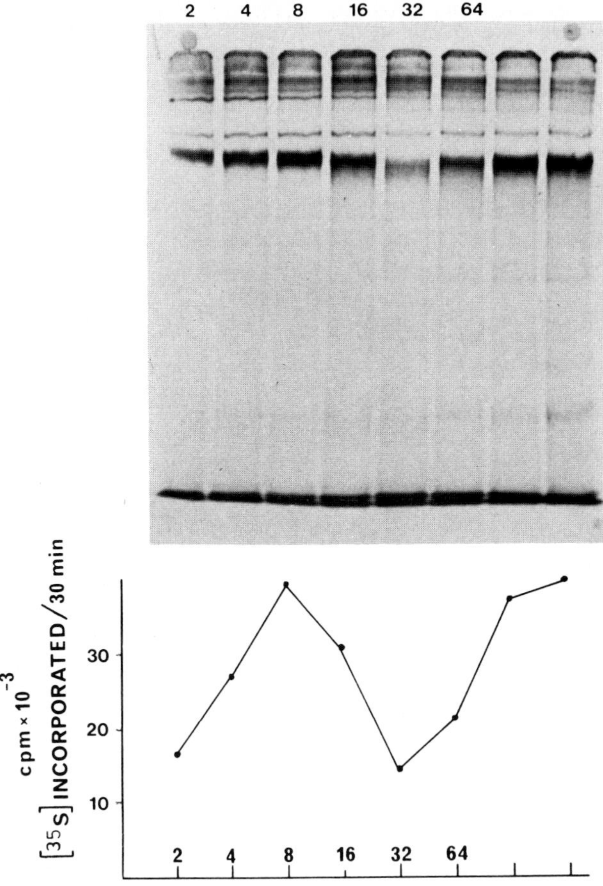

Fig. 9. Net $[^{35}S]O_4^{2-}$ incorporation into the 185 K component during embryogenesis. Volvox spheroids (female strain HK 10, sexually induced) were pulse-labeled in vivo for 30 min with $[^{35}S]O_4^{2-}$. After the initiation of gonidal cleavage a pulse-labeling experiment was performed at each stage of division. *Top*, the total membrane fractions were applied to 6% SDS polyacrylamide gels and visualized by fluorography. *Bottom*. Net incorporation of $[^{35}S]$-radioactivity into the 185 K component as a function of the developmental stage

However, one crucial property postulated by the model remains to be established for the 185 K component: Its capability to form cell-to-cell contacts.

How could the sulfation-desulfation of a surface component control differentiation. The hypothetical scheme in Fig. 10 gives a possible explanation in terms of our model: In this scheme the assumption is made that the desulfated degradation product of the 185 K component suppresses the unequal (differentiating) cleavage. The sulfated 185 K component is engaged in contact formation and thus becomes exhausted at a defined stage of division. Consequently, the successive cleavage divides the embry-

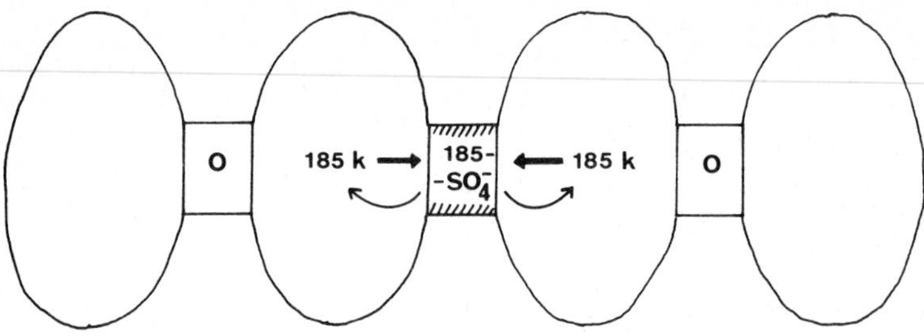

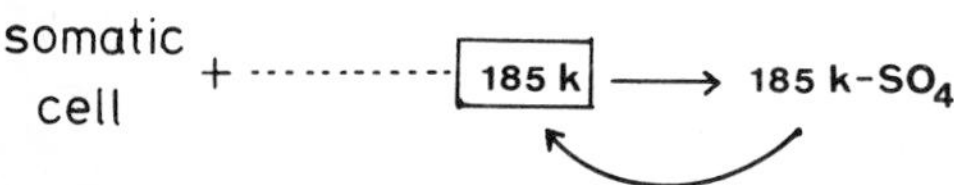

Fig. 10. A hypothetical scheme, showing two subclasses of cells: two cells contain the sulfated 185 K component in a cell-to-cell contact, whereas the other two cells lack this component. For details see Sect. 4

onic cells into two subclasses: One half of cells lacks the sulfated 185 K surface component. Therefore, in this subclass the differentiating cleavage is no longer suppressed.

During embryogenesis of Volvox the differentiating cleavage is usually seen at only one stage of division. This means that the suppression of unequal cleavage must be reestablished in all successive stages of division. The pulse-labeling experiment of Fig. 9 indeed demonstrates that the net production of the sulfated 185 K component is initiated again after the differentiating cleavage.

References

Barth LJ (1964) "Development: Selected topics". Addison-Wesley, Reading Massachusetts

Frazier W, Glaser L (1979) Surface components and cell recognition. Annu Rev Biochem 48: 491-523

Kochert G (1968) Differentiation of reproductive cells in Volvox carteri. J Protozool 15: 438-452

Powers JH (1907) New forms of Volvox. Trans Am Microsc Soc 27: 123-149

Powers JH (1908) Further studies in Volvox, with descriptions of three new species. Trans Am Microsc Soc 28: 141-175

Starr RC (1969) Structure, reproduction, and differentiation in Volvox carteri f. nagariensis Iyengar, strains HK 9 & 10. Arch Protistenkd 111: 204-222

Starr RC (1971) Control of differentiation in Volvox. Dev Biol Suppl 4: 59-100

Starr RC, Jaenicke L (1974) Purification and Characterization of the hormone initiating sexual morphogenesis in volvox carteri f. nagariensis Iyengar. Proc Natl. Acad Sci USA 71: 1050-1054

Sumper M (1979) Control of differentiation in Volvox carteri, a model explaining pattern formation during embryogenesis. FEBS Lett 107: 241-246

Sumper M, Wenzl S (1980) Sulphation-Desulphation of a membrane component proposed to be involved in control of differentiation in Volvox carteri. FEBS Lett 114: 307-312

Wenzl S, Sumper M (1979) Evidence for membrane-mediated control of differentiation during embryogenesis of Volvox carteri. FEBS Lett 107: 247-249

Discussion[2]

During the discussion with Drs. Helmreich, Fisher, Heilmeyer, Demaille, Hilz, Hofmann and Holzer, the following additional information on the properties and the metabolism of the sulfated cell surface component was obtained:

(a) The chemical nature of the sulfur linkage is unknown, but possibly involves carbohydrate moieties; (b) the precursor for the bound sulfate is assumed to be activated sulfate; (c) the sulfate label is metabolically labile with a fairly constant half life of 20 min throughout embryogenesis. This implies the existence of a very active, presumably plasma membrane-associated sulfatase in *Volvox*.

Structure, Possible Function, and Biosynthesis of VPg, the Genome-linked Protein of Poliovirus

C.J. Adler, B.L. Semler, P.G. Rothberg, N. Kitamura, and E. Wimmer[1]

1 Introduction

Poliovirus has attracted immense medical interest in the first 70 years of this century as it has caused epidemics of the dreaded paralytic disease, poliomyelitis. During this period poliovirus became a prototype for animal viruses and was subjected to numerous investigations at the molecular level. These studies revealed properties of the viral genome and mechanisms of viral replication that, when compared to other animal viruses, are unique. Two of the special features of poliovirus structure and replication will be dealt with in this paper: the genome-linked protein VPg, and post-translational, proteolytic processing of viral proteins (for recent reviews, see Rueckert 1976; Korant et al. 1980a; Kitamura et al. 1980a).

Poliovirus belongs to the genus enterovirus of the family Picornaviridae. Other genera of this family are the rhinoviruses (whose members cause the common cold in humans), the cardioviruses of rodents (e.g., encephalomyocarditis virus, EMC) and the aphthoviruses (foot-and-mouth disease virus of domesticated animals). Poliovirus is a small, naked (lacking a membraneous envelope), icosahedral particle consisting of 60 copies each of 4 coat proteins (VPI, VP2, VP3, and VP4) and a single-stranded genomic RNA that is 7485 ± 40 nucleotides long. The RNA is polyadenylated (60 ± 40 A residues) at the 3′ end (Yogo and Wimmer 1972); at the 5′ end it is covalently linked to a small polypeptide (Lee et al. 1976, 1977). This polypeptide has been named VPg, "VP" in keeping with the nomenclature for poliovirion proteins, and "g" for the genome to which it is bound. At the time of its discovery, a genome-linked protein of viral RNA was a novelty. Today, however, many RNA viruses with this feature have been identified: all genera of the picornaviridae, a calicivirus (another animal virus), and several different plant viruses (for references, see Rothberg et al. 1980; Kitamura et al. 1980b).

2 The Genome-linked Protein VPg

2.1 Properties of VPg

The genome-linked protein can be released from viral RNA by digestion with RNase T2. Owing to the specificity of this enzyme the polypeptide remains linked to a nuc-

1 Department of Microbiology, State University of New York, Stony Brook, NY 11794, USA

leotide that was identified as pUp (Lee et al. 1977; Flanegan et al. 1977). This nucleotide was found to be the 5′ end of the genome (Nomoto et al. 1977a). The nucleotidyl-polypeptide VPg-pUp released with RNase T2 migrates as a single band during electrophoresis in polyacrylamide gels, slightly faster than VP4 (mol. wt. 7,000) and elutes in the void volume when applied to a column of DEAE cellulose in low salt (Lee et al. 1977; Nomoto et al. 1977a). Moreover, VPg-pUp does not precipitate in 5%-20% trichloroacetic acid or in 10% acetic acid, and is rapidly eluted from polyacrylamide gels during standard fixation procedures (Lee et al. 1976, 1977). Finally, VPg-pUp migrates toward the cathode during paper electrophoresis on Whatman 3MM at pH 3.5 or pH 7 (Lee et al. 1976; Flanegan et al. 1977), and has an apparent pI of >9.4 (Adler and Wimmer, unpublished results). These properties suggest that VPg is a basic polypeptide of low molecular weight (<7,000), a conclusion recently confirmed by protein sequence studies (Kitamura et al. 1980b). In spite of its charged nature, VPg-pUp is insoluble in water and adsorbs rapidly to glass or plastic surfaces (Lee et al. 1976, 1977).

2.2 The Linkage Between VPg and Viron RNA

Work on VPg-pUp was complicated by its tendency to stick to glass or plastic from which it can be solubilized with 0.1% sodium dodecylsulfate (SDS). Initial studies on the linkage between VPg and RNA were therefore carried out with a nucleotidyl-peptide obtained from VPg-pUp by digestion with Pronase in 0.1% SDS (Lee et al. 1977; Nomoto et al. 1977a). The resistance of this nucleotidyl-peptide bond to 0.1 M HCl at 37°C for 2 h, to 1 M NaOH for 1 h at 100°C, or to nuclease Pl, made it unlikely that pUp was linked via a phosphoramidate bond or via a phosphodiester to Ser or Thr, nor was it likely that pUp was linked to a modified base (Nomoto et al. 1977a; Rothberg et al. 1978). On the other hand, the bond between pUp and the amino acid moiety was cleaved with snake venom exonuclease, an observation indicating a phosphodiester linkage (Lee et al. 1977; Nomoto et al. 1977a; Flanegan et al. 1977). Labeling studies with Tyr and the inability to radioiodinate VPg-pUp with [^{125}I] in the presence of chloramine T or lactoperoxidase suggested that VPg contains a single Tyr, the hydroxyl group of which is blocked. Based on studies of the sensitivity of Tyr-0^4-P to acid, Rothberg et al. (1978) then digested [^{32}P]VPg-pUp with 5.6 M HCl for 2 h at 110°C and obtained, among [^{32}P]Pi and incomplete digestion products, [^{32}P]0^4-(3′-phospho-5′-uridylyl)tyrosine (Tyr-pUp). The identity of [^{32}P]Tyr-pUp was confirmed by enzymatic digestion with snake venom exonuclease (yielding [^{32}P]pUp), with bacterial phosphomonoesterase (yielding [^{32}P]Tyr-pU and [^{32}P]Pi) or with micrococcal nuclease (yielding [^{32}P]Tyr-P and [^{32}P]Up). Thus the linkage group between VPg and poliovirus RNA is 0^4-(5′-uridylyl)tyrosine as shown in Fig. 1. Labeling studies of VPg with [^{3}H]Tyr have led Ambros and Baltimore (1978) to the same conclusion.

Nucleotidyl-tyrosine residues have previously been found in *E. coli* glutamine synthetase (Holzer and Wohlhueter 1972; Adler et al. 1975; and literature cited therein). In that case, the formation of 0^4-(5′-adenylyl)tyrosine residues in glutamine synthetase modulates the enzyme's activity, and a protein responsible for this modula-

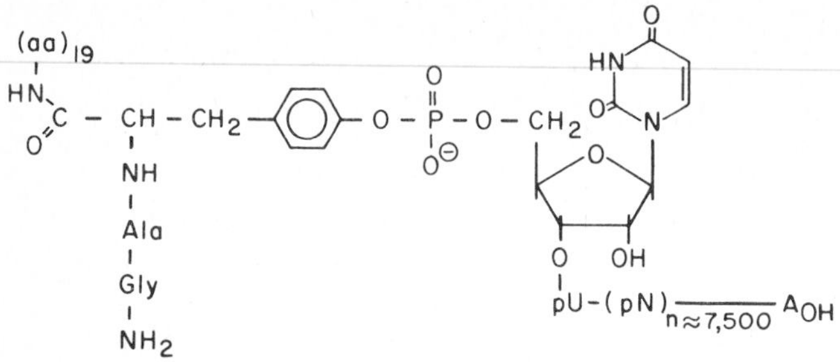

Fig. 1. 0^4-(5′-uridylyl)tyrosine is the linkage between VPg and poliovirus RNA. Tyrosine is the third amino acid from the NH_2-terminus; the protein is 22 amino acids in length

tion itself is derivitized to carry 0^4-(5′-uridylyl)tyrosine residues. Interestingly, the phosphodiester linkage of nucleotidyl-0^4-tyrosine is an "energy-rich" phosphate bond with a standard free energy of hydrolysis approx. -9.6 Kcal/mol (Holzer and Wohlhueter 1972).

In all cases studied so far, the genome-linked protein of viral RNA is always bound to a 5′-terminal pUp. These cases are poliovirus type 1 and type 2 (Nomoto et al. 1977a; Babich et al. 1980), EMC virus (Golini et al. 1978), aphthovirus (Harris 1979; Grubman 1980) and cow pea mosaic virus of plants (Stanley and Van Kammen 1979). In EMC viral RNA the linkage between VPg and RNA has also been shown to be Tyr-pUp (Vartapetian et al. 1979). It should be noted that a linkage between a tyrosine residue of a topoisomerase or gyrase of *E. coli* and a DNA substrate was recently observed (Tse et al. 1980). In contrast, the genome protein of adenovirus (Rekosh et al. 1977) is linked to the 5′ ends of double-stranded adenovirion DNA via a Ser-0-phosphodiester (Kelly, pers. comm.).

Specific cleavage of the linkage between the 0^4 of Tyr and the 5′-terminal phosphorus of the polynucleotide chain by an enzyme of host cell origin has been observed (Ambros et al. 1978). This "tyrosine polynucleotide phosphodiesterase", or "VPg-ase", is inhibited by RNA and heparin but not by a close analog of the cleavage site, thymidine 5′-monophospho-p-nitrophenyl ester, even when present in great molar excess (Rothberg and Wimmer, unpublished results). Moreover, work by Ambros and Baltimore (1980) has shown that the proteinase K digestion product of VPg-pUp [$(aa)_n$-Tyr-pUp] is not digested by the phosphodiesterase. These observations suggest that a polynucleotide moiety is required for the enzyme to be active. During the replicative cycle of poliovirus, cleavage of VPg from some newly synthesized viral RNA does occur (Nomoto et al. 1976, 1977a, b) but it remains to be shown whether or not (1) such cleavage is required for viral replication, and (2) whether the above mentioned phosphodiesterase is indeed involved in the hydrolytic cleavage of VPg from viral RNA.

The function of the "tyrosine polynucleotide phosphodiesterase" in the host cell is obscure to date. It is not even known whether or not a nucleotidyl-tyrosine bond is the natural substrate of this enzyme.

2.3 Possible Function of VPg

Any consideration of the function of VPg requires some familiarity with poliovirus replication. In the following we shall briefly review the replication cycle.

2.3.1 Replication Cycle of Poliovirus

Replication (see Fig. 2) begins with (1) virus attachment to the cell surface receptors, its penetration through the cell membrane, and the subsequent uncoating of the viral genome. (2) Immediately after uncoating, the viral genome engages in protein synthesis (see below). One or several of the virus-specific proteins then function to synthesize (3) viral RNA's in a membrane bound replication complex. Newly synthesized RNA may serve as a template in RNA synthesis, as mRNA in translation, or (4) combine with capsid proteins, VP1, VP2, and VP0 to form progeny virions (for a review, see Kitamura et al. 1980a).

Virion RNA and viral mRNA are thought to have identical nucleotide sequences, the only difference being the presence or absence of VPg (Nomoto et al. 1977b; Pettersson et al. 1977). Accordingly, naked virion RNA, free of detectable coat proteins, should be able to initiate an infection once it penetrates the cellular membrane. Indeed, virion RNA is taken up by cells and progeny virus is formed in normal yields. The efficiency of "transfection" (infectivity per molecule of RNA), however, is low (Koch 1973).

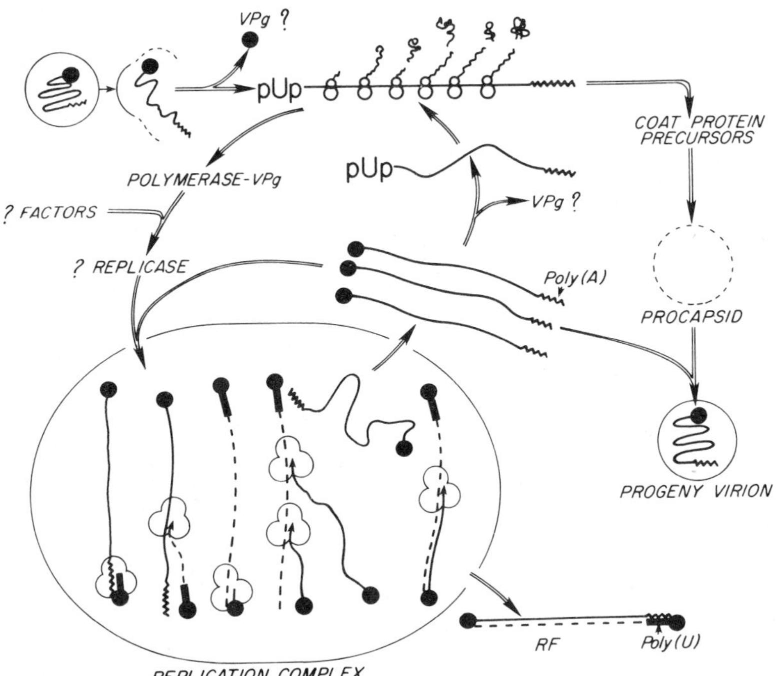

Fig. 2. Replication scheme of poliovirus. See text for details

2.3.2 Observations Suggesting that VPg is Involved in RNA Synthesis

The observation that VPg is absent from mRNA suggests that the polypeptide is not involved in translation. In fact, virion RNA, the VPg of which has been proteolytically degraded by proteinase K, retains its full specific infectivity (Nomoto et al. 1977b). It is therefore unlikely that VPg plays a role during the first two steps of viral replication. Analysis of viral RNAs isolated from infected cells has revealed the surprising fact that all newly synthesized RNA, both plus and minus strands, are VPg linked (Nomoto et al. 1977a; Pettersson et al. 1978). Moreover, even the nascent strands of the multistranded replicative intermediate within the membrane bound replication complex are terminated with VPg, whereas pppN- termini are absent (Nomoto et al. 1977a; Pettersson et al. 1978; Wimmer 1979). These observations have led to the suggestion that VPg is involved in the initation of RNA synthesis, possibly as a primer (Nomoto et al. 1977a). This hypothesis seemed to have been contradicted, however, by our failure to identify free VPg in infected cells (Golini and Wimmer, unpublished results). We therefore speculated that VPg is generated by proteolytic cleavage from a precursor polypeptide at the moment of initiation of RNA synthesis (Nomoto et al. 1977a; Wimmer 1979). Such a unique mechanism may be possible since poliovirus synthesizes most, if not all, viral proteins by proteolytic cleavage of a gigantic precursor, the polyprotein (see below).

3 Post-translational Processing of Poliovirus Proteins

3.1 General Outline

Among animal RNA viruses the picornaviruses may be unique in that the entire genetic information of the virus is translated from a genome-sized RNA, end-to-end into a polyprotein (Jacobson and Baltimore 1968). This polyprotein is cleaved by proteinases into numerous proteins (Fig. 3), a process called "post-translational processing" (Summers and Maizel 1968; Holland and Kiehn 1968; Jacobson and Baltimore 1968; see recent reviews by Rueckert et al. 1979; Kitamura et al. 1980a).

The initial processing of the polyprotein, referred to as "primary cleavages", probably occurs *in statu nascendi*. The products are three polypeptides P1-1a, P2-X (or possibly a slightly larger precursor P2-3b), and P3-1b. The primary products are subsequently processed into "secondary cleavage products", a mechanism that results in the complex scheme of polypeptides shown in Fig. 3. The polypeptides have been mapped into the genome with respect to their precursor/product relationships and time of appearance in translation (reviewed by Rueckert 1976; Rekosh 1977). The functions of the end products of cleavage are known for some polypeptides. For example, polypeptides VP1, VP2, VP3, and VP4 (products of P1-1a) are capsid proteins, polypeptide P3-7c is likely to be a proteinase (see below), and polypeptide P3-4 is an RNA-dependent RNA polymerase (reviewed by Kitamura et al. 1980a). A final cleavage, possibly distinct from secondary cleavage, is that of VP0 to VP4 and VP2 which occurs only during maturation of the virion ("maturation cleavage"; see Rueckert 1976).

POLIOVIRUS PROTEINS: SYNTHESIS AND PROCESSING

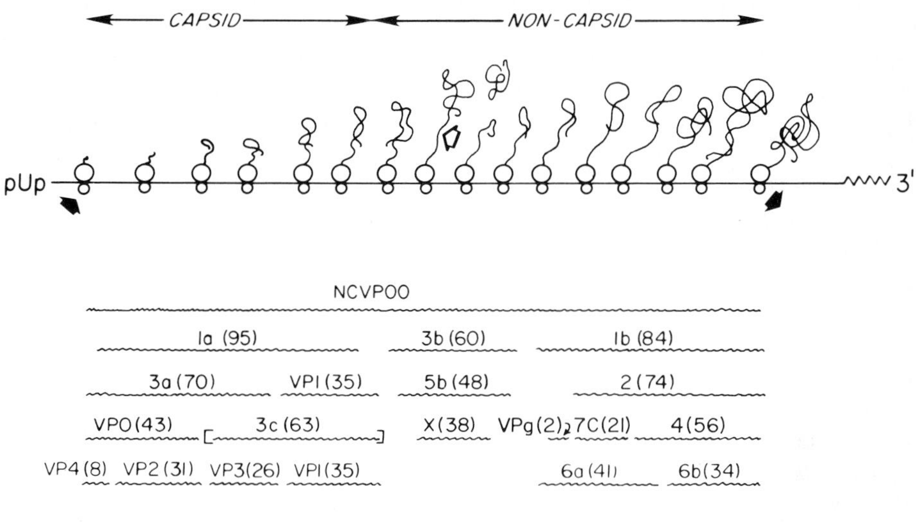

Fig. 3. Poliovirus protein synthesis and proteolytic cleavage scheme. The molecular weights of proteins (*in parentheses*) are given in kilodaltons. Proteins are not drawn to scale. Redrawn in part after Rueckert et al. (1979)

It should be emphasized that the scheme shown in Fig. 3 is simplified. Many small (mol wt. <12,000) polypeptides, normally detectable in poliovirus infected cells, have been omitted. Some of these small polypeptides may serve an important yet unknown function. There are also some other, larger cleavage intermediates that have been observed but have been omitted in Fig. 3 because their exact precursor/product relationships are obscure.

Establishing the processing scheme shown in Fig. 3 was complicated by the alternate pathways of cleavage of certain polypeptides. This phenomenon has been observed with rhinoviruses (McLean et al. 1976), aphthovirus (Doel et al. 1978), and poliovirus (Rueckert et al. 1979). In the latter case P3-2 is cleaved either into P3-7c and P3-4, or into P3-6a and P3-6b (Fig. 3). Whereas Rueckert and his colleagues (pers. comm.) have not observed the alternate pathway of cleavage of P3-2 with some poliovirus strains, Wiegers and Dernik (pers. comm.) have seen P3-6a and P3-6b with all strains analyzed. The biological significance of this phenomenon is unknown.

3.2 Proteinases Involved in Processing

The nature and the specificity of the proteinase(s) involved in post-translational processing of poliovirus are poorly understood. It has become apparent, however, that picornaviruses code for at least one proteinase.

3.2.1 Are Cellular Proteinases Involved?

Until recently it has been generally assumed that cellular proteinase(s) act on the polyprotein (see Korant et al. 1980a). This conclusion is based on the following observations: (1) With most picornaviruses studied, cleavage occurs on the nascent polypeptide chain (an exception is a report by Kiehn and Holland 1970, on Coxsackie virus); (2) there are differences in sensitivity of proteolytic enzymes toward chloromethyl ketones of lysine or phenylalanine (TLCK or TPCK), depending on the cell line, rather than the infecting virus; (3) there are small differences in the size of the cleavage products, depending on the host cell; and (4) cleavage appears to be sensitive to diisopropyl fluorophosphate. Accordingly, Korant et al. (1980b) have implicated a cellular, polysome-bound proteinase in the initial processing of the viral polyprotein.

Inasmuch as it seems attractive to speculate involvement of cellular proteinases in primary processing of the picornavirus polyprotein, the observations used to support this hypothesis may be explained in different ways. As the evidence is weak, the hypothesis becomes increasingly doubtful in light of two recent discoveries: (1) A large precursor protein of EMC virus is capable of autocatalytic cleavage yielding a proteinase (Palmenberg and Rueckert, pers. comm.), and (2) the pair of amino acids that is split to yield P2-3b and P3-1b (a "primary" cleavage site) is the same as the pairs split in "secondary cleavages" (Semler, Anderson, Rothberg, Wishart, Kitamura, and Wimmer, unpublished data; Kitamura et al. 1980b) (see Sect. 3.2.4).

Thus, during the first few rounds of viral translation, the synthesis of the polyprotein may proceed to completion. The polyprotein may then cleave autocatalytically. The viral proteinase thus liberated could subsequently process newly synthesized polyprotein *in statu nascendi*. As to other arguments, it is known that TLCK and TPCK have effects on host cell metabolism that may have nothing to do with viral protein processing (see Sect. 3.2.5). The size differences of viral polypeptides observed in different cell lines may be due to "trimming" by host cell enzymes (Rueckert 1976). Finally, Korant et al. (1980b) have observed that the cellular, ribosome-bound proteinase is inactivated during poliovirus infection, whereas polyribosomes of infected cells retain proteolytic activity when assayed with viral precursor polypeptides. This polyribosome-associated activity may be the viral proteinase.

3.2.2 Evidence for a Viral Proteinase

The following experiments have led to the identification of a virus-coded proteinase. It was shown initially that extracts of infected cells are capable of cleaving precursors of viral capsid polypeptides, whereas extracts of uninfected cells lacked such activity (Korant 1972; Korant 1973; Esteban and Kerr 1974; Lawrence and Thach 1975). The time course of induction of the cleaving activity parallels the progression of the virus' replication in vivo; the rate by which the cleaving activity is produced corresponds to the multiplicity of infection (Lawrence and Thach 1975; Korant et al. 1979). Moreover, the cleaving activity can be abolished completely and irreversibly in infected cells with carbobenzoxyleucyl chloromethyl ketone (CLCK) (see Sect. 3.2.5). The cleaving activity reappears only if excess CLCK is removed from the infected cells by extensive washing and if viral protein synthesis is allowed to proceed. Since host cell-specific translation is inhibited by poliovirus, the renewed production of proteinase must require synthesis of a viral gene product (Korant et al. 1979). Finally, the cleaving activity appears to be temperature-sensitive if produced in vivo by a mutant of poliovirus (Korant et al. 1979). These data strongly suggest that the proteinase itself is a viral gene product. They do not rule out the possibility, however, that a viral gene product serves to activate or redirect an endogenous cellular proteinase.

Direct evidence for the synthesis of a virus-specific proteinase was obtained from experiments of translation of picornavirus RNA in vitro.

An increase in proteolytic activity in L-cell extracts programmed to translate EMC virus RNA (Esteban and Kerr 1974) or the de novo synthesis of such activity in Krebs ascites cell extracts programmed by the same viral RNA (Lawrence and Thach 1975) indicated that the proteinase is virus coded.

These initial experiments were deficient due to an inherent inefficiency of the cell-free translation systems, a problem solved by Pelham and Jackson (1976). These investigators treated rabbit reticulocyte extracts with micrococcal nuclease (to destroy endogenous mRNA), and subsequently supplemented it with tRNA. The resulting cell-free translation system showed excellent response to added mRNA, and endogenous protein synthesis was nearly abolished.

Pelham (1978), Svitkin and Agol (1978), and Rueckert and his colleagues (Shih et al. 1978, 1979) have used the nuclease-treated recticulocyte extracts programmed with polio or EMC RNA to show extensive and faithful proteolytic processing of the viral proteins synthesized in vitro. Moreover, addition of inhibitors of initiation or translocation suggested that a viral gene product involved in proteolysis mapped within the 3′-terminal half of the viral genome.

It should be pointed out that in vitro only 10 min are required for a ribosome to traverse the poliovirus genome, and 30 min to traverse the EMC genome, although both RNAs have roughly the same chain length and coding capacity (Shih et al. 1978, 1979). The difference in translation time appears to be due to a (sterical?) block located approximately in the middle of the EMC genome that is absent from poliovirus RNA. The block in EMC RNA slows down the ribosome as if it were to pause. Paradoxically and for unknown reasons the EMC RNA is still translated in vitro four-fold more efficiently than poliovirus RNA. In any case, the relatively long time needed for a ribosome to traverse the picornavirus RNA allows one to accurately establish the time at which newly synthesized viral proteins appear one after the other (see Pelham 1978; Shih et al. 1979).

Finally, the proteolytic activity was identified among the in vitro translation products of EMC RNA (Palmenberg et al. 1979) or in extracts of EMC virus-infected ascites cells (Svitkin et al. 1979; Gorbalenya et al. 1979). It is the viral polypeptide p22 of EMC (mol. wt. 20,000) that maps adjacent to the putative viral RNA polymerase. By analogy, P3-7C of poliovirus (Fig. 3) corresponds to p22 of EMC. Experiments to correlate P3-7c with proteolytic activity are now in progress. Remarkably, the precursor to the EMC protease (a polypeptide corresponding to P3-2 in Fig. 3) is capable of autocatalytic cleavage yielding the proteinase p22 and the putative EMC RNA polymerase (Palmenberg and Rueckert pers. comm.).

Korant et al. (1979) have come to the conclusion, on the other hand, that the proteinase of poliovirus is P2-X (see Fig. 3). It is entirely possible that picornaviruses produce more than one proteinase, and that consequently both P2-X and P3-7c have proteolytic activity. An alternative explanation is that polypeptide P3-6a, containing the amino acid sequence of P3-7c, may have copurified with P2-X and thus been responsible for apparent proteinase activity in P2-X (Korant pers. comm.). Previously, Lawrence and Thach (1975) were also victims of copurification of viral proteins. They concluded that the third largest coat protein of EMC (polypeptide γ, corresponding to VP3) was a viral proteinase. In fact, γ and p22 copurify under a variety of conditions (see Shih et al. 1979). Moreover, p22 was originally thought to be of host cell origin (Butterworth and Rueckert 1972).

3.2.3 Properties of the Viral Proteinase

Studies on the properties of the picornaviral proteinase have all been carried out in vivo or with crude extracts of infected cells. Nevertheless, the following observations have been made (reviewed by Korant 1979; Korant et al. 1980b): (1) The enzyme has an apparent pH optimum of 7 that is lower than that of most cellular cytoplasmic proteinases and higher than that of the major lysozomal enzymes; (2) Enzyme activity requires an unblocked sulfhydryl group as it is inhibited by iodoacetamide (Korant 1973) or N-ethyl-maleimide (Pelham 1978). Clearly, these experiments should be repeated with more specific inhibitors for sulfhydryl proteinases (e.g., leupeptin). Other studies of inhibitors will be discussed in Sec. 3.2.5.

3.2.4 Cleavage Sites in Viral Proteins

Little is known about the sites in picornavirus polypeptides that are cleaved during post-translational processing. Terminal analyses of capsid proteins (see Bachrach 1977; Scraba 1979) or nonstructural proteins (Semler et al., manuscript in preparation; Kitamura et al. 1980b), however, have shown similarities suggesting common signals (Table 1). The predominant pair of amino acids that is cleaved is . . . Gln-Gly . . . (. . . Q-G . . .), but the viral proteinase may also act on . . . Gln-Thr . . . (. . . Q-T . . .), and . . . Gln-Ser . . . (. . . Q-S. . .). A common C-terminal amino acid of VP1 is Leu (L), but the N-terminal counterpart of the adjacent polypeptide (in polio P2-X) is as yet unknown.

Bachrach (1977) has pointed out that the presence of threonine (T) at the N-terminus of aphthovirus VP3, rather than serine (S) in VP3 of mengovirus, could be accounted for by a point mutation in a Ser codon and has also speculated that cleavage of picornavirus polypeptides may always occur between an α-helix maker (Gln, Leu) and an α-helix breaker (Gly, Ser, Pro, Asn) amino acid. It should be pointed out, however, that both mengovirus capsid proteins (Scraba 1979) as well as poliovirus capsid proteins (Dernik, pers. comm.) are very low in α-helical content.

Two points are worth mentioning: (1) A pair of amino acids exceptional for cleavage is . . . Asn-Ser . . . (. . . N-S . . .) (see Table 1). This pair occurs at the junction between VP4 and VP2. The precursor for these capsid polypeptides is VPO (Fig. 3) that is cleaved only during morphogenesis, when procapsid and virion RNA combine (review by Rueckert 1976). It is possible that this "maturation cleavage" of the provirion utilizes a proteinase distinct from that acting in all other processing. (2) Based on both poliovirus protein and RNA sequencing data, cleavage to produce P3-1b (see Fig. 3) from the polyprotein appears to have occurred between . . . Gln-Gly . . ., which is a signal of "secondary cleavage" (Semler et al., manuscript in preparation). Previously,

Table 1. NH_2 and COOH terminal amino acids of picornavirus proteins

Virus	Viral proteins								
						├── P3-1b			──┤
								├── P3-2	──┤
	VP4	VP2	VP3	VP1	P2-X	P-3?	VPg	P3-7c	P3-4
	NH_2-COOH	NH_2-COOH	NH_2-COOH	NH_2-COOH	NH_2-COOH	NH_2-COOH	NH_2-COOH	NH_2-COOH	NH_2-COOH
Poliovirus type 1 (M)	a — N	S	G	G — Y	G Q	G — Q	G — Q	G — Q	G — F[c]
Aptho O_1K	a	D	T	G					
Aptho O_1B	a	D	T	G					
Aptho A_{12}	a	D — Q[b]	T — Q	G — L					
Aptho C_3	a	D	T	G					
Mengo	a	D — Q	S — Q	G — L					

[a] Blocked NH_2-terminus; [b] Possibly glutamic acid; [c] COOH-terminal amino acid of the polyprotein

Abbreviations used: A = alanine, D = aspartic acid, F = phenylalanine, G = glycine, L = leucine, N = asparagine, Q = glutamine, S = serine, T = threonine, Y = tyrosine, For references see Bachrach 1977: Scraba 1979; Kitamura et al. (1981) Nature 291, pp 547-553

it was thought that P3-1b is a product of "primary cleavage" by host cell enzymes. As pointed out in Sect. 3.2.1, the . . .Gln-Gly . . . sequence at the junction between P2-X and P3-1b suggests cleavage by the viral proteinase. Alternatively, a precursor to P3-1b could have been cut from the polyprotein by a cellular proteinase and rapidly trimmed to P3-1b by the viral proteinase. In this regard, the C-terminus of P2-X is of great interest.

Whatever the specificity of the viral proteinase, its activity must be governed by stringent conditions, as proteins of HeLa cells are not degraded during the first 4 hr of infection with poliovirus (Korant et al. 1980b). During this period the bulk of viral translation and protein processing occurs. Moreover, the enzyme synthesized by EMC virus does not cleave the poliovirus capsid precursor protein (Shih et al. 1979) although both viral proteins may have similar cleavage signals. This latter observation may suggest involvement of secondary or tertiary structure of the substrate. Similarly, the temperature sensitivity to 41°C of EMC proteinase, reported by Pelham (1978) may reflect sensitivity of the proteinase to changes in conformation of the substrate rather than inactivation of the enzyme. Changes of secondary and/or tertiary structure of the substrate may also be the major reason for inhibition of processing if amino acid analogs are incorporated into the viral polyprotein. This phenomenon, first observed by Jacobson and Baltimore (1968), was utilized to accumulate large precursor proteins when fluorophenylalanine, canavanine (proline analog), ethionine (methionine analog) and azetidine-2-carboxylic acid (proline analog), were added to the infected cells (Jacobson et al. 1970). Interpretation of these data is complicated, however, in that the amino acid analogs were incorporated not only into the substrate polypeptides but into the viral proteinase itself.

3.2.5 Studies with Specific Inhibitors of the Viral Proteinase

Based upon the observation that the viral proteinase cleaves a peptide bond the carboxylic group of which is donated by Gln or Leu, Korant et al. (1980a) have synthesized a number of amino acid derivatives and oligopeptides and tested these in poliovirus infected cells. Carbobenzoxy leucyl chloromethyl ketone (CLCK) was found to inhibit poliovirus replication at 0.1 mM. The inhibitor, labeled with [^{3}H]Leu, bound mainly to two viral polypeptides of 40,000 and 23,000 mol. wt. It remains to be seen whether or not the 23K protein is P3-7c, the putative proteinase of poliovirus. The 40K protein could be P2-X or P3-6a, the product of alternate cleavage of P3-2 (Korant, pers. comm.; see Sect. 3.2.2).

A suitable, synthetic substrate for poliovirus proteinase is acetyl phenylalanyl glycyl alanyl leucyl thiobenzyl ester (Ac-Phe-Gly-Ala-Leu-SBz; Korant et al. 1980a). Owing to the chromogenic group in the substrate, enzyme action can be followed spectrophotometrically, a property that has been utilized to screen for viral proteinase mutants. The chloromethyl ketone derivative of the tetrapeptide (Ac-Phe-Gly-Ala-Leu-CMk) irreversibly inhibits viral replication presumably by covalent binding to the viral proteinase.

The viral proteinase is also inhibited reversibly by Zn^{2+} ions at 0.1 mM (Butterworth and Korant 1974; Korant and Butterworth 1976). It is thought that Zn^{2+} may change the conformation of viral polypeptides. Accordingly, the effect of Zn^{2+} on viral replication differs slightly with different picornaviruses (Korant et al. 1980a).

Difficult to interpret, and rather confusing, are results of inhibition of proteolysis with tosylsulfonyl phenylalanyl chloromethyl ketone (TPCK) and with the same derivative of lysine (TLCK). It has been claimed that TPCK has a preferential effect on chymotrypsin, and TLCK on trypsin (see Shaw and Ruscica 1971). Both inhibitors have an effect on polioprotein cleavage even in the same cell line (HeLa; Korant 1972; Summers et al. 1972). These results are not definitive as TPCK also inhibits initiation of protein synthesis in uninfected cells (Pong et al. 1975). Thus, polio proteins synthesized in the presence of TPCK may represent run-off molecules that could be chemically modified by the ability of the inhibitor to alkylate polypeptides.

4 Possible Regulation of Viral RNA Synthesis by Post-translational Processing

Clearly, the complex mechanism of protein synthesis of picornaviruses must influence, if not regulate, viral RNA replication. Unfortunately, there are no facts available to date that would hint at how this regulation occurs. Consequently, the field is open to speculation.

Any hypothesis must consider the fact that the genome-linked protein VPg, viral proteinase (P3-7c, by inference from EMC) and viral RNA polymerase (P3-4) all originate from a common precursor polypeptide (P3-1b; see Fig. 3). Moreover, VPg does not occur free in the infected cell (that is, unbound, and in quantities large enough to detect). Therefore, VPg must enter the RNA synthesizing machinery in the form of a precursor polypeptide. With EMC virus, such a precursor polypetide has been found (polypeptide "H", mol. wt. 12,000; Pallansch et al. 1980) but there is no evidence as yet that "H" is involved in RNA synthesis. There are several small, poliovirus-coded polypeptides (mol. wt. < 12,000) not shown in Fig. 3, that can be detected in poliovirus infected cells, one of which may be the precursor to VPg. How does this all fit together?

The following model may serve as a working hypothesis: polypeptide P3-1b is cleaved autocatalytically (Palmenberg and Rueckert, pers. comm.), or by P3-7c already formed, to "Pre-VPg" and to P3-2. The latter is processed further to P3-7c and P3-4. The viral RNA polymerase P3-4 Lundquist et al. 1974; Flanegan and Baltimore 1979) may combine with cellular factors (Dasgupta et al. 1979), and attach to a membrane. This complex has affinity to the 3′-terminal poly(A) of virion RNA. "Pre-VPg" (or another, VPg-containing polypeptide), together with proteinase P3-7c also enter the membrane-bound complex. Pre-VPg is cleaved to yield VPg onto which the RNA is then polymerized.

Clearly, an additional number of proteins and enzymes must participate in RNA synthesis and replication. For example, a specific protein (P2-X?) that combines with P3-4 that directs it to bind with membranes, or an enzyme that catalyzes the formation of the bond between the tyrosine of VPg and the 5′-terminal U of the RNA (VPg-tyr-OH + pppU → VPg-tyr-O-pU + PPi?), etc., may exist.

An approach to solving the fascinating problem of genome synthesis of picornavirus RNA is to reconstitute a crude system capable of replicating viral RNA. Such a system can subsequently be separated into components, the precise function of which can be identified.

Note Added in Proof: The complete structure of the poliovirus genome has now been elucidated and the precise cleavage signals within the polyprotein have been determined (Kitamura, Semler, Rothberg, Larsen, Adler, Emini, Hanecak, Lee, v.d. Werf, Anderson and Wimmer, Nature (1981) 291, 547-553; Semler, Anderson, Kitamura, Rothberg, Wishart and Wimmer (1981), Proc Nat Acad Sci USA 78, 3464-3468; Semler, Hanecak, Anderson and Wimmer (1981) Virology 114, 589-594; Larsen, Anderson, Dorner, Semler and Wimmer (1982) J Virol, in press). Partial sequence analysis of aphthovirus RNA has identified some cleavage signals within the precursor to the capsid polypeptides of aphthovirus (Küpper et al., Nature (1981) 289, 555-559; Boothroyd et al. (1981) Nature *290*, 800-802).

References

Adler S, Purich D, Stadtman E (1975) Cascade control of *Escherichia Coli* glutamine synthetase. Properties of the P_{II} regulatory protein and the uridylyltransferase-uridylyl-removing enzyme. J Biol Chem 250: 6264-6272

Ambros V, Baltimore D (1978) Protein is linked to the 5′ end of poliovirus RNA by a phosphodiester linkage to tyroseine. J Biol Chem 253: 5263-5266

Ambros V, Baltimore D (1980) Purification and properties of a HeLa cell enzyme able to remove the 5′-terminal protein from poliovirus RNA. J Biol Chem 255: 6739-6744

Ambros V, Pettersson R, Baltimore D (1978) An enzymatic activity in uninfected cells that cleaves the linkage between poliovirion RNA and the 5′-terminal protein. Cell 15: 1439-1446

Babich A, Wimmer E, Toyoda H, Nomoto A (1980) The genome-linked protein of poliovirus type 2 RNA is covalently linked to a nonanucleotide identical to that of poliovirus type 1 RNA. Intervirology 13: 192-199

Bachrach HL (1977) Foot-and-Mouth-Disease Virus: Properties, Molecular Biology, and Immunogenicity. In: Romberger JE (ed) Beltsville Symp Agric Res, vol I. Virology in agriculture. Allanheld Osmun & Co, Montclair, pp 3-32

Butterworth B, Korant B (1974) Characterization of the large picornaviral polypeptides produced in the presence of zinc ion. J Virol 14: 282-291

Butterworth B, Rueckert R (1972) Kinetics of synthesis and cleavage of encephalomyocarditis virus-specific proteins. Virology 50: 535-549

Dasgupta A, Baron M, Baltimore D (1979) Poliovirus replicase: A soluble enzyme able to initiate copying of poliovirus RNA. Proc Natl Acad Sci USA 76: 2679-2683

Doel T, Sangar D, Rowlands D, Brown F (1978) A re-appraisal of the biochemical map of Foot-and-Mouth disease virus RNA. J Gen Virol 41: 395-404

Esteban M, Kerr I (1974) The synthesis of encephalomyocarditis virus polypeptides in infected L-cells and cell-free systems. Eur J Biochem 45: 567-576

Flanegan J, Baltimore D (1979) Poliovirus polyuridylic acid polymerase and RNA replicase have the same viral polypeptide. J Virol 29: 352-360

Flanegan J, Pettersson R, Ambros V, Hewlett M, Baltimore D (1977) Covalent linkage of a protein to a defined nucleotide sequence at the 5′ terminus of the virion and replicative intermediate RNAs of poliovirus. Proc Natl Acad Sci USA 74: 961-965

Golini F, Nomoto A, Wimmer E (1978) The genome-linked protein of picornaviruses. IV. Difference in the VPgs of encephalomyocarditis virus and poliovirus as evidence that the genome-linked proteins are virus-coded. Virology 89: 112-118

Gorbalenya A, Svitkin Y, Kazachkov Y, Agol V (1979) Encephalomyocarditis virus-specific polypeptide p22 is involved in the processing of the viral precursor polypeptides. FEBS Lett 108: 1-9

Grubman M (1980) The 5′ end of Foot-and-Mouth disease virion RNA contains a protein covalently linked to the nucleotide pUp. Arch Virol 63: 311-315

Harris T (1979) The nucleotide sequence at the 5′ end of Foot-and-Mouth disease virus RNA. Nucleic Acids Res 7: 1765-1786

Holland J, Kiehn E (1968) Specific cleavage of viral proteins as steps in the synthesis and maturation of enteroviruses. Proc Natl Acad Sci USA 60: 1015-1022

Holzer H, Wohlhueter R (1972) (Glutamine Synthetase) tyrosyl-O-adenylate: A new energy-rich phosphate bond. In: Weber G (ed) Advances in enzyme regulation, vol X. Pergamon Press, Oxford, pp 121-132

Jacobson M, Baltimore D (1968) Polypeptide cleavages in the formation of poliovirus proteins. Proc Natl Acad Sci USA 61: 77-84

Jacobson M, Asso J, Baltimore D (1970) Further evidence on the formation of poliovirus proteins. J Mol Biol 49: 657-669

Kiehn E, Holland J (1970) Synthesis and cleavage of enterovirus polypeptides in mammalian cells. J Virol 5: 358-367

Kitamura N, Adler C, Wimmer E (1980a) Structure and expression of the picornavirus genome. In: Genetic variation of viruses, Ann NY Acad Sci (in press)

Kitamura N, Adler C, Rothberg P, Martinko J, Nathenson S, Wimmer E (1980b) The genome-linked protein of picornaviruses. VII. Genetic mapping of poliovirus VPg by protein and RNA sequence studies. Cell 21: 295-302

Koch G (1973) Interactions of poliovirus-specific RNAs with HeLa cells and *E. coli.* Curr Top Microbiol Immunol 62: 89-138

Korant B (1972) Cleavage of viral precursor proteins in vivo and in vitro. J Virol 10: 751-759

Korant B (1973) Cleavage of poliovirus-specific polypeptide aggregates. J Virol 12: 556-563

Korant B (1979) Role of cellular and viral proteases in the processing of picornavirus proteins. In: Perez-Bercoff R (ed) The molecular biology of picornaviruses. Plenum Publsihing Co Ltd, New York London

Korant B, Butterworth B (1976) Inhibition by zinc of rhinovirus protein cleavage: Interaction of zinc with capsid polypeptides. J Virol 18: 298-306

Korant B, Chow N, Lively M, Powers J (1979) Virus-specific protease in polio-infected HeLa cells. Proc Natl Acad Sci USA 76: 2992-2995

Korant B, Chow N, Lively M, Powers J (1980a) Proteolytic events in replication of animal viruses. Ann NY Acad Sci 343: 304-318

Korant B, Langner J, Powers J (1980b) Protein synthesis and cleavage in picornavirus infected cells. In: Koch G, Ricter D (eds) Biosynthesis, modification and processing of cellular and viral protein. Academic Press, London New York (in press)

Lawrence C, Thach R (1975) Identification of a viral protein involved in post-translational maturation of the encephalomyocarditis virus capsid precursor. J Virol 15: 918-928

Lee Y, Nomoto A, Wimmer E (1976) The genome of poliovirus is an exceptional eukaryotic mRNA. Prog Nucleic Acids Res Mol Biol 19: 89-95

Lee Y, Nomoto A, Detjen B, Wimmer E (1977) The genome-linked protein of picornaviruses. I. A protein covalently linked to poliovirus genome RNA. Proc Natl Acad Sci USA 74: 59-63

Lundquist R, Ehrenfeld E, Maizel J Jr (1974) Isolation of a viral polypeptide associated with poliovirus RNA polymerase. Proc Natl Acad Sci USA 71: 4773-4777

McLean C, Matthews T, Rueckert R (1976) Evidence of ambiguous processing and selective degradation in the noncapsid proteins of rhinovirus 1A. J Virol 19: 903-914

Nomoto A, Lee Y, Wimmer E (1976) The 5′ end of poliovirus mRNA is not capped with m^7G-(5′)ppp(5′)Np. Proc Natl Acad Sci USA 73: 375-380

Nomoto A, Detjen B, Pozzatti R, Wimmer E (1977a) The location of the polio genome protein in viral RNAs and its implication for RNA synthesis. Nature (London) 268: 208-213

Nomoto A, Kitamura N, Golini F, Wimmer E (1977b) The 5′-terminal structures of poliovirion RNA and poliovirus mRNA differ only in the genome-linked protein VPg. Proc Natl Acad Sci USA 74: 5345-5349

Pallansch M, Kew O, Palmenberg A, Golini F, Wimmer E, Rueckert R (1980) Picornaviral VPg sequences are contained in the replicase precursor. J Virol 35: 414-419

Palmenberg A, Pallansch M, Rueckert R (1979) Protease required for processing picornaviral coat protein resides in the viral replicase gene. J Virol 32: 770-778

Pelham H (1978) Translation of encephalomyocarditis virus RNA in vitro yields an active proteolytic processing enzyme. Eur J Biochem 85: 457-462

Pelham H, Jackson R (1976) An efficient mRNA-dependent translation system from reticulocyte lysates. Eur J Biochem 67: 247-256

Pettersson R, Flanegan J, Rose J, Baltimore D (1977) 5'-terminal nucleotide sequence of poliovirus polyribosomal RNA and virion RNA are identical. Nature (London) 268: 270-272

Pettersson R, Ambros V, Baltimore D (1978) Identification of a protein linked to nascent poliovirus RNA and to the polyuridylic acid of negative strand RNA. J Virol 27: 357-365

Pong S, Nuss D, Koch G (1975) Inhibition of initiation of protein synthesis in mammalian tissue culture cells by L-1-Tosylamido-2-phenylethyl chloromethyl ketone. J Biol Chem 250: 240-245

Rekosh D (1977) The molecular biology of picornaviruses. In: Nayak D (ed) The molecular biology of animal viruses, vol. I. Marcel Dekker, New York, pp 63-110

Rekosh D, Russell W, Bellet A, Robinson A (1977) Idenfication of a protein linked to the ends of adenovirus DNA. Cell 11: 283-295

Rothberg P, Harris T, Nomoto A, Wimmer E (1978) The genome-linked protein of picornaviruses. V. 0^4-(5'-Uridylyl)tyrosine is the bond between the genome-linked protein and the RNA of poliovirus. Proc Natl Acad Sci USA 75: 4868-4872

Rothberg P, Adler C, Kitamura N, Wimmer E (1980) VPg: The genome-linked protein of picornaviruses. In: Koch G, Ricter D (eds) Biosynthesis, modification, and processing of cellular and viral protein. Academic Press, London New York (in press)

Rueckert R (1976) On the structure and morphogenesis of picornaviruses. In: Fraenkel-Conrat H, Wagner R (eds) Comprehensive virology, vol 6. Plenum Press, New York, pp 131-213

Rueckert R, Matthews T, Kew O, Pallansch M, McLean C, Omilianowski D (1979) Synthesis and processing of picornaviral protein. In: Perez-Bercoff R (ed) The molecular biology of picornaviruses. Plenum Publishing Co Ltd, New York London, pp 113-125

Scraba D (1979) The picornavirion: Structure and assembly. In: Perez-Bercoff R (ed) The molecular biology of picornaviruses. Plenum Publishing Co Ltd, New York London, pp 1-23

Shaw E, Ruscica J (1971) The reactivity of His-57 in chymotrypsin to alkylation. Arch Biochem Biophys 145: 485-489

Shih D, Shih C, Kew O, Pallansch M, Rueckert R, Kaesberg P (1978) Cell-free synthesis and processing of the proteins of poliovirus. Proc Natl Acad Sci USA 75: 5807-5811

Shih D, Shih C, Zimmern D, Rueckert R, Kaesberg P (1979) Translation of encephalomyocarditis virus in reticulocyte-lysates: Kinetic analysis of the formation of virion proteins and a protein required for processing. J Virol 30: 472-480

Stanley J, Van Kammen A (1979) Nucleotide sequences adjacent to the proteins covalently linked to the cowpea mosaic virus genome. Eur J Biochem 101: 45-49

Summers D, Maizel J (1968) Evidence for large precursor proteins in poliovirus synthesis. Proc Natl Acad Sci USA 59: 966-971

Summers D, Shaw E, Stewart M, Maizel J Jr (1972) Inhibition of cleavage of large poliovirus-specific precursor proteins in infected HeLa cells by inhibitors of proteolytic enzymes. J Virol 10: 880-884

Svitkin Y, Agol V (1978) Complete translation of EMC virus RNA and faithful cleavage of virus-specific proteins in a cell-free system from Krebs-2 cells. FEBS Lett 87: 7-11

Svitkin Y, Gorbalenya A, Kazachkov Y, Agol V (1979) Encephalomyocarditis virus-specific polypeptide p22 possessing a proteolytic activity. FEBS Lett 108: 6-9

Tse Y, Kirkegard K, Wang J (1980) Covalent bonds between protein and DNA. Formation of phosphotyrosine linkage between certain DNA topoisomerases and DNA. J Biol Chem 255: 5560-5565

Vartapetian AB, Drygin Y, Chumakov K (1979) The structure of the covalent linkage between proteins and RNA in encephalomyocarditis virus. Bioorg Kim 5: 1876-1878

Wimmer E (1979) The genome-linked protein of picornaviruses: Discovery, properties and possible function. In: Perez-Bercoff R (ed) The molecular biology of picornaviruses. Plenum Publishing Co Ltd, New York London, pp 175-190

Yogo Y, Wimmer E (1972) Polyadenylic acid at the 3'-terminus of poliovirus RNA. Proc Natl Acad Sci USA 69: 1877-1882

Discussion[2]

Upon respective questions of Drs. Fischer, Hilz and Switzer Dr. Wimmer stated that (1) mutants of the RNA-Protein linkage in picornaviruses so far are not available, (2) the enzymology of the linkage of VPg to RNA is unknown, and (3) processing of the viral proteins can be prevented by a tetrapeptide isolated by Bruce Korant. In the presence of this protease inhibitor, however, RNA synthesis is also turned off.

2 Rapporteur: Heinrich Betz

The Acetylcholine Receptor: Control of Its Synthesis, Stability, and Cell Surface Distribution

H. Betz[1]

1 Introduction

Nicotinic acetylcholine receptors (AChR) mediate signal reception and transduction at neuromuscular and interneuronal synapses of vertebrates. Binding of the neurotransmitter acetylcholine or its agonists to AChR induces a conformational change in this receptor molecule which leads to the opening of transmembrane channels permeable to sodium and potassium ions. The resulting ion fluxes depolarize the postsynaptic membrane and thus induce a propagated action potential.

During the past 10 years, the structure, function and metabolism of the AChR have been investigated in great detail (for review see Heidmann and Changeux 1978; Fambrough 1979). This paper focuses on the regulation of AChR during synaptic development and discusses possible mechanisms for the transsynaptic control of the synthesis, degradation and cell surface distribution of this receptor.

2 Properties, Metabolism and Cell Surface Distribution of Extra- and Subsynaptic AChR

AChR is a major component of the postsynaptic membrane of neuromuscular junctions and of different cholinergic synapses in the central nervous system (Fambrough 1979; Vogel et al. 1977). In the adult motor end plate, its density is more than 1000-fold higher than in the "extrasynaptic" part of the sarcolemma. This is in marked contrast to the distribution of AChR on noninnervated myotubes and cholinoceptive neurons: there, receptors are uniformly dispersed over the entire cell surface, and their density in the differentiated postsynaptic membrane is only 10 to 100-fold higher. Synapse formation thus requires the localization of AChR under the innervating nerve terminal.

Because of their different distribution, the properties of extra- and subsynaptic AChR have been compared in great detail. As summarized in Table 1, both receptor species differ in their antagonist affinities, their isoelectric points, their lateral mobility in the plane of the membrane, the characteristics of their ion channels, and their immunological properties. Also, they exhibit a different metabolism in vivo: extrasynaptic receptors are subjected to continuous turnover, whereas subsynaptic receptors in the adult are essentially stable. Many of these differences have been related to a cova-

1 Max-Planck-Institut für Psychiatrie, Abt. Neurochemie, 8033 Martinsried, FRG

Table 1. Differences between extra- and subsynaptic AChR. (For references see Fambrough 1979)

	Extrasynaptic (Embryonic)	Subsynaptic (Adult)
Distribution	Diffuse (10^2 to 10^3 per μm^2)	Clustered (ca. 10^4 per μm^2)
K_D for d-tubocurarine	5.5×10^{-7} M	4.5×10^{-8} M
Isoelectric point	5.3	5.1
Lateral mobility (diffusion coefficient)	Large (5×10^{-11} cm^2/s)	Small ($<10^{-12}$ cm^2/s)
Mean channel opening time	3 to 5 ms	0.7 to 1 ms
Antigenic determinants detected by mayastenia gravis antibodies	++++	++
Metabolic half-life	15 to 30 h	5 to 15 days

lent modification of AChR during synaptogenesis. In membrane preparations from fish electric organ, a tissue extremely rich in nicotinic receptor sites, AChR could indeed be phosphorylated in vitro (Teichberg et al. 1977; Gordon et al. 1977). Furthermore, Vandlen et al. (1979) have reported that purified fish receptor contains phosphoserine.

Electrophysiological and autoradiographic studies have shown that the subsynaptic localization of AChR is achieved by at least two different processes: (1) the aggregation of receptors in the subsynaptic membrane, and (2) the removal of excess receptors from the extrasynaptic regions of the myofiber surface. Both processes are regulated by transsynaptic signals. As discussed in Sect. 3, the former presumably involves substances released by the motor nerve terminal. The latter depends on functional neurotransmission and results from a repression of extrasynaptic AChR synthesis by muscle electrical activity (Betz et al. 1977, 1980; Burden 1977a; Reiness and Hall 1977). Due to the continuing degradation ($t_{1/2}$ 20 to 30 h) of embryonic receptors, this repression causes a rapid drop in extrasynaptic acetylcholine sensitivity. Upon further development, subsynaptic AChR becomes protected against degradation (Burden 1977b; Betz et al. 1980). Possible mechanisms underlying the developmental control of AChR are discussed in the following section.

3 Models for AChR Aggregation and Regulation

As outlined in Sect. 2, the various changes in AChR properties and cell surface distribution occurring during synaptogenesis appear to be closely related to the innervation and electrical activation of the muscle cell, or postsynaptic neuron:

1. In a series of elegant experiments, Anderson et al. (1977) demonstrated that the addition of motor neurons to cultured Xenopus myocytes, whose AChR had previously been tagged with fluorescent α-bungarotoxin, caused an accumulation of fluorescence under sites of nerve-muscle contact. This suggests that motor neurons induce a redistribution of receptors in the plasma membrane of muscle cells. Furthermore, different groups reported that extracts from spinal cord, brain or nerve cell lines

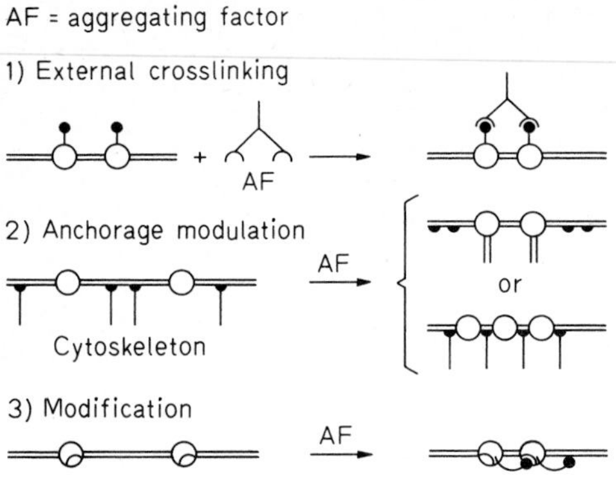

Fig. 1. AChR aggregation: possible mechanisms. For explanation see Sect. 3

contain proteinaceous factors which enhanced the aggregation of AChR on cultured myotubes (Podleski et al. 1978; Christian et al. 1978; Jessell et al. 1979). The mechanism of action of these factors is not yet clear (Fig. 1): recent data favor the concept that receptor clustering involves fixation of AChR to proteins anchoring the latter to th plasma membrane or even to the cytoskeleton itself (Prives et al. 1980; Bloch, pers. comm.). External crosslinking and convalent modifications of AChR have also been implicated (Changeux and Danchin 1976; Christian et al. 1978).

2. The processes underlying the metabolic stabilization of AChR are unknown. As in the case of receptor aggregation, both covalent modification and external cross-linking mechanisms have been considered (Teichberg et al. 1977; Prives et al. 1980). Evidence suggesting a cross-linking process comes from the work of Prives et al. (1980) who showed that the addition of concanavalin A to the culture medium decreased the rate of degradation of myotube AChR. By contrast, anti-AChR accelerated receptor degradation, a mechanism which is believed to contribute to the pathogenesis of myasthenia gravis, an autoimmune disease associated with muscle AChR (Fambrough 1979; Kao and Drachman 1977).

3. The regulation of extrasynaptic AChR synthesis has been extensively investigated in primary cultures of embryonic muscle (for review, see Fambrough 1979) and, also recently. neural retina (Betz 1980b). As depicted in Fig. 2, AChR induction most likely involves a cyclic AMP-dependent pathway (Betz and Changeux 1979; Blosser and Appel 1980). By contrast, the repression of receptor synthesis by membrane depolarization, i.e., from muscle or neuronal electrical activity, has been postulated to proceed via the opening of voltage-dependent Ca^{2+} channels (Shainberg et al. 1976) and subsequent stimulation of guanylate cyclase (Betz and Changeux 1979). Support for this hypothesis has come from several papers which report that levels of cyclic GMP are transiently raised upon nerve or muscle activity (Nestler et al. 1978; Study et al. 1978). Attempts to correlate cyclic GMP and AChR levels in depolarized embryonic myotubes in vitro have so far, however, been unsuccessful (Betz 1980a). This does not necessarily disprove the model of Fig. 2. Electrophysiological experiments revealed that the repression of AChR by electrical activity has "memory" properties: stimuli of 10 s every 5,5 h were sufficient to reduce significantly extrasynaptic acetylcholine sensitivity (Lomo and Westgaard 1976). A relatively stable control mechanism

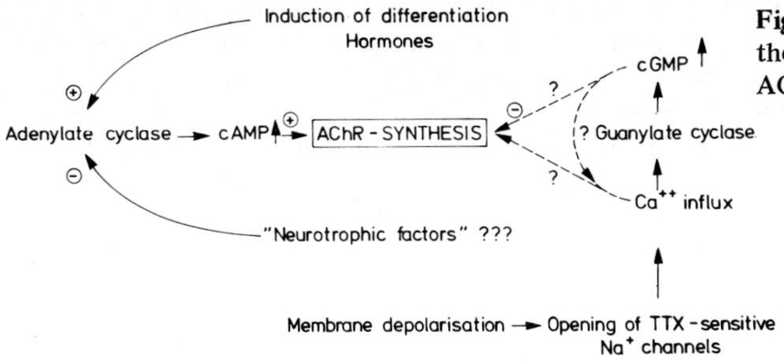

Fig. 2. Hypothetical scheme on the regulation of extrasynaptic AChR synthesis

thus must be the target of transient changes in intracellular Ca^{2+} and/or cyclic GMP. Reversible modifications of membrane and/or nuclear regulatory proteins appear a likely possibility.

4 Conclusions

AChR is an integral membrane protein which, during development, is highly regulated with respect to its metabolism and its cell surface distribution. Some of these regulations are likely to involve an interconversion, most probably a phosphorylation, of the receptor molecule itself. In addition, modifications of regulatory proteins are thought to participate in the repression of AChR synthesis by membrane depolarisation.

Acknowledgements. The author thanks Drs. J.P. Bourgeois and J.P. Changeux for collaboration, Dr. D. Graham for a critical reading of the manuscript, Mrs. U. Müller for technical assistance, and Mrs. H. Macher and C. Bauereiss for help during the preparation of this paper. Own work cited in this article was supported by the Deutsche Forschungsgemeinschaft. H.B. holds a Heisenberg award from the Deutsche Forschungsgemeinschaft.

References

Anderson MJ, Cohen MW, Zorychta E (1977) Effects of innervation on the distribution of acetylcholine receptors on cultured muscle cells. J Physiol 268: 731-756

Betz H (1980a) Effects of drug-induced paralysis and depolarisation on acetylcholine receptor and cyclic nucleotide levels of chick muscle cultures. FEBS Lett 118: 289-292

Betz H (1980b) Regulation of acetylcholine receptor. Adv Physiol Sci Vol. 36: 313-322

Betz H, Bourgeois JP, Changeux JP (1977) Evidence for degradation of the acetylcholine (nicotinic) receptor in skeletal muscle during the development of the chick embryo. FEBS Lett 77: 219-224

Betz H, Bourgeois JP, Changeux JP (1980) Evolution of cholingergic proteins in developing slow and fast skeletal muscles in chick embryo. J Physiol (London) 302: 197-218

Betz H, Changeux JP (1979) Regulation of muscle acetylcholine receptor synthesis in vitro by cyclic nucleotide derivatives. Nature (London) 278: 749-752

Blosser JC, Appel SH (1980) Regulation of acetylcholine receptor by cyclic AMP. J Biol Chem 253: 3088-3093

Burden S (1977a) Development of the neuromuscular junction in the chick embryo; the number, distribution, and stability of acetylcholine receptors. Dev Biol 57: 317-329
Burden S (1977b) Acetylcholine receptors at the neuromuscular junction: developmental change in receptor turnover. Dev Biol 61: 79-85
Changeux JP, Danchin A (1976) Selective stabilisation of developing synapses as a mechanism for the specification of neuronal networks. Nature (London) 264: 705-712
Christian CN, Daniels MP, Sugiyama H, Vogel Z, Jacques L, Nelson PC (1978) A factor from neurons increases the number of acetylcholine receptor aggregates on cultured muscle cells. Proc Natl Acad Sci USA 75: 4011-4015
Fambrough D (1979) Control of acetylcholine receptors in skeletal muscle. Physiol Rev 59: 165-227
Gordon AS, Davis CG, Diamond I (1977) Phosphorylation of membrane proteins at a cholingergic synapse. Proc Natl Acad Sci USA 74: 263-267
Heidmann T, Changeux JP (1978) Structural and functional properties of the acetylcholine receptor protein in its purified and membrane-bound states. Annu Rev Biochem 47: 317-357
Jessell TM, Siegel RE, Fischbach GD (1979) Induction of acetylcholine receptors on cultured skeletal muscle by a factor extracted from brain and spinal cord. Proc Natl Acad Sci USA 76: 5397-5401
Kao I, Drachman DB (1977) Myasthenic immunoglobulin accelerates acetylcholine receptor degradation. Science 196: 527-529
Lomo T, Westgaard RH (1976) Control of ACh sensitivity in rat muscle fibers. Cold Spring Harbor Symp Quant Biol 40: 263-274
Nestler EJ, Beam KG, Greengard P (1978) Nicotinic cholinergic stimulation increases cyclic GMP levels in vertebrate skeletal muscle. Nature (London) 27: 451-453
Podleski TR, Axelrod D, Ravdin P, Greenberg I, Johnson MM, Salpeter MM (1978) Nerve extract induces increase and redistribution of acetylcholine receptors on cloned muscle cells. Proc Natl Acad Sci USA 75: 2035-2039
Prives J, Christian C, Penman S, Olden K (1980) Neuronal regulation of muscle acetylcholine receptors: role of the muscle cytoskeleton and receptor carbohydrate. In: Giacobini E (ed) Tissue culture in neurobiology. Raven Press, New York, pp 35-52
Reiness CG, Hall ZW (1977) Electrical stimulation of denervated muscle reduces methionine inorporation into ACh receptor. Nature (London) 268: 655-657
Shainberg A, Cohen SA, Nelson PG (1976) Induction of acetylcholine receptors in muscle cultures. Pflueger's Arch 361: 255-261
Study RE, Brakefield YO, Bartfai T, Greengard P (1978) Voltage-sensitive calcium channels regulate guanosine 3′, 5′-cyclic monophosphate levels in neuroblastoma cells. Proc Natl Acad Sci USA 75: 6295-6299
Teichberg V, Sobel A, Changeux JP (1977) In vitro phosphorylation of the acetylcholine receptor. Nature (London) 267: 540-542
Vandlen LR, Wu WCS, Eisenach JC, Raftery MA (1979) Studies of the composition of purified Torpedo californica acetylcholine receptor and of its subunits. Biochemistry 18: 1845-1854
Vogel Z, Maloney GJ, Ling A, Daniels MP (1977) Identification of synaptic acetylcholine receptor sites in retina with peroxidase-labeled α-bungarotoxin. Proc Nat Acad Sci USA 74: 3268-3272

Discussion[2]

With Drs. Demaille, Hofmann, Helmreich, Heilmayer and Walter as participants, the following statements were made:

(1) The currently available data on the development of the elements of the calcium cycle do not fit with the postulated role of calcium in acetylcholine receptor

---

2 Rapporteur: Heinrich Betz

repression. This may, however, be explained by the existence of different calcium compartments. Indeed, from dysgenic mutants of mouse indirect evidence has been obtained that the increase in cytoplasmic calcium during contraction is not required for receptor regulation; (2) the stoichiometry and the physiological significance of acetylcholine receptor phosphorylation are still unknown. The kinase involved is insensitive to both cyclic AMP and cyclic GMP, but has not been further characterized; (3) during the differentiation into slow and fast muscle fibers, one observes striking differences in acetylcholine receptor expression and clustering. These are presumably related to different patterns of electrical and/or contractile activity (cf. Betz et al. 1980).

β-Catecholamine-Stimulated Adenylate Cyclase; an Associating-Dissociating System

A. Bakardjieva[1], R. Peters[2], M. Hekman[1], H. Hornig[1], W. Burgermeister[1], and E.J.M. Helmreich[1]

Hormonally regulated adenylate cyclase is a complex membrane-bound and vectorially oriented enzyme system of which at present three components are fairly well characterized: The hormone receptor, the guanylnucleotide binding subunit, and the catalytic moiety. The stoichiometry is assumed to be 1:1:1 but this is not proven (Fig. 1) (see Helmreich and Bakardjieva 1980). In the case of β-catecholamine-stimulated adenylate cyclase it became apparent from recent kinetic and biochemical evidence that activity is controlled by reversible interactions between the receptor, the guanylnucleotide binding protein, and the enzyme. Recent data which bear on these interactions are discussed here. Although the presently available information is quite persuasive, it is not yet conclusive.

1 Coupling of Receptor with Adenylate Cyclase

1.1 Kinetic Evidence

The kinetics of adenylate cyclase activation by β-catecholamines have been studied extensively by A. Levitzki and his group (see Tolkovsky and Levitzki 1978). The evidence may be summarized as follows: When hormone and GTP are saturating, the system at steady state is described by the equation:

$$E \underset{k_{off}}{\overset{\text{GTP, hormone, } k_{on}}{\rightleftharpoons}} E^* \qquad (1)$$

Thus, the total enzyme concentration in the membrane is:
$[E_T] = [E] + [E^*]$
and at steady state:
$k_{on}\,[E] = k_{off}\,[E^*]$
Therefore one obtains:
$[E^*] = [E_T] \,/\, (1 + k_{off}/k_{on})$;

1 Department of Physiological Chemistry, University of Wuerzburg, School of Medicine, Koellikerstrasse 2, 8700 Wuerzburg, FRG
2 Center of Biological Chemistry, University of Frankfurt and Max Planck Institute for Biophysics, 6000 Frankfurt, FRG

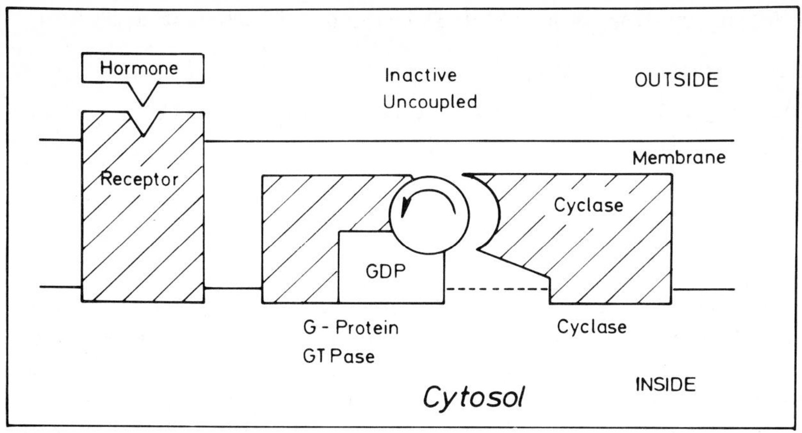

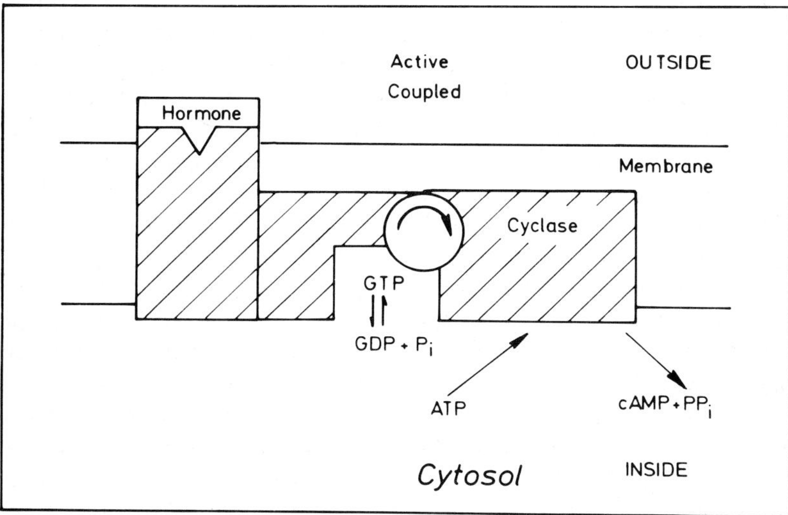

Fig. 1. Schematic presentation of a hormonally stimulated adenylate cyclase system and its activation-deactivation by association-dissociation. (Reproduced from the doctoral thesis of Michal Svoboda, by courtesy of Dr. Michal Svoboda and Professor Jean Christophe, Department of Biochemistry and Nutrition, Medical School, Université Libre de Bruxelles, Brussels, Belgium)

Hence, at steady state the part of the total adenylate cyclase which is in the active state is determined by the ratio of the rate constants k_{on} for activation and k_{off} for deactivation. With hormone and nonhydrolyzable GppNHp instead of natural hydrolyzable GTP, k_{off} becomes zero and all adenylate cyclase molecules will finally be converted to the active state, i.e., $[E_T] = [E^*]$ (Tolkovsky and Levitzki 1978; see also for further literature Levitzki and Helmreich 1979).

The available binding and kinetic data can be described by the following mechanism:

$$H + R \underset{K_H}{\rightleftharpoons} HR + E \cdot GTP \underset{k_4}{\overset{k_3}{\rightleftharpoons}} [HRE \cdot GTP \rightleftharpoons HRE^* \cdot GTP]$$

$$\overset{k_5}{\longrightarrow} HR + E^* \cdot GTP \overset{k_6}{\longrightarrow} E + GDP + P_i$$

where K_H is the hormone-receptor dissociation constant. It should be noted that the committed step which leads to activation is the encounter of HR and E · GTP, since $k_5 \gg k_4$.

The bimolecular rate constant governing the formation of HRE · GTP is k_3. Under physiological conditions in the presence of saturating concentrations of GTP, the activated form E* will decay to its inactive form E concomitantly with the hydrolysis of GTP at the regulatory site. When GTP is substituted for by Gpp(NH)p a nonhydrolyzable GTP analog, $k_6 \ll k_5$ and k_6 becomes ratelimiting preventing the decay of E* → E. This model has been developed and referred to as collision coupling mechanism by Levitzki and co-workers (Tolkovsky and Levitzki 1978).

In accordance with this model, progressive inactivation of β-receptors causes a decrease in the maximal binding capacity but no change in the maximal level of activity when decay of the active species E* is prevented by Gpp(NH)p (Tolkovsky and Levitzki 1978). To give an example: Bakardjieva et al. (1979a, b) have decreased by experimental manipulation, in that case by enrichment of membranes of Chang liver cells in culture with 30 mol % dimyristoyl- or dioleoyl-phosphatidylcholine, the number of functional receptors by about 50% (Fig. 2). This results in a decrease of (–)isoproterenol stimulated activity at steady state (Table 1). However, on incubation with (–)isoproterenol and GppNHp, when k_{off} becomes zero, finally nearly all the adenylate cyclase molecules become activated despite the reduced number of functioning receptors, although at a slower rate (Bakardjieva et al. 1979a, b and unpublished 1980). This behavior is readily explained by the collision coupling mechanism (Tolkovsky and Levitzki 1978) which predicts that the first-order rate constant k_{on} of hormonal activation of adenylate cyclase is dependent on the number of receptors, whereas the number of enzyme molecules finally activated is independent of the number of receptors.

1.2 Biochemical Evidence

There are other aspects of changes in β-receptor-like adenylate cyclase activity in phospholipid enriched membranes which deserve attention:

At 17°C (see Table 2) adenylate cyclase (and $[Na^+, K^+]$-activated ATPase) activities are very low (comp. Table 1 with Table 2, and Bakardjieva et al. 1979b). The abrupt decrease in activity below the phase transition temperature of the membrane, T_c, is expressed in discontinuities in Arrhenius plots (Bakardjieva et al. 1979b). This may be interpreted to mean that at lower temperatures the protein components are squeezed out of the rigid gel-like membrane lattice into a greatly diminished fluid solvent space. The as-

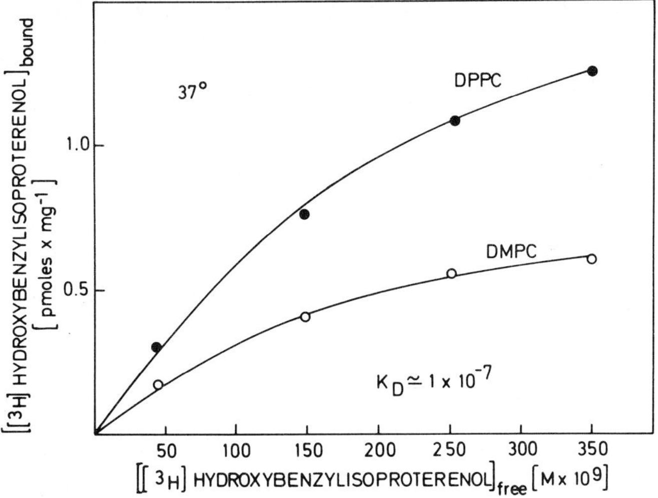

Fig. 2. β-receptor sites in Chang liver cells at 37°. The receptor sites were titrated with a β-agonist, [^{3}H] hydroxybenzylisoproterenol, which is capable of activating adenylate cyclase. DPPC refers to membranes enriched with 30 mol % dipalmitoyl phosphatidylcholine. Control membranes not enriched with phospholipids gave results which did not differ significantly from the DPPC enriched membranes. DMPC refers to membranes enriched with 30 mol % dimyristoyl phosphatidylcholine. For experimental details the paper by Bakardjieva et al. (1979b) should be consulted. A publication of the data reported here is in preparation

Table 1. Adenylate cyclase activities at 37° in membranes from cells enriched with phospholipids by fusion with liposomes

	Adenylate cyclase activities (pmol cAMP · mg^{-1} · min^{-1})			
Lipid suppl.	Basal, non-stimulated	NaF	(−)iso-proterenol	(−)iso-proterenol + Gpp (NH)p
None	30 ± 3.0	82 ± 9.0	191 ± 25	n.d. [c]
DPPC [a]	32 ± 2.8	80 ± 7.0	195 ± 32	357 ± 41
DMPC [b]	10 ± 5.1	50 ± 9.5	71 ± 11	280 ± 34

[a, b] Dipalmitoyl- and dimyristoyl phosphatidylcholine, respectively. Values are mean ± S.D. of four experiments. Assays were carried out at 1×10^{-2} M NaF, 1×10^{-5} M (−)isoproterenol and 1×10^{-4} M Gpp(NH)p

[c] Not determined. For details see: Bakardjieva et al. (1979b)

sumption that in protein-rich membranes the fluid lipid phase can actually become nearly saturated with protein has been shown to apply to bacteriorhodopsin in halobacterium membranes (Cherry et al. 1978). Experiments which support that contention are shown in Fig. 3. In these experiments, the phospholipid to protein ratio in Chang liver cell membranes has been changed drastically by enrichment with up to 70 mol % dipalmitoyl phosphatidylcholine (Bakardjieva et al. unpublished experiments 1980). The endothermic

Table 2. Adenylate cyclase activities at 17° in membranes from cells enriched with phospholipids by fusion with liposomes

	Adenylate cyclase activities (pmol cAMP · mg^{-1} · min^{-1})		
Lipid suppl.	Basal, nonstimulated	NaF	(–)isoproterenol
None	6 ± 0.9	21 ± 2.7	34 ± 2.5
DPPC [a]	5.5 ± 1.1	20 ± 3.5	32 ± 2.9
DMPC [b]	5.5 ± 1.0	18 ± 2.0	28 ± 3.4

[a, b] Dipalmitoyl- and dimyristoyl phosphatidylcholine, respectively. Values are mean ± S.D. of four experiments. Assays were carried out at 1 × 10^{-2} M NaF, and 1 × 10^{-5} M (–)isoproterenol. For details see: Bakardjieva et al. (1979b)

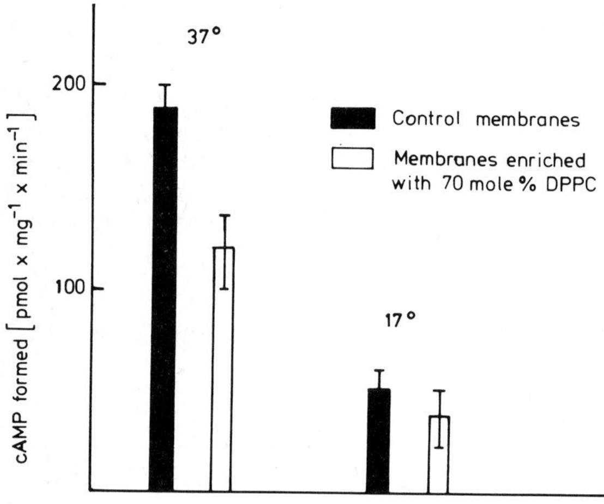

Fig. 3. (–)isoproterenol stimulated adenylate cyclase activity in membranes from Chang liver cells and in membranes from the same cells enriched with 70 mol % DPPC at 37° and 17°. DPPC refers to dipalmitoyl phosphatidylcholine. The cells were enriched with the phospholipid by fusion with liposomes. Assays were carried out with 1 × 10^{-5} M (–)-isoproterenol. For experimental details the paper by Bakardjieva et al. (1979b) should be consulted. A publication of the data reported here is in preparation

phase transition in pure dipalmitoyl phosphatidylcholine liposomes is around 41°C. As may be seen from Fig. 3, there is a marked decrease in β-catecholamine-stimulated adenylate cyclase activity already at 37° in the membranes enriched with 70% dipalmitoyl phosphatidylcholine. A comparably low activity is found in normal membranes only at lower temperature, 17°, where as expected, however, the differences in activities between enriched and normal membranes are much smaller. Interestingly, there was a similar (30%-40%) decrease in β-adrenergic binding sites at 37° in the membranes rigidified with 70% dipalmitoyl phosphatidylcholine like in the membranes fluidized with 30% dimyristoyl phosphatidylcholine (Fig. 2).

There are therefore two important variables in membranes with high protein content, i.e., where the ratio of protein lipid is 2:1 or greater: The size of the available solvent space and its viscosity.

When phospholipids are increased and the ratio of protein to phospholipids is shifted in favor of the latter and the phospholipids introduced are fluidizers, such as dimyristoyl or dioleoyl phosphatidylcholines which have phase transition temperatures (T_c) well below 37°, the membrane becomes more fluid at 37° (Bakardjieva et al. 1979a, b). Consequently, receptors become more mobile. This in turn reduces the stochastic probability of coupling with adenylate cyclase which expresses itself in reduced specific activity (Table 1). This is interpreted to mean that on increasing the fluid solvent space and reducing the solute concentration in that space dissociation is favored like in the case of oligomeric proteins in aqueous solutions on dilution.

On the other hand, when "rigid" phospholipids are introduced, such as dipalmitoyl phosphatidylcholine, in concentrations sufficiently high to elevate the phase transition temperature, a different mechanism may account for the observed reduction of β-catecholamine-stimulated activity (see Table 2 and Fig. 3). Under these conditions the available fluid solvent space may be reduced and the protein concentration in that space increased. As a consequence, association of membrane proteins is favored. This needs to be clarified and is currently studied (Bakardjieva, unpublished experiments, 1980).

Taken as a whole, the data strongly argue that receptors which couple within a highly organized membraneous oligometric system must remain confined to a solvent space whose size and fluidity is critical. There is little doubt that membrane structure (i.e., the phospholipid to protein ratio), and membrane dynamics (i.e., the fluidity of the membrane) profoundly affect the activity of membraneous multicomponent systems such as adenylate cyclase. This interpretation is based on the experiments of Cherry et al. (1977) who have shown that membrane viscosity in phospholipid layers is dependent on protein concentration.

Mechanisms operative in regulation of membrane structure and fluidity are under active study (Hirata et al. 1979). But, at present, we do not know what keeps the components of hormonally stimulated adenylate cyclase in place. In human blood platelets a small but significant increase in the temperature-phase-transition profile appears to be a necessary accompaniment for adenylate cyclase activation by small concentrations of prostaglandin E_1 that is at molar ratios of prostaglandin E_1 to phospholipids in the range of 1:5,000 to 1:10,000 (Mruk, Burgermeister, and Bakardjieva, unpublished experiments, 1980).

2 Receptor Mobility

There is overwhelming evidence that hormone receptors including the β-adrenergic receptor become desensitized and nonfunctional upon prolonged exposure to agonists (Lefkowitz and Williams 1978; Perkins et al. 1978; Terasaki et al. 1978). In frog erythrocytes (Mukherjee et al. 1975), human astrocytoma cells (Johnson et al. 1977), and S49 lymphoma cells (Shear et al. 1976) a principal feature of desensitization is a decrease in the number of functional β-adrenergic receptors, which are no longer accessible either because of a transition to an inactive receptor conformation which binds the agonist more tightly (Williams and Lefkowitz 1977), or because of receptor internalization (Chuang and Costa 1979), or both. Desensitization only occurs with β-adrenergic agonists and subsequent to adenylate cyclase activation. Moreover, it requires interaction with the guanylnucleotide binding protein as shown by Sternweis and Gilman (1979) with a coupling-deficient (Unc) mutant of S49-lymphoma cells. But at present the molecular basis of desensitization remains obscure. Precise information is lacking at what step in the coupling reaction desensitization occurs and whether it involves covalent modification (Simpson and Pfeuffer 1980a; Terasaki et al. 1978).

We have initiated experiments using the technique of fluorescence recovery after photo bleaching that might provide information relevant to the mechanism of desensitization of β-receptors. To our knowledge these experiments provide for the first time data on the mobility of β-catecholamine receptors. In order to make possible such experiments, new fluorescent ligands with high affinity for β-adrenergic receptors were synthesized (M. Hekman and H. Hornig, unpublished experiments, 1980).

Fluorescent β-Adrenergic Ligands

$O-CH_2-CH(OH)-CH_2-NH-CH_2-C(CH_3)_2-NH$; ring substituent: NO_2 ; NBD ring: N, O, N, NO_2

MH – 53

$O-CH_2-CH(OH)-CH_2-NH-CH_2-C(CH_3)_2-NH$; ring substituent: $CH_2-CH=CH_2$; NBD ring: N, O, N, NO_2

Alprenolol – NBD

These 4-nitrobenzo-2-oxa-1,3-diazolyl(=NBD)-derivatives of potent β-adrenergic antagonists bind specifically with dissociation constants in the order of 10^{-10} M. Chang liver cells were incubated in Petri dishes with alprenolol-NBD in presence or absence of an excess of the nonfluorescent antagonist, carazolol, for 10 min at 37°C (see legends to Figs. 4 and 5). The cells were carefully washed with buffer, and photo bleaching was carried out at room temperature with an argon-ion laser at 476 nm. The radius of the bleached area was 2.0 μm and the time of bleaching was 1/60-1/15 s. Recovery of fluorescence in the bleached area was measured as a function of time and recorded. Fluorescence recovery is due to the lateral diffusion of unbleached fluorescent molecules. The diffusion coefficient, D is calculated according to:

$$D = \frac{\omega^2}{4 \cdot t_{1/2}} \cdot \gamma \, [cm^2 \cdot sec^{-1}] \qquad (2)$$

where ω (= 2 μm) is the radius of the laser beam, γ (= 0.88) is an experimental parameter determined by the extent of bleaching and the profile of the laser beam, and $t_{1/2}$ is the time constant of fluorescence recovery. Experiments with high concentrations (2×10^{-6} M) of alprenolol-NBD served as control (not shown here). As expected, the unbound or lipid-bound label gave $D \approx 1 \times 10^{-8}$ cm$^2 \cdot$ s^{-1} and $t_{1/2} \approx 1$ s typical for the lateral diffusion of lipids in membranes. Moreover fluorescence recovery was not affected by preincubation with competitive inhibitors of NBD-alprenolol binding, propranolol or carazolol.

In Fig. 4 an experiment with Chang liver cells at a 100 times lower (2×10^{-8} M) alprenolol-NBD concentration is shown. Fluorescence recovery was only 60%, $t_{1/2}$ = 80-100 s and accordingly $D \approx 1 \times 10^{-10}$ cm$^2 \cdot$ s^{-1}, a value characteristic for the lateral diffusion of membrane proteins.

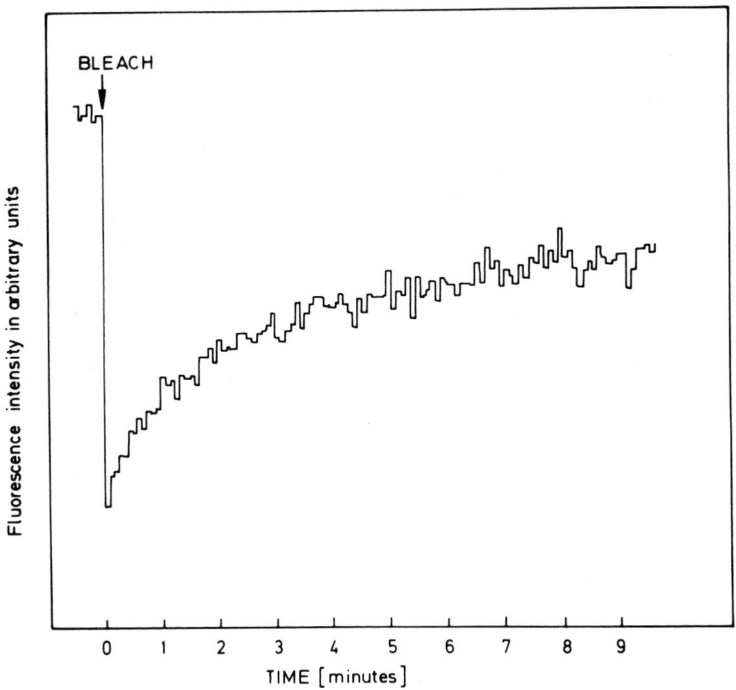

Fig. 4. Time dependence of fluorescence recovery after photo bleaching. Chang liver cells were incubated with 2×10^{-8} M alprenolol-NBD at room temperature, washed and bleached for 1/15 s $t_{1/2}$ = 80 s. For details see text. Unpublished experiments of Peters et al. (1980). Hornig, diploma in chemistry (1980)

In Fig. 5 an experiment is shown with cells which were preincubated with the powerful β-blocker, carazolol, before alprenolol-NBD was added also with an excess of the blocker. In this experiment, $t_{1/2}$ was again about 90 s. Recovery of fluorescence after bleaching was nearly 100%, but the initial fluorescence intensity of the cells treated with the blocker was only 40% of the fluorescence intensity of the cells shown in Fig. 4, where

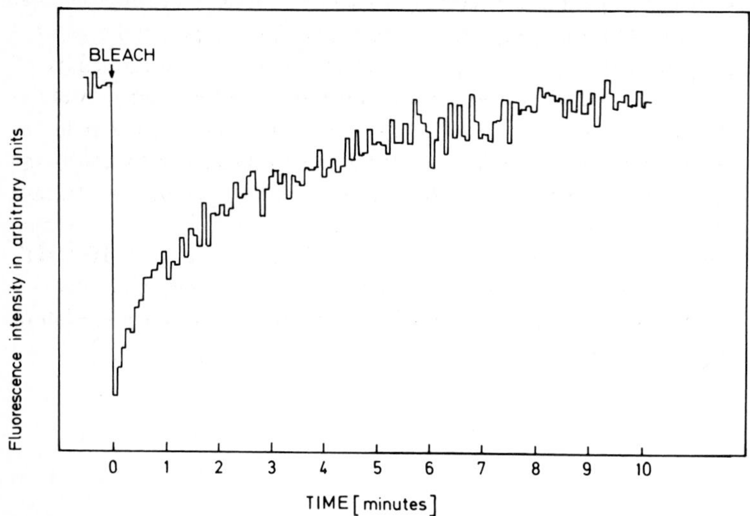

Fig. 5. Time dependence of fluorescence recovery after photo bleaching. Chang liver cells were preincubated with 2×10^{-5} M carazolol and incubated with 2×10^{-8} M alprenolol-NBD together with 2×10^{-5} M carazolol at room temperature, washed and bleached for 1/15 s $t_{1/2} = 90$ s. For details see text. Unpublished experiments of Peters et al. (1980)

the nonfluorescent blocker was absent. Hence displacement of the fluorescent ligand by the nonfluorescent ligand resulted in a drastic reduction of the fluorescence intensity but nearly complete recovery.

In Fig. 6 the experiments shown in Figs. 4 and 5 have been schematically replotted using a normalized scale of fluorescence intensity. In the absence of the blocker carazolol

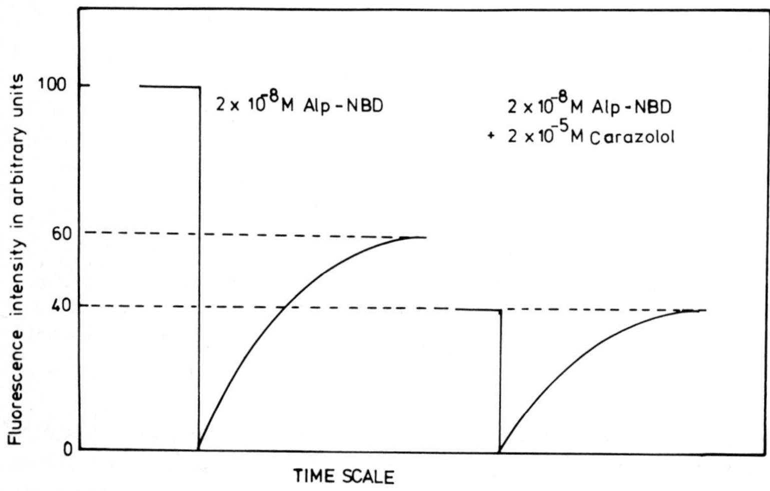

Fig. 6. Schematic representation of the photo bleaching experiments in Figs. 4 and 5

(Fig. 6, *left*) the initial fluorescence intensity (before photo bleaching) represents the sum of specifically and nonspecifically bound alprenolol-NBD. In the presence of carazolol (Fig. 6, *right*) the initial fluorescence intensity is reduced to 40%. If we assume that only the specifically bound alprenolol-NBD is displaced at the applied concentration of carazolol and the nonspecific binding of the fluorescent ligand is unaffected by the blocker, the 60% difference of initial fluorescence intensities in both experiments can be ascribed to alprenolol-NBD bound to receptors. Since fluorescence recovery in the absence of carazolol was between 40%-60%, either all or at least two-thirds of the β-receptor-antagonist complex are immobile. We are presently investigating the validity of this assumption with a much greater concentration range of blocker to make sure that unspecifically bound fluorescence is not displaced by β-blocking agents. If the assumption should prove valid the β-receptor-antagonist complex would behave strikingly different from the insulin receptor complex. When the photo bleaching data published by Schlessinger et al. (1978) for the rhodamin-insulin-receptor complex are plotted in an analogous fashion (Fig. 7), one readily sees that in this case and in the presence of an ex-

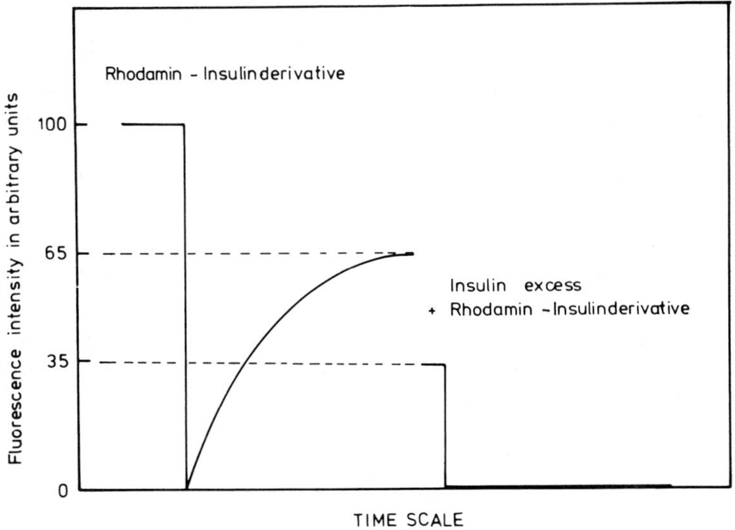

Fig. 7. Schematic representation of photo bleaching experiments by Schlessinger et al. (1978)

cess of nonfluorescent insulin (Fig. 7, *right*) all the unspecific binding, which amounts also to about 35% of the total fluorescence intensity in the absence of nonfluorescent insulin, is immobile because it does not recover after bleaching. Hence the specific insulin receptor-bound fluorescence (Fig. 7, *left*), which again has a comparable recovery with the β-antagonist-receptor complex of about 65%, is completely mobile in the membrane of 3T3 fibroblasts.

We are now extending this work by synthesizing and testing β-adrenergic fluorescent agonists with high affinity for the β-receptor, in order to compare the mobility of the β-adrenergic receptor in "active" and "inactive" conformations. The β-receptor-agonist complex is in contrast to the antagonist β-receptor complex capable of activating adenylate cyclase, although an agonist binds to the same site on the β-receptor to which the antagonist binds. Moreover, in contrast to the antagonist-receptor complex only the agonist-receptor complex becomes desensitized. We would not be surprised if the β-agonist-receptor complex were more mobile than the β-antagonist-receptor complex and more like the insulin-receptor complex (Schlessinger et al. 1978). Activation of adenylate cyclase by the agonist-receptor complex may be followed by processes which involve transport of receptors. For cyclase activation, mobility of the receptor in the membrane as a whole might not be required; unrestricted mobility might even interfere with activation as discussed before. On the other hand, mobility of the receptor in the whole membrane is quite likely a prerequisite for transport, cryptization and recycling of receptors (cf. Schlessinger et al. 1978; Su et al. 1980)[3]. To allow for these processes, which may be required to initiate desensitization, the receptor must acquire a new property. What is the nature of this change in the receptor which profoundly alters its structure and prepares the receptor for inactivation and removal: covalent modification, a modifier protein? We do not know.

3 Role of GTP-GDP-exchange

A few years ago, in a review in FEBS-Letters, one of us (Helmreich 1976) pointed out that hormone-receptor complexes are rather long lived and not readily reversible allowing for several discrete cycles of activation and deactivation of adenylate cyclase. A mechanism which would account for that property is collision coupling of which the kinetics were extensively studied by A. Levitzki and his group (Tolkovsky and Levitzki 1978; Rimon et al. 1978). Moreover, Helmreich postulated regulatory steps necessary for deactivation, and pointed out that the points of control instrumental in modulation of coupling between receptor and adenylate cyclase might be beyond the hormone-receptor level. In the past few years work of several laboratories (cf: Ross and Gilman 1980) has clarified the role of a guanylnucleotide binding protein (G-protein) required for activation of adenylate cyclase (Pfeuffer 1977, 1979) and probably also for modulation of hormone-receptor interactions (Sternweis and Gilman 1979). This and the studies of Cassel and Selinger have led to the proposal of the so-called GTP-GDP-cycle (Cassel and Selinger 1976, 1977, 1978) (Fig. 8).

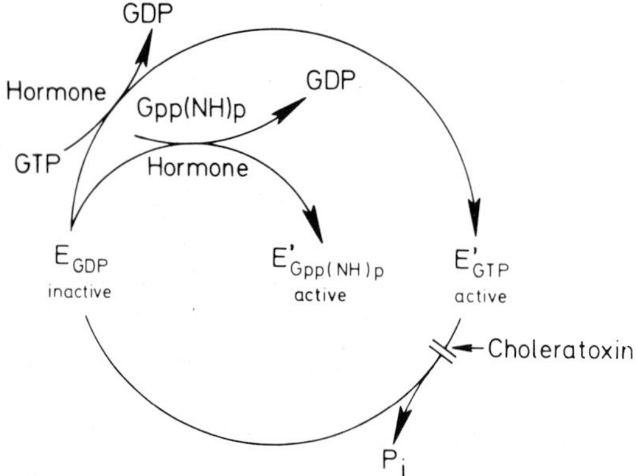

Fig. 8. The GTP-GDP cycle of Cassel and Selinger. (Reproduced with permission from Cassel and Selinger 1978; Proc Natl Acad Sci USA 75:4155)

3 Note Added in Proof: These studies have in the meantime been completed: cf: Henis YI et al. (1981) Proc Natl Acad Sci USA (in press)

It is assumed that transition of adenylate cyclase from an inactive to an active state requires simultaneous binding of hormone and guanylnucleotide at specific sites on separate proteins. A cycle of hormonal activation is terminated when the GTP is hydrolyzed, and a new cycle of adenylate cyclase activation can occur only if a new molecule of GTP is bound and GDP, the product of the GTPase reaction, discharged. GTP-GDP exchange governs adenylate cyclase activity and Pfeuffer (1979) could show with solubilized intact components that only the GTP- or GTP-γ-S-form of the G-protein binds to the catalytic moiety of adenylate cyclase, whereas the GDP-form does not form an active holoenzyme complex. Exchange is interrupted when either a nonhydrolizable GTP-analog is bound at the G-protein or GTP-hydrolysis is blocked by ADP-ribosylation catalyzed by choleratoxin (Cassel and Pfeuffer 1978; Gill and Meren 1978). In all these instances adenylate cyclase remains in a permanently active state. More recently, Limbird et al. (1980) could show that binding of the β-adrenergic agonist [^{3}H]-hydroxybenzyl-isoproterenol to the β-adrenergic receptor of rat reticulocyte membranes results in the coupling of the isolated components, receptor and G-protein associated with the adenylate cyclase system. The β-receptor-G-protein complex was not formed when β-receptors were unoccupied or occupied by an antagonist. Moreover, incubation of the receptor with a β-agonist in the presence of guanylnucleotides reversed or prevented formation of the receptor-G-protein complex. According to Cassel and Selinger (1978) the only function of the hormone in adenylate cyclase activation is to facilitate exchange of GDP for GTP. But if that is so, what controls and adjusts the rates of GTP hydrolysis to the rates of the association/dissociation cycle of the guanylnucleotide binding protein and the other components, catalytic moiety and receptor?

4 Control of Adenylate Cyclase

One is tempted to speculate that control of adenylate cyclase activity might involve a regulatory bicyclic cascade similar to that which regulates glutamine synthetase activity in E. coli. Stadtman and co-workers (Chock et al. 1980) have shown that there are two interlinked cyclic cascades, one is the adenylylation cycle and the other is the uridylylation cycle. Since the same enzyme, an adenylyltransferase, catalyzes both adenylylation and deadenylylation, the two reactions would be a futile cycle in the absence of additional regulation. Futile cycling is however avoided when the adenylylation cycle is linked to the uridylylation cycle by the P2 protein, discovered by Shapiro in Stadtman's laboratory, which exists in an uridylylated and a deuridylylated form and acts as modifier protein. The unmodified form of P2 activates the adenylylation reaction of glutamine synthase, whereas the uridylylated form activates the deadenylylation reaction catalyzed by the same adenylyltransferase (Fig. 9).

Adenylate cyclase has features which make this kind of regulation attractive. GTP-GDP exchange occurs only in the presence of hormone (Cassel and Selinger 1978). Is the hormone-receptor complex itself the modifier protein linked to the adenylate cyclase system via the guanylnucleotide binding protein? But if this were so, what converter enzymes are involved in covalent control of the receptor and where could a bicyclic cascade be operative in the modification of the adenylate cyclase-guanylnucleotide binding protein complex? Cassel and Pfeuffer (1978), Gill and Meren (1978), Johnson et al.

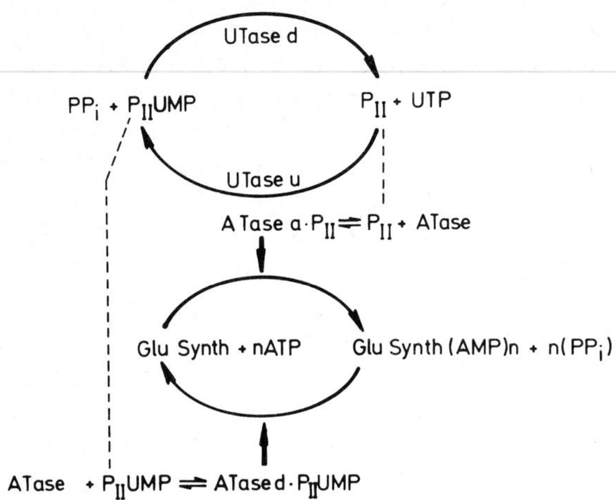

Fig. 9. Closed bicyclic cascade of glutamine synthetase regulation. *Glu Synth*, glutamine synthetase. *UTase u* and *UTase d*, uridylyl- and deuridylyltransferase activities, respectively, which catalyze auto-uridylyl – deuridylylation of the P_{II} regulatory protein. *ATase a* and *ATase d*, adenylyl- and deadenylyltransferase enzyme. Both activities are under control of a variety of effectors not shown here (Chock et al. 1980)

(1978) have shown that GTPase activity is blocked when the guanylnucleotide binding protein with Mr 42,000 (Pfeuffer 1977) is ADP-ribosylated by choleratoxin. Moss and Vaughan (1978), Moss et al. (1979) have purified a mono-ADP-ribosyltransferase activity in the cytosol of turkey erythrocytes, which decreases adenylate cyclase activity, but the physiological substrate, its location relative to the membraneous adenylate cyclase system, its presence in other cells, and its biological importance are still unknown. Furthermore, an ADP-ribosyl removing enzyme activity has not yet been found. Ezra and Salomon (1980) recently reported on a GTP-dependent phosphotransferase reaction in rat ovarian plasma membranes and proposed that desensitization of adenylate cyclase response to luteinizing hormone (lutropin) and GTP is due to a covalent modification of membrane proteins catalyzed by that phosphotransferase activity.

Undoubtedly, a covalent regulatory cascade would be an elegant and refined way of controlling steady-state concentrations of active and inactive adenylate cyclase in response to hormones and other effectors (Ca^{2+}). This type of control would also be in keeping with a characteristic property of some hormone actions, because hormone activations of adenylate cyclase are greatly amplified: Activation via a few thousand receptors per cell results in the formation of cyclic AMP whose action again is greatly magnified by activation of protein kinases, which in turn act on other enzymes catalyzing a variety of metabolic reactions.

At present, a discussion of regulation of hormonally stimulated adenylate cyclase merely lends itself to more or less plausible speculations. The reasons for that are obvious: Purification and characterization of the components of this membraneous enzyme system are slow and tedious, but progress is being made in several laboratories, notably in Dr. Gilman's laboratory (Northup et al. 1980). It does not seem too optimistic, therefore, to predict that in future symposia on metabolic interconversion of enzymes adenylate cyclase may find a legitimate place in the program.

Acknowledgements. Work reported here was supported by a grant (He 22/30) of the Deutsche Forschungsgemeinschaft to A. Levitzki and Ernst J.M. Helmreich and by the Fonds der Chemischen Industrie, e.V.. We gratefully acknowledge the able assistance of Ms. H. Dietrich, Mrs. I. Löffler, and Ms. G. Meyer.

References

Bakardjieva A, Galla H-J, Helmreich EJM, Levitzki A (1979a) Horm Cell Regul 3:11-27
Bakardjieva A, Galla H-J, Helmreich EJM (1979b) Biochemistry 18:3016-3023
Cassel D, Pfeuffer T (1978) Proc Natl Acad Sci USA 75:2669-2673
Cassel D, Selinger Z (1976) Biochim Biophys Acta 452:538-551
Cassel D, Selinger Z (1977) Proc Natl Acad Sci USA 74:3307-3311
Cassel D, Selinger Z (1978) Proc Natl Acad Sci USA 75:4155-4159
Cherry RJ, Müller U, Schneider G (1977) FEBS Lett 80:465-469
Cherry RJ, Müller U, Henderson R, Heyn MP (1978) J Mol Biol 121:283-298
Chock PB, Rhee SG, Stadtman ER (1980) Annu Rev Biochem 49:813-843
Chuang DM, Costa E (1979) Proc Natl Acad Sci USA 76:3024-3025
Ezra E, Salomon V (1980) Abstract S2-P6. Abstr 13th Fed Eur Biochem Soc Meet August 24-29, 1980, Jerusalem, Israel, p 38
Gill DM, Meren R (1978) Proc Natl Acad Sci USA 75:3050-3054
Helmreich EJM (1976) FEBS Lett 61:1-5
Helmreich EJM, Bakardjieva A (1980) Biosystems 12:295-304
Henis YI, Hekman M, Elson EL, Helmreich EJM (1981) Proc Natl Acad Sci USA (in press)
Hirata F, Strittmatter WJ, Axelrod J (1979) Proc Natl Acad Sci USA 76:368-372
Hornig H (1980) Work performed in partial fulfillment of the requirements for Dipl. chem. of the Faculty of Chemistry and Pharmacy, University of Würzburg
Johnson G, Kaslow HR, Bourne HR (1978) J Biol Chem 253:7120-7123
Johnson GL, Wolfe BB, Harden TK, Molinoff PB, Perkins JP (1977) J Biol Chem 252:1472-1480
Lefkowitz RJ, Williams LT (1978) In: George WJ, Ignarro LJ (eds) Advances in cyclic nucleotide research, vol 9. Raven Press, New York, pp 1-17
Levitzki A, Helmreich EJM (1979) FEBS Lett 101:213-219
Limbird LE, Gill DM, Lefkowitz R (1980) Proc Natl Acad Sci USA 77:775-779
Moss J, Vaughan M (1978) Proc Natl Acad Sci USA 75:3621-3624
Moss J, Stanley SJ, Oppenheimer NJ (1979) J Biol Chem 254:8891-8894
Mukherjee C, Caron MC, Lefkowitz RJ (1975) Proc Natl Acad Sci USA 72:1945-1949
Northup JK, Sternweis PC, Smigel MD, Schleifer LS, Ross EM, Gilman AG (1980) Proc Natl Acad Sci USA 77:6516-6520
Perkins JP, Johnson GL, Harden KT (1978) In: George WJ, Ignarro LJ (eds) Advances in cyclic nucleotide research, vol 9. Raven Press, New York, pp 19-32
Pfeuffer T (1977) J Biol Chem 252:7224-7234
Pfeuffer T (1979) FEBS Lett 101:85-88
Rimon G, Hanski E, Braun S, Levitzki A (1978) Nature (London) 276:394-396
Ross EM, Gilman AG (1980) Annu Rev Biochem 49:533-564
Schlessinger J, Willingham MC, Pastan I (1978) Proc Natl Acad Sci USA 75:5353-5357
Shear M, Insel PA, Melmon KL, Coffino P (1976) J Biol Chem 251:7572-7576
Simpson IA, Pfeuffer T (1980a) Eur J Biochem 111:111-116
Simpson IA, Pfeuffer T (1980b) FEBS Lett 115:113-117
Sternweis PC, Gilman AG (1979) J Biol Chem 254:3333-3340
Su Y, Harden TK, Perkins JP (1980) J Biol Chem 255:7410-7419
Terasaki WL, Brooker G, de Vellis J, Inglisch D, Hsu Ch-Yi, Moylan RD (1978) In: George WJ, Ignarro LJ (eds) Advances in cyclic nucleotide research, vol 9. Raven Press, New York, pp 33-52
Tolkovsky AM, Levitzki A (1978) Biochemistry 17:3795-3810
Williams LT, Lefkowitz RJ (1977) J Biol Chem 252:7207-7213

The references are incomplete. For further references the recent reviews by Ross and Gilman (1980) and of Chock et al (1980) in Annu Rev Biochem should be consulted.

Discussion[3]

Upon respective questions of Drs. Ballard, Demaille, Fischer and Heilmeyer, Dr. Helmreich made the following statements in the discussion:

(1) Evidence for a recycling of β-adrenergic receptors so far comes only from desensitisation-resensitization experiments in the presence of cycloheximide. Neither the internalization of receptors in the presence of agonist nor their insertion into the surface membrane after agonist removal are blocked by the drug. This excludes a de novo synthesis of receptors to be involved in resensitization. (2) Calmodulin-stimulated and -insensitive adenylate cyclases can be distinguished not only by their different distribution, the former being found in brain and pancreas exclusively, but also by their different sensitivity to adenosine stimulation and calcium inhibition. (3) The contribution of the presented rate constants to the different steps of adenylate cyclase activation is so far not well investigated. Only in the case of the erythrocyte membrane, estimates of the individual rate constants of the collision coupling model have been presented (Tolkovsky AM, Levitzki A (1978) Biochemistry 17:3795). See also: Citri Y, Schramm M (1980) Nature (London) 287:297-300.

3 Rapporteur: Heinrich Betz

Subject Index

acetylcholine receptor (AChR) 372 f
– model for
– – aggregation 373
– – regulation 373
– synthesis control 372
acetyl-CoA-carboxilase 28, 76
– inactivation 28
acetyl-transferase 141
adenylate cyclase 378 ff
– biochemical evidence 380
– β-catecholamine stimulated 378
– control 389
– coupling of receptor 378
– kinetic evidence 378
adenylisation 181, 264
adipocytes 77
adipose tissue 28, 134
ADP-ribosylated proteins
– separation by HPLC molecular filtration 287
ADP-ribosylation 270 ff
– of actin 280
– during rat liver development 274
– effects of growth 280
– – hormonal influences 280
– histone H1 of Hela cells, of, in vitro 273
– – in vivo 273
– histones 271
– in vitro, actin-like proteins of isolated nuclei 286
– in vitro, added actin 286
– non histone chromatin proteins, of 280
– non histone proteins 271
– ribonuclease 294
– treatment with phosphodiesterase 271
ADP-ribosylation, nuclear
– acceptor proteins 270
ADPR-RNAase (ADP-ribosolated RNAase) 298 f
– affinity chromatography 298
– cellulose acetate electrophoresis 299
adrenaline 30, 34
allosteric effectors 165
3-Aminobenzamide, analog of nicotinamide 6
aminopeptidase I 174
aminotransferase 160
angiotensin I, II 248
angiotensin converting enzyme (ACE) 248
ascites tumor 116
aspartate kinase 260
ATP-analogs 47 f
– adenosine chlormethane pyrophosphonate 47
– adenosine 5'chlormethyl-pyrophosphonate 48
– adenosine 5'-(p-fluorosulfophenyl-phosphatate 48
ATP-citrate lyase 77 f
ATP-derivative 47 f
– fluorosulfophenyl, analog of adenylic acid 47

bacillus subtilis 159
bovine factor X 150
brush border membranes 18 f

Ca^{2+} 28 ff
– calmodulin 28, 30
– intracellular receptors 28
caffeine 60
calcisome 310
– convalent regulation within 310
– properties 310
calcium-binding proteins, evolutionary tree 305
calmodulin 71, 306, 321
– activation of c.-dependent enzymes 306
– effect of the activity 71
carbon starvation 259
carboxipeptidase A 148, 154 f, 248
carcinogenic agent 6
catalase 266
cathepsin C 189
cathepsin D 189
cathepsin G 148
cell differentiation 44
cell proliferation 44
cellular aging 241
cellular refractoriness 25

chemical modification of sulfhydrils 13
chemotactic peptides 155
chloroquine 179
cholesterol esterase 76
chromatin 55
chymotrypsin 149
conformational change 165, 266
converting enzymes 186 ff
cross-reactive enzyme 163 ff, 175, 181
cyclic adenosine 3',5'-monophosphate (cAMP) 10, 18 ff, 53 ff, 82, 93 f
– dibutyryl-3',5'- 92
– intercellular 23
– subunit C 10
– subunit R 10
cyclic adenosine 3',5'-monophosphate dependent protein kinase (cAMPdPK) 17 ff, 30
– active site catalytic subunit 17
– degradation 20
– degradative inactivation of subunit C 18
– intracellular sensor 21
– subunit C 20
cytochrome C oxidase 228 ff
– in vitro synthesis 232
– precursor processing 228
– precursor for subunit IV 232
– – tryptic fingerprints 235
– preparation of antibodies 229
– purification by hydrophobic chromatography 229

2-deoxiglucose 6-phosphate 77 f
diaphragm skeletal muscle 77
DNA, damaged 7
DNA repair 3 ff
– automodification of the synthease 5
– ligation 8
double immunoprecipitation 163

elastase 149
embryogenesis, biochemical model explaining pattern formation 346
enzyme activity, principles of regulation 111
enzyme degradation 159, 174, 217, 259
– regulation of 266
enzyme synthesis 217, 259
epinephrine 82
erythrocytes 116, 239
Escherichia coli 181, 259
esterase 146
exocytosis 156

fructose-1,6-bisphosphatase 168, 186
fructose-1,6-bisphosphatase, interconversion 179

generator-sensor-suppressor cycle (GSS cycle) 304
genetic disorder 239
glucagon 92, 95, 100, 102 f
gluconeogenesis 101
glucose-6-phosphate dehydrogenase (G-6-PD) 77, 209, 239
– defiency of 239
glutamine phosphoribosyl-pyrophosphate amidotransferase (PRPP) 159 f
glutamine synthetase 181, 259 ff
– inactivation 262
– turnover 266
γ-glutamyltranspeptidase 199 f
glutathione, oxidized 64
glutathione peroxidase 138, 141
glutathione reductase 141
glycerol phosphate acyl transferase 28
glycogen 28 ff
– breakdown 28
– metabolism 28, 34
– – skeletal muscle, mammalian 30
glycogen phosphorylase, reversible covalent modification 28
glycogen synthase 28 ff, 59 f, 74 ff, 82
– b_1; b_2 32
– inactivation 28
– kinases 32, 34
glycogen synthase phosphatase 32 f
glycogen synthesis 28
glycolysis 100
granulocytes 146

heparin 148
histidine decarboxilase 335 ff
– *Lactobacillus* sp., from 335
– subunit structure 342
histone H1 4 f, 32, 44, 53
histone H2B 13 ff, 19, 32, 44
– ATP-binding site of C 15
– modification 17
– modification of sulfhydryls 15
– subunit C 15
hydroximethyl glutaryl coenzyme A reductase (HMG CoA R) 76
hypertension 248

inactivation 159 f, 168, 179, 208, 259
– catabolite 168, 174, 179
– glucose, by 168
– intermediate 175 f
– oxigen, by 160
– selective 159
inhibitor 1 34 f, 38, 41, 60
inhibitor 1 phosphatase 35

inhibitor 2 38, 41, 60
insulin 35, 74 ff, 131 ff
– control of metabolism 75
insulin mediators 74, 83
– anti-insulin receptor antibody 83
– control of covalent phosphorylation 74
– pituitary peptide 83
integral membrane proteins (IMPs) 221
interconversion 134, 179, 181
iron-sulfur enzyme 161

kinase
– cAMP-dependent 22
– intracellular sensor for cAMP 22
– type I 16
Klebsiella aerogenes 259

leucocytes 146
lipase, hormone sensitive 76
lipoate dehydrogenase 141
liver, peptide stabilizing factor 116
liver phosphofructokinase (PFK) 68, 100 ff, 117
– control by fructose 2,6-biphosphate 100
liver phosphorylase kinase, regulation 68
lymphocytes 6 ff
– ADP-ribosylated 7
– N-methyl-N'-nitro-N-nitroso-guanidine
lysosomal proteins 223
– recognition marker 223
lysosomes 186 f, 213

mast cell carboxipeptidase 154 f
mast cells 145 ff
– atypical 146
– degranulation 156, 158
membrane proteins 199
membrane transport 218
microcomplement fixation 163, 175
mitochondria, inner membrane-matrix compartment 128
mixed function oxidase 261
Mono ADP-ribosylation of proteins 269
Mono(ADP-ribosyl)protein conjungate 272 ff
– levels in intact tissue 272
– in lymphocytes from normal donors 276
– in lymphocytes from patients with chronic lymphocytic leucemia 276
mouse diaphragma 76
mouse thymocytes 21
muscle phosphofructokinase 117
muscle protein inhibitor 116
muscle sarcolemnal protein 78
myosin kinase 28
myosin-[^{31}P]NMR spectra 322

NAD(P)$^{+}$-transhydrogenase 141
Nbs_2 (5'5-dithiobis[2-nitrobenzoic]acid) 10 ff
– ATP site in C 11
– binding of R to C 16
– catalytic function 13
– chemical modification 10
– inactivation of C 11
– modification of the SH groups 11
– γ-P subsite 16
– SH groups 10
– subunit C 11, 16
– subunit R 16
– type II R subunit 16
neoplastic transformation 56
Neurospora crassa 60
nitrogen starvation 259
nuclear poly(ADP-R)synthetase 289

peroxide 134
phosphatase complex, sarcoplasmic 62
phosphoenolpyruvate carboxikinase (PEPCK) 174, 210
phosphofructokinase (PFK) 91 ff, 100 ff, 113ff
– *Bacillus stearothermophilus* 120
– bacterial (*E. coli*) 118 f
– "futile cycling" 91
– inactivity enzyme 97
– kinetic properties 95
– multimodulation 111, 113, 115, 117, 119
– stimulation by fructose 2,6-bisphosphate 105 f
– yeast (*saccharomyce cerevisiae*) 118
phosphorylase 76
– [^{31}P]NMR spectra 322
phosphorylase a 32, 38, 59, 60
phosphorylase a phosphatase 59
phosphorylase a + b 69 f
phosphorylase b 59
phosphorylase b kinase 76
phosphorylase kinase 28, 30, 32 ff, 38, 59 f, 70
– Ca^{2+} binding to calmodulin 325
– liver 70 f
– phosphorylation 329
phosphorylase kinase phosphatase 38
α-phosphorylase kinase phosphatase 32, 41
β-phosphorylase kinase phosphatase 32 f
phosphorylase phosphatase 32 f, 38, 41, 60, 70
– adrenal cortex (bovine) 36
– ATP-Mg-dependent 37
– – dog liver 36 f
– – heart (rats, rabbits) 37
– – liver (rats, rabbits) 37 f, 40
– – skeletal muscle (rats, rabbits, pigeon) 36 f
– inhibition by glutathione 64
– inhibitor 2 38

– latent 108 f
– manganese stimulation 61
– Neurospora crassa 36
– purification of a complex 61
phosphorylation 172, 181
Physarum polycephalum 52
– regulatory role of protein kinases 53
planaria *Dugesia tigrina* 54
plasma membranes 209
poliovirion proteins genome (VPg) 356 ff
– biosynthesis 356
– linkage betw. VPg and virion RNA 357
– structure 356
poliovirus 356 ff
– replication cycle 359
poliovirus proteins 356 ff
– post-translational processing 360
– – proteinase 361
poly(ADP-ribose) (ADPR) 3 ff
– nicotinamide-adenine dinucleotide (NAD) 3
poly ADPR content
– in normal hamster liver 284
– in precancerous hamster liver 284
poly(ADP-ribosyl)ation 5
– of proteins 269
poly(ADP-ribosyl)protein, enzymatic synthesis 3
poly(ADP-R)synthesis 6
– aldosterone, in vitro effect of 288
– in cardiocyte nuclei; function of age and organspecific hormonal influences 288
poly(ADP-ribosyl)protein conjungate, levels in intact tissue 272
poly(ADP-R)synthetase 3, 6ff
– cortisol and aldosterone treatment of hypophysectomized rats 290
– effect of hypophysectomy 290
– inhibitors 6 f
ppGpp 165
processing proteinase 236 f
– mitochondria 236
– rat liver 236
– yeast 236
prohistidine decarboxilase, kinetics of activation 336
protease 174, 203
– contaminating 158
– human erythrocytes, of 239
– mast cell 145
– membrane 240
– mitochondrial membrane, inner 235
– physiological function 154
– serine 145, 151
– substrate specifity 152 f
– viral, clearage sites in 364
– zinc 248
protease-deficient mutants 160
protein breakdown 208, 215
protein kinase 10 ff, 28, 30, 44, 82
– active site 46
– – structure 10, 44
– autophosphorylation 49 ff
– cAMP-dependent (cAMPdPK) 10 f, 13, 18, 22 ff, 28, 49, 52, 96
– – action with insulin 80
– – brain 44
– – brain (pig) 44
– cAMP-independent 18, 34
– chromatin template activity 54
– intracellular cAMP, indicator of level 22
– mechanism of
– – action 44
– – cAMP-binding 49
– – dissociation of the holoenzymes 44
– nuclear translocation 54 f
– phosphorylation 34
– regulation of hormonal response 10
– restricted degradation 10
– subunit(s)
– – biological role of 44
– – inhibition of catalytic 314
– – regulatory 54, 314
– – regulatory, phosphorylation of native 317
– – regulatory, proteolytic fragments of 316
protein phosphatase(s) 30, 33, 35
– activity in myosin 41
– ATP-Mg-dependent 36, 41
– classification 28
– factor eIF-2 40
– mammalian liver in 38
– protein phosphorylation 28
– substrate specifities 36
protein phosphatase(s) – type 1 (PrP-1) 30, 33, 35 f, 39 ff
– ATP-Mg-dependent 36
– substrate specifities 36
protein phosphatase – type 1a (PrP-1a) 33
protein phosphatase – type 1b (PrP-1b) 33
protein phosphatase(s) – type 2 (PrP-2) 30, 39, 41
– separation 38
protein phosphatase – type 2a (PrP-2a) 40 f
protein phosphatase – type $2b_1$ (PrP-$2b_1$) 39 ff
– subunit structures 40
protein phosphatase – type $2b_2$ (PrP-$2b_2$) 39 ff
– subunit structures 40
protein phosphatase C 38
– separation (mammalian liver) 37
proteolysis 19, 177, 239, 243
– limited 145, 186, 199
– lysosomal 208
– selective 159, 165, 259

– susceptibility 165, 186, 216, 267
proteolytic degradation 260, 265 f
pyruvate dehydrogenase 76, 134
– converter enzymes 130
– kinase 130
– phosphatase 33, 130
pyruvate dehydrogenase complex (PDC) 124 ff
– dihydrolipoyl transacetylase 126
– PDH_3 kinase 135
– phosphorylation 135
– structure 125
– subunit composition 125
pyruvate dehydrogenase complex, mammalian 124 f
– dihydrolipoyl dehydrogenase 124
– function 124, 127, 129, 131
– kinase 131
– multisite phosphorylation 129
– phosphatase 131
– pyruvate dehydrogenase 124
– regulation 124, 127 ff, 131
– – covalent; pyruvate dehydrogenase component 128
– – monophosphorylated nonapeptide 128
– – phosphorylated tetradecapeptide 128
– structure 124, 127, 129, 131

rabbit muscle 60
rat liver 92 f, 95, 108
– purification of
– – enzyme 3
– – poly(ADP-R')synthetase 4
receptor mobility 383
ribosomal S6-protein 78

saccharomyces cerevisiae 168, 174, 179
second messenger 136
signal sequence 220
– clerage of 220
skeletal muscle 33, 146
– inhibitor 1, inhibitor 2 33
– rabbit 34, 59
– – glycogen synthease phosphatase 32
skeletal muscle phosphatases regulation
– by Mn^{2+} interchange 59
– by sulfhydril-disulfide interchange 59
small intestine 18, 146
subtilisin 186
sulfation 345, 351
– cell surface component 345, 351
– embryogenesis, during 345
– 185 K molecule 351
– sulfate residues 351
– turnover rate 351
superoxide dismutase 263

thalassemia 239, 243 f
thermolysin 248
thiols 212
thrombin 150
"topogenesis" 217
"topogenic" sequences 217, 224
– insertion sequences 218, 222
– signal sequences 218, 222
– sorting sequences 218, 223
– stop-transfer sequences 218, 221
N^{α}-tosyl-L-lysine chloromethyl ketone (TLCK) 23 ff
triglyceride 28
– breakdown 28
– lipase 28
– synthesis 28
troponin
– [^{31}P]NMR spectra 322
– phosphorylation 329
troponin I 59 f
troponin I phosphorylation, deinhibition 329
troponin C 321
tryptophan hydroxilase kinase 28
turkey gizzard smooth muscle 41
0^4-(5'-uridylyl)tyrosine 358